Energetics of Organometallic Species

NATO ASI Series

Advanced Science Institutes Series

A Series presenting the results of activities sponsored by the NATO Science Committee, which aims at the dissemination of advanced scientific and technological knowledge, with a view to strengthening links between scientific communities.

The Series is published by an international board of publishers in conjunction with the NATO Scientific Affairs Division

A **Life Sciences**	Plenum Publishing Corporation
B **Physics**	London and New York
C **Mathematical**	Kluwer Academic Publishers
and Physical Sciences	Dordrecht, Boston and London
D **Behavioural and Social Sciences**	
E **Applied Sciences**	
F **Computer and Systems Sciences**	Springer-Verlag
G **Ecological Sciences**	Berlin, Heidelberg, New York, London,
H **Cell Biology**	Paris and Tokyo
I **Global Environmental Change**	

NATO-PCO-DATA BASE

The electronic index to the NATO ASI Series provides full bibliographical references (with keywords and/or abstracts) to more than 30000 contributions from international scientists published in all sections of the NATO ASI Series.
Access to the NATO-PCO-DATA BASE is possible in two ways:

– via online FILE 128 (NATO-PCO-DATA BASE) hosted by ESRIN,
Via Galileo Galilei, I-00044 Frascati, Italy.

– via CD-ROM "NATO-PCO-DATA BASE" with user-friendly retrieval software in English, French and German (© WTV GmbH and DATAWARE Technologies Inc. 1989).

The CD-ROM can be ordered through any member of the Board of Publishers or through NATO-PCO, Overijse, Belgium.

Series C: Mathematical and Physical Sciences - Vol. 367

Energetics of Organometallic Species

edited by

J. A. Martinho Simões
Departmento de Engenharia Química,
Instituto Superior Técnico,
Lisbon, Portugal

Springer Science+Business Media, B.V.

Proceedings of the NATO Advanced Study Institute on
Energetics of Organometallic Species
Curia, Portugal
September 3–13, 1991

Library of Congress Cataloging-in-Publication Data

NATO Advanced Study Institute on Energetics of Organometallic Species
  (1991 : Curia, Portugal)
    Energetics of organometallic species : proceedings of the NATO
  Advanced Study Insitute on Energetics of Organometallic Species held
  at Curia, Portugal, September 3-13, 1991 / edited by J.A. Martinho
  Simões.
        p.   cm. -- (NATO ASI series. Series C, Mathematical and
  physical sciences ; no. 367)
    Includes bibliographical references and index.
    ISBN 978-0-7923-1707-4   ISBN 978-94-011-2466-9 (eBook)
    DOI 10.1007/978-94-011-2466-9
    1. Organometallic compounds--Congresses. 2. Thermochemistry-
  -Congresses.   I. Simões, J. A. Martinho. II. Title. III. Series.
  QD410.N355 1991
  547'.05'0456--dc20   92-9043

ISBN 978-0-7923-1707-4

Printed on acid-free paper

CONTENTS

Foreword

An overview of modern organometallic thermochemistry, made by some
of the most active scientists in the area, is offered in this book.
The contents correspond to the seventeen lectures delivered at the NATO
ASI *Energetics of Organometallic Species* (Curia, Portugal, September
1991), plus three other invited contributions from participants of that
summer school. These papers reflect a variety of research interests,
and discuss results obtained with several techniques. It is therefore
considered appropriate to add a few preliminary words, attempting to
bring some unity out of that diversity.

In the first three chapters, results obtained by classical
calorimetric methods are described. Modern organometallic
thermochemistry started in Manchester, with Henry Skinner, and his
pioneering work is briefly surveyed in the first chapter. The
historical perspective is followed by a discussion of a very actual
issue: the trends of stepwise bond dissociation enthalpies. Geoff
Pilcher, another Manchester thermochemist, makes, in chapter 2, a
comprehensive and authoritative survey of problems found in the most
classical of thermochemical techniques - combustion calorimetry -
applied to organometallic compounds. Finally, results from another
classical technique, reaction-solution calorimetry, are reviewed in the
third chapter, by Tobin Marks and coworkers. More than anybody else,
Tobin Marks has used thermochemical values to define synthetic
strategies for organometallic compounds, thus indicating an application
of thermochemical data of which too little use has been made so far.

Calorimetry is not the only experimental approach to obtain
information on the energetics of chemical reactions in solution:
equilibrium and kinetic studies have been providing a wealth of
important data, as illustrated in chapters 4 and 5. Bill Jones and
coworkers, in chapter 4, include a description of the energy profiles
of hydrocarbon activation by an unsaturated rhodium complex and discuss
the kinetic and thermodynamic selectivities of aromatic C-H activation

versus η^2-arene coordination. Brad Wayland, in chapter 5, makes a brief survey of the thermochemical results obtained for rhodium (II) porphyrin complexes. One of the highlights of these data is the exothermic carbonyl insertion into a Rh-H bond.

The energetics of short-lived species cannot be probed by the traditional calorimetric techniques. Chapter 6, by Ted Burkey, deals with the application of photoacoustic calorimetry to thermochemical studies of transient organometallic unsaturated complexes in solution. Data obtained by the photoacoustic technique can often be coupled to redox potentials of radicals and yield values of heterolytic bond dissociation enthalpies, pK_a's, etc.. These relationships are conveniently summarized in a thermochemical mnemonic included in Dan Wayner's chapter 7. Also included in this chapter are a brief description of photomodulation voltammetry (a technique used to determine redox potentials of transient radicals) and its application to the measurement of bond energies in radical cations. Mats Tilset, in chapter 8, illustrates the use of another electrochemical technique, derivative cyclic voltammetry, to probe the energetics of metal-hydrogen bonds in a number of transition metal-hydride complexes.

Ideally, the energetics of chemical bonds should be referred to the perfect gas phase. Solution data are extremely useful, since most important chemical processes occur in solution, and many *may* provide a good approximation of gas phase data. Unfortunately, in the case of organometallic compounds, the evidence for this last assumption is rather scarce. We know too little about solvation energetics, in particular about sublimation enthalpies. Efforts to measure or to estimate these quantities are addressed in chapters 9 and 10. In chapter 9, by Alexander Carson, the sublimation enthalpies of several organometallic molecules, obtained with a Knudsen cell, are combined with combustion calorimetry results to derive bond enthalpy data. Jim Chickos, Donald Hesse, and Joel Liebman describe, in chapter 10, a method for the estimation of sublimation enthalpies of hydrocarbons.

Most results covered by Robin Walsh's contribution, chapter 11, do not require any assumptions regarding solvation enthalpies, since they were obtained in the gas phase. Yet, other assumptions are often necessary to extract thermochemical results from kinetic and equilibrium experiments. A brief description of these methods and hypotheses, illustrated with data for silanes and alkanes, is made in chapter 11.

Chapters 12 and 13 are a sort of an interlude in our exploration of organometallic energetics. In chapter 12, Joe Connor introduces and, through a set of selected data, discusses the idea of "bond enthalpy transferability" as applied to transition metal organometallic compounds. Although we know many of the rules that allow us to predict if two similar bonds in different organic compounds have similar "strengths" (e.g. by comparing the bond orders and the bond lengths), other considerations are required when the exercise is attempted for d-element molecules (e.g. the oxidation state of the metal and the coordination number). The transferability of bond enthalpies is a very relevant issue for predicting the energetics of reactions where no experimental data are available. The estimation of thermochemical data for organometallic compounds is considered in chapter 13, where a critical survey of the methods that have been (or can be) used for that purpose, is made.

Several mass spectrometry techniques provided valuable information on metal-ligand bonding energetics in neutral and ionic coordinatively saturated compounds and organometallic fragments. Furthermore, gas phase data such as ionization and electron attachment energies can be compared with their solution counterparts, oxidation and reduction potentials, providing information on solvation energetics. This bridge between gas phase and solution data is analyzed by David Richardson (chapter 14) and illustrated for several organometallic and coordination complexes. It is worth pointing out that Richardson's basic data, ionization and electron attachment energies, were obtained from equilibrium and bracketing studies of ion-molecule reactions in a

Fourier transform ion cyclotron mass spectrometer cell. FTICR, one of the favorite techniques for probing gas phase chemistry and thermochemistry, has also been used by Amy Stevens Miller and Thomas Miller, to determine the gas phase acidities of a number of transition metal-hydride complexes (chapter 15). Trends in these acidities are explained in terms of metal-hydrogen bond dissociation enthalpies and electron affinities of the organometallic fragment radicals. Solvation effects on acidities are also addressed by comparing the gas phase acidities with the solution pK_a's of the hydrides.

Some of Amy Stevens Miller's acidity data were obtained by studying ion-molecule reactions in a selected-ion flow tube (SIFT). An apparatus from the same family, a flowing afterglow-triple quadrupole mass spectrometer, was used by Lee Sunderlin and Bob Squires to measure the energy thresholds for CO dissociation in several transition metal carbonyl anions (chapter 16). In some cases, these results were coupled with electron affinity data to yield M-CO stepwise bond dissociation enthalpies in the neutral species. Jack Beauchamp and Petra van Koppen, in chapter 17, review the contribution of several other gas phase techniques (ICR, ion beam and the measurement of kinetic energy release distributions) and theoretical methods to investigate the thermochemistry and the mechanisms of organometallic reactions. The measurement of kinetic energy release distributions is covered in some detail and its application to probe the energetics of large molecules (such as co-enzyme B_{12}) is emphasized. The survey is illustrated with many examples, and includes a "periodic table" of the trends of atomic metal ion reactivity.

Part of the beauty of the experiments by Peter Armentrout and David Clemmer (chapter 18) lies in the simplicity of the systems analyzed. For example, by probing the energetics of "diatomic" species like ML or ML^+, they are able to examine the details of M-L (or M^+-L) bonding by varying M or L. As many of these species are sufficiently "simple" to be handled by sophisticated theoretical methods, the experimental data can also be used for the calibration of these methods. The guided ion beam mass spectrometric technique also provides stepwise metal-ligand bond dissociation enthalpies in multi-ligand species, ML_n^+, enabling a detailed discussion of the bond cleavage energetics as a function of the degree of unsaturation of the metal. All this is addressed in chapter 18, together with a discussion of the mechanism of hydrocarbon activation by atomic metal cations.

Theoretical methods have had a large impact in all fields of Chemistry, and Thermochemistry is no exception. The two final chapters are on *quantitative* theoretical methods, i.e. methods that not only provide qualitative help to understand chemical reactions or molecular structures, but they are also able to supply reliable data for a reaction enthalpy, a bond dissociation enthalpy, bond lengths, and bond angles. In chapter 19, Tom Ziegler surveys the applications of Density Functional Theory to several processes relevant to the energetics of organometallic species, including bond cleavages, ionization energies, electron affinities, etc.. The use of Density Functional Theory to probe profiles of many reactions involving organometallic complexes is also illustrated in chapter 19. The other theoretical contribution, chapter 20, by Margareta Blomberg, Per Siegbahn, Mats Svensson, and Ian Wennerberg, addresses simpler organometallic systems, in particular reactions of bare second row metal atoms and ions with hydrocarbons. The paper includes, for example, reaction profiles of C–C and C–H activation as a function of the metal and its oxidation state. The stepwise Ni–CO dissociation energies in nickel tetracarbonyl were also calculated and compared to experimental data.

A final comment on nomenclature, units, and the like. There was great editorial flexibility on these matters. Some authors, usually American, favor *kcal/mol*, others *kJ/mol*; some authors use *free energy* instead of *Gibbs energy*; some authors use the standard deviation of the mean (σ) for the uncertainty interval (which, by the way, looks better in kcal/mol), others adopt the old thermochemical convention (2σ); some authors do not use the new IUPAC recommended designation for the groups in the Periodic Table; some authors use ΔH_f^o and not

$\Delta_f H_m^o$ for enthalpies of formation. While most of these choices do not damage the set of stimulating papers now published, some standardization wouldn't do any harm either... Perhaps more serious, particularly for the casual reader, is the proliferation of designations and symbols, such as bond *strengths*, bond *enthalpies*, bond *energies*, bond *dissociation enthalpies*, bond *dissociation energies*, bond *disruption enthalpies*, D, $\bar{D}$, E, BDE, and so on. Although it is believed that in the present book there will not be any misunderstand of what each author is talking about, the field has reached a stage where some agreement on this would be desirable. It will certainly be easier than deciding for kcal or kJ.

A word of acknowledgement for the generous support and advice from NATO Scientific Affairs Division (Dr. Luís Veiga da Cunha, Director, ASI Programme), and the support from several other organizations, including Secretaria de Estado do Ensino Superior, Portugal, Junta Nacional de Investigação Científica e Tecnológica, Portugal, Instituto Nacional de Investigação Científica, Portugal, National Science Foundation, U.S.A., Luso-American Foundation for Development, Portugal, The British Council, and Instituto Superior Técnico, Portugal. Very special thanks to Professors J. A. Connor and J. L. Beauchamp, from the Scientific Committee of the ASI, and to my students and colleagues who have worked so hard to make the ASI possible, in particular Mr. Hermínio P. Diogo.

J. A. Martinho Simões

January 1992.

HISTORICAL PERSPECTIVE OF ORGANOMETALLIC THERMOCHEMISTRY

H.A. SKINNER
Chemistry Department
University of Manchester
Manchester, M13 9PL
England

ABSTRACT. My involvement with thermochemistry and calorimetry goes back to 1946, at which time simple organic compounds containing C, H, O and N were the main interest of thermochemists. Organometallic compounds were then of minor interest, and few such data had been measured. The development later of rotating bomb calorimetry, the use of combustion aids, advances in solution reaction calorimetry, and in particular, the adaptation of the Calvet twin-microcalorimeter, opened the door to study organometallic compounds in general, but it was well into the 60's before this happened. From 1970 onwards, however, more laboratories became involved with organometallics in general, particularly in Portugal at Lisbon and Oporto, in Spain, and in Russia. In more recent times, non-calorimetric developments such as photoionization mass-spectroscopy are being applied to measure individual bond-dissociation energies in specific MX_n compounds. The results indicate that these, $D_1, D_2 \ldots D_n$, vary considerably from the mean value, $\bar{D} = \frac{1}{n}(D_1 + D_2 \ldots + D_n)$.

The main reason for variability in the individual D values in MX_n compounds is that the valence-state of the metal changes in moving along the series MX_n, MX_{n-1}, $MX_{n-2} \ldots M$. In the case of SiH_4, e.g. this change occurs at the step $SiH_3 \rightarrow SiH_2$; in CH_4, however, the valence change occurs at the step $CH_2 \rightarrow CH$. The analysis of these, and other examples is attempted in this paper, to indicate the complexity of the overall problem, and the need for more experimental data than is provided by thermochemical studies alone.

1. HISTORICAL REVIEW

The thermochemistry of organometallic compounds has become subject to more extensive research within the last twenty years. Prior to 1960, thermochemists were mainly engaged in obtaining reliable enthalpy of formation data for organic compounds from precision measurements of heats of combustion by bomb-calorimetry. The techniques of static bomb calorimetry - perfected by Rossini and his collaborators in the 1940's - were applied to hydrocarbons, and to organic compounds of C, H, O and N so successfully that in due course empirical bond-energy additive schemes, such as that of Allen[1], enabled accurate prediction of unknown values, and virtually removed the need for further direct experimental measurements. The development of the rotating bomb calorimeter in the 1950's by Sunner, Good, Waddington and others extended the scope of combustion calorimetry to cover organic compounds of S, P, halogens, and of organo-silicon[2] and

1

J. A. Martinho Simões (ed.), Energetics of Organometallic Species, 1–8.
© 1992 *Kluwer Academic Publishers.*

organo-boron[3] compounds. The method has also been applied to a few organo-metallic compounds of germanium[4,5], lead[6], arsenic[7], bismuth[8] and selenium[9], and to one transition-metal complex, i.e. $Mn_2(CO)_{10}$, by Good[10]. Thermochemical studies of organo-*transition* metal compounds have mainly involved non-combustion studies, in particular by microcalorimetric methods.

My own involvement with calorimetry, and with organometallic compounds, began when I was employed as a lecturer in physical chemistry under Professor Michael Polanyi, at Manchester University. In 1946, the War being over, I became free for the first time to choose my own research programme. I sought the advice of Michael Polanyi and he suggested that I might try to measure the heats of formation of alkyl halides - for which available data at that time were of dubious value. But money for research was scarce in 1946, and the cost of setting up a conventional bomb-combustion calorimeter ruled out that method. I had to find a less expensive procedure and a simple solution calorimeter proved to be the answer. My Ph.D. research student at that time, A.S. Carson[11], used this to measure both the heat of iodination (in solution) and the heat of hydrolysis (in acid solution) of cadmium dimethyl. Combined together, these measurements led to ΔH_f^0 values for both CH_3I and for $Cd(CH_3)_2$:

(i) $Cd(CH_3)_2 + 2I_2(soln) \rightarrow CdI_2 + 2CH_3I$

(ii) $Cd(CH_3)_2 + H_2SO_4(soln) \rightarrow CdSO_4 + 2CH_4$

Later similar studies were made using zinc alkyls[12] and mercury alkyls[13], again with the primary objective of establishing ΔH_f^0 values for alkyl and aryl halides, but the subsidiary determination of ΔH_f^0 values for the zinc and mercury alkyls opened the door, as it were, on metal-carbon bond-energies, and this problem became the major one as time moved forward.

In 1953, IUPAC held its Conference in Uppsala, Sweden, at which I was present as a member of the Thermodynamics Commission. Edward Calvet was present at the same meeting, as a French representative. We became acquainted and I learnt from him of the development of the Tian-Calvet microcalorimeter and of its capability to study processes at elevated temperatures. I was very interested to learn more of this and Calvet persuaded the British Council in 1955 to send me to visit his laboratory in Marseille to see the microcalorimeter in operation. At that time, the microcalorimeter was being used to measure specific heats, thermal conductivities, radioactive transformations, catalytic surface activities and in several biochemical and zoological studies[14]. I was sure that the microcalorimeter offered scope to study thermal decomposition and other simple chemical reactions, but I had to wait until the mid-sixties before finances allowed purchase of the equipment and prove my foresight to be correct. The advantage of the microcalorimeter for studies on transition-metal complexes is that only small quantities of compound are required for each measurement and often the compounds of interest are not readily available, nor easily prepared.

The traditional static-bomb combustion calorimeter, although generally not suited to studies on transition organometallic compounds, can be successfully applied to metal carbonyls since on combustion in dry O_2, no water is formed, and there are no complications from hydration of the metal oxide formed[15]. It has also been adapted successfully by Rabinovitch[16] by the use of combustion aids; a relatively small amount of the metal complex is admixed with a large amount of combustion aid (benzoic acid, or polyethylene) to ensure complete combustion. Telnoi[17] has applied a similar procedure, by coating the sample with paraffin wax, or by enclosing it in a polyethylene capsule.

The various calorimetric techniques now available provide the means to measure enthalpies of formation of organometallic compounds in general. More novel techniques based on ion-beam studies, pulsed laser pyrolysis, photoionization and guided ion-beam mass spectroscopy[18,19,20] can provide <u>individual</u> bond dissociation energies of specific M-X bonds in MX_n, and this offers the scope for a much deeper analysis of the nature of the bonding.

2. METAL BONDING CHARACTERISTICS

For several groups of organometallic compounds the available thermochemical data are sufficient to seek correlations within the group, and in particular to examine the variation in <u>mean</u> bond energy levels, $\bar{D}(M-X)$, in MX_n compounds on changing the metal. Examinations of this type have been described by Pilcher[21] and Connor[22] for metal hydrides, metal alkyls and aryls, and for metal carbonyls, metal arenes and cyclopentadienyls. Correlation was made with $\Delta H_f^o(M,g)$, indicating in each case that the $\bar{D}(M-X)$ values <u>increase</u> along a given series with <u>increasing</u> $\Delta H_f^o(M,g)$ values.

> Figure 1. $\bar{D}(M-H)$, $\bar{D}(M-R)$ $v.$ $\Delta H_f^o(M)$ for Group IV metals.
> Figure 2. $\bar{D}(M-H)$, $\bar{D}(M-Ph)$, $\bar{D}(M-Me)$, $\bar{D}(M-Et)$ for Bi, Sb, As, P.
> Figure 3. $\bar{D}(M-Neopentyl)$, $\bar{D}(M-benzyl)$, $\bar{D}(M-NEt_2)$, $\bar{D}(M-O^iPr)$
> for Ti, Zr, Hf.
> Figure 4. $\bar{D}(M-CO)$ for metal carbonyls.

The linear correlation in Fig.1 implies that $\bar{D}(M-X)$ values are directly proportional to $\bar{D}(M-M)$, since these metals have the tetrahedral diamond-like structure, for which $\bar{D}(M-M) = 0.5 \Delta H_f^o(M)$. For other metals, the ratio of $\bar{D}(M-M)$ to $\Delta H_f^o(M)$ is variable and usually > 0.5.

However, a direct correlation of $\bar{D}(M-X)$ in MX_n with $\Delta H_f^o(M)$ ignores the fact that the <u>valence</u>-state of the metal in MX_n is rarely the same as in the ground-state of the free metal atom. The effect is most clearly shown by the variations in the stepwise dissociation energies in MX_n, as, for example, in CH_4[23] and in SiH_4[19,20]. The relevant experimental data (at 0 K) are summarized below:-

MH_n	r_e Å	H–M–H angle	ΔH^o kJ mol^{-1}	$\bar{D}$	D_1	D_2	D_3	D_4
							kJ mol^{-1}	
CH_4	1.094	109.5°	1642	410.5	432			
CH_3	1.079	120°				457		
$CH_2, {}^3B_1$	1.078	~136°	753				419	
$CH, {}^2\pi$			334					334
SiH_4	1.455	109.5°	1266	316.5	373			
SiH_3	~1.49	~112°	893			292		
$SiH_2, {}^1A_1$	1.516	92°	601				312	
$SiH, {}^2\pi$	1.520		289					289

4

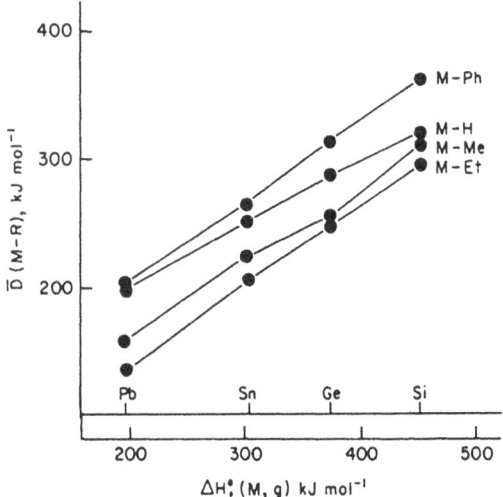

Figure 1. $\bar{D}(\text{M-H})$, $\bar{D}(\text{M-R})$ v $\Delta H_f^O(\text{M})$ for Group IV metals

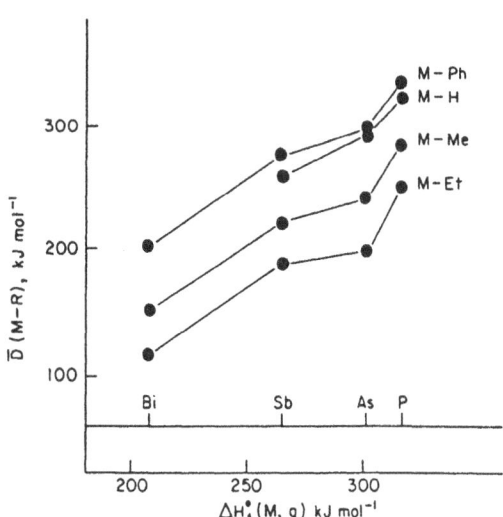

Figure 2.

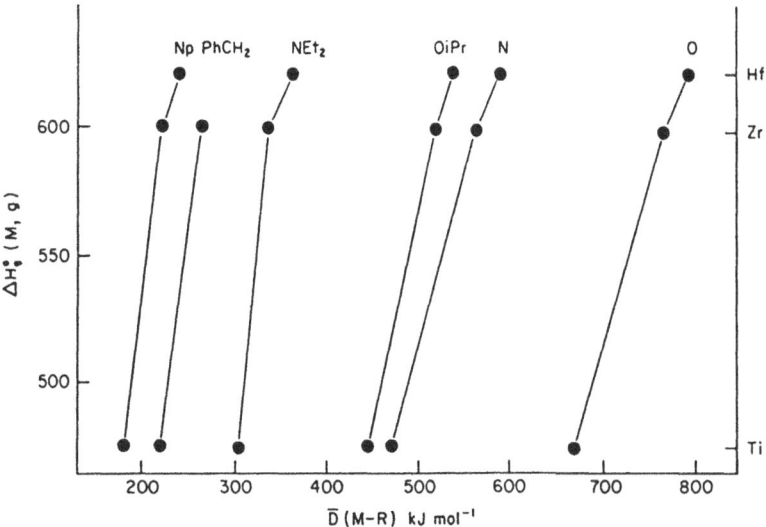

Figure 3. $\bar{D}$(M-R) values v ΔH_f^o(M) for Ti, Zr, Hf.

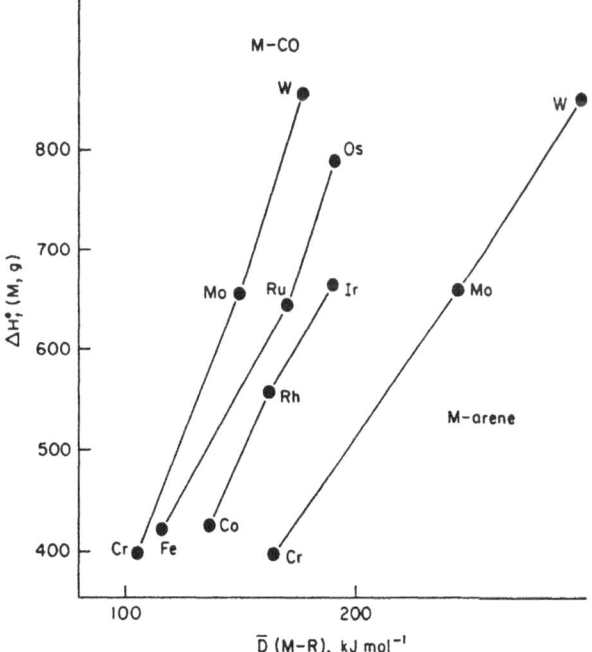

Figure 4. $\bar{D}$(M-CO) values plotted against ΔH_f^o(M, g)

The differences in the fluctuation patterns, $D_1 \rightarrow D_2 \rightarrow D_3 \rightarrow D_4$ in CH_4 and SiH_4 are reflected in structural differences between .CH_3 and .SiH_3, and between :CH_2 and :SiH_2. The methyl radical is planar, and derives from trigonal carbon, $t^3\pi$, whereas the .SiH_3 radical is pyramidal, with bond-angles only slightly larger than tetrahedral. The planar structure of the .CH_3 radical has been attributed[24] to the increase in the overlap of C-trigonal orbitals with $H(S = 0.7)$ relative to that of C–tetrahedral orbitals $(S = 0.68)$. In the case of Si–H bonding there is no similar overlap advantage and the silyl radical retains the tetrahedral structure. The difference between CH_2 and SiH_2 are even more marked - the ground-state of CH_2 being a triplet $^3B_1$, deriving from 4-valent carbon, whereas SiH_2 is a singlet, $^1A_1$, deriving from the 2-valent ground-state of Si, ^3P. The bond-angle of ~ 140° in CH_2 suggests a valence state for C in between trigonal and the digonal state, $d_1 d_2 \pi \pi$, whereas the 92° bond-angle in SiH_2 is compatible with essentially p-bonding.

Calculated valence-state excitation energies[25] in C and Si in defined tetravalent configurations are listed below and used to evaluate mean bond-energy values, $\bar{D}^*(M–H)$, measured with respect to the valence-state of M in each case.

V.S. energy

C,	V_4, T^4	632 kJ mol^{-1}	Si,	V_4, T^4	484 kJ mol^{-1}
	V_4, $t^3\pi$	648 kJ mol^{-1}		V_4, $t^3\pi$	496 kJ mol^{-1}
	V_4, $d^2\pi\pi$	694 kJ mol^{-1}		V_4, $d^2\pi\pi$	523 kJ mol^{-1}

	ΔH^a_o	$\bar{D}$	V.S.	ΔH^*_a	$\bar{D}^*$
CH_4	1642	410.5	T^4	2274	568.5*
CH_3	1210	403.3	$t_3\pi$	1858	619.3*
SiH_4	1266	316.5	T^4	1750	437.5*
SiH_3	893	297.7	T^3T	1377	(T = 121)
				1256*	418.7*

The $\bar{D}^*$ for tetrahedral C–H in CH_4 is calculated to be less than $\bar{D}^*$ for trigonal C–H in CH_3, as reflected by the shorter bond-length in the free radical. By contrast, the tetrahedral $\bar{D}^*$ (Si–H) in SiH_4 is larger than in the .SiH_3 free radical, for which $\bar{D}^*$ was calculated on the basis that the free electron occupies a tetrahedral sp^3 orbital, (excitation energy = $\frac{1}{4} T_4 = 121$ kJ). In this case, the bond-length is longer in the radical than in SiH_4.

In $^3B_1$ CH_2, the bond-angle implies a valence-state in between trigonal and digonal, although nearer to the trigonal state. For the trigonal state, $t^2 t\pi$, (with the free electrons in one of the t-orbitals and in the π-orbital) we have:

	ΔH_a^o	$\bar{D}$	V.S.	$\bar{D}^*$
$CH_2, {}^3B_1$	753	376.5	$\underline{t}^2 t\pi$	$(t = 216)$
				592.5^*

giving a $\bar{D}^*$ value slightly less than in $.CH_3$. However, since the bond-angle in CH_2 implies some digonal character, the true $\bar{D}^*$ should be increased and the bond-length would imply a resultant $\bar{D}^*$ very close to that in the methyl radical.

Detailed analysis of the type given for CH_4 and SiH_4 can be made for other MR_n compounds of Group II, III and IV, but is not yet applicable to transition metal compounds. The necessary structural and experimental data are not available, but the new techniques now being applied by Berkowitz, Armentrout and several others offer the scope to do this in due course.

3. REFERENCES

1. Allen, T.L. (1959) *J. Chem. Phys.*, **31**, 1039.
2. Good, W.D., Lacina, J.L., De Prater, B.L., McCullough, J.P. (1964) *J. Phys. Chem.*, **68**, 579.
3. Good, W.D., Månsson, M. (1966) *J. Phys. Chem.*, **70**, 97.
4. Bills, J.L., Cotton, F.A. (1964) *J. Phys. Chem.*, **68**, 806.
5. Carson, A.S., Carson, E.M., Laye, P.G., Spencer, J.A., Steele, W.V. (1970) *Trans. Faraday Soc.*, **66**, 2459.
6. Good, W.D., Scott, D.W., Waddington, G. (1956) *J. Phys. Chem.*, **60**, 1090.
7. Mortimer, C.T., Sellers, P. (1964) *J. Chem. Soc.*, 1965.
8. Steele, W.V. (1979) *J. Chem. Thermo.*, **11**, 187.
9. Barnes, D.S., Mortimer, C.T. (1963) *J. Chem. Thermo.*, **5**, 371.
10. Good, W.D., Fairbrother, D.M., Waddington, G. (1958) *J. Phys. Chem.*, **62**, 853.
11. Carson, A.S., Hartley, K., Skinner, H.A. (1949) *Proc. Roy. Soc.*, A195, 500.
12. Carson, A.S., Hartley, K. Skinner, H.A. (1949) *Trans. Faraday Soc.*, **45**, 1159.
13. Hartley, K., Pritchard, H.O., Skinner, H.A. (1950) *Trans. Faraday Soc.*, **46**, 1019.
14. Calvet, E. (1962) "Recent Progress in Microcalorimetry", H.A. Skinner (ed.), in Experimental Thermochemistry, Vol.2, Interscience Publishers, New York and London, pp.385-410.
15. Barnes, D.S., Pittam, D.A., Pilcher, G., Skinner, H.A., Virmani, Y. (1974) *J. Less Common Met.*, **36**, 177; **38**, 53.
16. Tel'noi, V.I., Rabinovitch, I.B., Gribov, G.B., Pashinkin, A.S., Salamatin, B.A., Chernova, V.I. (1972) *Russ. J. Phys. Chem.*, **46**, 465.
17. Tel'noi, V.I., Rabinovitch, I.B., Latyneva, V.N., Lineva, A.N. (1971) *Dokl. Akad. Nauk, SSSR*, **197**, 353.
18. Lewis, K.E., Golden, D.M., Smith, G.P. (1984) *J. Amer. Chem. Soc.*, **106**, 3905-12.
19. Berkowitz, J., Greene, J.P., Cho, H., Ruscic, B. (1987) *J. Chem. Phys.*, **86**, 1235-48.
20. Boo, B.H., Armentrout, P.B. (1987) *J. Amer. Chem. Soc.*, **109**, 3549-59.
21. Pilcher, G., Skinner, H.A. (1982) "Thermochemistry of organometallic compounds", in F.R. Hartley and S. Patai (eds), "The Chemistry of the Metal-Carbon Bond", J. Wiley and Sons, Limited, pp. 43-90.
22. Skinner, H.A., Connor, J.A. (1985) *Pure and App. Chem., I.U.P.A.C.*, **57**, 79-88.

23. Glyshko, V.P., Gurvich, L.V. (1979) "Thermochemical Constants of Chemical Compounds", Vol. 2, *Akad. Nauk.*, Moscow.
24. Murrell, J.N., Kettle, S.F.A., Tedder, J.M. (1965) in "Valence Theory", Wiley and Sons Limited, London, p. 54, 361.
25. Hinze, J., Jaffé, H.H. (1962) *J. Amer. Chem. Soc.*, **84**, 540; Pilcher, G., Skinner, H.A. (1962) *J. Inorg. Nucl. Chem.*, **24**, 937-952.

COMBUSTION CALORIMETRY OF ORGANOMETALLIC COMPOUNDS

G.PILCHER

Department of Chemistry
University of Manchester
Manchester
M13 9PL
U.K.

ABSTRACT

The application of combustion calorimetry for determining enthalpies of formation of organometallic compounds is examined critically. The static-bomb method has been widely used but the results are generally disappointing whereas the rotating-bomb method is successful but has been used in only a few cases.

INTRODUCTION

Enthalpies of combustion of organic compounds are the major source of their enthalpies of formation but for organometallic compounds the combustion method has been only partially successful and its contribution to this topic has been disappointing. In principle the complete oxidation of an organometallic compound to produce carbon dioxide, water, and the metallic oxide would appear to be the most promising general reaction to use to determine the enthalpy of formation, i.e. one could imagine that this would be applicable in a straightforward manner to all such compounds. In practice however, the situation is quite different and the main reasons for this are:-

(1) Enthalpies of combustion are large and are required to high precision and accuracy hence a high degree of purity of the sample is essential. Only a limited number of organometallic compounds can be obtained in the quantities (several g) and in a state of purity (> 99.9%) necessary for combustion measurements.

(2) Organometallics usually exhibit problems with incompleteness of combustion. Many of these compounds explode on ignition in oxygen but the use of "combustion aids" such as benzoic acid or hydrocarbon oil can assist in the attainment of complete combustion of the organic part of the molecule. The auxiliary combustion aid often acts as a moderating influence on the combustion reaction, but a price is paid in the precision of measurement because the compound contributes a

J. A. Martinho Simões (ed.), Energetics of Organometallic Species, 9–34.
© 1992 *Kluwer Academic Publishers.*

smaller fraction than usual to the measured enthalpy change. Incomplete oxidation of the metal is often a problem: as the metal oxide is a solid product it has the facility when forming to trap unburned compound and unoxidized metal within the grains of oxide produced. The seriousness of this problem depends on the element concerned and it appears to be particularly severe in the combustion of aluminium and silicon compounds.

(3)   The products of combustion must be capable of being thermodynamically defined. Some oxides can be produced in different crystalline forms or in an amorphous form and in some cases the energy differences between these forms are significant. X-ray examination can provide clues but a small proportion of crystallinity in an amorphous form can falsely indicate complete crystallinity.

Any of these problems may arise on combustion of an organometallic ompound in oxygen, however some of these difficulties may be overcome by applying rotating-bomb calorimetry.

It is difficult to set criteria by which the reliability of combustion results on organometallics can be judged, but certain tests can be applied.

(a)   The completeness of combustion of the organic part of the molecule can be established by measuring the amount of carbon dioxide produced.

(b)   Examination of the oxide, or oxides, produced should be made to try to establish the completeness of oxidation of the metal and the crystalline form of the oxide.

(c)   If the enthalpies of formation for one or more compounds of a particular metal obtained by combustion calorimetry agree with results obtained by other methods, e.g. reaction calorimetry, this gives confidence in the combustion technique for other compounds of that particular metal.

(d)   Occasionally there are sufficient results for compounds of a particular metal for application of a bond-energy scheme and if the results seem sensible and fit with the scheme, this gives some confidence in the experimental results. If however, the results do not fit in with our expectations, we should not automatically cast suspicion on the experimental results as this would imply that we have a complete understanding of the subject which is not the case.

STATIC-BOMB COMBUSTION CALORIMETRY

Most of the measurements on organometallic compounds have been made using static-bomb calorimeters. Some satisfactory results have been obtained but far too many are of doubtful significance. We will look at the present situation critically considering the metals in order of the Periodic classification.

Group IA Alkali metals

Lebedev *et al.* [1] burned ethyl-lithium and n-butyl lithium in a static-bomb but gave few experimental details. The samples were contained in collodion and benzoic acid was used as a combustion aid. The product Li_2O contained some Li_2O_2 and Li_2CO_3 and thermal corrections made for these but no mention was made of the possible hydration of the lithium oxide: as the oxide is formed at a high temperature, hydration may be a slow process. The results seem satisfactory and that for n-butyllithium agreed with the value by Fowell and Mortimer [2] from the enthalpy of hydrolysis.

	$\Delta_f H_m^0/(kJ \cdot mol^{-1})$	
C_2H_5Li (cr)	-58.6 ± 5.4	(combustion)
n-C_4H_9Li (l)	-133.1 ± 7.1	(combustion)
n-C_4H_9Li (l)	-132.2 ± 3.3	(hydrolysis)

Whilst static-bomb measurements are feasible for organolithium compounds, use of reaction calorimetry would seem to be a more sensible approach.

There are no other static-bomb measurements on organometallic compounds of the alkali metals, nor on compounds of the group 1B metals, Cu, Ag, Au.

Group II

No measurements have been reported for compounds of the group II A metals, Be, Mg, Ca, Sr, Ba but there have been many studies on compounds of the group II B metals, Zn, Cd, Hg.

Zinc and cadmium

Long and Norrish [3] made 50 combustion experiments on diethylzinc with 20 variations in technique and from visual inspection of the products for incomplete combustion, they concluded that none of the experiments were fully satisfactory. They accepted results from 11 of the experiments which were considered to be the least bad, and their combustion result was confirmed by Lautsch *et al.* [4] who reported no experimental details. The enthalpies of formation from the combustion studies agree with values derived from enthalpies of hydrolysis in water and in aqueous acid, and enthalpies of iodination. [5]

	$\Delta_f H_m^0(l)/(kJ \cdot mol^{-1})$	
	combustion	reaction calorimetry
$(CH_3)_2Zn$	28.8 ± 6.3 [3]	22.6 ± 8.4 [5]
$(C_2H_5)_2Zn$	16.7 ± 13.4 [3]	22.2 ± 8.4 [5]
	20.5 ± 2.9 [4]	15.4 ± 8.4 [5]
		15.4 ± 7.5 [5]
$(n\text{-}C_3H_7)_2Zn$	-57.7 ± 23.4 [3]	
$(n\text{-}C_4H_9)_2Zn$	-104.2 ± 23.8 [3]	

Although the static-bomb combustion method can be applied to alkylzinc compounds, the reaction-solution calorimetric method is superior. For alkylcadmium compounds this situation is more marked, the combustion and reaction-solution calorimetric results disagree and the latter are those accepted.

	$\Delta_f H_m^0(l)/(kJ \cdot mol^{-1})$	
	combustion	reaction calorimetry
$(CH_3)_2Cd$	88.3 ± 12.5 [3]	67.4 ± 4.2 [6]
		69.9 ± 1.2 [6]
$(C_2H_5)_2Cd$	90.3 ± 3.8 [4]	60.7 ± 1.7 [5]
		56.9 ± 5.0 [5]

Mercury

The first static-bomb combustion measurements on organomercury compounds were made by Berthelot in 1899 who obtained nearly correct results, within 25 $kJ \cdot mol^{-1}$ for dimethylmercury and diethylmercury: modern measurements have proved successful. On combustion, the metal does not oxidize to a marked extent and only small corrections are required for the formation of HgO and $Hg_2(NO_3)_2$. Carson et al. [7] for dimethylmercury and diethylmercury established completeness of combustion from the carbon dioxide produced and the derived enthalpies of formation were in agreement with those from reaction calorimetric studies. For diphenylmercury, the combustion results of Fairbrother and Skinner [8] agree with those of Carson and Wilmshurst [9]: the latter authors used hydrocarbon oil as a combustion aid and obtained CO_2 recoveries of 99.85%.

	$\Delta_f H_m^0/(\text{kJ·mol}^{-1})$	
	combustion	reaction calorimetry
$(CH_3)_2Hg(l)$	58.9 ± 0.4 [7]	56.0 ± 3.3 [10]
		51.5 ± 1.7 [10]
$(C_2H_5)_2Hg(l)$	27.2 ± 0.8 [7]	29.3 ± 3.3 [11]
		31.8 ± 3.8 [11]
$(C_6H_5)_2Hg(cr)$	279.9 ± 7.5 [8]	274.1 ± 8.4 [12]
	299.6 ± 1.3 [7]	
	285.8 ± 6.2 [9]	
$(C_6H_5CH_2)_2Hg(cr)$	189.3 ± 4.1 [13]	
$(C_6H_5C{\equiv}C)_2Hg(cr)$	621.3 ± 5.2 [13]	

Static-bomb combustion calorimetry of organomercury compounds has been developed into a reliable technique and is probably preferable to reaction calorimetry as it is easier to carry out.

The enthalpies of formation in the condensed phase can be converted to gas phase values using enthalpies of vaporization or sublimation and these lead to sensible values for the mean bond dissociation enthalpies, $<D>/(\text{kJ·mol}^{-1})$: Hg-CH$_3$, 130.0; Hg-C$_2$H$_5$, 102.7; Hg-C$_6$H$_5$, 160.1; Hg-CH$_2$C$_6$H$_5$, 80.0; Hg-C$\equiv$CC$_6$H$_5$, 312.0.

Group III A Scandium, Yttrium and the Lanthanides

The only static-bomb combustion study in this area is an admirable piece of work by Rabinovich *et al.* [14] on the tricyclopentadienyls of Sc, Y, La, Pr, Tm, and Yb. The samples were judged to be 99.99% pure by mass spectrometry, and the samples were sealed under paraffin wax to protect them from oxidation before ignition and the completeness of combustion was established by CO_2 analyses. With selected enthalpies of sublimation and with $\Delta_f H_m^0(Cp,g)/(\text{kJ·mol}^{-1}) = 264.4 \pm 9.0$ [15], $<D(M-Cp)>$ values in these compounds can be derived.

	$\Delta_f H^0_m(cr)$	$\Delta_f H^0_m(g)$	$\Delta_f H^0_m(M,g)$	$<D(M-Cp)>$
	$kJ \cdot mol^{-1}$	$kJ \cdot mol^{-1}$	$kJ \cdot mol^{-1}$	$kJ \cdot mol^{-1}$
$YbCp_3$	29.3 ± 5.1	138.1 ± 6.2	151.9	269.0
$TmCp_3$	-49.5 ± 5.1	61.8 ± 6.2	247.3	326.2
$PrCp_3$	-28.2 ± 8.7	97.3 ± 9.2	372.8	356.2
$LaCp_3$	35.7 ± 5.1	150.3 ± 6.5	431.0	358.0
$ScCp_3$	-13.6 ± 4.6	83.5 ± 5.8	381.6	363.8
YCp_3	-45.2 ± 4.6	66.5 ± 5.8	424.7	383.8

The compounds have been ordered here according to increasing $<D(M-Cp)>$ to show that $<D(M-Cp)>$ is roughly proportional to $\Delta_f H^0_m(M,g)$ as to be expected.

Group III B Aluminium and Gallium

For alkylaluminium compounds, bomb calorimetry appears to be a complete failure; (a) repeated measurements are discrepant, and (b) values of $\Delta_f H^0_m$ from combustion calorimetry show large differences from values obtained by reaction calorimetry. The following few examples are typical of the very many such differences in combustion calorimetry results that could be quoted.

	$\Delta_f H^0_m(l)/(kJ \cdot mol^{-1})$		$\delta/(kJ \cdot mol^{-1})$
Et_2AlH	-307.5 ± 18.8 [16]	-204.2 ± 7.1 [17]	103
$(i-Bu)_2AlH$	-402.1 ± 16.7 [16]	-289.1 ± 4.6 [17]	104
nPr_3Al	-322.2 ± 4.6 [17]	-251.0 ± 15.8 [18]	71

The comparison between combustion calorimetric results and those from reaction calorimetry are no more encouraging.

	$\Delta_f H_m^0(l)/(kJ \cdot mol^{-1})$	
	combustion	reaction calorimetry
$(CH_3)_3Al$	-120.0 ± 10.0 [3]	-150.6 ± 6.7 [19]
$(C_2H_5)_3Al$	-217.1 ± 9.2 [16]	-156.9 ± 20.9 [20]
$(n\text{-}C_3H_7)_3Al$	-251.0 ± 15.8 [18]	-297.5 ± 18.8 [18]

Combustion calorimetry of aluminium alkyls is not a topic worth pursuing: reaction calorimetry would be the correct approach and the small number of results obtained so far require confirmation. The combustion method would probably be suitable for compounds such as aluminium alkoxides as it has been successfully applied to aluminium acetylacetonate. [21]

In contrast, combustion measurements on gallium alkyls appear to be successful, the main solid product was Ga_2O_3 mixed with traces of free metal and carbon for which corrections were made. Measurement of the enthalpy of iodination of $(CH_3)_3Ga(l)$ gave $\Delta_f H_m^0 = -80.3 \pm 12.5$ kJ·mol^{-1} [22] in agreement with the value from combustion measurements.

	$\Delta_f H_m^0/(kJ \cdot mol^{-1})$
$(CH_3)_3Ga$	-79.9 ± 8.2 [23]
$(C_2H_5)_3Ga$	-99.9 ± 5.4 [24]
$(n\text{-}C_4H_9)_3Ga$	-284.6 ± 5.0 [24]
$(i\text{-}C_4H_9)_3Ga$	-293.0 ± 5.0 [24]

It is probable that further combustion measurements on gallium and indium compounds should prove successful.

Group IVA Titanium, Zirconium, Hafnium

Rabinovich et al. [25,26] burned several titanium compounds in a static bomb using paraffin wax coated samples and measuring the total CO_2 produced. For those compounds containing chlorine, a glass wool liner holding hydrazine hydrochloride solution was placed in the bomb to reduce chlorine to chloride. The work was done carefully and the results reported were; (Cp = cyclopentadienyl)

$\Delta_f H_m^0 (cr)/(kJ \cdot mol^{-1})$			
$TiCp_2$	-70.7 ± 9.5	$Cp_2Ti(OCOC_6H_5)$	-967.3 ± 13.6
Cp_2TiCl_2	-384.8 ± 8.7	$CpTiCl_3$	-610.7 ± 7.9
$Cp_2Ti(CH_3)_2$	-29.0 ± 9.5	$Cp_2Ti(C_6H_5)_2$	70.8 ± 13.6

In spite of the care taken, some of these results are grossly in error as shown by reaction-calorimetric studies carried out in Lisbon [27]. From the enthalpies of reactions such as

$$Cp_2Ti(CH_3)_2(cr) + 2HCl(soln) = Cp_2TiCl_2(soln) + 2CH_4(g)$$

the difference in the enthalpies of formation of $Cp_2Ti(CH_3)_2$ and Cp_2TiCl_2 can be accurately determined. These differences are:

	combustion	reaction
	$kJ \cdot mol^{-1}$	
$Cp_2Ti(CH_3)_2$-Cp_2TiCl_2	355.8 ± 12.9	356.6 ± 12.7
$Cp_2Ti(C_6H_5)_2$-Cp_2TiCl_2	455.6 ± 16.1	645.4 ± 11.7
$Cp_2Ti(O_2CC_6H_5)_2$-Cp_2TiCl_2	-582.5 ± 16.1	-406.5 ± 11.9

The enthalpies of formation of the Cp_2TiL_2 derivatives derived from reaction calorimetry were based on the value of $\Delta_f H_m^0 (Cp_2TiCl_2, cr)$ as determined by combustion calorimetry: if the latter is in error there would be a constant error in the enthalpies of formation of all the Cp_2TiL_2 derivatives. From the manner of interpretation of the results such a constant error would have no effect on the derived values of E(Ti-L) and $\overline{D}$(Ti-L), nevertheless it would be desirable to obtain a reliable value for $\Delta_f H_m^0$(Cp$_2$TiCl$_2$,c), probably by rotating-bomb calorimetry. This would be a difficult project but would lead to more certainty in the values of $\Delta_f H_m^0$ for Cp$_2$TiL$_2$ complexes. Other combustion results on organotitanium compounds merely serve to demonstrate that the combustion method is unsuitable for compounds of this element.

	$\Delta_f H_m^0$(l)/(kJ·mol^{-1})	
	combustion	reaction calorimetry
Ti(OEt)$_4$	-1475.9 ± 6.6 [28]	-1460.1 ± 3.5 [29]
Ti(O-iPr)$_4$	-1579.3 ± 3.9 [30]	-1702.9 ± 8.4 [31]
Ti(NMe$_2$)$_4$	-455.0 ± 6.4 [32]	-382.0 ± 5.4 [31]
Ti(NEt$_2$)$_4$	-488.4 ± 8.6 [33]	-619.2 ± 4.2 [31]
	-504.5 ± 7.0 [32]	

One concludes from this comparison that whenever there is an alternative method available, the combustion method should be avoided for organotitanium compounds.

Combustion measurements made using techniques similar to those employed for the titanium compounds have been made for zirconium and hafnium compounds [34]: the results are listed below.

$\Delta_f H_m^0$(cr)/(kJ.mol^{-1})	
Cp$_2$Zr(CH$_3$)$_2$	-44.4 ± 2.1
Cp$_2$Zr(C$_6$H$_5$)$_2$	275.7 ± 10.9
Cp$_2$ZrCl$_2$	-538.1 ± 2.9
Cp$_2$HfCl$_2$	-536.0 ± 2.5

It is likely that the criticisms made of combustion results for titanium compounds also apply to measurements made on zirconium and hafnium compounds and it would be unwise to rely unduly on these results.

Group IVB: Germanium, Tin and Lead

Germanium

Static-bomb combustion measurements on organogermanium compounds seem to be reasonably successful although the rotating-bomb method is much superior. A problem arises concerning the state of the solid GeO_2 produced because the energies of transition between the three forms are large [35]:

$$GeO_2(am) \rightarrow GeO_2(c, hexagonal) \quad \Delta H = -15.7 \text{ kJ·mol}^{-1}$$

$$GeO_2(am) \rightarrow GeO_2(c, tetragonal) \quad \Delta H = -41.1 \text{ kJ·mol}^{-1}$$

Powder X-ray photographs on the products of combustion of $Ge(C_2H_5)_4$ in Manchester [36] indicated the hexagonal form whereas for combustion of the same compound in Gorky [37], the tetragonal form was found. It seems sensible to assume that the form actually produced is amorphous, the same state as that produced in the combustion of germanium metal admixed with paraffin wax as combustion aid, [38] although that was originally considered to refer to the hexagonal form. It is this uncertainty concerning the form of the GeO_2 that casts some doubt on static-bomb combustion results on organogermanium compounds. The static-bomb combustion results reported so far are:

	$\Delta_f H_m^o/(\text{kJ·mol}^{-1})$
$(CH_3)_4Ge(l)$	-98.3 ± 9.4 [39]
	-131.1 ± 8.3 [40]
$(C_2H_5)_4Ge(l)$	-189.7 ± 3.3 [36]
	-205.0 ± 7.1 [37]
$(n-C_3H_7)_4Ge(l)$	-288.3 ± 4.9 [36]
$((C_2H_5)_3Ge)_2(l)$	-372.9 ± 11.9 [37]
$(C_2H_5)_3GeOO(t-C_4H_9)$ (l)	-486.2 ± 7.0 [41]
$((C_2H_5)_3Ge)_2O$ (l)	-611.3 ± 11.9 [41]
$((CH_3)_2GeO)_4(c)$	-1514.5 ± 25.9 [42]
$((C_2H_5GeO)_4(c)$	-1519.3 ± 27.6 [43]
$(C_2H_5)_3GeN(C_2H_5)_2$ (l)	-342.5 ± 6.1 [44]
$((C_2H_5)_3Ge)_2Hg$ (l)	-101.0 ± 9.5 [45]
$((i-C_3H_7)_2Ge)_2Hg$ (l)	-273.4 ± 9.7 [45]

The disagreements between the results for $(CH_3)_4Ge$ and $(C_2H_5)_4Ge$ do show that the static-bomb combustion method is not the ideal method for organogermanium compounds.

Tin

Static-bomb combustion calorimetry of organotin compounds has been a great success: combustions are carried out at a higher than usual pressure of oxygen (45 atm), CO_2 analyses have demonstrated the complete combustion of the organic part of the molecule and only minor corrections were required for the presence of traces of free metal and of the monoxide in the solid residue. There is excellent agreement between independent measurements on $Sn(CH_3)_4$, $Sn(C_2H_5)_4$, $Sn(n-C_4H_9)_4$, $Sn(C_6H_5)_4$ and $Sn_2(C_6H_5)_6$. In the following table are listed the best values for the enthalpies of formation of organotin compounds as obtained by combustion, enthalpies of vaporization and the enthalpies of formation in the gaseous state.

	$\Delta_f H_m^0 (l,cr)$	$\Delta_{cr\ l}^g H_m^0$	$\Delta_f H_m^0 (g)$
		kJ·mol^{-1}	
$(CH_3)_4Sn(l)$	-52.3 ± 1.9 [46]	33.1 ± 1.3	-19.2 ± 2.5
$(CH_3)_3Sn(C_2H_5)(l)$	-67.2 ± 2.4 [46]	37.7 ± 1.7	-29.5 ± 2.9
$(CH_3)_3Sn(i-C_3H_7)(l)$	-87.4 ± 4.2 [47]	40.6 ± 2.1	-46.8 ± 4.7
$(CH_3)_3Sn(t-C_4H_9)(l)$	-120.9 ± 4.6 [47]	54.0 ± 4.2	-66.9 ± 6.2
$(CH_2=CH)_4Sn(l)$	300.8 ± 7.6 [48]		
$(C_2H_5)_3Sn(CH=CH_2)(l)$	36.0 ± 2.9 [48]		
$(C_2H_5)_4Sn(l)$	-95.7 ± 1.6 [46]	51.0 ± 2.1	-44.7 ± 2.6
$(CH_2=CHCH_2)_4Sn(l)$	-170.2 ± 7.1 [48]		
$(n-C_3H_7)_4Sn(l)$	-211.3 ± 5.0 [46]	66.9 ± 2.1	-144.4 ± 5.4
$(i-C_3H_7)_4Sn(l)$	-184.7 ± 5.0 [47]	64.9 ± 4.2	-122.5 ± 6.5
$(n-C_4H_9)_4Sn(l)$	-304.6 ± 6.7 [46]	82.8 ± 2.1	-221.8 ± 7.0
$(i-C_4H_9)_4Sn(l)$	-331.0 ± 5.9 [48]		
$(C_6H_5)_4Sn(cr)$	411.6 ± 3.7 [49]	161.1 ± 4.2	572.7 ± 5.6
$(C_2H_5)_6Sn_2(l)$	-217.6 ± 8.4 [50]	62.8 ± 4.2	154.8 ± 9.4
$(C_6H_5)_6Sn_2(cr)$	661.4 ± 15.6 [50]	188.3 ± 4.2	849.7 ± 16.2
$(C_6H_5)_3Sn(CH=CH_2)(cr)$	411.4 ± 6.6 [51]	114.1 ± 2.5	525.5 ± 7.1
$(C_6H_5)_3Sn(C\equiv CC_6H_5)(cr)$	596.2 ± 7.8 [51]	137.6 ± 2.0	733.8 ± 8.1
$(C_6H_5)_2Sn<(CH_2)_4(cr)$	195.0 ± 5.2 [52]	106.8 ± 5.4	301.8 ± 7.5
$(C_6H_5)_2Sn<(CH_2)_5(l)$	214.0 ± 6.8 [52]	75.0 ± 1.5	289.0 ± 6.3

There are sufficient enthalpies of formation of organotin compounds for the values in the gaseous state to be tested by application of a bond-energy scheme. In the following table the Laidler scheme has been used with the parameters given in reference [53].

| | $\Delta_f H_m^0(g)/(kJ \cdot mol^{-1})$ | | |
	observed	calculated	Δ
$(CH_3)_4Sn$	-19.2 ± 2.5	-19.2	0.0
$(CH_3)_3Sn(C_2H_5)$	-29.5 ± 2.9	-28.4	-1.1
$(CH_3)_3Sn(i-C_3H_7)$	-46.8 ± 4.7	-46.0	-0.8
$(CH_3)_3Sn(t-C_4H_9)$	-66.9 ± 6.2	-104.2	37.3
$(C_2H_5)_4Sn$	-44.7 ± 2.6	-55.5	10.8
$(n-C_3H_7)_4Sn$	-144.4 ± 5.4	-137.9	-6.5
$(n-C_4H_9)_4Sn$	-221.8 ± 7.0	-220.3	1.5
$(CH_3)_3Sn(C_6H_5)$	113.1 ± 5.2	116.9	-3.8
$(C_6H_5)_4Sn$	572.7 ± 5.6	525.5	47.2
$(CH_3)_6Sn_2$	-26.9 ± 8.4	-26.9	0.0
$(C_6H_5)_6Sn_2$	849.7 ± 16.2	790.2	59.5
$(C_6H_5)_2Sn{<}(CH_2)_4$	301.8 ± 7.5	270.9	30.9
$(C_6H_5)_2Sn{<}(CH_2)_5$	289.0 ± 6.3	250.3	38.7

It is apparent that the application of a bond-energy scheme gives sensible results in that for those molecules in which steric strain is expected, $(CH_3)_3Sn(t-C_4H_9)$, $(C_6H_5)_4Sn$, $(C_6H_5)_6Sn_2$ and in the cyclic compounds $(C_6H_5)_2$ $Sn{<}(CH_2)_4$, $(C_6H_5)_2Sn{<}(CH_2)_5$, it is in fact observed. For the last two molecules the strain probably arises from interaction between the phenyl groups and that the five and six-membered rings appear to be strain-free. As independent measurements for some particular compounds are in agreement and a bond-energy scheme produces a reasonable outcome, we can be confident in the results of static-bomb combustion calorimetry for organotin compounds.

Group VA Vanadium, Niobium, Tantalum

Tel'noi *et al.* [25] using the technique of paraffin wax coated samples and measuring the total carbon dioxide produced determined the enthalpy of combustion of dicyclopentadienyl vanadium. Bisbenzene vanadium was measured by Fischer and Reckziegel [54], and in both investigations the solid product was found to be $V_2O_5(cr)$. A series of niobium and tantalum alkoxides were studied using static-bomb combustion calorimetry by Tel'noi *et al.* [55]: the samples were enclosed in polythene capsules to protect them from water vapour prior to ignition because of the ease of their hydrolysis and this suggests that in this case that reaction-solution calorimetry would have been a satisfactory experimental method. The results however, appear to be sound as from the enthalpies of formation in the gaseous state the mean bond dissociation enthalpies <D(M-L)> can be derived and are listed below. The constancy of <D(Nb-O)> and <D(Ta-O)> suggests that the combustion experiments were satisfactory for these compounds and incomplete combustion of the metal would be unlikely because in the molecule the metal atom is surrounded by oxygen atoms.

	$\Delta_f H_m^0/(kJ \cdot mol^{-1})$	<D(M-L)>/(kJ·mol^{-1})
$V(C_5H_5)_2(cr)$	144.9 ± 8.8	420.0 ± 10.0
$V(C_6H_6)_2(cr)$	37.2 ± 15.3	286.3 ± 10.0
$Nb(OCH_3)_5(l)$	-1413.8 ± 2.5	419.2 ± 10.9
$Nb(OC_2H_5)_5(l)$	-1584.1 ± 2.5	417.1 ± 11.3
$Nb(O-n-C_3H_7)_5(cr)$	-1701.6 ± 4.2	418.4 ± 10.5
$Ta(OCH_3)_5(l)$	-1417.5 ± 2.5	438.9 ± 11.7
$Ta(OC_2H_5)_5(l)$	-1638.5 ± 2.9	440.1 ± 10.9
$Ta(O-n-C_3H_7)_5(cr)$	-1759.4 ± 4.2	439.7 ± 10.9

Group VB Arsenic, Antimony, Bismuth

Static-bomb combustion calorimetry has been applied to compounds of arsenic, antimony and bismuth. Long and Sackman [56] studied the trimethyls or arsenic, antimony and bismuth; Lautsch *et al.* [48] the corresponding triethyl derivatives; and Birr [57] the triphenyls of arsenic and bismuth. The static-bomb method does not appear to be very satisfactory. For the arsenic compound combustions, the solid product consisted of the three oxides As_2O_3, As_2O_4 and As_2O_5 plus some elemental arsenic and carbon. The solution in the bomb contained As(III) and As(V), hence a complex analytical procedure was required, nevertheless, the final state could not be thermodynamically well defined. For the organoantimony combustions the solid product contained Sb_2O_3, Sb_2O_4 and some unburned antimony and carbon, for the bismuth compounds the solid product was Bi_2O_3 containing a substantial proportion of elemental bismuth. The static-bomb combustion results are given in the following table but because of the difficulties in analysing the final state, these results should be treated with caution. A further demonstration that the static-bomb combustion method is unsatisfactory for compounds of these elements is that triphenylarsine and triphenylbismuth have been studied by rotating-bomb combustion calorimetry: for $AsPh_3(cr), \Delta_f H_m^o/(kJ \cdot mol^{-1}) = 310.0 \pm 6.7$ [58], differing by 83.3 $kJ \cdot mol^{-1}$ from the static-bomb combustion value, and for $BiPh_3(cr), \Delta_f H_m^o/(kJ \cdot mol^{-1}) = 489.7 \pm 5.2$ [59] differing by 20.7 $kJ \cdot mol^{-1}$ from the static-bomb value

$\Delta_f H_m^o/(kJ \cdot mol^{-1})$		
As(CH$_3$)$_3$(l)-16.3 ± 10.0	As(C$_2$H$_5$)$_3$(l) 13.0 ± 16.7	As(C$_6$H$_5$)$_3$(cr) 393.3 ± 16.7
Sb(CH$_3$)$_3$(l) 0.8 ± 25.1	Sb(C$_2$H$_5$)$_3$(l) 5.0 ± 10.5	Sb(C$_6$H$_5$)$_3$(cr) 329.3 ± 16.7
Bi(CH$_3$)$_3$(l) 158.2 ± 14.2	Bi(C$_2$H$_5$)$_3$(l) 169.9 ± 16.7	Bi(C$_6$H$_5$)$_3$(cr) 469.0 ± 16.7

Group VI A Chromium, Molybdenum; Tungsten

Hexacarbonyls of Chromium, Molybdenum and Tungsten

Several measurements of the enthalpies of combustion of the hexacarbonyls of the group VIA metals have been reported. The enthalpies of formation are listed below and are compared with the values determined by thermal decomposition using hot-zone calorimetry, these latter values can be considered reliable.

	$\Delta_f H^0_m(cr)/(kJ \cdot mol^{-1})$	
	combustion	decomposition
$Cr(CO)_6$	-1043.5 ± 1.7 [60]	-978.2 ± 2.1 [61]
	-1077.8 ± 4.6 [62]	
	-982.2 ± 6.2 [63]	
	-999.6 ± 18.8 [61]	
$Mo(CO)_6$	-990.8 ± 4.2 [64]	-989.1 ± 1.5 [64]
	-986.6 ± 6.2 [60]	
	-982.8 ± 6.2 [62]	
$W(CO)_6$	-961.1 ± 4.2 [65]	-959.4 ± 2.1 [65]
	-942.7 ± 6.2 [60]	

Barnes *et al.* [61,64,65] measured the energies of combustion in dry oxygen, as the carbonyls contain no hydrogen there could be no complication from hydration of the oxides produced. For the molybdenum and tungsten hexacarbons the carbon dioxide recoveries were of the order of 99.97 per cent and the oxide residues contained only trace amounts of unburned metal. The combustion measurements on chromium hexacarbonyl were less successful, the carbon dioxide recovery was 99.74 per cent and difficulties were experienced in analysing the oxide residue for unburned metal. Shuman *et al.* [63] burned chromium hexacarbonyl with a relatively large amount of polythene and benzoic acid as combustion aids and obtained complete conversion to chromium oxide with no unburned metal and no evidence of hydration of the oxide. The measurements on these hexacarbonyls demonstrate that with care, the static-bomb combustion method can produce successful results for compounds of these metals.

Chromium

Tel'noi, Rabinovich *et al.* [66,67,68] have measured the enthalpies of combustion of a series of organochromium compounds using a technique similar to that in their successful study of chromium hexacarbonyl. The results are listed below.

$$\Delta_f H_m^0/(\text{kJ·mol}^{-1})$$

$(C_5H_5)_2Cr(cr)$ 178.1 ± 2.6 [66]	$(C_6H_6)_2CrCl(cr)$ -43.9 ± 12.7 [67]
$(C_6H_6)_2Cr(cr)$ 146.4 ± 8.4 [67]	$(C_6H_6)_2CrBr(cr)$ -3.6 ± 8.6 [67]

$(C_6H_5C_2H_5)_2Cr(l)$ 59.4 ± 4.0 [68] $(1,3-(CH_3)_2C_6H_4)_2CrI(cr)$ -188.5 ± 8.8 [67]

$(1,2-(C_2H_5)_2C_6H_4)_2Cr(l)$ -55.6 ± 7.6 [68]

$(1,3,5-(CH_3)_3C_6H_3)_2CrI(cr)$ -188.5 ± 8.8 [67]

$(1,2-(iC_3H_7)_2C_6H_4)_2Cr(l)$ -204.4 ± 8.4 [68]

$(C_6H_5C_6H_5)_2CrI(cr)$ 151.8 ± 9.0 [67]

$(i-C_3H_7C_6H_5)(1,2-(iC_3H_7)_2C_6H_4Cr(l)$ -132.4 ± 7.5 [68]

Further evidence in support of these results is provided by the fact that the microcalorimetric study of the thermal decomposition of bis-benzene chromium by Connor et al. [69] gave $\Delta_f H_m^0/(C_6H_6)_2Cr,cr)/(\text{kJ·mol}^{-1}) = 141.9 ± 4.9$, in agreement with the combustion value.

Molybdenum and Tungsten

Tel'noi et al. [70] applied their static-bomb combustion technique to some dicyclopentadienyl derivatives of molybdenum and tungsten: the enthalpies of formation are listed below.

$$\Delta_f H_m^0(cr)/(\text{kJ·mol}^{-1})$$

Cp_2MoH_2	195.0 + 2.5	Cp_2WH_2	250.6 ± 2.5
Cp_2MoCl_2	-95.8 ± 2.5	Cp_2WCl_2	-71.1 ± 2.5
Cp_2MoI_2	30.1 ± 2.1	Cp_2WI_2	117.7 ± 2.9

Dicyclopentadienyl tungsten dichloride was studied by rotating-bomb calorimetry by Minas da Piedade et al. [71] who obtained -63.5 ± 7.7 kJ·mol⁻¹ for its enthalpy of formation in reasonably good agreement with the static-bomb value. The results of these static-bomb measurements are not however correct to within the limits of experimental uncertainty because, from reaction-solution calorimetric studies made in Lisbon [72,73], differences between the enthalpies of formation of some of these derivatives were accurately determined, and comparison of the differences is given below.

	combustion	reaction
	$kJ \cdot mol^{-1}$	
$Cp_2MoH_2\text{-}Cp_2MoCl_2$	290.8 ± 3.5	306.1 ± 6.2 [72]
$Cp_2MoI_2\text{-}Cp_2MoCl_2$	125.9 ± 3.3	165.6 ± 8.2 [73]
$Cp_2WH_2\text{-}Cp_2WCl_2$	321.7 ± 4.0	285.9 ± 5.0 [72]
$Cp_2WI_2\text{-}Cp_2WCl_2$	188.1 ± 4.3	128.9 ± 7.6 [73]

This comparison shows that the errors in the combustion results are very much smaller than was apparent in the corresponding comparisons for titanium compounds suggesting that further work to refine the combustion method may ultimately result in successful measurements for molybdenum and tungsten compounds.

Group VIIA Manganese, Rhenium

The enthalpy of combustion of dicyclopentadienylmanganese was measured by Rabinovich $et\ al.$ [24] with the sample protected by paraffin wax and with satisfactory carbon dioxide recoveries: the solid product was $(MnO(10\text{-}24\%)+Mn_3O_4)$. There are two independent measurements on cyclopentadienylmanganese tricarbonyl, in not very good agreement.

	$\Delta_fH_m^o(cr)/(kJ \cdot mol^{-1})$
Cp_2Mn	200.8 ± 4.2
$CpMn(CO)_3$	-525.5 ± 4.2 [74]
	-542.2 ± 8.4 [48]
$Re_2(CO)_{10}$	-1558.5 ± 6.7 [75]

Combustion measurements on rhenium compounds are limited to a study of dirhenium decacarbonyl by Chemova *et al.* [75] but the enthalpy of formation is in very poor agreement with that derived from microcalorimetric measurements of the enthalpy of decomposition [76],

$$\Delta_f H_m^0(cr)/(kJ \cdot mol^{-1}) = -1660 \pm 11.$$

Group VIII Iron, Cobalt, Nickel

The enthalpies of combustion of iron pentacarbonyl and of nickel tetracarbonyl were measured by Cotton *et al.* [77,78] and the derived enthalpy of formation of nickel tetracarbonyl is in reasonable agreement with that derived from equilibrium studies by Sykes and Townshend, [79]

$$\Delta_f H_m^0(l)/(kJ \cdot mol^{-1}) = -622.6 \pm 5.0$$

	$\Delta_f H_m^0/(kJ \cdot mol^{-1})$
$Fe(CO)_5(l)$	-765.1 ± 6.6 [77]
$Ni(CO)_4(l)$	-633.5 ± 4.6 [78]
$Cp_2Fe(cr)$	168.2 ± 2.6 [66]
$Cp_2Co(cr)$	236.6 ± 2.5 [66]
$Cp_2Ni(cr)$	284.8 ± 4.6 [66]

The dicyclopentadienyls of iron, cobalt and nickel were measured by Rabinovich *et al.* [66] with the samples protected by paraffin wax and with satisfactory carbon dioxide recoveries. In each experiment careful analysis of the solid product was required as these were found respectively to be, $[Fe_2O_3(7-14\%)+Fe_3O_4]$, $[Co(7-25\%)+CoO(54-71\%)+Co_3O_4(12-25\%)]$ and $[Ni(16-50\%)+NiO]$.

General assessment of static-bomb combustion calorimetry applied to Organometallics

The work described so far represents a major effort on the part of many experimentalists to determine enthalpies of formation of organometallic compounds by static-bomb combustion calorimetry. The quantity of work and the care with which it has been done are both admirable which makes greater the disappointment of the general quality of the final results. It is impossible however, to fail at any task without first making an attempt. Summarizing the overall situation is obviously a personal judgement, the view of this author is as follows.

(1) The method has had great success and is probably superior to reaction-solution calorimetric methods for compounds of Hg and Sn.

(2) There is apparent success but reaction-solution calorimetric methods are easier and superior for compounds of Li, Ga and Zn. For Ge compounds, the rotating-bomb method is to be preferred.

(3) For compounds of Cr, Mo and W it is possible to obtain successful results but these compounds do present very great experimental difficulties.

(4) Results for compounds of the following elements should be regarded with caution because satisfactory confirmation of the results using other experimental methods is required: Zr, Hf, Fe, Co, Ni, Mn, V, Nb, Ta, Sc, Y, La, Pr, Tm, Yb.

(5) The method appears to be unsatisfactory for compounds of Cd, Ti, Re, As, Sb, Bi so that reaction-solution calorimetric or rotating-bomb methods should be employed.

(6) The results for compounds of Si, Al and Pb should be rejected and it does not seem at all worthwhile for such measurements on compounds of these elements to be attempted in the future.

ROTATING-BOMB COMBUSTION CALORIMETRY

The rotating-bomb technique involves placing a suitable solution in the bomb so that shortly after combustion of the sample rotating the bomb can ensure that solid residues are dissolved. Moreover, rotation ensures that the final solution is of uniform concentration and that thermodynamic equilibrium is established between the liquid and gaseous phases. Although the final state can be defined, many of the correction terms needed to convert the observed energy change to the standard state value are unknown. Comparison experiments, by burning a sample of known enthalpy of formation in such amount as to produce a final solution of the same composition enables these corrections to be cancelled.

Rotating-bomb measurements on organometallic compounds have been successful in producing reliable results. Unfortunately, only a few such measurements have been reported because rotating-bomb calorimeters are not as widely available as static-bomb calorimeters.

Silicon

The rotating-bomb technique for silicon compounds was introduced by Good et al. [80]. The compound was mixed with a fluorine containing derivative so after combustion and rotation, the solution in the bomb was [H_2SiF_6+HF]aq. For hexamethyldisiloxane with $\alpha\alpha\alpha$-trifluorotoluene as combustion auxiliary, the combustion reaction was $[Si(CH_3)_3]_2O(l) + 5C_6H_5CF_3(l) + 49.5O_2(g) + 406H_2O(l) \rightarrow 41CO_2(g) + 2\{[H_2SiF_6+1.5HF](212H_2O)\}(l)$.

Comparison experiments were made by burning elemental silicon with polyvinylidene fluoride to produce the same final solution. The technique is difficult and has the drawback that combustion of the fluorine containing auxiliary contributes about 75 per cent to the total energy quantity. Nevertheless, this is the only reliable method for measuring enthalpies of combustion of organosilicon compounds and measurements made by static-bomb calorimetry should be ignored. The reported values that can be considered most reliable are:

$\Delta_f H_m^0/(kJ \cdot mol^{-1})$	
$[Si(CH_3)_3]_2O(l)$	-814.6 ± 5.4 [80]
$Si(CH_3)_4(l)$	-257.9 ± 3.2 [81]
$Si(C_6H_5)_4(cr)$	184.9 ± 5.9 [82]

Germanium

The rotating-bomb method for germanium compounds was introduced by Bills and Cotton [83], who measured the energy of combustion of germanium tetraethyl encapsulated in polyester bags. Aqueous hydrofluoric acid was used to dissolve the germanium oxide produced by the combustion. Comparison experiments were made by burning benzoic acid and simultaneously dissolving a sample of $GeO_2(cr, hexagonal)$ in the HF solution. All the other rotating-bomb measurements on organogermanium compounds have been made by Carson et al. who placed KOH (aq) in the bomb to dissolve the GeO_2 and made the comparison experiments in a similar way. The rotating-bomb results are:

$\Delta_f H_m^0/(kJ \cdot mol^{-1})$			
$Ge(C_2H_5)_4(l)$	-206.4 ± 7.5 [83]	$(C_6H_5)_3GeCH=CH_2(cr)$ 263.9 ± 7.0 [86]
$Ge(C_6H_5)_4(cr)$	288.6 ± 23.6 [84]	$(C_6H_5)_3GeC \equiv CC_6H_5(cr)$ 471.4 ± 7.9 [86]
$Ge(CH_2C_6H_5)_4(cr)$	223.3 ± 11.9 [85]	$[(C_6H_5)_3Ge]_2O(cr)$ 161.1 ± 10.6 [86]
$(C_6H_5Ge)_2(cr)$	453.7 ± 14.2 [85]	$(C_6H_5)_3Ge<(CH_2)_4(cr)$ 34.3 ± 10.0 [87]

Lead

The first application of rotating-bomb calorimetry to organometallic compounds was for tetraethyllead by Good *et al.* [88] because these authors were so suspicious of their static-bomb results on this compound. The solution placed in the bomb was nitric acid containing some arsenious oxide to ensure that the dissolved lead was in the Pb^{2+} state. For comparison experiments, hydrocarbon oil was burned in the presence of $Pb(NO_3)_2(cr)$ to produce the same final solution, hence the enthalpy of formation was derived from the enthalpy of the reaction

$$(C_2H_5)_4Ph(l) + 13.5\ O_2(g) + 2[HNO_3\ 30H_2O](l) \rightarrow 8CO_2(g) + 11\ H_2O(l) + Pb(NO_3)_2(cr)$$

The rotating-bomb results for lead compounds that have been reported are:

	$\Delta_fH_m^0/(kJ \cdot mol^{-1})$
$Pb(CH_3)_4(l)$	98.1 ± 4.4 [89]
$Pb(C_2H_5)_4(l)$	53.1 ± 5.0 [88]
$Pb(C_6H_5)_4(cr)$	515.3 ± 15.4 [90]

The static-bomb results were for $Pb(CH_3)_4$ -24.7, and for $Pb(C_2H_5)_4$ 219.7 $kJ \cdot mol^{-1}$ showing that method to be completely unsatisfactory.

Arsenic, Bismuth

Rotating-bomb calorimetry has been applied to only one arsenic compound, triphenylarsine, by Mortimer and Sellers, [58] with a solution of NaOH (aq) in the bomb, the final solution contained sodium arsenite, sodium arsenate, sodium carbonate and the excess sodium hydroxide. Comparison experiments consisted of burning benzoic acid in the presence of NaOH (aq) and separate experiments were made to determine the energy of oxidation of arsenite to arsenate. The result $\Delta_fH_m^0/(C_6H_5)_3As,cr)/(kJ \cdot mol^{-1}) = 310.0 \pm 6.7$ differed from the static-bomb value by 83.3 $kJ \cdot mol^{-1}$. Steele [59] measured $\Delta_fH_m^0$ $((C_6H_5)_3Bi,cr)/(kJ \cdot mol^{-1}) = 489.7 \pm 5.2$ by rotating-bomb calorimetry using a method similar to that for the arsenic compound, the result differed from the static-bomb value by 20.7 $kJ \cdot mol^{-1}$

Dimanganesedecacarbonyl

Good *et al.* [91] measured $Mn_2(CO)_{10}$ by rotating-bomb calorimetry, with a bomb solution of HNO_3 (aq) containing some H_2O_2 to ensure the dissolved manganese was in the Mn^{2+} state. Solution of the manganese could only be obtained when combustions were done under a low pressure, 5 atm, of oxygen so relatively large quantities of carbon were present in the final state. Comparison experiments were made by burning hydrocarbon oil in the presence of $Mn_2(NO_3)_2$ 10.3 H_2O solution, hence the enthalpy of formation was derived from the enthalpy of the reaction

$$Mn_2(CO)_{10}(cr) + 6O_2(g) + 4[HNO_3\ 16H_2O](l) \rightarrow 10CO_2(g) + 45.4\ H_2O(l) + 2[Mn(NO_3)_2\ 10.3\ H_2O](l)$$

The result, $\Delta_f H_m^o$ $(Mn_2(CO)_{10},cr)/(kJ\cdot mol^{-1}) = -1677 \pm 3.3$ is in agreement with a value determined by a microcalorimetric determination of the enthalpy of iodination, which gave for $\Delta_f H_m^o(cr)$, $-1674.4 \pm 8.4\ kJ\cdot mol^{-1}$ [69]

CONCLUSION

It is apparent that the contribution of combustion calorimetry to the thermochemistry of organometallic compounds has been much less than would reasonably have been anticipated. The static-bomb method has been especially disappointing in its outcome whereas the rotating-bomb method has been successful in all cases where it has been applied, unfortunately these are too few. The alternative procedure of reaction-solution calorimetry is well recognised as an excellent method for deriving enthalpies of formation. It cannot however, be applied as a universal method.

Combustion calorimetry will continue to play a role in this field. The rotating-bomb method should be developed to extend to compounds of many more elements and the investigator should be selective in the compounds to be studied, so that combustion results can be combined profitably with those from reaction-solution calorimetry. A simple example would suffice: if the rotating-bomb method could be successfully applied to compounds such as Cp_2TiCl_2, Cp_2MoCl_2 as we believe has been done for Cp_2WCl_2[71], then the extensive results from reaction calorimetry on Cp_2ML_2 compounds could be placed in a firmer setting.

REFERENCES

1. Yu. A. Lebedev, E.A. Miroshnickenko and A.M. Chaikin, *Dokl. Akad. Nauk. SSSR*, **145**, 751 (1962).

2. P.A. Fowell and C.T. Mortimer, *J. Chem. Soc.* 3793 (1961).

3. L.H. Long and R.G.W. Norrish, *Phil. Trans. Roy. Soc. London*, **A241**, 587 (1949).

4. W.F. Lautsch, P. Erzberger and A. Tröber, *Wiss. Z. Tech. Hochsch. Chem. Leuna-Merseburg*, **1**, 31 (1958).

5. A.S. Carson, K. Hartley and H.A. Skinner, *Trans. Faraday Soc.* **45**, 1159 (1949).

6. A.S. Carson, K. Hartley and H.A. Skinner, *Proc. Roy. Soc. London*, **A195**, 500 (1949).

7. A.S. Carson, E.M. Carson and B. Wilmshurst, *Nature*, **170**, 320 (1952).

8. D.M. Fairbrother and H.A. Skinner, *Trans. Faraday Soc.* **52**, 956 (1956).

9. A.S. Carson and B. Wilmshurst, *J. Chem. Thermodynamics*, **3**, 251 (1971).

10. K. Hartley, H.O. Pritchard and H.A. Skinner, *Trans. Faraday Soc.* **46**, 1019 (1950).

11. K. Hartley, H.O. Pritchard and H.A. Skinner, *Trans. Faraday Soc.* **47**, 254 (1951).

12. C.L. Chernick, H.A. Skinner and I. Wadsö, *Trans. Faraday Soc.* **52**, 1088 (1956).

13. A.S. Carson and J.A. Spencer, *J. Chem. Thermodynamics*, **16**, 423 (1984).

14. C.G. Devyatykh, I.B. Rabinovich, V.I. Tel'noi, G.K. Borisov and L.F. Zyazina, *Dokl. Akad. Nauk. SSSR*, **217**, 673 (1974).

15. D.J. De Frees, R.T. McIver and W.J. Hehre, *J. Am. Chem. Soc.* **102**, 3334 (1980).

16. Yu. Kh. Shaulov, G.O. Shmyreva and V.S. Tubyanskaya, *Russ. J. Phys. Chem.* **39**, 51 (1965).

17. S. Pawlenko, *Chem. Ber.* **100**, 3591 (1967).

18. C.O. Shmyreva, G.B. Sakharovskaya, A.F. Popov, N.N. Korneev and A.A. Smolyaninova, *Russ. J. Phys. Chem.* **45**, 260 (1971).

19. C.T. Mortimer and P. Sellers, *J. Chem. Soc.* 1978 (1963).

20. P.A. Fowell, Ph.D. Thesis, *University of Manchester* (1961).

21. K. Cavell and G. Pilcher *J. Chem. Soc. Faraday Trans. 1*, **73**, 1590 (1977).

22. P.A. Fowell and C.T. Mortimer, *J. Chem. Soc.* 3734 (1958).

23. L.H. Long and J. F. Sackman, *Trans. Faraday Soc.* **54**, 1797 (1958).

24. G.M. Kol'yakova, I.B. Rabinovich and E.N. Zorina, *Dokl. Akad. Nauk. SSSR*, **197**, 353 (1971).

25. V.I. Tel'noi, I.B. Rabinovich, V.N. Latyaeva and A.N. Lineva, *Dokl. Akad. Nauk. SSSR*, **197**, 353 (1971).

26. V.I. Tel'noi, I.B. Rabinovich, V.D. Tikhonov, V.N. Latyaeva, L.L. Vyshmskaya and G.A. Razuvaev, *Dokl. Akad. Nauk. SSSR*, **174**, 1374 (1967).

27. M.J. Calhord, A.R. Dias, M.E. Minas da Piedade, M.S. Salema and J.A. Martinho Simões, *Organometallics*, **6**, 734 (1987).

28. E.A. Volchkova, D.D. Smolyaninova, V.G. Genchel, L.L. Lopatkina and Yu. Kh. Shaulov, *Russ. J. Phys. Chem.* **46**, 1053 (1972).

29. D.C. Bradley and M.J. Hillyer, *Trans. Faraday. Soc.* **62**, 2367 (1966).

30. V.G. Genchel, E.A. Volchkova, R.M. Aizatullova and Yu. Kh. Shaulov, *Russ. J. Phys. Chem.* **47**, 643 (1973).

31. M.F. Lappert, D.S. Patil and J.B. Pedley, *J. Chem. Soc. Chem. Comm.* 830 (1975).

32. V.E. Mikhailov, A.K. Baev, A.I. Sachek and V.M. Al'khimovich, *Russ. J. Phys. Chem.* **60**, 1382 (1986).

33. G.M. Kol'yanov, I.B. Rabinovich and N.S. Vyazanken, *Dokl. Akad. Nauk. SSSR*, **200**, 735 (1971).

34. K.V. Kir'yanov, V.I. Tel'noi, G.A. Vasil'eva and I.B. Rabinovich, *Dokl. Akad. Nauk. SSSR*, **231**, 1021 (1976).

35. P. Gross, C. Haymann and J.T. Bingham, *Trans. Faraday Soc.* **62**, 2338 (1966).

36. A.E. Pope and H.A. Skinner, *Trans. Faraday Soc.* **60**, 1404 (1964).

37. I.B. Rabinovich, V.I. Tel'noi, N.V. Karyakin and G.A. Razuvaev, *Dokl. Akad. Nauk. SSSR*, **149**, 217 (1963).

38. A.D. Mah and L.H. Adami, *U.S. Bureau of Mines Rpt. Invest. No.* 6034 (1962).

39. Yu. Kh. Shaulov, A.K. Federov and V.G. Genchel, *Russ. J. Phys. Chem.* **43**, 744 (1969).

40. L.H. Long and C.I. Pulford, *J. Chem. Soc. Faraday Trans. II*, **82**, 567 (1986).

41. I.B. Rabinovich, V.I. Tel'noi, N.V. Karyakin and G.A. Razuvaev, *Dokl. Akad. Nauk. SSSR*, **200**, 842 (1971).

42. E.A. Volchkova, D.D. Smolyaninova, V.G. Genchel, K. Lapatkina and Yu. Kh. Sohaulov, *Russ. J. Phys. Chem.* **46**, 1053 (1972).

43. V.G. Genchel, A.I. Toporkova, Yu. Kh. Shaulov and D.D. Smolyaninova, *Russ. J. Phys. Chem.* **48**, 1085 (1974).

44. G.M. Kol'yakova, I.B. Rabinovich and N.S. Vyazankin, *Dokl. Akad. Nauk. SSSR*, **200**, 735 (1971).

45. G.M. Kol'yakova, I.B. Ravinovich, E.N. Gladyshev and N.S. Vyazankin, *Dokl. Akad. Nauk. SSSR*, **204**, 419 (1972).

46. J.V. Davies, A.E. Pope and H.A. Skinner, *Trans. Faraday Soc.* **59**, 2233 (1963).

47. D.J. Coleman and H.A. Skinner, *Trans. Faraday Soc.* **62**, 1721 (1966).

48. W.F. Lautsch, A. Tröber, W. Zimmer, L. Mehner, W. Linck, H-M. Lehmann, H. Brandenberger, H. Korner, H-J. Metschker, K. Wagner and R. Kaden, *Z. Chem.* **3**, 415 (1963).

49. G.P. Adams, A.S. Carson and P.G. Laye, *J. Chem. Thermodynamics*, **1**, 393 (1969).

50. V.I. Tel'noi and I.B. Rabinovich, *Russ. J. Phys. Chem.* **40**, 842 (1966).

51. A.S. Carson, P.G. Laye and J.A. Spencer, *J. Chem. Thermodynamics* **17**, 277 (1985).

52. A.S. Carson, E.H. Jamea, P.G. Laye and J.A. Spencer, *J. Chem. Thermodynamics* **20**, 923 (1988).

53. G.Pilcher and H.A. Skinner, *"The Chemistry of the Metal-Carbon Bond"*, F.R. Hartley and S. Patai, editors: pp 43-90, Wiley, New York (1982).

54. E.O. Fischer and A. Reckziegel, *Chem. Ber.* **94**, 2204 (1961).

55. V.I. Tel'noi, I.B. Rabinovich, B.I. Kozykin, B.A. Salamantin and K.V. Kiranov, *Dokl. Akad. Nauk. SSSR*, **205**, 364 (1972).

56. L.H. Long and J.F. Sackman, *Trans. Faraday Soc.* **50**, 1177 (1954), **51**, 1062 (1955), **52**, 1201 (1956).

57. K.H. Birr, *Z. Anorg. Allg. Chem.* **30**, 621 (1960); **31**, 192 (1961).

58. C.T. Mortimer and P. Sellers, *J. Chem. Soc.* 1965 (1964).

59. W.V. Steele, *J. Chem. Thermodynamics* **11**, 187 (1979).

60. K.A. Sharifov and T.N. Rezhukhina, *Tr. Inst. Fiz. i. Mat. Akad. Nauk. Azerb. SSR, Ser. Fiz.* **6**, 53 (1953).

61. D.A. Pittam. G. Pilcher, D.S. Barnes, H.A. Skinner and D. Todd, *J. Less Common Met.* **42**, 217 (1975).

62. F.A. Cotton, A.K. Fischer and G. Wilkinson, *J. Am. Chem. Soc.* **78**, 5168 (1956).

63. M.S. Shuman, V.I. Chernova, V.V. Zakharov and I.B. Rabinovich, *Tr. Khim. Khim. Tekhnol., Gorky*, **1**, 78 (1974).

64. D.S. Barnes, G. Pilcher, D.A. Pittam, H.A. Skinner, D. Todd and Y. Virmani, *J. Less Common Met.* **36**, 177 (1974).

65. D.S. Barnes, G. Pilcher, D.A. Pittam, H.A. Skinner and D. Todd, *J. Less Common Met.* **38**, 53 (1974).

66. V.I. Tel'noi, K.V. Kirynov, V.I. Ermolaev and I.B. Rabinovich, *Dokl. Akad. Nauk. SSSR*, **220** 1088 (1975).

67. V.I. Tel'noi, I.B. Rabinovich, B.G. Gribov, A.S. Pashinkin, B.A. Salamatin and V.I. Chernova, *Russ. J. Phys. Chem.* **46**, 465 (1972).

68. V.I. Tel'noi, I.B. Rabinovich and V.A. Umilin, *Dokl. Akad. Nauk. SSSR*, **209**, 197 (1973).

69. J.A. Connor, H.A. Skinner and Y. Virmani, *J. Chem. Soc. Faraday Trans. I*, **69**, 1218 (1973).

70. V.I. Tel'noi, I.B. Rabinovich, K.V. Kir'yanov and A.S. Smirnov, *Dokl. Akad. Nauk. SSSR*, **231**, 733 (1976).

71. M.E. Minas da Piedade, Li Shaofeng and G. Pilcher, *J. Chem. Thermodynamics*, **19**, 195 (1987).

72. J.C.G. Calado, A.R. Dias, J.A. Martinho-Simões and M.A.V. Ribeiro da Silva, *J. Organometallic Chem.* **142**, 321 (1979).

73. J.C.G. Calado, A.R. Dias, J.A. Martinho-Simões and M.A.V. Ribeiro da Silva, *Rev. Port. Quim.* **21**, 129 (1979).

74. E.V. Evstigneeva and G.O. Shmyreva, *Russ. J. Phys. Chem.* **39**, 529 (1965).

75. V.I. Chernova, M.S. Shuman, I.B. Rabinovich and V.G. Syrkin, *Tr. Khim. Khim. Tekhnol., Gorky*, **2**, 43 (1973)..

76. D.L.S. Brown, J.A. Connor and H.A. Skinner, *J. Organometallic Chem.* **81**, 403 (1974).

77. F.A. Cotton, A.K. Fischer and G. Wilkinson, *J. Am. Chem. Soc.* **81**, 800 (1959).

78. A.K. Fischer, F.A. Cotton and G. Wilkinson, *J. Am. Chem. Soc.* **79**, 2044 (1957).

79. K.W. Sykes and S.C. Townshend, *J. Chem. Soc.* 2528 (1955).

80. W.D. Good, J.L. Lacina, B.L. DePrater and J.P. McCullough, *J. Phys. Chem.* **68**, 579 (1964).

81. W.V. Steele, *J. Chem. Thermodynamics*, **15**, 595 (1983).

82. W.V. Steele, *J. Chem. Thermodynamics*, **10**, 445 (1978).

83. J.L. Bills and F.A. Cotton, *J. Phys. Chem.* **68**, 806 (1964).

84. G.P. Adams, A.S. Carson and P.G. Laye, *Trans. Faraday Soc.* **65**, 113 (1969).

85. A.S. Carson, E.M. Carson, P.G. Laye, J.A. Spencer and W.V. Steele, *Trans. Faraday Soc.* **66**, 2459 (1970).

86. A.S. Carson, E.H. Jamea, P.G. Laye and J.A. Spencer, *J. Chem. Thermodynamics*, **20**, 1223 (1988).

87. A.S. Carson, J. Dyson, P.G. Laye and J.A. Spencer, *J. Chem. Thermodynamics*, **20**, 1423 (1988).

88. W.D. Good, D.W. Scott and G. Waddington, *J. Phys. Chem.* **60**, 1090 (1956).

89. W.D. Good, D.W. Scott, J.L. Lacina and J.P. McCullough, *J. Phys. Chem.* **63**, 1139 (1959).

90. A.S. Carson, P.G. Laye, J.A. Spencer and W.V. Steele, *J. Chem. Thermodynamics*, **4**, 783 (1972).

91. W.D. Good, D.M. Fairbrother and G. Waddington, *J. Phys. Chem.* **62**, 853 (1958).

ORGANO-f-ELEMENT THERMOCHEMISTRY. IMPLICATIONS FOR REACTIVITY AND BONDING FROM METAL-LIGAND BONDING ENERGETICS.

MICHAEL A. GIARDELLO, WAYNE A. KING, STEVEN P. NOLAN,
MARINA PORCHIA, CHAND SISHTA, AND TOBIN J. MARKS*
Department of Chemistry
Northwestern University
Evanston, IL 60208-3113
U.S.A.

ABSTRACT. This article surveys the thermochemistry of organolanthanide and organoactinide complexes in several interrelated areas. For trivalent 4f and tetravalent 5f as well as tetravalent (d°) group 4 complexes, patterns in metal-ligand σ bond enthalpies are rather similar for a diverse range of ligands. This information is directly applicable to the analysis and invention of a variety of stoichiometric and catalytic transformations, including olefin hydroamination, hydrophosphination, and hydrosilylation, as well as dehydrogenative silane polymerization. Studies of metal-ligand bonding energetics in zero-valent lanthanide and group 4 bisarene sandwich complexes reveal metal-arene bonding which is both similar and rather strong.

1. Introduction

The developing sophistication of contemporary organometallic chemistry has raised increasing numbers of questions about metal-ligand bonding energetics as we attempt to better understand bonding, to rationalize the courses of observed reactions, and to invent new transformations. For newly developing areas such as the organometallic compounds of the early transition and f-elements, unusual bonding modes and reaction patterns vis-à-vis more conventional organometallics have underscored the need for such information. In this article, we review recent results in our studies of organo-f-element bonding energetics and show how these impact upon the development of new reaction chemistry and upon better understanding bonding in unusual types of molecules. We illustrate the former with new f-element-catalyzed hetero-atom/metalloid transformations and the latter with the bonding energetics in zero-valent lanthanide bisarene sandwich complexes.

2. Experimental Approach and f-Element Trends

Our initial studies focussed on determining *relative* metal hydride, hydrocarbyl, alkoxide, thiolate, amide, and phosphide bond disruption enthalpies in broad series of tetravalent organoactinide and trivalent organolanthanide complexes. These series included $Cp'_2 AnR_2(Cp'$ $= \eta^5-Me_5C_5$; An = Th,U),[1] $Cp_3AnR/(Me_3SiC_5H_4)_3AnR$,[2] and $Cp'_2LnR/Me_2SiCp''_2 LnR/Et_2Si-$ $CpCp''LnR$[3] ($Cp'' = \eta^5-Me_4C_5$; Ln = La,Nd,Sm,Lu; $Cp = \eta^5-C_5H_5$) complexes. The

35

thermochemical technique employed was anaerobic isoperibol batch titration calorimetry using either iodine or protonolytic reagents as shown in eqs.(1)-(4). The derived bond enthalpy

$$L_nM\text{-}R + I_2 \longrightarrow L_nM\text{-}I + RI \qquad\qquad \Delta H_{rxn(1)} \qquad (1)$$

$$\Delta H_{rxn(1)} = D(L_nM\text{-}R) + D(I_2) - D(L_nM\text{-}I) - D(R\text{-}I) \qquad (2)$$

$$L_nM\text{-}R + HX \longrightarrow L_nM\text{-}X + RH \qquad\qquad \Delta H_{rxn(3)} \qquad (3)$$

$$\Delta H_{rxn(3)} = D(L_nM\text{-}R) + D(H\text{-}X) - D(L_nM\text{-}X) - D(R\text{-}H) \qquad (4)$$

information is extremely informative for comparing $L_nM\text{-}R/L_nM\text{-}R'$ bond enthalpies within a homologous series of complexes or for predicting heats of reactions interconnecting members of the series. Converting such relative bond enthalpies to an absolute scale is less straightforward, however, and requires a judiciously chosen "anchor point" (e.g., an accurate estimate of $D(L_nM\text{-}I)$ or $D(L_nM\text{-}X)$). Rigorously absolute values of $D(L_nM\text{-}R)$ are accessible via one-electron oxidation sequences as shown in eqs.(5)-(10), provided suitable $L_nM/L_nM\text{-}R/L_nM\text{-}X$

$$L_nM\text{-}R + X_2 \longrightarrow L_nM\text{-}X + RX \qquad\qquad \Delta H_{rxn(5)} \qquad (5)$$

$$L_nM\text{-}X \longrightarrow L_nM + 1/2\ X_2 \qquad\qquad \Delta H_{rxn(6)} \qquad (6)$$

$$X\cdot \longrightarrow 1/2\ X_2 \qquad\qquad 1/2D(X_2) \qquad (7)$$

$$R\text{-}X \longrightarrow R\cdot + X\cdot \qquad\qquad D(R\text{-}X) \qquad (8)$$

$$\overline{L_nM\text{-}R \longrightarrow L_nM + R\cdot} \qquad\qquad\qquad\qquad (9)$$

$$D(L_nM\text{-}R) = \Delta H_{rxn(5)} + \Delta H_{rxn(6)} - 1/2\ D(X_2) + D(R\text{-}X) \qquad (10)$$

ensembles are available.[2b] This approach has recently been applied to an actinide(III)/(IV) and several lanthanide(II)/(III) systems: $(Me_3SiC_5H_4)_3U/(Me_3SiC_5H_4)_3U\text{-}X/R^{[2b]}$ and $Cp'_2Ln/Cp'_2LnX/R,^{[3]}$ where $Ln = Sm, Eu, Yb$. Here $X_2 = I_2$, $t\text{-}BuOOt\text{-}Bu$, and $n\text{-}PrSSn\text{-}Pr$.

The general f-element bond disruption enthalpy patterns which emerge from these investigations differ greatly from those of middle and late transition elements and can be summarized as follows:

1. $D(M\text{-}H) - D(M\text{-}alkyl)$ is generally small.

2. $D(M\text{-}heteroelement) > D(M\text{-}H), D(M\text{-}alkyl)$.

3. $D(M\text{-}Cl) > D(M\text{-}Br) > D(M\text{-}OR) > D(M\text{-}SR) >$

    $D(M\text{-}I) > D(M\text{-}H) > D(M\text{-}NR_2) > D(M\text{-}alkyl) \geq$

    $D(M\text{-}M') > D(M\text{-}PR_2)$.

4. $D(M-C\equiv CR) > D(M\text{-aryl})$, $D(M\text{-vinyl}) > D(M\text{-alkyl})$.

5. $D(M\text{-halogen}) \approx D_1(M(\text{halogen})_n)$ for M in the same oxidation state.

6. Metallacyclobutanes are highly strained; metallacyclopentanes are not.

7. $D(Yb^{II}\text{-acetylene}) > D(U^{III}\text{-CO}) >> D(Eu^{II}\text{-olefin})$; all are very weak.

To first order, trends 1-3 above can be rationalized on the basis of electronegativity models, where the electropositive character of the f-elements (and early transition elements) leads to enhancement of metal-ligand bond enthalpies involving more electronegative ligands.[4] This behavior is especially evident in plots of D(H-X) versus D(M-X), which exhibit a large dispersion in data points for group 4 and f-element complexes (e.g., Figure 1). In contrast, such plots for middle and late transition elements evidence a greater clustering of data points about

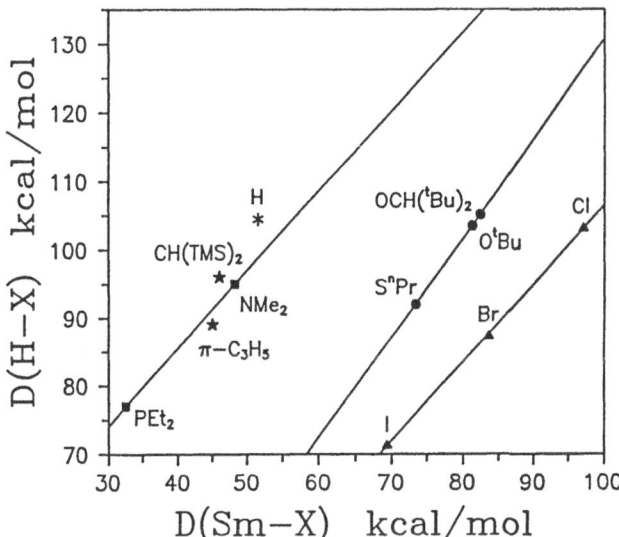

Figure 1. Correlations between D(H-X) values and the corresponding experimental $D(Cp_2'Sm\text{-R/X})$ values. Ligands of the same general type are denoted by separate symbols and the lines represent least-squares fits to the data points.

a single line.[5] Trend 7 largely reflects the relatively weak π bonding ability of these ions for reasons involving both orbital energies and spatial extensions.

In regard to trends within the lanthanide series, the data acquired to date indicate that absolute $D(L_nLn\text{-X/R})$ values generally parallel known Ln(III) → Ln(II) redox couples,[3a] while trends as a function of X/R exhibit little metal sensitivity (e.g., Figure 2). Perhaps more striking is the close congruency between the bond disruption enthalpies in a series of 4f organosamarium complexes[3] and those in a 5f $(Me_3SiC_5H_4)_3U\text{-R/X}$ series with similar ligands (Figure 3),[2] as well as in a $Cp_2'ThR_2/X_2$ series with similar ligands (Figure 4).[1] Clearly the metal-ligand bonding in these 4f and 5f organometallics is very similar. Figure 5 compares the aforementioned organosamarium data with those for an archetypical early transition metal Zr(IV) series.

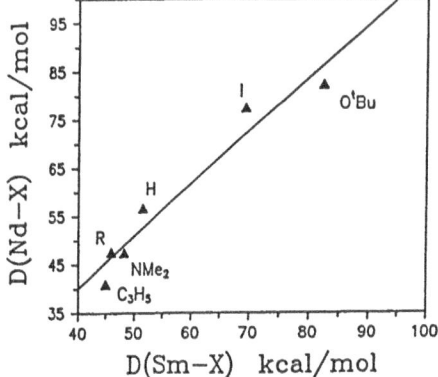

Figure 2. Comparison of D(Cp$_2'$Sm-R/X) data to the corresponding D(Cp$_2'$Nd-R/X) data. R = CH(SiMe$_3$)$_2$ and C$_3$H$_5$ = η^3-allyl.

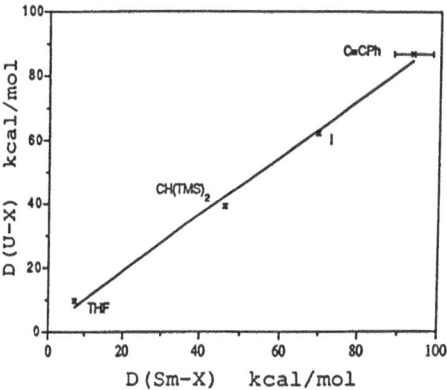

Figure 3. Comparison of D(Cp$_2'$Sm-R/X) and D(Cp$_2'$Sm-L) data to the corresponding D[(Me$_3$SiC$_5$H$_4$)$_3$U-R/X] and D[(Me$_3$SiC$_5$H$_4$)$_3$U-THF] data. The uncertainty in D(Sm-C≡CPh) reflects a competing isomerization process.

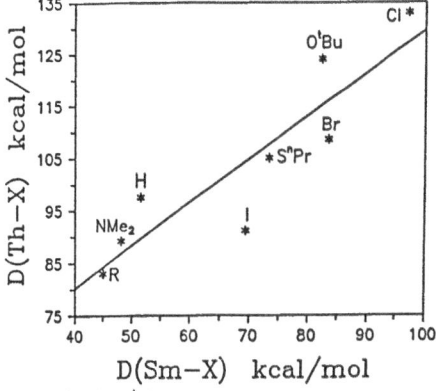

Figure 4. Comparison of D(Cp$_2'$Sm-R/X) data to those of the corresponding Cp$_2'$Th(R/X)-R/X series (refs. 1a,b; using an approximate thermodynamic anchor). For Sm, R = CH(SiMe$_3$)$_2$ while for Th, R = Me. The X = H and n-PrS data for Th are averages of D[Cp$_2'$Th(X)-X] and D[Cp$_2'$Th(OtBu)-X] data, and the X = halide data are from D$_1$(ThX$_4$).

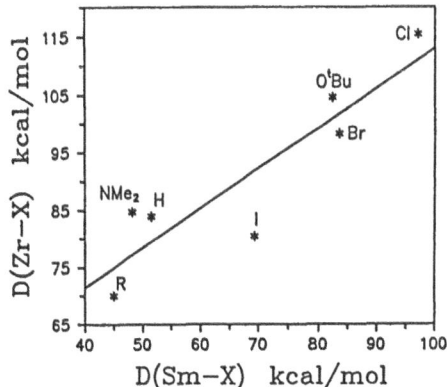

Figure 5. Comparison of D(Cp$_2'$Sm-R/X) data to those of the corresponding Cp$_2'$Zr(R/X)-R/X series (ref. 4c, anchored to D$_1$(ZrCl$_4$); average of D[Cp$_2'$Zr(R/X)-R/X] and D[Cp$_2'$Zr(Cl/I)-R/X]). For Sm, R = CH(SiMe$_3$)$_2$ while for Zr, R = CH$_3$.

Again, the congruency in bond enthalpies is rather striking. Besides conveying important information on similarities in metal-ligand bonding, the foregoing linear trends also have a practical ramification. Unknown D(M-R/X) values should be predictable with some confidence if they are known for members of other f-element or early transition metal series.

3. f-Element-Heteroatom Bonds. Energetics and Related Catalytic Chemistry

Information about metal-ligand bonding energetics can be especially useful in examining the thermodynamics of known or hypothetical transformations involving a particular metal center. Such transformations can be components of catalytic cycles, and it is then possible to ask, as part of the invention process for catalytic processes, whether there are major thermodynamic impediments in an unexplored cycle. In the case of labile metal centers where cognate reactions are known to be facile, this type of analysis is a major asset.

In the case of Cp$_2'$LnR complexes, R = hydride or simple alkyl, it is known that olefin insertion processes are kinetically very rapid (e.g., eq.(11)).[6] If insertions into other types of

$$Cp_2'LnR + CH_2=CH_2 \longrightarrow Cp_2'LnCH_2CH_2R \qquad (11)$$

Ln-X bonds could be effected(eq.(11a)) and if this process could be coupled with subsequent

$$Cp_2'LnX + CH_2=CH_2 \longrightarrow Cp_2'LnCH_2CH_2X \qquad (11a)$$

protonolysis of the resulting Ln-C bond (eq. (11b)), a cycle would be generated for the catalytic

$$Cp_2'LnCH_2CH_2X + HX \longrightarrow Cp_2'LnX + CH_3CH_2X \qquad (11b)$$

addition of HX functionalities to olefins. Both intramolecular and intermolecular variants can be envisioned (Figure 6). Moreover, thermochemical data are available to predict, along with

40

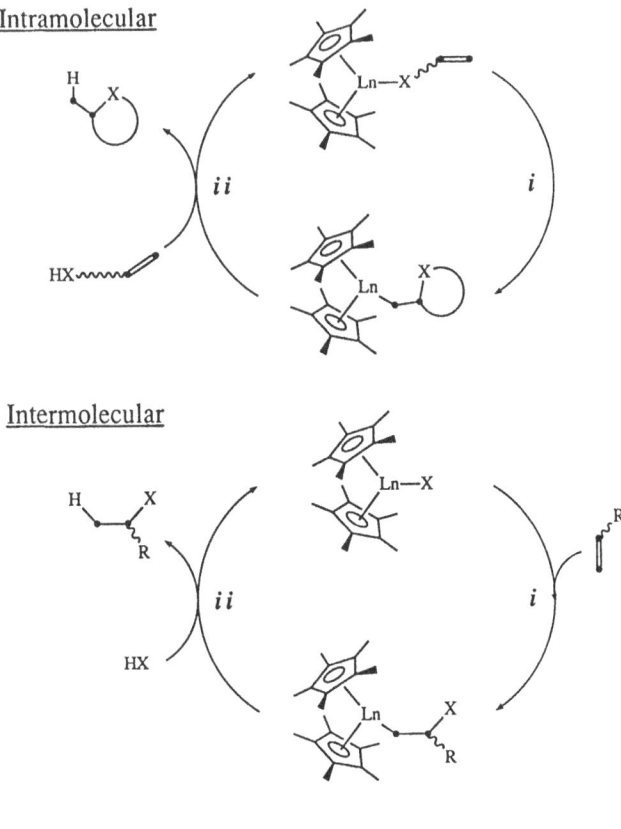

Intramolecular

Intermolecular

Expectations $\Delta S, \Delta S^{\ddagger}$ Favor Intramolecular Processes

$\Delta H_{ii} < \Delta H_i$

$k_{ii} > k_i$

$\Delta H_i (X) : CH_3 \leq H < PR_2, NR_2 < SR, OR$

Figure 6. Intramolecular and intermolecular sequences for the organolanthanide-catalyzed addition of HX functionalities bonds to olefins.

existing kinetic information, which cycles are most promising for exploration. As summarized in Figure 6, protonolysis is invariably estimated to be more exothermic than olefin insertion,[1-3] while chemical information suggests that it will also be more rapid.[1,7] Initial experiments have focussed on intramolecular catalytic transformations since it was anticipated that entropic factors would be far more favorable. Space does not permit an exhaustive discussion of the hydroamination/cyclization chemistry,[8] however Table I conveys an appreciation for the broad scope and excellent (> 95%) regioselectivity. In agreement with thermodynamic and kinetic expectations, olefin insertion is turnover-limiting.[8a] The X = PR$_2$ system has also been explored.[9] Although results are not presently as extensive as for amino-olefins, it is apparent that catalytic hydrophosphination/cyclization is both feasible and rapid (Figure 7). Further exploration of the scope and mechanisms of these thermochemically designed catalytic heteroatom transformations is in progress.

Table I. Scope and turnover frequencies for organolanthanide-catalyzed olefin hydroamination/cyclization

Organolanthanide-Catalyzed Olefin Hydroamination / Cyclization Scope

Entry	Substrate	Product	$N_t (h^{-1})$ °C
1	H₂N⌷⌷⌷	pyrrolidine, N-H	140 (60°C)
2	H₂N⌷⌷⌷	piperidine, N-H	5 (60°C)
3	NH₂ branched	pyrrolidine, N-H	45 (25°C)
4	H₂N branched	pyrrolidine, N-H	36 (25°C)
5	H₂N branched	pyrrolidine, N-H	125 (25°C)
6	MeNH⌷⌷⌷	pyrrolidine, N-Me	11 (25°C)*
7	aryl-NH₂ allyl	indoline, N-H	12 (80°C)
8	H₂N branched	azepane, N-H	0.3 (60°C)*

Catalyst = Cp'₂LaR
* = Me₂SiCp''₂NdR

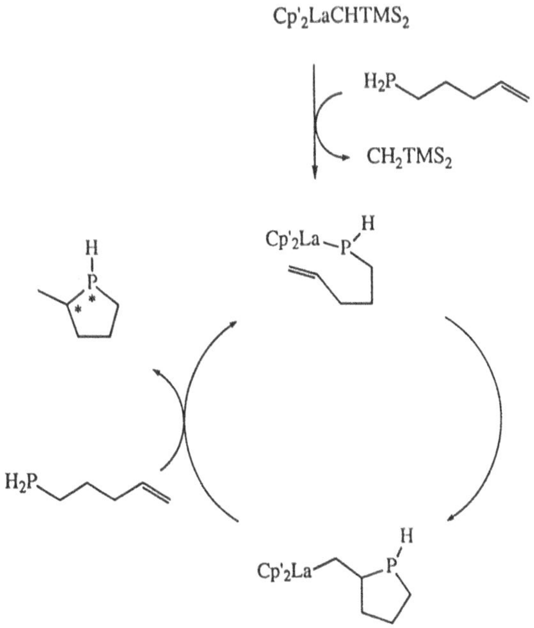

Cp'$_2$LaCHTMS$_2$

- $N_t = 30\text{-}40\ h^{-1}$ at $60\ ^\circ C$

- Zero-order in olefin over two half-lives

- 2:1 diastereomer product ratio

Figure 7. Organolanthanide-catalyzed hydrophosphination/cyclization.

4. f-Element-Metalloid/Metal Bonds. Energetics and Related Catalytic Chemistry

Although there is great interest in the synthesis, bonding, and reactivity of unusual hetero-bimetallic bonds, essentially nothing is known about the bonding energetics. An iodinolytic study of absolute metal-ligand bond disruption enthalpies in the series Cp$_3$UM(CO)$_2$Cp (M = Fe,Ru) and Cp$_3$UM'Ph$_3$ (M' = Si,Ge,Sn) provides the first thermochemical information on "early-late" and "early-metalloid" metal-metal bonds.[2a] It is found that the bond enthalpies are comparatively weak and fall in a relatively narrow range. In a D(U-R/X) vs. D(H-R/X) plot, the data deviate considerably from the R = hydrocarbyl line (Figure 8). An important implication of these bond enthalpy relationships is that trends in the thermodynamics of a great many transformations will depend largely upon the nature of the M/M' reagents and products, since U-M/M' is relatively constant. This results in monotonic trends in ΔH_{rxn} patterns which

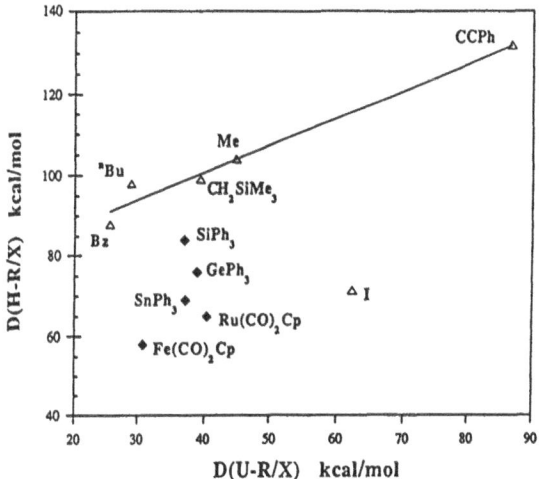

Figure 8. Correlation between measured $(Me_3SiC_5H_5)U$-R/X/M/M' bond disruption enthalpies and the corresponding literature D(R-H)/D(X-H)/D(M-H)/D(M'-H) values.

are illustrated in eqs.(12)-(18) for several classes of metal-metal bond formation and bond scission processes (the uncertainties in estimated ΔH_{rxn} values is ± 6 kcal/mol). Eqs.(12)-(16) illustrate why amine elimination routes are most effective for the synthesis of metal-metal bonds

$$Cp_3U\text{-}NMe_2 + HSiPh_3 \longrightarrow Cp_3U\text{-}SiPh_3 + HNMe_2 \qquad (12)$$

$$\Delta H_{calcd} \approx \quad 0 \text{ kcal/mol.}$$

$$Cp_3U\text{-}NMe_2 + HGePh_3 \longrightarrow Cp_3U\text{-}GePh_3 + HNMe_2 \qquad (13)$$

$$\Delta H_{calcd} \approx -10 \text{ kcal/mol.}$$

$$Cp_3U\text{-}NMe_2 + HSnPh_3 \longrightarrow Cp_3U\text{-}SnPh_3 + HNMe_2 \qquad (14)$$

$$\Delta H_{calcd} \approx -15 \text{ kcal/mol.}$$

$$Cp_3U\text{-}NMe_2 + HFe(CO)_2Cp \longrightarrow Cp_3U\text{-}Fe(CO)_2Cp + HNMe_2 \qquad (15)$$

$$\Delta H_{calcd} \approx -20 \text{ kcal/mol.}$$

$$Cp_3U\text{-}NMe_2 + HRu(CO)_2Cp \longrightarrow Cp_3U\text{-}Ru(CO)_2Cp + HNMe_2 \qquad (16)$$

$$\Delta H_{calcd} \approx -22 \text{ kcal/mol.}$$

involving the heavier metalloids.[10] Eq.(17) illustrates an alternative synthetic pathway and

$$Cp_3U\text{-}H + HM'Ph_3/HM(CO)_2Cp \longrightarrow Cp_3U\text{-}M'Ph_3/M(CO)_2Cp + H_2 \quad (17)$$

$$\Delta H_{calcd} \approx \quad -7 \text{ kcal/mol } (M' = Si)$$

$$-17 \text{ kcal/mol } (M' = Ge)$$

$$-22 \text{ kcal/mol } (M' = Sn)$$

$$-27 \text{ kcal/mol } (M = Fe)$$

$$-29 \text{ kcal/mol } (M = Ru)$$

$$Cp_3U\text{-}CH_3 + HM'Ph_3/HM(CO)_2Cp \longrightarrow Cp_3U\text{-}M'Ph_3/M(CO)_2Cp + CH_4 \quad (18)$$

$$\Delta H_{calcd} \approx \quad -12 \text{ kcal/mol } (M' = Si)$$

$$-22 \text{ kcal/mol } (M' = Ge)$$

$$-27 \text{ kcal/mol } (M' = Sn)$$

$$-33 \text{ kcal/mol } (M = Fe)$$

$$-35 \text{ kcal/mol } (M = Ru)$$

indicates that metal-metal bond hydrogenolysis will be very endothermic. Eq.(18) provides yet another synthetic pathway to these element-element bonds and also predicts that these complexes will not be effective agents for the activation of saturated alkanes. Indeed, hydrocarbon activation only appears promising for substrates with weak C-H bonds, which form exceptionally strong U-C bonds, or which are coupled to an exothermic (e.g., U-O bond-forming) follow-up reaction (eqs.(19)-(22)).[11]

$$Cp_3U\text{-}SiPh_3 + \diagup\!\!\!\!\diagdown \longrightarrow Cp_3U(allyl) + HSiPh_3 \quad (19)$$

$$\Delta H_{calcd} \approx \quad -1 \text{ kcal/mol}$$

$$Cp_3U\text{-}SiPh_3 + HC\equiv CPh \longrightarrow Cp_3U\text{-}C\equiv CPh + HSiPh_3 \quad (20)$$

$$\Delta H_{calcd} \approx \quad -4 \text{ kcal/mol}$$

$$Cp_3U\text{-}SiPh_3 + C_6H_6 \longrightarrow Cp_3U\text{-}C_6H_5 + HSiPh_3 \tag{21}$$

$$\Delta H_{calcd} \approx +8 \text{ kcal/mol}$$

$$Cp_3U\text{-}Ru(CO)_2Cp + \text{(acetone)} \longrightarrow \text{(enol-}O\text{-}UCp_3) + HRu(CO)_2Cp \tag{22}$$

$$\Delta H_{calcd} \approx -3 \text{ kcal/mol}$$

The olefin insertion chemistry of early metal-metalloid bonds is of considerable interest in the discussion of pathways by which metalloid-element bonds might be catalytically added to olefins (e.g., hydrosilylation[12]) and other unsaturated substrates. Eqs.(23)-(26) estimate the thermodynamic constraints under which such insertion processes can be effected at f-element and

$$Cp_3U\text{-}SiPh_3 + \| \longrightarrow Cp_3U\overline{}SiPh_3 \tag{23}$$

$$\Delta H_{calcd} \approx -15 \text{ kcal/mol}$$

$$Cp_3U\text{-}GePh_3 + \| \longrightarrow Cp_3U\overline{}GePh_3 \tag{24}$$

$$\Delta H_{calcd} \approx 0 \text{ kcal/mol}$$

$$Cp_3U\text{-}SnPh_3 + \| \longrightarrow Cp_3U\overline{}SnPh_3 \tag{25}$$

$$\Delta H_{calcd} \approx +11 \text{ kcal/mol}$$

$$Cp_3U\text{-}Ru(CO)_2Cp + \| \longrightarrow Cp_3U\overline{}Ru(CO)_2Cp \tag{26}$$

$$\Delta H_{calcd} \approx +13 \text{ kcal/mol}$$

by extrapolation, early transition element centers.[12a] Eq.(23) also represents one component of a plausible cycle for the organo-f-element-catalyzed hydrosilylation of olefins, with the second step being an exothermic metal alkyl protonolysis reaction (Figure 9a). An alternative cycle involving initial f-element metal alkyl formation, followed by transmetallation can also be envisioned (Figure 9b). Interestingly, neither catalytic pathway is predicted to have significant thermodynamic barriers, both are estimated to be almost equally exothermic, and the estimated exothermicities are in good agreement with ΔH_{rxn} values estimated for R_3SiH + ethylene from standard tabulations[13] ($\sim$ -29 kcal/mol).

Scheme II

a)

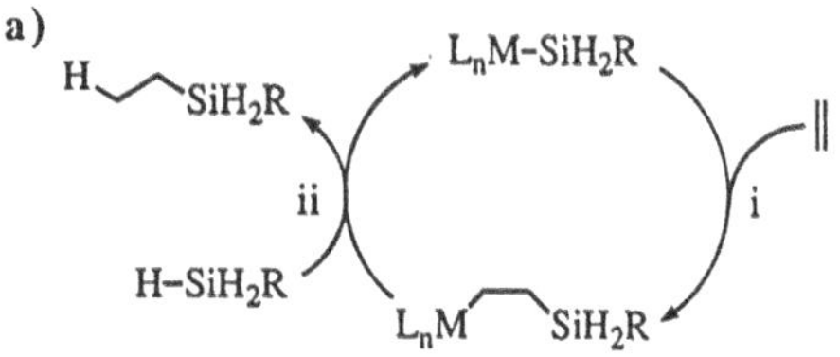

$$\Delta H_i \approx -15 \text{ kcal/mol} \; ; \; \Delta H_{ii} \approx -12 \text{ kcal/mol}$$

b)

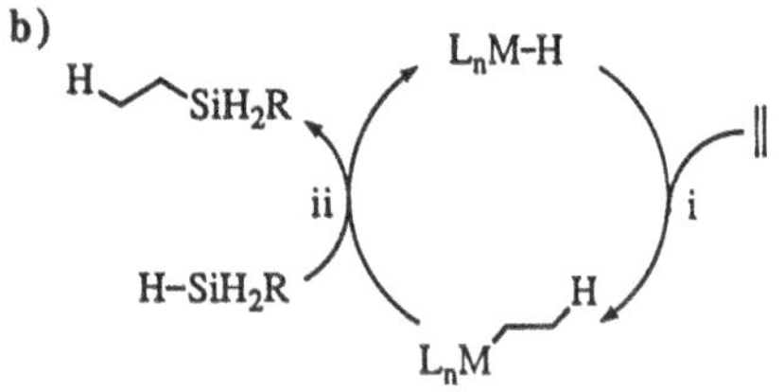

$$\Delta H_i \approx -20 \text{ kcal/mol} \; ; \; \Delta H_{ii} \approx -9 \text{ kcal/mol}$$

Figure 9. Thermodynamic analysis of two possible pathways for the organo-f-element-catalyzed hydrosilylation of olefins.

Coupling eq.(17) to the metalloid-metalloid bond-forming process of eq.(27) models what may be a key transformation in the catalytic dehydrogenative polymerization[14] of group 14 hydrides

$$Cp_3U\text{-}M'Ph_3 + HM'Ph_3 \longrightarrow Cp_3UH + Ph_3M'\text{-}M'Ph_3 \quad (27)$$

$$\Delta H_{calcd} \approx \; -4 \text{ kcal/mol (M'=Si)}$$

$$-4 \text{ kcal/mol (M'=Ge)}$$

$$0 \text{ kcal/mol (M'=Sn)}$$

(e.g., eq.(28)). There has been much speculation concerning the mechanism by which early

$$n \; RSiH_3 \xrightarrow{\text{catalyst}} H\text{-}(\underset{H}{\overset{R}{Si}})_n H + (n\text{-}1)H_2 \quad (28)$$

transition metal complexes catalyze eq.(28), and oxidative addition/reductive elimination, silylene-mediated, free radical, and "heterolytic" four-center processes have all been proposed.[14-18] The cycle in Figure 10 portrays a mechanistic scenario for dehydrogenative silane polymerization which proceeds via a four-center hetrolytic bond transposition mechanism. A thermodynamic analysis of this cycle using our Cp_3USiPh_3 data indicates that both components

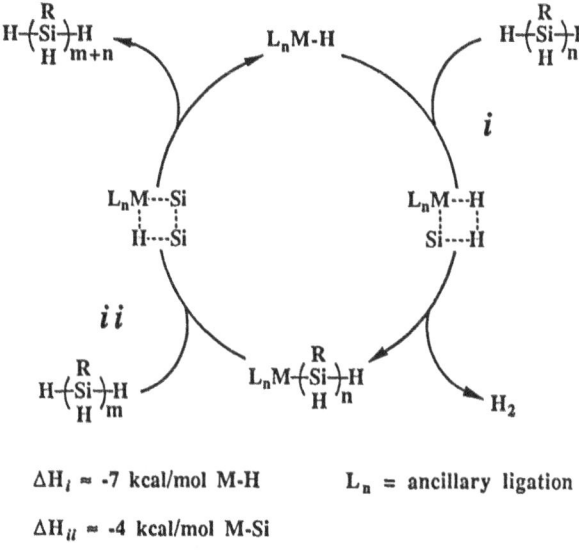

$\Delta H_i \approx$ -7 kcal/mol M-H L_n = ancillary ligation

$\Delta H_{ii} \approx$ -4 kcal/mol M-Si

Figure 10. Catalytic cycle for the dehydrogenative polymerization of silanes based upon four-center bond transpositions and U-Si bond enthalpy data.

of the cycle, analogous to eqs.(17) and (27), respectively, are exothermic. If Figure 10 were indeed a viable pathway, then metal centers incapable of oxidative addition/reductive elimination sequences, which possess only one possible Si/H σ functionality per metal center, and for which other four-center processes are generally facile, would be good candidates for active catalysts. Indeed, we have recently shown that labile, electrophilic organolanthanide complexes of the type $R_2SiCp_2''LnR$ and $Cp_2'LnR$ are extremely active catalysts for eq.(28).[19]

Further studies of early metal-metalloid bonding energetics are in progress.

5. Bonding Energetics in Zero-Valent Bisarene Lanthanide Sandwich Complexes

The traditional description of metal-ligand bonding in organolanthanides has been one of largely electrostatic interactions necessarily involving metals in relatively high oxidation states and anionic hydrocarbyl/polyene ligands.[20] Chemical and thermochemical evidence suggests that the metal-ligand bonding in the few known simple lanthanide complexes with neutral olefin,[21] alkyne,[3a] and arene[22] ligands is rather weak. In marked contradiction to this picture are recently synthesized zero-valent lanthanide bisarene sandwich complexes of the formula $LnAr_2$ (Ar = η^6-1,3,5-t-Bu$_3$C$_6$H$_3$), several of which exhibit impressive thermal stability (sublimable at 100°C).[23] This characteristic and physicochemical evidence suggesting both zero-valent metal character and significant covalency in metal-ligand bonding[23b] raises intriguing questions about the magnitudes of the metal-arene bond strengths, especially vis-à-vis the more common transition element sandwich complexes.

Iodinolytic batch titrational calorimetry has been carried out on $LnAr_2$ complexes in toluene solution. Parallel NMR titrations verify the rapidity and reaction stoichiometry of eq.(29). These results can be combined with tabulated LnI_3 heat of formation, $Ln°$ heat of sublimation, and I_2 heat of solution data in the thermodynamic cycle of eqs.(29)-(33) (s = solution; c = crystal; g = gas). In essence, eq.(33) describes the metal-arene bond disruption process in

$$LnAr_{2(s)} + 3/2\ I_{2(s)} \longrightarrow LnI_{3(c)} + 2\ Ar_{(s)} \qquad \Delta H_{rxn}(29) \qquad (29)$$

$$3/2\ I_{2(c)} \longrightarrow 3/2\ I_{2(s)} \qquad 3/2\ \Delta H_{soln} \qquad (30)$$

$$LnI_{3(c)} \longrightarrow Ln°_{(c)} + 3/2\ I_{2(c)} \qquad -\Delta H_f° \qquad (31)$$

$$Ln°_{(c)} \longrightarrow Ln°_{(g)} \qquad \Delta H°_{sub} \qquad (32)$$

$$LnAr_{2(s)} \longrightarrow Ln°_{(g)} + 2\ Ar_{(s)} \qquad 2\bar{D}(Ln\text{-}Ar) \qquad (33)$$

solution, where $Ln°_{(g)}$ is taken to be "unsolvated" (the enthalpy required to strip both arene ligands from the "bare" metal atom). This approach is preferable to attempted calculation of gas phase $\bar{D}(Ln\text{-}Ar)$ parameters which requires unavailable (and possibly unobtainable) heats of sublimation. Furthermore, most organometallic chemistry of interest occurs in the solution phase. The heats of solution of the $LnAr_2$ complexes and Ar are expected to be rather small($\leq$ 3-5 kcal/mol)[1-3] and to largely cancel. For example, we find ΔH_{soln} for YAr_2 to be 3.5(8) kcal/mol.[24] For comparative purposes, $\bar{D}(M\text{-}Ar)$ values for zero-valent MAr_2 group 4 sandwich complexes[25] have also been determined via iodinolytic calorimetry and an $M° {\rightarrow} MI_4$ cycle analogous to eqs.(29)-(33).

Thermochemical data for the $LnAr_2$ and MAr_2 complexes[26] as well as literature data[27] for several "classical" group 6 arene sandwich complexes are set out in Table II. Several trends are

Table II. Metal-arene bond disruption enthalpies for zero-valent sandwich complexes

Compound	$\bar{D}(Ln/M\text{-}Ar),$[a] kcal/mol
YAr_2	67(1)[b]
$GdAr_2$	67(2)[b]
$DyAr_2$	46(2)[b]
$TiAr_2$	50(1)[b]
$HfAr_2$	66(2)[b]
$Cr(mesitylene)_2$	36(1)[c]
$Mo(C_6H_6)_2$	59(1)[c]
$W(C_6H_5CH_3)_2$	73(1)[c]

[a] $Ar = \eta^6\text{-}1,3,5\text{-}t\text{-}Bu_3C_6H_3$

[b] Defined in eq.(33); quantities in parentheses are 95% confidence limits (3σ).

[c] Gas phase data of ref. 27.

immediately evident. The lanthanide-arene bond enthalpies are rather strong -- comparable to those of the zero-valent group 4 congeners and comparable to or greater than those of the classical group 6 complexes. These results suggest that the bonding in all three classes of compounds is similar and that the electronic structure/metal-ligand orbital overlap in the LnAr$_2$ complexes deviates drastically from that in traditional organolanthanide complexes. Measurements on additional LnAr$_2$ complexes are currently in progress and should better define the details of the metal-ligand bonding.

Conclusions

In this brief overview, we have focussed on several themes emerging from our studies of organo-f-element thermochemistry. First, differences in metal-ligand σ bond enthalpy patterns between trivalent 4f and tetravalent 5f complexes, and also between 4f,5f and group 4 transition metal complexes, are rather small for a wide range of ligands. In contrast, differences between these bond enthalpy patterns and those of traditional middle and late transition element complexes are rather large. These energetic patterns serve not only to provide invaluable, quantitative information on metal-ligand bonding, but also to understand and to design chemical transformations at these metal centers. Examples include a range of early transition metal/f-element-centered bond formation and bond scission reactions, with the most challenging being those which constitute new and useful catalytic cycles. These include catalytic olefin hydroamination, hydrophosphination, and hydrosilylation, as well as dehydrogenative silane polymerization. Thermochemical measurements also provide a unique insight into the metal-ligand bonding energetics of unusual species such as zero-valent lanthanide and group 4 bisarene sandwich complexes. In both cases, the bond enthalpies are found to be prodigious.

Acknowledgments

We thank the NSF for generous support of this research under grants CHE8800813 and CHE9104112. M. P. thanks CNR for a postdoctoral fellowship. We thank Prof. F. G. N. Cloke for samples of arene complexes.

References

1. (a) Bruno, J. W.; Marks, T. J.; Morss, L. R. J. Am. Chem. Soc. **1983**, 105, 6824-6832.
    (b) Bruno, J. W.; Stecher, H. A.; Morss, L. R.; Sonnenberger, D. C.; Marks, T. J. J. Am. Chem. Soc. **1986**, 108, 7275-7280.
    (c) Smith, G. M.; Suzuki, H.; Sonnenberger, D. C.; Day, V. W.; Marks, T. J. Organometallics **1986**, 5, 549-561.
2. (a) Nolan, S. P.; Porchia, M.; Marks, T. J. Organometallics **1991**, 10, 1450-1457.
    (b) Schock, L. E.; Seyam, A. M.; Marks, T. J. Polyhedron **1988**, 7, 1517-1530.
    (c) Sonnenberger, D. C.; Morss, L. R.; Marks, T. J. Organometallics **1985**, 4, 352-355.
3. (a) Nolan, S. P.; Stern, D.; Hedden, D.; Marks, T. J. ACS Sympos. Series **1990**, 428, 159-174.
    (b) Nolan, S. P.; Stern, D.; Marks, T. J. J. Am. Chem. Soc. **1989**, 111, 7844-7853.
4. (a) Gagné, M. R.; Nolan, S. P.; Seyam, A. M.; Stern, D.; Marks, T. J. in Fackler, J. P., Ed., "Metal-Metal Bonds and Clusters in Chemistry and Catalysis," Plenam Press: New York,

1990, pp. 113-125.

(b) Marks, T. J.; Gagné, M. R.; Nolan, S. P.; Schock, L. E.; Seyam, A. M.; Stern, D. Pure Appl. Chem. **1989**, 61, 1665-1672.

(c) Schock, L. E.; Marks, T. J. J. Am. Chem. Soc. **1988**, 110, 7701-7715.

5. Bryndza, H. E.; Fong, L. K.; Paciello, R. A.; Tam, W.; Bercaw, J. E. J. Am. Chem. Soc. **1987**, 109, 1444-1456.

6. (a) Mauermann, H.; Swepston, P. N.; Marks, T. J. Organometallics **1985**, 4, 200-202.

(b) Jeske, G.; Lauke, H.; Mauermann, H.; Swepston, P. N.; Schumann, H.; Marks, T. J. J. Am. Chem. Soc. **1985**, 107, 8091-8103.

(c) Jeske, C.; Schock, L. E.; Mauermann, H.; Swepston, P. N.; Schumann, H.; Marks, T. J. J. Am. Chem. Soc. **1985**, 107, 8103-8110.

(d) Jeske, G.; Lauke, H.; Mauermann, H.; Schumann, H.; Marks, T. J. J. Am. Chem. Soc. **1985**, 107, 8111-8118.

(e) Watson, P. L.; Parshall, G. W. Acc. Chem. Res. **1985**, 18, 51-56.

7. (a) Fagan, P. J.; Manriquez, J. M.; Maatta, E. A.; Seyam, A. M.; Marks, T. J. J. Am. Chem. Soc. **1981**, 103, 6650-6667.

(b) Fagan, P. J.; Manriquez, J. M.; Vollmer, C. H.; Day, C. S.; Day, V. M.; Marks, T. J. J. Am. Chem. Soc. **1981**, 103, 2206-2220.

8. (a) Gagné, M. R.; Stern, C. L.; Marks, T. J. J. Am. Chem. Soc., in press.

(b) Gagné, M. R.; Nolan, S. P.; Marks, T. J. Organometallics **1990**, 9, 1716-1718.

(c) Gagné, M. R.; Marks, T. J. J. Am. Chem. Soc. **1989**, 111, 4108-4109.

9. Giardello, M. A.; Sishta, C.; Marks, T. J., manuscript in preparation.

10. (a) Porchia, M.; Brianese, N.; Casellato, U.; Ossola, F.; Rossetto, G.; Zanella, P.; Graziani, R. J. Chem. Soc., Dalton Trans **1989**, 677-681.

(b) Porchia, M.; Ossola, F.; Rossetto, G.; Zanella, P.; Brianese, N. J. Chem. Soc., Chem. Commun. **1987**, 550-551.

(c) Porchia, M.; Cassellato, V.; Ossola, F.; Rossetto, G.; Zanella, P; Graziani, R. J. Chem. Soc., Chem. Commun. **1986**, 1034-1035.

11. (a) Sternal, R. S.; Sabat, M.; Marks, T. J. J. Am. Chem. Soc. **1987**, 109, 7920-7921.

(b) Sternal, R. S.; Marks, T. J. Organometallics **1987**, 6, 2621-2623.

12. (a) Tilley, T. D. in Patai, S.; Rappoport, Z., Eds. "The Chemistry of Organic Silicon Compounds," John Wiley: Chichester, 1989, Chapt. 24.

(b) Ojima, I. in ref. 12a, pp. 1479-1526.

13. Walsh, R. in ref. 12a, pp. 371-391.

14. (a) Walzer, J. F., Woo, H. G., Tilley, T. D., Polymer Preprints **1991**, 32, 441-442.

(b) Tilley, T. D. Comments Inorg. Chem. **1990**, 10, 37-51.

(c) Woo, H. G.; Tilley, T. D. J. Am. Chem. Soc. **1989**, 111, 8043-8044.

15. (a) Gauvin, F., Harrod, J. F. Polymer Preprints **1991**, 32, 439-440.

(b) Mu, Y., Aitken, C., Cote, B., Harrod, J. F., Samuel, E. Canad. J. Chem. **1991**, 69, 264-276.

(c) Harrod, J. F.; Mu, Y.; Coté, B. Polymer Preprints **1990**, 31, 228-229.

(d) Aitken, C.; Barry, J.-P.; Gauvin, F.; Harrod, J. F.; Malek, A.; Rousseau, D. Organometallics **1989**, 8, 1732-1736.

(e) Harrod, J. F. ACS Sympos. Series **1988**, 360, 89-100.

16. (a) Campbell, W. H.; Hilty, T. K.; Yurga, L. Organometallics **1989**, 8, 2615-2618.

(b) Chang, L. S.; Corey, J. Y. Organometallics **1989**, 8, 1885-1893.

17. (a) Corey, J. Y., Zhu, S. H., Bedard, T. C., Lange, L. D. Organometallics **1991**, 10, 924-

930.

(b) Brown-Wensley, K. A. Organometallics **1987**, <u>6</u>, 1590-1591.

18. (a) Harrod, J. F.; Ziegler, T.; Tschinke, V. Organometallics **1990**, <u>9</u>, 897-902.

    (b) Harrod, J. F.; Yun, S. S. Organometallics **1987**, <u>6</u>, 1381-1387.

19. Forsyth, C. M.; Nolan, S. P.; Marks, T. J. Organometallics **1991**, <u>10</u>, 2543-2545.

20. (a) Bursten, B. E.; Strittmatter, R. J. Angew Chem. Int. Ed. Engl., in press, and references therein.

    (b) Edelstein, N. in Marks, T. J.; Fragalá, I., Eds. Fundamental and Technological Aspects of Organo-f-Element Chemistry, D. Reidel: Dordrecht, Holland, 1985, Chapt. 7.

    (c) Rösch, N. Inorg. Chim. Acta **1984**, <u>94</u>, 297-299.

    (d) Raymond, K. N.; Eigenbrot, C. W., Jr. Acc. Chem. Res. **1980**, <u>13</u>, 276-283.

21. Nolan, S. P.; Marks, T. J. J. Am. Chem. Soc. **1989**, <u>111</u>, 8538-8540.

22. Cotton, F. A.; Schwotzer, W. J. Am. Chem. Soc. **1986**, <u>108</u>, 4657-4658.

23. (a) Brennan, J. G.; Cloke, F. G. N.; Sameh, A. A.; Zalkin, A. J. Chem. Soc., Chem. Commun. **1987**, 1668-1669.

    (b) Anderson, D. M.; Cloke, F. G. N.; Cox, P. A.; Edelstein, N.; Green, J. C.; Pang, T.; Sameh, A. A.; Shalimoff, G. J. Chem. Soc., Chem. Commun. **1989**, 53-55.

24. King, W. A.; Marks, T. J., unpublished results.

25. (a) Cloke, F. G. N.; Lappert, M. F.; Lawless, G. A.; Swain, A. C. J. Chem. Soc., Chem. Commun. **1987**, 1667-1668.

    (b) Cloke, F. G. N.; Courtney, K. A. E.; Sameh, A. A.; Swain, A. C. Polyhedron **1989**, <u>8</u>, 1641-1648.

26. King, W. A.; Marks, T. J.; Cloke, F. G. N., manuscript in preparation.

27. (a) Connor, J. A.; El-Sayed, N. I.; Martinho-Simoes, J. A.; Skinner, H. A. J. Organometal. Chem. **1981**, <u>212</u>, 405-410.

    (b) Connor, J. A.; Martinho-Simoes, J. A.; Skinner, H. A.; Zafrani-Mottar, M. F. J. Organometal. Chem. **1979**, <u>179</u>, 331-356.

THE ROLE OF BOND ENERGIES IN HYDROCARBON ACTIVATION BY TRANSITION METAL CENTERS

William D. Jones[*], R. Martin Chin, Lingzhen Dong, Simon B. Duckett, and Edward T. Hessell
Department of Chemistry
University of Rochester
Rochester, New York 14627
USA

ABSTRACT. This paper describes the fundamental thermodynamic and kinetic factors that influence carbon-hydrogen bond activation at homogeneous transition metal centers. Advances have been made in both understanding the interactions of hydrocarbons with metals and in the energetics of these interactions. We have examined reactions of a series of arenes with $(C_5Me_5)Rh(PMe_3)PhH$ and begun to map out the kinetic and thermodynamic preferences for arene coordination. The effects of resonance, specifically the differences in the Hückel energies of the bound vs free ligand, are now believed to fully control the C-H activation/η^2-coordination equilibria. We have begun to examine the reactions of rhodium isonitrile pyrazolylborates for alkane and arene C-H bond activation. A new, labile, carbodiimide precursor has been developed for these studies.

1. Introduction

The reactions of coordinatively unsaturated metal complexes with hydrocarbons has been the subject of intense study during the past decade.[1] The basic thermodynamic parameters that dictate the feasibility of the oxidative addition of an alkane to a metal have not been generally well established, making predications about reactivity an uncertain business. The initial studies undertaken in our laboratory were aimed at using the same metal/ligand system to study a series of hydrocarbon activation reactions with the goal of extracting some basic thermodynamics about the additions.

J. A. Martinho Simões (ed.), Energetics of Organometallic Species, 53–67.
© 1992 *Kluwer Academic Publishers.*

2. Thermodynamics of C-H Bond Activation and Metal-Carbon Bond Energies

The initial basic discoveries made with the metal fragment $[(C_5Me_5)Rh(PMe_3)]$. Three reactions were studied in order to map out the thermodynamics allowing the comparison of the oxidative addition of benzene and propane (Scheme I). The barrier (Free energy) for loss of benzene was determined by measuring the rate at which $(C_5Me_5)Rh(PMe_3)PhH$ loses C_6H_6 in C_6D_6 as a function of temperature. Similarly, synthesis of the complex $(C_5Me_5)Rh(PMe_3)(n\text{-Pr})H$ and monitoring the rate of reductive elimination of propane allowed measurement of this barrier also. Finally, a measurement of the kinetic selectivity of the coordinatively unsaturated intermediate $[(C_5Me_5)Rh(PMe_3)]$ for benzene vs propane was measured to be 4.2 : 1 by irradiation of $(C_5Me_5)Rh(PMe_3)H_2$ in a benzene/propane solvent mixture. The combination of these values as shown in Scheme II allowed the calculation of the Free energy difference separating $(C_5Me_5)Rh(PMe_3)PhH$ + propane from $(C_5Me_5)Rh(PMe_3)(n\text{-Pr})H$ + benzene. The value of 8.7 Kcal/mole corresponds to an equilibrium constant of 2.2×10^6 favoring benzene activation.[2]

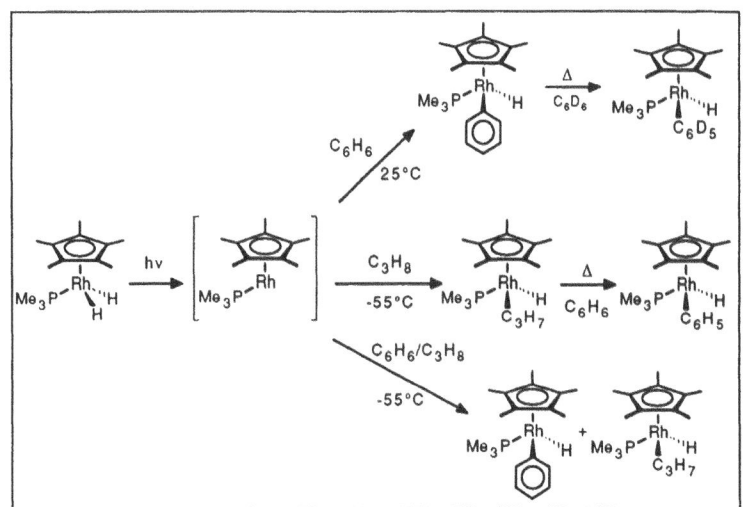

Scheme I. Reactions used to establish thermodynamics of $[(C_5Me_5)Rh(PMe_3)]$.

It is worth pointing out that since the primary C-H bond in propane is 8 Kcal/mole weaker than the C-H bond in benzene, it might have been expected that propane activation should have been more facile. However, the determination that the energetics lie 8.7 Kcal/mole in the opposite direction can only be accounted for in terms of the *difference* in metal-carbon bond strengths for Rh-Ph vs Rh-*n*-propyl. *i.e.*, the rhodium-phenyl bond must be 16.7 Kcal/mole stronger than the rhodium-*n*-propyl bond. The greater strength of the product M-C bond dominates the C-H bond activation thermodynamics, leading to the conclusion that the strong C-H bond is broken because this leads to the strongest M-C bond. This basic observation appears to be general, in that there has been widespread observation of a preference (both kinetic and thermodynamic) for cleavage of stronger C-H bonds in the presence of weaker C-H bonds.[3]

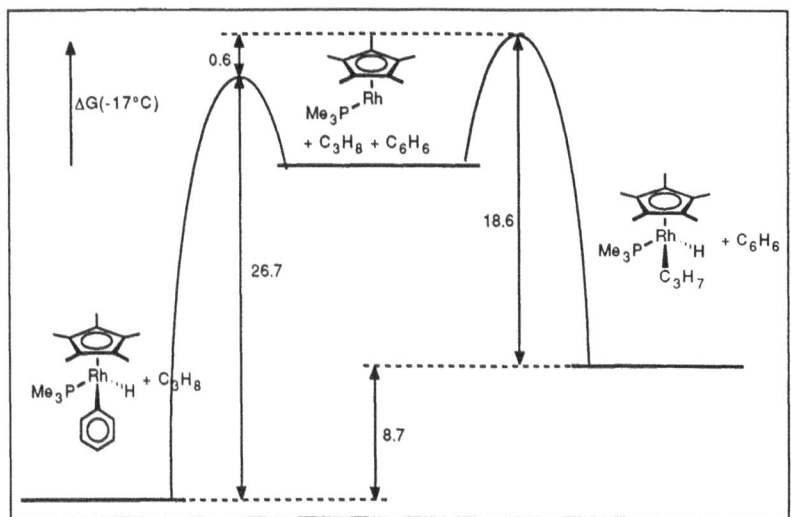

Scheme II. Free Energy diagram for benzene vs propane.

3. Polycyclic Fused η²-Arene Studies

The relative thermodynamic and kinetic selectivities of C-H bond activation vs η²-arene coordination have been measured for $[(C_5Me_5)Rh(PMe_3)]$ by thermolysis of $(C_5Me_5)Rh(PMe_3)(Ph)H$ in hexane solution in the presence of an arene. With naphthalene, a 2:1 equilibrium is set up between $(C_5Me_5)Rh(PMe_3)(\eta^2$-naphthalene) and $(C_5Me_5)Rh(PMe_3)(2$-naphthyl)H, both of which are preferred over $(C_5Me_5)Rh(PMe_3)(Ph)H$. Furthermore, the rate of loss of naphthalene from $(C_5Me_5)Rh(PMe_3)(\eta^2$-naphthalene) was measured by monitoring the kinetics of the conversion back to $(C_5Me_5)Rh(PMe_3)(Ph)H$ in neat benzene, giving the overall barrier for naphthalene loss. Combining these kinetic and thermodynamic data with those previously established for benzene gives the free energy diagram shown in Scheme III.[4]

The barrier for interconversion of the naphthyl hydride complex and the η²-naphthalene complex was determined by ^{31}P spin saturation transfer (SST) experiments. For the rate of loss of the naphthyl hydride complex, the activation parameters were determined ($\Delta H^{\ddagger} = 16.9(7)$ Kcal/mole, $\Delta S^{\ddagger} = -8.1(2)$ e.u.),[5] which compare favorably with the previously determined values for the formation of $(C_5Me_5)Rh(PMe_3)(\eta^2$-p-xylene) from $(C_5Me_5)Rh(PMe_3)(p$-xylyl)H ($\Delta H^{\ddagger} = 16.3(2)$ Kcal/mole, $\Delta S^{\ddagger} = -6.3(8)$ e.u.),[2] indicating that dropping the energy of the η²-naphthalene complex has little effect upon the transition state for aryl-hydride to η^2-arene interconversion. The depth of the well for $(C_5Me_5)Rh(PMe_3)(\eta^2$-benzene) has been independently determined in flash photolysis experiments done in collaboration with Dr. Robin Perutz of the University of York.[5] These studies showed that the intermediate $[(C_5Me_5)Rh(PMe_3)]$ reacts rapidly (<200 ns) to give the η^2-C_6H_6 complex, which then goes on more slowly ($\tau \cong 150\ \mu s$) to insert into the C-H bond. The activation parameters for this step of the reaction were also determined, giving $\Delta H^\ddagger = 11.15$ (0.31) Kcal/mole and $\Delta S^\ddagger = -4.9$ (1.1) e.u.

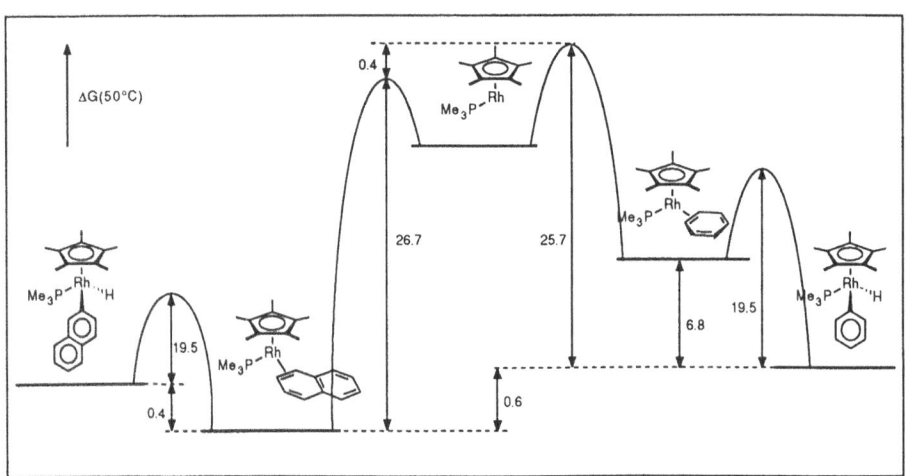

Scheme III. Free Energy diagram for benzene and naphthalene complexes.

As can be seen, the unique feature about the polycyclic aromatic lies in the stability of the η^2 complex relative to the naphthyl hydride complex. That is, for benzene, the η^2-complex is destabilized relative to the η^2-naphthalene complex, whereas the C-H activation complexes are of comparable energy. *It is becoming clear that it is the energy of the η^2 complex that determines whether or not C-H oxidative addition will occur, and that the bond strengths should not be the only factor considered.* The origin of this destabilization can be directly attributed to the interruption of the resonance of the aromatic ring, which is minimized in the naphthalene complex. Table I summarizes calculations of the energies of the free and bound arene in terms of the p-orbital interaction integral β.[6] The balance point for C-H activation vs η^2 coordination occurs at $\Delta E_r \cong 1.25\ \beta$. Values larger than this give C-H activation only, and smaller values result in only η^2-arene complexation (*vide infra*). Other examples support this hypothesis, as outlined in Scheme IV and summarized below.

Table I. Resonance Energies for η^2-Arene Complexes (in units of β).

Complex	Resonance Energy of		ΔE_r
	Free Arene	η^2-Arene	β (Kcal/mol)
(structure, M)	2	0.47	1.53 (30.6)
(structure, M) IIa	3.68	2.42	1.26 (25.2)
(structure, M) IIb	3.68	1.95	1.73 (34.6)
(structure, M, M) IIc	3.68	2.00	1.68 (33.6)
(structure, M, M) IId	3.68	0.99	2.69 (53.8)
(structure, M, OMe)	4.43	3.39	1.04 (20.8)
(structure, M, OMe)	4.43	3.19	1.24 (24.8)
(structure, M, OMe)	4.43	3.14	1.29 (25.8)
(structure, M) IVa	5.31	4.11	1.20 (24.0)
(structure, M) IVb	5.31	3.53	1.82 (36.4)
(structure, M) Va	5.45	4.38	1.07 (21.4)
(structure, M) Vb	5.45	4.13	1.32 (26.4)
(structure, M) Vc	5.45	4.11	1.34 (26.8)

58

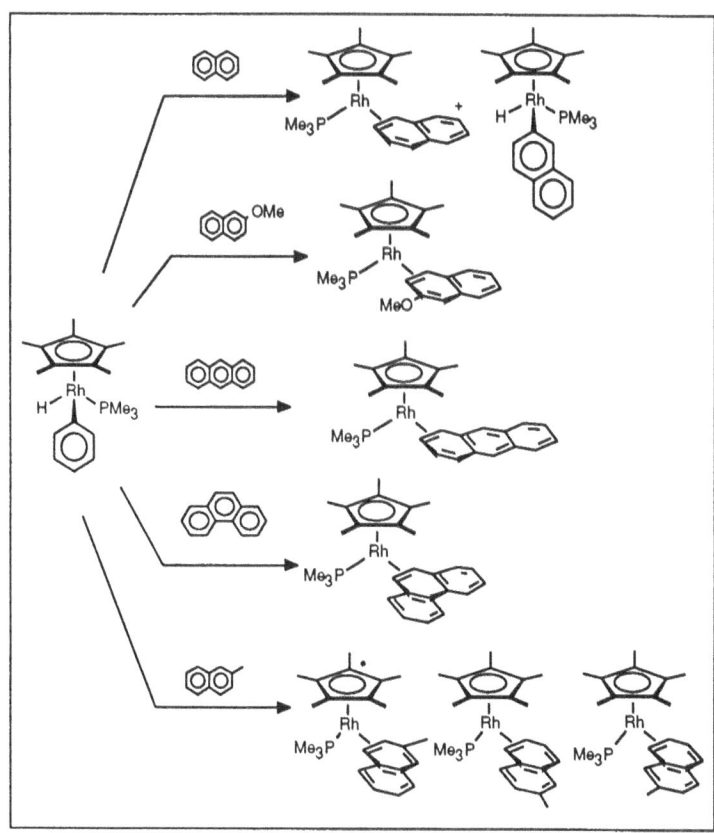

Scheme IV. Reactions of polycyclic aromatics with $(C_5Me_5)Rh(PMe_3)PhH$.

With 2-methoxynaphthalene, only the η^2-complex is observed in the 3,4 position of the ring. No C-H activation is seen. In this example, the loss of resonance energy is minimal ($\Delta E_r = 1.04$ β, Table I), so that the aromatic behaves as an olefin! The reactivity is similar to that of styrene, which also forms only an η^2 complex with the vinylic portion of the molecule. Other simple arene selectivities (thermodynamic, since the arenes exchange under the reaction conditions) are indicated in Scheme IV. With phenanthrene and anthracene, only a single η^2 complex is seen, again in accord with the ΔE_r calculations in Table I . With 2-methylnaphthalene, however, three different $\eta2$ complexes are observed since they all have comparable ΔE_r energies.

In the case of phenanthrene, the η^2 complex was crystallized and structurally characterized (Figure 1).[4] Two structural features are to be noted: 1) only 2 carbons interact symmetrically with the metal and 2) the arene ring is oriented away from the phosphine and towards the C_5Me_5 ring. The UV-visible spectrum (Figure 2) of the red complex is interpreted in terms of metal to ligand π^* charge transfer bands that are shifted to lower energy than the $\pi \longrightarrow \pi^*$ bands in free phenanthrene.

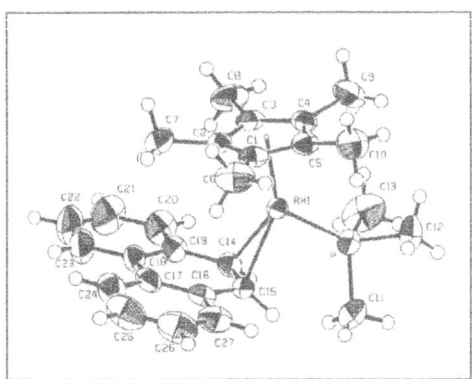

Figure 1. X-ray of $(C_5Me_5)Rh(PMe_3)(\eta^2\text{-phenanthrene})$.

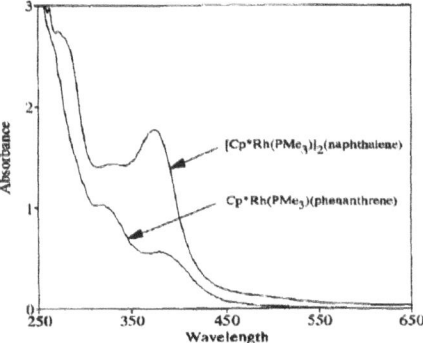

Figure 2. UV of $(C_5Me_5)Rh(PMe_3)(\eta^2\text{-phenanthrene})$.

The 2-methoxynaphthalene derivative also offered the opportunity to study the kinetic isomers of the reaction. Preparation of $(C_5Me_5)Rh(PMe_3)[6\text{-}(2\text{- methoxy-naphthalene})]H$ was performed by way of hydridic reduction of the corresponding bromide complex. The initially formed aryl hydride complex rapidly isomerized between four species by way of rhodium-hydride $\longrightarrow \eta^2$-arene equilibria, assigned as shown in Scheme V. This result is to be expected, as the ΔE_r values from Table I are 1.24 and 1.29. Over several days time at room temperature, the four isomers were converted into the single more stable thermodynamically preferred isomer. This isomerization required migration past the carbon at the ring juncture, and consequently occurred more slowly.

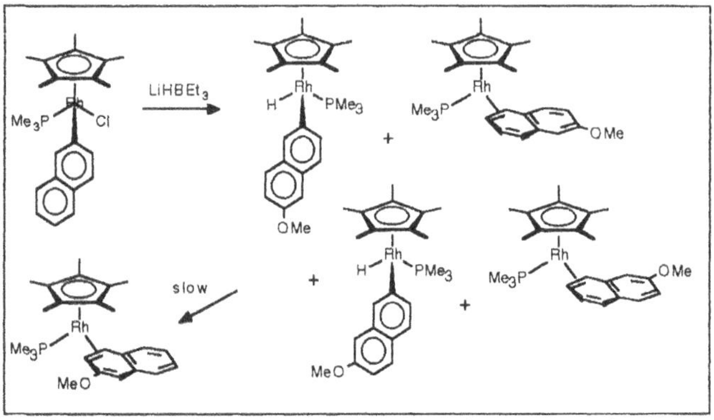

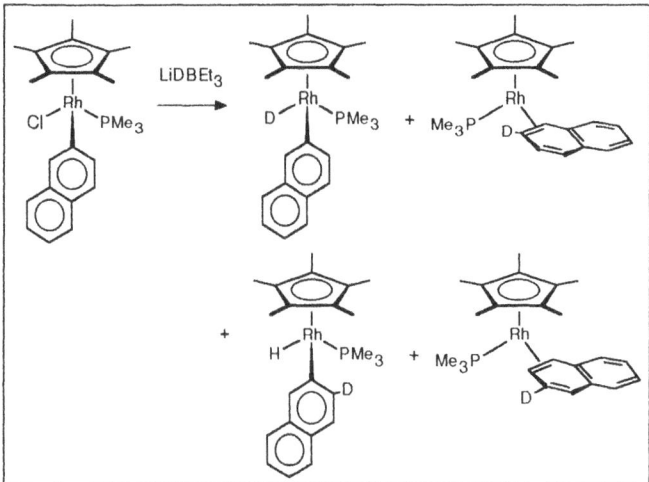

Scheme V. 2-methoxynaphthalene interconversions.

Similar rearrangements have been probed in the parent naphthalene system. Reaction of the complex $(C_5Me_5)Rh(PMe_3)(2$-naphthyl$)Br$ with $LiDBEt_3$ gives a mixture of 4 isomers in which the deuterium has scrambled over the two possible 2-naphthyl derivatives and two possible η^2 isomers (Scheme VI). Slowly, the initial isomers equilibrate with the four isomers in which the rhodium has migrated across to the unsubstituted ring. Since this rearrangement occurs more rapidly than the rate at which naphthalene dissociates from $(C_5Me_5)Rh(PMe_3)(\eta^2$-naphthalene$)$, the reaction must be intramolecular.

Scheme VI. Naphthalene-d_1 interconversions.

Further heating of the reaction of $(C_5Me_5)Rh(PMe_3)PhH$ with naphthalene ultimately leads to a new thermodynamically preferred product in which two rhodiums are attached to the aromatic ligand. Periodic removal of naphthalene helps to shift this equilibrium towards the binuclear adduct, allowing its isolation and characterization in pure form.

COSY NMR studies indicate that both metals are on one aromatic ring. The X-ray structure, shown in Figure 3, confirms that both metals are on the same aromatic ring but on opposite faces. The preference for keeping the aromatic ring pointing up towards the Cp* ring can also be seen here, as in the phenanthrene complex.[7]

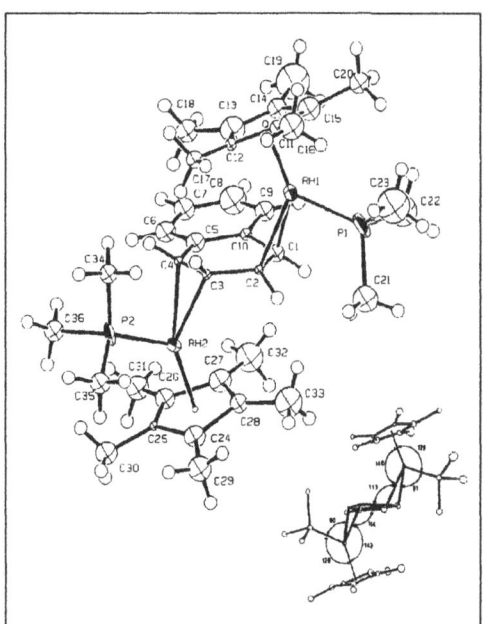

Figure 3. ORTEP drawing of [(C$_5$Me$_5$)Rh(PMe$_3$)]$_2$(naphthalene).

The ΔE_r calculation for the possible binuclear adducts given in Table I are in agreement with the observed isomer. While the large value of ΔE_r (1.68) would suggest that only C-H activation should be observed, this energy cost should be divided over two metal-arene interactions, giving a ΔE_r of 0.84 per metal, which is well within the η^2 regime.

A variety of other larger fused polycyclic aromatics have also been examined. These include pyrene, perylene, fluoranthene, and triphenylene. Preliminary studies indicate that in most cases η^2 arene complexation is strongly preferred, but that in others C-H activation is preferred (Scheme VII). As in the earlier cases, examination of the resonance energy costs based on simple Hückel energies allows one to account for the observations. The results of these calculations are summarized in Table II.

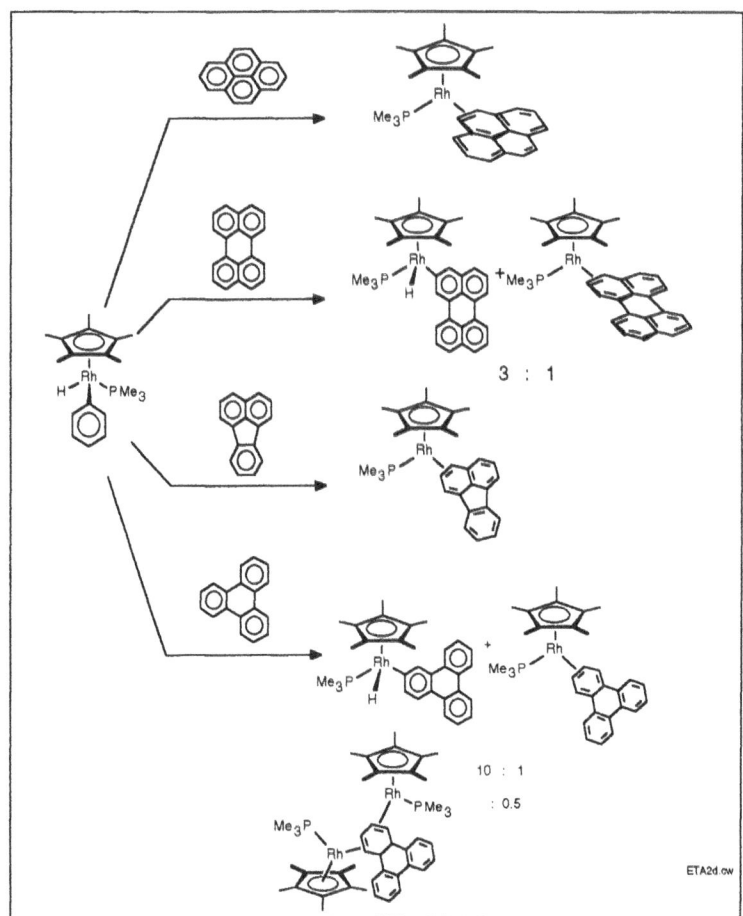

Scheme VII. Reactions with larger fused polycyclic aromatics.

Table II. Resonance Energies for η^2-Arene Complexes (in units of β).

Complex	Resonance Energy of		ΔE_r
	Free Arene	η^2-Arene	β (Kcal/mol)
	6.51	5.45	1.06 (21.2)
	6.51	5.09	1.42 (28.4)
	6.50	5.22	1.28 (25.6)
	6.50	5.03	1.46 (29.2)
	6.50	4.97	1.53 (30.6)
	8.25	7.02	1.23 (24.6)
	8.25	6.68	1.57 (31.4)
	7.27	5.68	1.60 (32.0)
	7.27	5.89	1.38 (27.6)

4. Rhodium Pyrazolylborate Studies

Our earlier studies with $(C_5Me_5)Rh(CNR)_2$ complexes indicated that they were inefficient (substoichiometric) in functionalizing aromatic C-H bonds with isonitriles.[8] The recent work in C-H activation by Bill Graham using $(HBPz^*_3)Rh(CO)_2$ complexes led us to investigate the reactions of the analogous bis isonitrile complexes.

The complexes $(HBPz^*_3)Rh(CNR)_2$ where R = neopentyl and 2,6-xylyl have been prepared. Structural characterization of these complexes revealed an unexpected feature.[9] The trispyrazolylborate rings were η^2, not η^3 (Figure 4)! The rhodium prefers a 16-electron square planar geometry over an 18-electron trihapto geometry. An attempt to protonate the free pyrazole ligand led to protonation of the metal, inducing η^3 coordination of the tris-pyrazolylborate ligand to the newly formed cationic Rh(III) center (Figure 5).

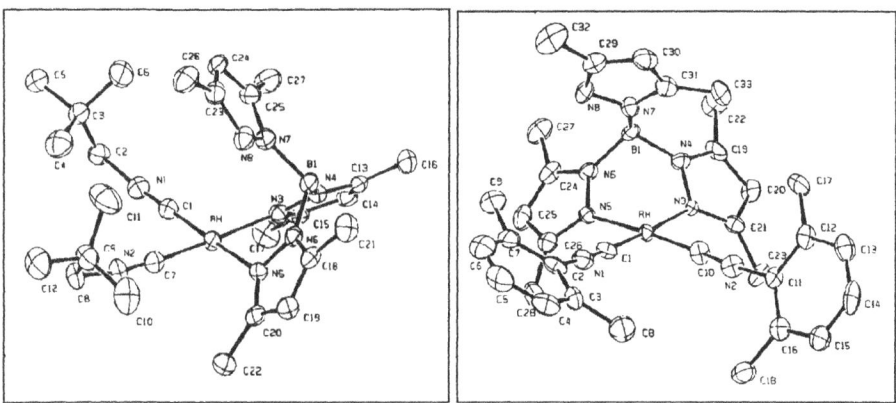

Figure 4. ORTEP drawings of $(HBPz^*_3)Rh(CNneopentyl)_2$ and $(HBPz^*_3)Rh(CNxylyl)_2$.

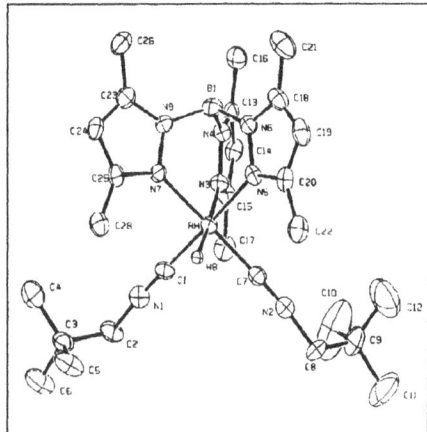

Figure 5. ORTEP drawing of $[(HBPz^*_3)Rh(CNneopentyl)_2H]^+$.

Irradiation of the complexes in benzene solution slowly results in C-H activation of the hydrocarbon solvent (Eq 1).[10] In this case, the product is η^3 and the formation of an extra bond in the product makes the oxidative addition adduct quite stable. Benzene exchange with C_6D_6 is very sluggish, even at high temperatures. Attempts to force isonitrile insertion into the Rh-Ph bond have failed, attesting to the stability of the η^3 adduct.

We have discovered that reaction of the bis-isocyanide complexes with phenyl azide (PhN_3) leads via [1,3] dipolar addition to the new carbodiimide complexes (Eq 2). Irradiation of these compounds leads to the rapid and efficient loss of carbodiimide and activation of benzene solvent. Irradiation in alkane solvent also gives clean oxidative addition products. Some selectivities are given in Table III. Comparison of the [Tp*Rh(CNR)] fragment with [(C$_5$Me$_5$)Rh(PMe$_3$)] shows the former to be slightly more selective.

These oxidative addition reactions are reversible in the presence of added CNR. The rates appear to be first order in both rhodium complex and in CNR. Activation parameters are consistent with a bimolecular reaction. Several possible mechanisms are indicated in Scheme VIII, all of which involve loss of benzene in or prior to the rate determining step with CNR.

Table III. Preferences for alkane activation.

substrate	k_{rel} Tp*Rh(CNR)	k_{rel} Cp*Rh(PMe$_3$)
C_6H_{12}	1.0	1.0
C_6H_6	69	19.5
mesitylene, 1°	36	-
mesitylene, 2°	25	-
pentane, 1°	14.6	5.9
pentane, 2°	0	0
propane, 1°	14.6	2.6
propane, 2°	0	0

(per H, relative to C_6H_{12})

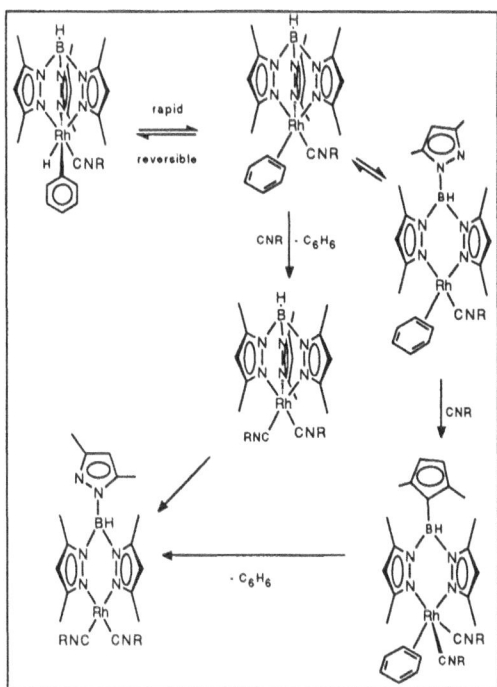

Scheme VIII. Possible mechanisms for CNR induced elimination of benzene.

5. Summary

In summary, the chemistry of a variety of rhodium complexes has been explored with regard to their fundamental interactions with hydrocarbons and their mechanisms of reactions. The factors that control the stability of η^2-arene complexation relative to C-H bond activation. Knowledge of these factors will lead to better predictive ability for the reaction of a given hydrocarbon or functionalized species with a low valent metal center.

6. References

(1) For recent reviews, see: Hill, C. L. *Activation and Functionalization of Alkanes;* Wiley: New York, 1989. Davies, J. A.; Watson, P. L.; Liebman, J. F.; Greenberg, A. *Selective Hydrocarbon Activation*; VCH Publishers, Inc.: New York, 1990. Shilov, A. E. *Activation of Saturated Hydrocarbons by Transition Metal Complexes*; D. Reidel: Boston, 1984. Crabtree, R. H. *Chem. Rev.* **1985**, *85*, 245-269. Halpern, J. *Inorg. Chim. Acta* **1985**, *100*, 41-48.

(2) Jones, W. D.; Feher, F. J. *J. Amer. Chem. Soc.*, **1984**, *106*, 1650-1663. Jones, W. D.; Feher, F. J. *Acc. Chem.Res.***1989**, *22*, 91-100.

(3) Bergman, R. G. *Science* **1985**, *223*, 902-908. Buchanan, J. M.; Stryker, J. M.; Bergman, R. G. *J. Amer. Chem. Soc.* **1986**, *108*, 1537-1550. Jones, W. D.; Maguire, J. A. *Organometallics* **1985**, *4*, 951-953.

(4) Jones, W. D.; Dong, L. *J. Am. Chem. Soc.* **1989**, *111*, 8722-8723.

(5) Belt, S. T.; Dong, L.; Duckett, S. B.; D. Jones, W. D.; Partridge, M. G.; and Perutz, R. N. *J. Chem. Soc., Chem. Commun.* **1991**, 266-269.

(6) This method of analyzing π-bonding energies has been presented previously in: Brauer, D. J.; Kruger, C. *Inorg. Chem.* **1977**, *16*, 884-891,

(7) Chin, R. M.; Dong, L.; Duckett, S. B.; Jones, W. D. *Organometallics*, in press.

(8) Jones, W. D.; Feher, F. J. *Organometallics*, **1983**, *2*, 686-687.

(9) Jones, W. D.; Hessell, E. T. *Inorg. Chem.* **1991**, *30*, 778-783.

(10) Jones, W. D.; Duttweiler, R. P.; Feher, F. J.; Hessell, E. T. *Nouv. J. Chem.*, **1989**, *13*, 725-736.

Rh-C Bond Dissociation Enthalpies for Organometallic Derivatives of Rhodium Porphyrins

Bradford B. Wayland
Department of Chemistry
University of Pennsylvania
Philadelphia, Pennsylvania 19104-6323
U.S.A.

ABSTRACT. Solution equilibrium studies of rhodium porphyrin reactions with carbon monoxide, aldehydes, alkanes and alkenes that result in formation of Rh-C bonds are used in deriving estimates of Rh—C bond dissociation enthalpies. Maximum Rh—C D values occur for the methyl and formyl derivatives $(D(Rh-CH_3 \sim 242$ kJ mol^{-1}(58 kcal mol^{-1}); $D(Rh-CHO=249$ kJ mol^{-1} (59 kcal mol^{-1})) where the steric demands of the organic fragments are minimized.

INTRODUCTION

Rhodium (II) porphyrin complexes manifest an unusual breadth of organometallic reactivity including thermodynamically difficult transformations such as methane activation,[1,2] CO reductive coupling[3-5] and formation of metallo formyl complexes from reaction with H_2 and CO.[6-8] In the course of evaluating the scope and patterns of rhodium porphyrin reactivity, measurable equilibria have been observed for several reactions that have been used in providing information on Rh-C bond strengths.[9,10] Thermodynamic values for a series of rhodium porphyrin reactions that result in formation Rh-C bonds are given in Table1. Equilibrium constants for these reactions at several temperatures were determined in benzene solution by integration of the ^1H NMR and used in obtaining the experimental thermodynamic parameters. It is assumed that the solvation enthalpies of reactants and products for reactions in hydrocarbon media virtually cancel, and thus the reaction enthalpies in solution approximate the gas phase values. Enthalpy changes for each of these Rh-C bond forming reactions is obviously dependent on a Rh-C bond dissociation enthalpy (BDE), but derivation of Rh-C BDE values from reaction enthalpies requires additional information such as Rh-Rh and Rh-H bond dissociation enthalpies and thermochemical data for the organic radical fragments. Fortunately the dissociation enthalpies of $[(OEP)Rh]_2$ ($\Delta H° \sim 67$ kJ mol^{-1}) and $[(TXP)Rh]_2$ ($\Delta H° \sim 50$ kJ mol^{-1}) have been estimated from the ^1H NMR line broadening that is associated with exchange of the diamagnetic dimer with the paramagnetic monomers, and the (OEP)Rh—H BDE of 258 ± 8 kJ mol^{-1} (61.7 kcal mol^{-1}) has been derived from the enthalpy for the reaction of $[(OEP)Rh]_2$ with H_2 ($\Delta H° = -13\pm3$ kJ mol^{-1}) and the dissociation enthalpies for $[(OEP)Rh]_2$ (67 kJ mol^{-1})[9] and H_2 (436 kJ mol^{-1}). Gas phase $\Delta H°$ values for the required organic fragment reactions (Table 2) have been calculated from tabulations of ΔH_f (298K) values[11-14] or estimated by procedures described in the text. This article describes the construction of thermochemical cycles to extract estimates of Rh-C bond dissociation enthalpies from observed enthalpies for reactions 1-7 (Table 1).

J. A. Martinho Simões (ed.), Energetics of Organometallic Species, 69–74.

Table 1: Experimental thermodynamic values for reactions relevant to the estimation of Rh-C bond dissociation enthalpies in rhodium porphyrin organometallic derivatives. *

Reaction		$\Delta H°$ (kJ mol^{-1})	$\Delta S°$ (J K^{-1}mol^{-1})
1) (OEP) Rh-H + CO $\rightleftharpoons$ (OEP) Rh-CHO	$\Delta H_1°$	-54±4	-121±10
2) (OEP)Rh-H + (C$_4$H$_9$)CHO $\rightleftharpoons$			
(OEP)Rh-CH(OH)(C$_4$H$_9$)	$\Delta H_2°$	-42±4	-96±10
3) 2(TMP)Rh• + CH$_4$ $\rightleftharpoons$			
(TMP)Rh-H+(TMP)Rh-CH$_3$	$\Delta H_3°$	-54±4	-79±10
4) [(TXP)Rh]$_2$ + CH$_4$ $\rightleftharpoons$			
(TXP)Rh-H+(TXP)Rh-CH$_3$	$\Delta H_4°$	0±4	-29±10
5) [(OEP)Rh]$_2$ + CH$_2$=CH(CH$_3$) $\rightleftharpoons$			
(OEP)Rh-CH$_2$CH(CH$_3$)-Rh(OEP)	$\Delta H_5°$	-41±4	-84±10
6) [(OEP)Rh]$_2$ + CO $\rightleftharpoons$ (OEP)Rh-C(O)-Rh(OEP)	$\Delta H_6°$	-50±8	-129±10
7) [(OEP)Rh]$_2$ + 2CO $\rightleftharpoons$			
(OEP)Rh-C(O)-C(O)-Rh(OEP)	$\Delta H_7°$	-88±8	-259±20

* Determined by ^{1}H NMR in C$_6$D$_6$ solvent (reactions 1-5) and CD$_3$C$_6$D$_5$ (reactions 6,7).
  (OEP)=octaethylporphyrin dianion
  (TXP)=tetra(3,5dimethylphenyl)porphyrin dianion
  (TMP)=tetra(2,4,6trimethylphenyl)porphyrin dianion

Table 2: Enthalpy changes for reactions used in estimating Rh-C bond dissociation enthalpies.*

reaction	$\Delta H°$ kJ mol^{-1}	reaction	$\Delta H°$ kJ mol^{-1}
8) (OEP)Rh-H → (OEP)Rh• + H•	$\Delta H_8° = 259$	14) CH$_3$CH$_3$ → •CH$_2$CH$_3$ + H•	$\Delta H_{14}° = 422$
9) [(OEP)Rh]$_2$ → 2(OEP)Rh•	$\Delta H_9° = 67$	15) CH$_3$C$_6$H$_5$ → •CH$_2$C$_6$H$_5$ + H•	$\Delta H_{15}° = 369$
10) [(TXP)Rh]$_2$ → 2(TXP)Rh•	$\Delta H_{10}° = 50$	16) CH$_2$ = CH(CH$_3$) → •CH$_2$-CH(CH$_3$)•	$\Delta H_{16}° \sim 268$
11) H• + CO → •CHO	$\Delta H_{11}° = -64$	17) CO → CO*	$\Delta H_{17}° \sim 294$
12) H• + (Bu)CHO → •CH(OH)(Bu)	$\Delta H_{12}° = -116$	18) 2CO → (O)C-C(O)*	$\Delta H_{18}° \sim 305$
13) CH$_4$ → •CH$_3$ + H•	$\Delta H_{13}° = 439$		

*Thermodynamic data used in calculating these values is found in the following references: reactions 1-3 ref. 9; reactions 11-15 ref. 11-14; reactions 16-18 are explained in the text.

RESULTS AND DISCUSSION

(OEP)Rh—CHO BDE

Rhodium porphyrins achieved a measure of prominence as organometallic reagents through the observed reaction of H$_2$ and CO to produce metalloformyl (M-CHO) complexes. Octaethylporphyrinatorhodium(III) hydride ((OEP)Rh-H) reacts reversibly with CO in benzene to produce an equilibrium distribution with the formyl complex ((OEP)Rh-CHO) (equation 1). Thermodynamic values for reaction 1 were obtained by measurement of the

1) (OEP)Rh-H+CO $\rightleftharpoons$ (OEP)Rh-CHO

temperature dependence of the equilibrium constant using integration of the ^{1}H NMR (ΔH_1° =-54±4 kJ mol^{-1}; ΔS_1°= -121±10 J K^{-1}mol^{-1}) (table 1).[7] A thermochemical cycle to estimate the (OEP)Rh—CHO bond dissociation energy (BDE) is obtained by combining ΔH_1° with the (OEP)Rh—H BDE (259 kJ mol^{-1}),[9] and thermodynamic parameters for the formyl radical[11] (D(Rh-CHO)= $-\Delta H_1^{\circ}$+ ΔH_8°+ ΔH_{11}° = 249 kJ mol^{-1} (59.5 kcal mol^{-1})). The derived Rh-CHO BDE of 249 kJ mol^{-1} (59.5 kcal mol^{-1}) for

(OEP)Rh-CHO → (OEP)Rh• + •CHO $\quad\quad$ $\Delta H°$=249 kJ mol^{-1}(59.5 kcal mol^{-1})

(OEP)Rh-CHO is relatively large in absolute magnitude and nearly equal to the Rh-H BDE of 259 kJ mol^{-1} for (OEP)Rh-H.

Estimation of the (OEP)Rh-CH$_2$(OH)(Bu) BDE

(OEP)Rh-H reacts with aldehydes (RCHO) to form α-hydroxyalkyl complexes (OEP)Rh-CH(OH)R. In the specific case of pentanal (reaction 2) thermodynamic parameters for the reaction are available from solution equilibrium studies (ΔH_2°= -42±4 kJ mol^{-1}, ΔS_2° = -96±10 J K^{-1} mol^{-1}).[9] An estimate of 184 kJ mol^{-1} (44 kcal mol^{-1}) is obtained for

2) (OEP)Rh-H + (Bu)CHO ⇌ (OEP)Rh-CH(OH)(Bu)

the Rh—CH(OH)(Bu) BDE from a thermochemical cycle (D(Rh-CH(OH)(Bu) = $-\Delta H_2^{\circ}$ + ΔH_8° + ΔH_{12}°).

(OEP)Rh-CH(OH)Bu → (OEP)Rh• + •CH(OH)(Bu) $\quad$ $\Delta H°$=184 kJ mol^{-1}

Estimation of the Rh-CH$_3$ BDE for (TMP)Rh-CH$_3$ and (TXP)Rh-CH$_3$

(TMP)Rh• and [(TXP)Rh]$_2$ react with methane in C$_6$D$_6$ to produce measurable equilibria with the corresponding hydride and methyl derivatives (equations 3, 4).[1,2] The measured enthalpy changes for reaction 3 (ΔH_3° = -54±4 kJ mol^{-1}) along with the CH$_4$ BDE

3) 2 (TMP)Rh• + CH$_4$ ⇌ (TMP)Rh-H + (TMP)Rh-CH$_3$

4) [(TXP)Rh]$_2$ + CH$_4$ ⇌ (TXP)Rh-H + (TXP)Rh-CH$_3$

(439 kJ mol^{-1}) yields a value of 493 kJ mol^{-1} (118 kcal mol^{-1}) for the sum of the Rh-H and Rh-CH$_3$ bond dissociation energies ((D(TMP)Rh-CH$_3$ + D(TMP)Rh-H) = $-\Delta H_3^{\circ}$ + ΔH_{13}° = 493 kJ mol^{-1}). The enthalpy change for reaction 4 (ΔH_4° = 0±4 kJ mol^{-1}) along with the dissociation energy for [(TXP)Rh]$_2$ (50 kJ mol^{-1}) and D(CH$_4$) (439 kJ mol^{-1}) gives 489 kJ mol^{-1} (117 kcal mol^{-1}) for the sum of the Rh-H and Rh-CH$_3$ BDE ((D(TXP)Rh-H + D(TXP)Rh-CH$_3$) = $-\Delta H_4^{\circ}$ + ΔH_{10}° + ΔH_{13}° = 489 kJ mol^{-1}). The Rh-H BDE for (TMP)Rh-H (ν_{Rh-H}=2095cm^{-1}) and (TXP)Rh-H (ν_{Rh-H}=2094cm^{-1}) have not been directly measured, but judging from the difference in ν_{Rh-H} values they are probably slightly smaller than the measured value of 258 kJ mol^{-1} for (OEP)Rh-H (ν_{Rh-H}=2220cm^{-1}). Using an estimate of 251 kJ mol^{-1} (60 kcal mol^{-1}) for the Rh-H BDE places the (TMP)Rh—CH$_3$ BDE and the (TXP)Rh—CH$_3$ BDE at approximately 242 kJ mol^{-1} (57.8 kcal mol^{-1}) and 238 kJ mol^{-1} (56.9 kcal mol^{-1}). The experimental errors are undoubtedly larger than this difference Rh-CH$_3$ D values. The absolute magnitude and the small difference in the Rh-CH$_3$ and Rh-H D values are similar to observations on gas phase methyl and hydride derivatives of transition metal atoms and monocations.[15]

Preliminary studies for the corresponding reactions of (TMP)Rh• with ethane and toluene (2(TMP)Rh• + R-H→(TMP)Rh-H + (TMP)Rh-R (R=C$_2$H$_5$, CH$_2$C$_6$H$_5$)) yield equilibrium constants $(K(C_2H_6)(373K)=20\pm5; K(CH_3C_6H_5)(353K)=250\pm50)$.[16] Assuming that $\Delta S°$ is similar to that for the methane reaction $(\Delta S°_3=-79$ J K^{-1} mol$^{-1})$ yields estimates of -33 kJ mol^{-1} and -46 kJ mol^{-1} for the enthalpy changes for the C_2H_6 and $CH_3C_6H_5$ reactions. Using C-H BDE values of 422 kJ mol^{-1} for C_2H_6 and 368 kJ mol^{-1} for $CH_3C_6H_5$ yields a preliminary estimate of 204 kJ mol^{-1} (49 kcal mol^{-1}) for the $(TMP)Rh-CH_2CH_3$ BDE and 163 kJ mol^{-1} (39 kcal mol^{-1}) for the $(TMP)Rh-CH_2C_6H_5$ BDE. The combined steric and electronic effects associated with substitution of a methyl group for hydrogen lowers the Rh-C BDE by ~ 38 kJ mol^{-1} (~9 kcal mol^{-1}). The large difference between the estimated BDE values for $Rh-CH_2CH_3$ and $Rh-CH_2C_6H_5$ is dominated by the difference in the hydrocarbon D(C-H) values and not by the intrinsic Rh-C bond enthalpies in the alkyl complexes.

Estimation of the average effective Rh-C BDE for $(OEP)Rh-CH_2CH(CH_3)-Rh(OEP)$

$[(OEP)Rh]_2$ reacts with propene to form a bridged alkyl complex (equation 5).[17]

5) $[(OEP)Rh]_2 + CH_2=CH(CH_3) \rightleftharpoons (OEP)Rh-CH_2CH(CH_3)-Rh(OEP)$

The average Rh-C bond dissociation energy of 54 kJ mol^{-1} (13 kcal mol^{-1}) is obtained from one half the sum of $-\Delta H°_5$ (41 kJ mol^{-1}) and $D[(OEP)Rh]_2$ (67 kJ mol^{-1}). This $\overline{Rh-C}$ BDE is not directly comparable with those for rhodium alkyls because it includes a large contribution from conversion of the hypothetical diradical $(•CH_2CH(CH_3)•)$ to propene. Using Benson's parameters for radicals[18] provides an estimate 264 kJ mol^{-1} for $\Delta H°_f$ $(•CH_2CH(CH_3)•)$ and a $\Delta H°$ of 244 kJ mol^{-1} (58.4 kcal mol^{-1}) for conversion of propene to the diradical in its relaxed form. The dissociation of $(OEP)Rh-CH_2CH(CH_3)-Rh(OEP)$ into $2(OEP)Rh•$ and $•CH_2CH(CH_3)•$ evaluated by this procedure is 352 kJ mol^{-1} which yields an effective average Rh-C BDE of 176 kJ mol^{-1} (42 kcal mol^{-1}). Alternately assuming that dissociation of 2H• from propane to form $•CH_2CH(CH_3)•$ occurs with a $\Delta H°$ of ~828 kJ mol^{-1} (198 kcal mol^{-1}) along with $\Delta H°_f$ for $CH_2=CH(CH_3)(20$ kJ mol$^{-1})$, $CH_3CH_2CH_3(-104$ kJ mol$^{-1})$ and 2H• (436 kJ mol^{-1}) yields a $\Delta H°$ of ~268 kJ mol^{-1} (64 kcal mol^{-1}) for conversion of $CH_3CH=CH_2$ to $•CH_2CH(CH_3)•$. This procedure results in an estimate of 188 kJ mol^{-1} (45 kcal mol^{-1}) for the average Rh-C bond dissociation enthalpy for $(OEP)Rh-CH_2CH(CH_3)-Rh(OEP)$ to form $2(OEP)Rh•$ and $•CH_2CH(CH_3)•$. The two approximate approaches give similar estimates for $\overline{Rh-C}$, but the later model which assumes transferability of C-H BDE values is more widely applicable and easy to visualize. The average $\overline{Rh-C}$ dissociation energy by this method corresponds to forming a diradical where the structure is relaxed but the π bond is not formed and thus should be directly comparable to the other Rh-alkyl (D(Rh-C)) values. The two inequivalent Rh-C dissociation enthalpies cannot be individually evaluated, but assuming that the $Rh-CH_2CH(X)^-$ bond is similar to that in $(OEP)Rh-CH_2CH_3$ $(D(Rh—CH_2CH_3)$ ~203 kJ mol^{-1}) suggests that $D(Rh-CH_2-)$ is ~203 kJ mol^{-1} (48 kcal mol^{-1}) and that $D(Rh-CH(CH_3)-)$ is ~176 kJ mol^{-1} (42 kcal mol^{-1}). These estimates provide a plausible sequence of D(Rh-C) values which parallels the number of alkyl groups substituted for hydrogen on the α carbon.

Estimation of the average Rh-C(O)- bond enthalpies for (OEP)Rh-C(O)-Rh(OEP) and (OEP)Rh-C(O)-C(O)-Rh(OEP)

[(OEP)Rh]$_2$ reacts with CO to form dimetal ketone and dimetal α-diketone complexes (equations 6, 7).[3] The formal average Rh-C BDE in (OEP)Rh-C(O)-Rh(OEP) is

6) [(OEP)Rh]$_2$ + CO $\rightleftharpoons$ (OEP)Rh-C(O)-Rh(OEP)

7) [(OEP)Rh]$_2$ + 2CO $\rightleftharpoons$ (OEP)Rh-C(O)-C(O)-Rh(OEP)

59 kJ mol^{-1} (14 kcal mol^{-1}) and 78 kJ mol^{-1} (18.6 kcal mol^{-1}) in (OEP)Rh-C(O)-C(O)-Rh(OEP), which are not directly comparable with D((OEP)Rh—CHO) (249 kJ mol^{-1}). In reactions 6 and 7 conversion of the diradicals (:C=O, •C(O)-C(O)•) to CO makes a large contribution to the average D values. Setting the average C(O)-H bond dissociation enthalpy for H$_2$CO and HC(O)C(O)H at 366 kJ mol^{-1} (87.5 kcal mol^{-1}) and using the ΔH_f° (298K) for H$_2$CO, HC(O)C(O)H, CO and H• provides estimates for the enthalpy change in converting CO to the hypothetical diradicals (CO $\rightarrow$ CO* (ΔH°~294 kJ mol^{-1}); 2CO $\rightarrow$ C(O)-C(O)*(ΔH°~305 kJ mol^{-1})). This approach places the average Rh-C(O)-bond enthalpies for (OEP)Rh-C(O)-Rh(OEP) and (OEP)Rh-C(O)-C(O)-Rh(OEP) at ~203 kJ mol^{-1} (48.5 kcal mol^{-1}) and 232 kJ mol^{-1} (55.4 kcal mol^{-1}) respectively. These estimates produce a plausible sequence of effective bond dissociation enthalpies in comparison with (OEP)Rh-CHO (D(Rh-CHO) = 249 kJ mol^{-1} (59.5 kcal mol^{-1})).

Table 3: Derived Rh-C bond dissociation enthalpies (D) values for organometallic derivatives of rhodium porphyrins.

	D(Rh-C) kJ mol^{-1} (kcal mol^{-1})
(OEP)Rh-CHO $\rightarrow$ (OEP)Rh• + •CHO	249 (59.5)
(OEP)Rh-CH(OH)(Bu) $\rightarrow$ (OEP)Rh• + •CH(OH)(Bu)	184 (44.0)
(TMP)Rh-CH$_3$ $\rightarrow$ (TMP)Rh• + •CH$_3$	242 (57.8)
(TXP)Rh-CH$_3$ $\rightarrow$ (TXP)Rh• + •CH$_3$	238 (56.9)
(TMP)Rh-CH$_2$CH$_3$ $\rightarrow$ (TMP)Rh• + •CH$_2$CH$_3$	203 (48.5)
(TMP)Rh-CH$_2$C$_6$H$_5$ $\rightarrow$ (TMP)Rh• + •CH$_2$C$_6$H$_5$	138 (39.0)
(OEP)Rh-CH$_2$CH(CH$_3$)-Rh(OEP) $\rightarrow$ 2(OEP)Rh• + •CH$_2$CH(CH$_3$)•	188 (44.9)
(OEP)Rh-C(O)-Rh(OEP) $\rightarrow$ 2(OEP)Rh• + CO*	203 (48.5)
(OEP)Rh-C(O)-C(O)-Rh(OEP) $\rightarrow$ 2(OEP)Rh• + C(O)C(O)*	232 (55.4)

References

1) Wayland, B.B.; Ba,S.; Sherry, A.E. *J. Am. Chem. Soc.* **1991**, *113,* 5305.

2) Sherry, A.E.; Wayland, B.B. *J. Am. Chem. Soc.* **1990**, *112*, 1259.

3) Coffin, V.L.; Brennen, W.; Wayland, B.B. *J. Am. Chem. Soc.* **1988**, *110*, 6063.

4) Wayland, B.B.; Woods, B.A.; Coffin, V.L. *Organometallics* **1986**, *5*, 1059.

5) Wayland, B.B.; Sherry, A.E.; Coffin, V.L. *J. Chem. Soc., Chem. Commun.* **1989**, 662.

6) Wayland, B.B.; Woods, B.A.; Pierce, R. *J. Am. Chem. Soc.* **1982**, *104*, 302.

7)   Farnos, M.D.; Woods, B.A.; Wayland, B.B. *J. Am. Chem. Soc.* **1986**, *108*, 3659.

8)   Wayland, B.B.; Farnos, M.D.; Coffin, V.L. *Inorg. Chem.* **1988**, 27, 2745.

9)   Wayland, B.B. *Polyhedron* **1988**, 7, 1545.

10) Wayland, B.B.; Coffin, V.L.; Sherry, A.E.; Brennen, W.R. *ACS Symposium Series #428*, Marks, T. (Ed.), chapter 10, pp. 148-158.

11) Wagman, D.D.; Evans, W.H.; Parker, V.B.; Schumm, R.H.; Halow, I; Bailey, S.M.; Churney, K.L.; Nuttall, R.L. The NBS Tables of Chemical Thermodynamic Properties. *J. Phys. Chem. Ref. Data* **1982**, *11*, Suppl. no. 2.

12) Pedley, J.B.; Naylor, R.D.; Kirby, S.P. *Thermochemical Data of Organic Compounds*; Chapman and Hall: London, **1986**.

13) McMillen, D.F.; Golden, D.M. *Annu. Rev. Phys. Chem.* **1982**, *33*, 493.

14) Martinho Simões, J.A.; Beauchamp, J.L. *Chem. Rev.* **1990**, *90*, 629.

15) Armentrout, P.B. ACS Symposium Series #428, Marks, T. (Ed.), chapter 2, pp18-33.

16) Ba, S. Ph.D. Dissertation, The University of Pennsylvania, **1991**.

17) Wayland, B.B.; Ba, S. *Organometallics*, **1989**, *8*, 1438.

18) O'Neal, H.; Benson, S. in "Free Radicals" Vol.II Kochi, J.K., Ed.; Wiley: New York, **1973**, p343.

Acknowledgement

We gratefully acknowledge support of this work by the National Science Foundation and the Department of Energy, Division of Chemical Sciences, Offices of Basic Energy Sciences, through Grant DE-FG5-86ER13615.

SOLUTION THERMOCHEMISTRY OF ORGANOMETALLIC COMPOUNDS
USING PHOTOACOUSTIC CALORIMETRY: ENTHALPIES OF LIGAND
EXCHANGE, METAL-LIGAND BONDS, AND METAL-SOLVENT
INTERACTIONS

T. J. Burkey
Memphis State University
Department of Chemistry
Memphis, Tennessee USA 38152

ABSTRACT. This article discusses the application of photoacoustic calorimetry (PAC) to the study of organometallic compounds. Studies include the determination of the enthalpy of CO substitution on $M(CO)_6$ (M = Cr, Mo, W) by heptane, CO substitution on $Cr(CO)_6$ by amines (piperidine, n-butylamine, di-n-butylamine, tri-n-butylamine) and by weakly coordinating solvents (alkane, aromatic, and chlorocarbon solvents), and CO substitution on various metal carbonyls ($M(CO)_6$ (M = Cr, Mo, W), η^6-$C_6H_6Cr(CO)_3$, and η^5-$CpMn(CO)_3$ (Cp = cyclopentadienyl)) by triethylsilane. From known metal-CO bond energies estimates of metal-ligand bond energies are calculated. The relation of structure and bond energies is discussed. PAC was also used to investigate the quantum yields of photosubstitution of each metal carbonyl in perfluorodecalin (PFD). In each case the quantum yield decreased a result attributed to the weak metal-PFD interaction. The PAC studies provide insight into the behavior of organometallic species, and the mechanistic significance of these results is considered.

1. Introduction

Gas-phase techniques provide some of the most precise methods for determining thermodynamic data such as bond energies. Unfortunately many organometallic compounds are not thermally stable or sufficiently volatile to be used in the gas-phase. As an alternative, we have utilized photoacoustic calorimetry (PAC) for the study of organometallic species. This technique allows the determination of the heats of reaction in solution at ambient temperatures. Like gas-phase studies, this technique avoids complications of solid phase reactions where there may be interactions between two or more molecules of the starting compound. Solvation provides the gas phase equivalent of vaporization which allows the molecule of interest to react without interaction with a like molecule. Nevertheless, a solution technique does require consideration of the role that solvent may play in the reaction. Indeed PAC is an ideal method for examining the interaction of the solvent with the reaction species. An advantage over bomb calorimetry is that products of reaction need not be stable species; in fact, the technique is often used to study the thermochemistry of transient species.

The essential details of PAC will be discussed in Section 2, but two requirements of the technique are important to emphasize. First, the reactions of interest must be initiated by light, and second, there must be some knowledge about the quantum yield for the reaction. Despite these restrictions the technique continues to evolve. The development of deconvolution techniques allows the determination of reaction rates, while temperature

75

J. A. Martinho Simões (ed.), Energetics of Organometallic Species, 75–94.
© 1992 *Kluwer Academic Publishers.*

dependence studies in aqueous solutions have been used to study reaction volume changes. [1,2] PAC has also been used to determine quantum yields and extinction coefficients. [3] We are currently developing a high pressure PAC system to study reaction volume changes in organic solvents. Nonetheless, we believe the most exciting application of PAC is the determination of enthalpies of reactions generating transient species.

In this article, both recent published and unpublished work are discussed. Due to improvements in quantum yield measurements, published results have been updated and have resulted in significant changes in the interpretation of some results. Studies include the displacement of CO by heptane on $M(CO)_6$, (M = Cr, Mo, and W). This allows an estimate of the energy of the metal-heptane bond, an interaction of great interest due to its close relation to metal insertion into C-H bonds and agostic bonds. [4-6] Studies of the enthalpy of CO substitution on $Cr(CO)_6$ by amines of differing steric bulk provide an estimate of intrinsic Cr-amine bond energies and their dependence on steric effects. The enthalpy of reaction of various metal carbonyls with triethylsilane has been measured. For the metal centers studied, insertion into the Si-H bond is preferred over insertion into alkyl C-H bonds, and the results allow the determination of the magnitude of the metal-silane reaction. Results of PAC studies of the quantum yields for photosubstitution of metal carbonyls in a perfluorocarbon solvent are used to investigate the interaction of metal centers with a perfluorocarbon solvent. Finally the enthalpy of CO dissociation from $Cr(CO)_6$ in several "inert" solvents was determined to examine the Cr-solvent bond strength.

2. Experimental Methods

2.1 PAC THEORY AND INSTRUMENTATION

Photoacoustic calorimetry has been used to study the thermochemistry of biochemical, organic, inorganic and organometallic systems. [7-9] Light is used to initiate a reaction of interest, and the heat that is liberated from the reaction is detected acoustically. A photoacoustic calorimeter is diagramed in Figure 1. A laser light pulse is weakly focused on a quartz flow cuvette. The light initiates chemical and physical processes that liberate heat resulting in a pressure increase. The propagation of that pressure increase is an acoustic wave the amplitude of which is detected by a piezoelectric transducer. Typically the transducer face is clamped to a side of the cuvette that is parallel to the light beam. To enhance acoustic coupling silicon grease is applied between the transducer and the cuvette. The transducer signal is amplified, stored on a digital oscilloscope, and transferred to a computer. Waveforms can be rejected if a misfired laser pulse is detected. The signal-to-noise ratio is enhanced by averaging 30-100 waveforms. An important feature of the calorimeter is the frequency response of the transducer: it responds to heat liberated at rates much faster or near its frequency response but not to heat liberated much slower. This allows the heat liberated by transient species to be detected and the heat of fast reactions to be resolved from that of slow reactions. The observed waveform is a damped sine wave the amplitude of which can depend on the laser pulse duration, beam construction, beam position relative to the transducer, cell shape, solvent, and on the amount of heat liberated after the laser pulse. If the beam is too wide the wave generated at one side of the beam will arrive at a significantly different time than the wave generated at the other side of the beam, resulting in destructive interference and distortion of the acoustic signal. A similar consideration arises for the time duration of the light pulse. Allowable beam widths and pulse durations are determined by the transducer frequency response. For a 1 Mhz transducer, a 2 mm beam width and 5 ns pulse duration have been used. [1a] In terms of beam position, the farther the beam is from the transducer the more the wave is attenuated.

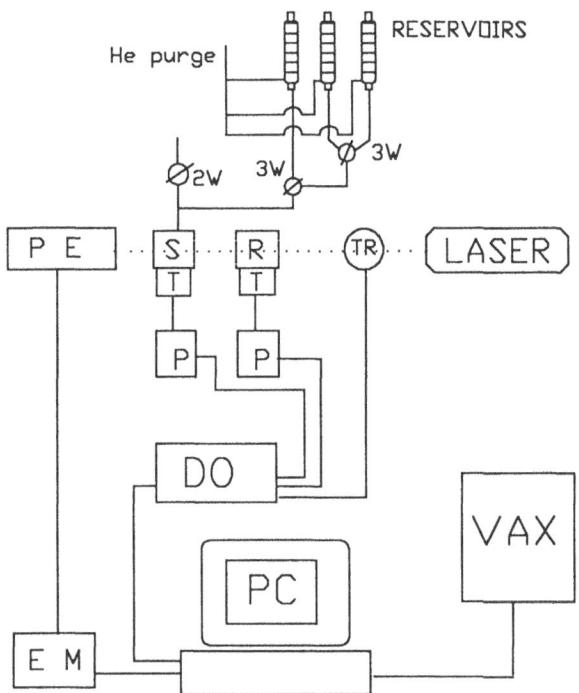

FIGURE 1. Photoacoustic Calorimeter. Laser = PTI LP 300 N_2 laser/PL 202 Dye laser, TR = PRA optical trigger, R = quartz cuvette (reference), S = quartz flow cuvette (sample), PE = Laser Precision RJP-735 pyroelectric probe, EM = Laser Precision RJ-7620 energy meter, RT = Panametrics 2.25 Mhz transducer , ST = Panametrics 1.0 Mhz transducer, P = Panametrics ultrasonic preamplifier, DO = Tektronix 2430 digital oscilloscope, PC = Zenith 386 computer, VAX = minicomputer, 2W = two-way Teflon stopcock, and 3W = three-way Teflon stopcock.

Finally, different solvents have different coefficients of thermal expansion and acoustic velocities, and these result in different wave amplitudes and transport times for the acoustic wave to reach the transducer.

To account for these variables, the calorimeter is calibrated with a reference compound dissolved in the same solvent used for the sample. In practice, signals are obtained for both sample and reference solutions without altering the positions of the PAC optical components. Photolysis of the reference compound must yield a known amount of heat for the amount of light absorbed. For convenience we use compounds that deposit all the absorbed light energy as heat in less than a ns. In organic solvents, we typically use ferrocene and o-hydroxybenzophenone.[10]

For the simple case where reactions only liberate heat much faster or slower than the response of the transducer, the signal amplitude will be proportional only to the heat liberated by the faster reactions. Thus the ratio of the signal amplitudes for the sample (S_s) and reference (S_r) solutions will be equal to the ratio of the heats liberated. The signals are corrected for differences in light intensity and the absorbance of the solutions to give the observed fractional heat (α), which is the ratio of the heats released from the sample and reference solutions per quanta. This is calculated from equation 1 where P_s and P_r are the

$$\alpha = (S_s/S_r)(P_r/P_s)(1-T_r)/(1-T_s) \tag{1}$$

pulse energies while collecting the sample and reference signals from the sample cuvette, and T_s and T_r are the corresponding light transmissions. The PAC signal for pure solvent is not always negligible, so it is routinely substracted from the signals for reference and sample solutions.

The ratio P_r/P_s can be conveniently determined by collecting a portion of the light beam and irradiating a reference solution in the reference cuvette. The transmission of solutions can be determined with a spectrophotometer, but we find it convenient to make this determination *in situ* during the experiment. The intensity of light that passes through the sample cell is determined using an energy meter and probe, so the transmission is simply the ratio of the intensity measured for a sample or reference solution to that for pure solvent. The *in situ* measurement avoids errors caused by alteration of the solution. For example, sampling for a spectrophotometric measurement may result in contamination or evaporation of the solution. In addition, the spectrophotometric measurement must be obtained at precisely the same temperature as that of the PAC experiment. While enthalpies are not very sensitive to temperature, extinction coefficients and sample volumes are, and a temperature change between the time the absorbance is measured and the acoustic signal is collected can introduce an error.

For certain experiments it is desirable to use other than ambient temperatures, but use of a flow system would require regulating the temperature of the solution handling apparatus. This includes a manifold connecting reservoirs to the sample cuvette (Figure 1). We have solved this problem by enclosing the flow cell and solvent handling apparatus in an air-temperature controlled chamber and allowing the components to come to thermal equilibrium.

The observed heat evolved from the sample solution (αE_v) is related by eq 2 to the enthalpy of reaction (ΔH), the quantum yield (Φ), and the energy of a mole of photons (E_v). The first term represents the fraction of light that does not result in reaction but is instead deposited as heat. The second term corresponds to the light energy used to drive the reaction minus the enthalpy of the reaction. If there is luminescence, an additional term must be included, and the quantum yield, the energy, and lifetime for luminescence must be known. Rearrangement of eq 2 yields eq 3, our working equation for determining the enthalpy of a reaction that can be initiated by light. Possible α values encountered are

$$\alpha E_v = E_v(1 - \Phi) + \Phi(E_v - \Delta H) \tag{2}$$

$$\Delta H = (1 - \alpha)E_v/\Phi \tag{3}$$

illustrative. If the quantum yield is unity and the reaction is endothermic, then α is positive and less than unity, but it is not less than zero since only E_v is available to drive an endothermic process.[11] If the quantum yield is unity and the reaction is exothermic, then α is greater than unity. Note for endothermic reactions, the smaller the quantum yield the larger α will be since more heat is given off when there is no net reaction (because of the $E_v(1 - \Phi)$ term). In terms of error analysis the accuracy of Φ is less critical the closer α is to unity.

2.2 SAMPLE PREPARATION AND MANIPULATION

Solutions are prepared air-free using Schlenck and cannula techniques. The absorbance of solutions is kept low, between 0.08 and 0.11. This ensures that the intensity of the beam along the path of the cell is nearly constant, and hence the heat emanating at each point along the light path is nearly constant. Therefore a small change in light transmission or light intensity will yield proportional changes in the transducer signal amplitude. At higher absorbances, a greater fraction of the light would be absorbed near the front of the sample, and since the transducer response to every point along the light path may not be equal, a non-linear transmission dependence could result. Higher absorbances can be used if the absorbance of the reference solution is carefully matched to that of the sample solution. Then the relative contribution to the acoustic signal at each point of the light path will be the same for the sample and reference solutions.

Reservoirs containing pure solvent, reference solution, and a sample solution are purged and pressurized (1-3 psi) with helium (Figure 1). Helium is used because its low solubility reduces the formation of bubbles as the solution passes through the sample cuvette. The reservoirs are connected to the cuvette by a glass manifold with Teflon stopcocks. The cuvette can be emptied and flushed with fresh solution from any of the reservoirs so that a signal can be collected from a new solution in 1-2 minutes. To ensure that products are not photolyzed and reactants are not depleted, the rate of solution flow is increased until the acoustic signal is independent of the flow rate. Alternatively the laser pulse rate can be decreased. To check for multiphoton effects the pulse intensity is varied 4-10 fold, typically in the range of 4 to $100 \mu J$. Pulse intensities are normally $20\text{-}30 \mu J$ for data collection.

2.3 OTHER CONSIDERATIONS

Should the heats of reaction approach the time response of the transducer then the amplitude of the acoustic signal will decrease and its phase will shift. Recent studies show that deconvolution of such waveforms provides not only the magnitude of heat evolved but also the rate of heat evolution so that enthalpies and rates of reactions can be determined.[1] Rather than using the limiting rate conditions described above, various deconvolution techniques are now commonly used to analyze PAC data.

It is worth pointing out that heat is always observed in a PAC experiment, and PAC is a poor tool for identifying intermediates in solution. At best PAC can be used to demonstrate that a heat decay depends on the concentration of a reagent. It is therefore important to characterize the photochemistry by other techniques. This should include isolation and characterization of photoproducts, determination of quantum yields for conditions applicable to the PAC experiment, and flash photolysis studies confirming the kinetics of transient species.

3. Coordination of Heptane to $M(CO)_5$ (M = Cr, Mo, and W)

3.1 BACKGROUND

The ligand exchange reaction of group VI-B metal hexacarbonyls have been extensively studied in hydrocarbon solvents. The kinetics are described by a two-term rate law (eq 4) which can be attributed to the competing dissociative (eq 5 and 6) and associative processes (eq 7).[12] The intermediate in the dissociative process is postulated to be coordinatively unsaturated, but more recent flash photolysis studies clearly demonstrate that $M(CO)_5$ is rapidly coordinated, even by alkanes.[13] There is not a consensus on the lifetime of uncoordinated $Cr(CO)_5$ in solution, and values of less than a ps to 100 ps have been

$$Rate = k_1[M(CO)_6] + k_2[L][M(CO)_6] \tag{4}$$

$$M(CO)_6 \dashrightarrow M(CO)_5 + CO \tag{5}$$

$$M(CO)_5 + L \dashrightarrow M(CO)_5L \tag{6}$$

$$M(CO)_6 + L \dashrightarrow [M(CO)_6L] \dashrightarrow M(CO)_5L + CO \tag{7}$$

$$M(CO)_6 \xrightarrow{h\nu} M(CO)_5 + CO \xrightarrow{S} M(CO)_5S \tag{8}$$

M = Cr, Mo, W; S = solvent

reported. No coordinatively-unsaturated $M(CO)_5$ is observed at longer timescales, only solvent coordinated $Cr(CO)_5$. Although the assignment is controversial, a transient infrared spectrum has been assigned to coordinatively-unsaturated $Cr(CO)_5$ after the photolysis of $Cr(CO)_6$ in solution.[14] These studies raise the question: what is the relative importance of coordinatively-unsaturated $M(CO)_5$ and solvent coordinated $M(CO)_5$ in ligand exchange reactions? Certainly a determining factor is the M-S bond energy (BE_{M-S}).

If the BE_{M-S} is significant, the enthalpy for CO dissociation in solution will be significantly less than BE_{M-CO}. It has been demonstrated that the enthalpies of organometallic reactions are insensitive to non-stereospecific solvation by hydrocarbon solvents;[15] therefore ΔH_8 will be equal to the difference in BE_{M-CO} and BE_{M-S}. Thus knowledge of the first two will allow calculation of the third.

3.2 ENTHALPY OF CO DISSOCIATION FROM $M(CO)_6$ IN HEPTANE

Using PAC, we determined the enthalpy of CO dissociation (ΔH_8) in heptane.[16] At low concentrations of dispersed nucleophile, the signal observed by PAC corresponds to eq 8. For $\Phi = 0.67$, we obtained 27.2 kcal/mol for ΔH_8 in heptane in agreement with a previous PAC study.[17] However, we have confirmed a recent finding that $\Phi = 0.72(2)$ in heptane, and this yields $\Delta H_8 = 25.2$ kcal/mol.[18] The activation enthalpy for CO dissociation from $Cr(CO)_6$ in the gas phase is 37(2)[19] and is postulated to be equal to BE_{M-CO} since the reverse reaction is near the gas kinetic limit.[20] This value is within the experimental error of that for CO dissociation in solution, the first step of dissociative ligand exchange (eq 5)[12] and suggests that the $Cr(CO)_5$ fragment is stabilized by nearly 12 kcal/mol relative to the transition state for CO dissociation. Abundant evidence indicates the stabilization involves a stereospecific bond between heptane and chromium.[21,22]

Similarly we obtained ΔH_8 for M = Mo and W which are significantly smaller than the gas-phase activation enthalpies for CO dissociation (ΔH_5^*, TABLE 1). As with chromium, the ΔH_8 is smaller then ΔH_5^* for the gas phase indicating there is considerable stabilization of the $M(CO)_5$ fragment in the alkane solvent. Solvent coordination may also explain differences in gas-phase and solution activation enthalpies for CO dissociation. The activation enthalpies in alkane solvents are 30 and 40 kcal/mol, for Mo and W, respectively, considerably smaller than those recently reported in the gas phase (TABLE 1).[12] The CO dissociation in solution may involve an interchange process where coordination of the solvent begins before complete CO dissociation, and a partial M-alkane bond is formed to the metal in the transition state. The behavior of the molybdenum and tungsten complexes may be contrasted to $Cr(CO)_6$ where the activation enthalpy for CO dissociation in decalin is the same as in the gas phase.[12] Presumably there is little chromium-solvent bonding in

TABLE 1. Enthalpies of CO Dissociation from $M(CO)_6$ in Heptane, and in the Gas Phase and Metal-Heptane bond Energies[a]

M	Φ[b]	α	ΔH_8	ΔH_5^{*}[c]	BE_{M-S}
Cr	0.72(2)	0.786(9)	25.2(13)	36.8(20)	11.6(26)
Mo	0.93(1)	0.750(10)	23.2(13)	40.5(20)	17.3(27)
W	0.79(1)	0.723(3)	31.3(6)	46.0(20)	14.7(21)

[a]energy in kcal/mol [b]Reference 18 [c]gas-phase values, references 20,23-25.

this transition state. The predilection of $Mo(CO)_6$ and $W(CO)_6$ to undergo an interchange process has also been observed for the substitution of CO by stable ligands such as phosphines.[26]

3.3 COMPARISON OF LITERATURE DATA

The BE_{M-S} are surprisingly large and there is little data for comparison. The NRC Laser Chemistry group has recently studied the equilibria between $W(CO)_5$, S, and $W(CO)_5S$ in the gas phase and from the temperature dependence determined BE_{W-S}.[27] For S = n-hexane and cyclohexane they obtained values of 10.8(30) and 11.6(30) kcal/mol. These results are somewhat smaller than the PAC results but still within experimental error.

Activation enthalpies for alkane dissociation from $M(CO)_5(RH)$ have been determined by Dobson's group. In one study, mixtures of heptane and 1-hexene were used to investigate the rates of heptane displacement from $W(CO)_5(n\text{-}C_7H_{16})$ by 1-hexene.[28] The reaction exhibits dissociative kinetics consistent with the mechanism shown in eq 9 and 10. For heptane dissociation (eq 9), the authors reported 8.4(5) kcal/mol for the activation enthalpy and 1.8(17) eu for the activation entropy. The small activation entropy is

$$W(CO)_5(n\text{-}C_7H_{16}) \xrightarrow{} W(CO)_5 + n\text{-}C_7H_{16} \qquad (9)$$

$$W(CO)_5 + C_6H_{12} \dashrightarrow W(CO)_5(\eta^2\text{-}C_6H_{12}) \qquad (10)$$

consistent with residual bonding in the transition state. If this were the case then the activation enthalpy would be less than the W-heptane bond dissociation energy, and this is what is observed. Evidence of residual W-alkane bonding in a transition state was also observed in a kinetically different system. The reaction of 4-acylpyridine with $W(CO)_5(CH)$ (CH = cyclohexane) obeys a single-term second-order rate law where $\Delta H^* = 3.4$ kcal/mol and $\Delta S^* = -13.0$ (6) cal/(deg mol).[29] The dependence is first order in 4-acylpyridine and $W(CO)_5(CH)$ implicating an associative process. The activation enthalpy is much lower than BE_{W-CH} (assuming the bond energy for CH is the same as for heptane, See 7.1) again indicating a stabilization of the transition state for CH dissociation. The low activation enthalpy and negative activation entropy for CH displacement are consistent with an associative process where there is bond formation to 4-acylpyridine in the transition state.

When comparing the same nucleophiles and alkane solvents, Dobson's group has shown that $Mo(CO)_5S$ and $W(CO)_5S$ have a greater propensity than $Cr(CO)_5S$ to utilize an interchange process for alkane displacement by a nucleophile.[30] Inspection of TABLE 1 shows that stronger M-S bonds may be responsible for this trend. While the origin of the stronger bonds for $Mo(CO)_5S$ and $W(CO)_5S$ might be attributed in part to electronic effects, we have shown in studies of amine substitution (See Section 4) that the $Cr(CO)_5$ fragment is very sensitive to steric effects. Steric effects may weaken Cr-S bonds and make an interchange process unfavorable in certain cases.

It is tempting to postulate that the large metal-alkane bond energies and rapid reaction of metal fragments with solvent preclude the formation of coordinatively-unsaturated complexes as reaction intermediates in solution. There is excellent data indicating this is not the case. Recent studies of ligand exchange of $Cr(CO)_5L$ in Dobson's group plus their results mentioned above for $W(CO)_5$ (n-C_7H_{16}) and 1-hexene demonstrate that for certain solvents and ligands there must be an intermediate the composition of which does not involve a coordinated solvent molecule or ligand.[31] Entropy may account for the formation of small but kinetically significant amounts $M(CO)_5$ in equilibrium with $M(CO)_5$ S. Furthermore, in many cases an associative displacement of coordinated solvent will be incumbered sterically thus increasing the free energy of activation and making the dissociative process more competitive.

4. Studies of Amine Exchange with $Cr(CO)_6$

4.1 BACKGROUND

A conflict in the interpretation of data in the literature prompted our investigation of the exchange of amines with $Cr(CO)_6$. Metal-amine bonds are generally believed to be weaker than the corresponding metal-CO bonds primarily because amines are more readily displaced than CO in solution.[12] Activation enthalpies for ligand dissociation in solution are used as a measure of bond energies,[32] and the activation enthalpies for amine dissociation are less than the corresponding activation enthalpies for CO dissociation.[12] In the case of thermal decomposition of $Cr(CO)_5$ (c-HNC_5H_{10}) in hexane, piperidine dissociation is the rate determining step with an activation enthalpy of 25.7 (12) kcal/mol.[33] This is 11-12 kcal/mol less than the activation enthalpy for CO dissociation from $Cr(CO)_6$ in decalin.[12] If it is assumed that the activation enthalpies of dissociative processes are equal to bond energies this suggests the Cr-CO bond energy is about 12 kcal/mol greater than the corresponding Cr-piperidine bond and that the enthalpy of CO exchange for piperidine is 12 kcal/mol. In contradiction with these results and/or interpretation Adamson's group reported 2.4 (10) kcal/mol for the enthalpy of piperidine exchange with $Cr(CO)_6$ in cyclohexane.[34] They suggested that solvation of the transition state for piperidine dissociation results in a low activation enthalpy and therefore ΔH^* is not a good measure of the bond energy. They concluded that the Cr-piperidine bond is nearly as strong as the Cr-CO bond. In further support of this argument, the activation entropy for piperidine dissociation is small (3.8 (48) eu) consistent with an associative process where solvent coordination could make the ΔH^* and ΔS^* low.[33] We found their arguments appealing and set out to confirm their results.

4.2 ENTHALPY OF AMINE EXCHANGE WITH $Cr(CO)_6$

Using high amine concentrations with $Cr(CO)_6$ in heptane, we determined the enthalpy of CO exchange with amines having different steric bulk (eq 5 and 6, L = amine).[8] NMR experiments demonstrate that the product hydrogen bonds to free amine via the coordinated amine proton (eq 11) and that the PAC measurements include the heat of hydrogen bond formation (ΔH_{11}) as well as ligand exchange ($\Delta H_{5,6}$). From temperature-dependent studies of the equilibria in eq 11, the enthalpy of hydrogen bond formation was determined allowing the calculation of the enthalpy of ligand exchange from eq 12. The results are reported in

$$Cr(CO)_5L + L \longrightarrow Cr(CO)_5L\text{-}L \tag{11}$$

L = piperidine, n-butylamine, di-n-butylamine

$$\Delta H_{5,6,11} = \Delta H_{5,6} + \Delta H_{11} \tag{12}$$

TABLE 2. For di-n-butylamine, we determined that there was incomplete hydrogen bonding for the concentrations used in the PAC experiments. The value for $\Delta H_{5,6,11}$ is only an upper limit. The $\Delta H_{5,6}$ was calculated from the contribution of ΔH_{11} to $\Delta H_{5,6,11}$ (65% based on equilibrium calculations).

TABLE 2. Enthalpy of CO Substitution on $Cr(CO)_6$ by Amines in Heptane and Chromium-Amine Bond Energies[a]

amine (L)	$\Delta H_{5,6,11}$	$-\Delta H_{11}$	$\Delta H_{5,6}$	BE(Cr-L)
Piperidine	5.5(17)	4.7(7)	10.2(19)	26.6(28)
n-butylamine	2.6(14)	3.6(3)	6.2(14)	30.6(24)
Di-n-butylamine	≤7.2(6)	5.4(15)	10.1(11)	26.7(23)
Tri-n-butylamine	[b]	[b]	14.5(14)	22.3(24)

[a]energy in kcal/mol, $\Phi = 0.72(2)$. [b]hydrogen bonding not possible

As pointed out previously, solvation differences of reactants and products tend to cancel, so $\Delta H_{5,6}$ should equal the difference between the Cr-CO and Cr-L bond energies. The Cr-L bond energies (BE$_{Cr\text{-}L}$) can then be calculated from $\Delta H_{5,6}$ and BE$_{Cr\text{-}CO}$ (TABLE 1).[35] The result for piperidine is in surprisingly good agreement with the activation enthalpy of 25.7 (12) kcal/mol reported for piperidine dissociation from $Cr(CO)_5(o\text{-}NHC_5H_{10})$.[33] Inspection of TABLE 2 reveals that BE$_{Cr\text{-}L}$ decreases as alkyl substitution for hydrogen increases. This trend is the opposite of that expected if differences in amine basicity were an important factor. The results are consistent with a steric effect, which increases by about 4 kcal/mol per alkyl group. The regularity of the increase further suggests that the steric interactions are essentially an alkyl-$Cr(CO)_5$ and not an alkyl-alkyl interaction. This being the case, the Cr-NH$_3$ bond energy may be estimated yielding a value of about 34 kcal/mol. This calculation leads to the conclusion that the intrinsic Cr-N bond for aliphatic amines is nearly as strong as the Cr-CO bond. In many cases it is believed that amine-metal bonds are weaker because the amines are incapable of backbonding. Our results show that steric hindrance may be responsible for weak metal-amine bonds in other metal complexes. It is interesting that the analogous phosphines are often utilized because of their greater apparent bond strengths. It would be instructive to determine if phosphines have a comparable sensitivity to steric effects.

5. Studies of Metal-Silane Interactions

5.1 BACKGROUND

We have ongoing studies of the interaction of silanes and metal centers. Our interest in metal-silane interactions originated in their similarity to metal-alkane interactions. Some metal centers completely insert into the Si-H bond of silanes while others form σ-complexes.[36] However, many metal complexes that insert into Si-H bonds cannot insert into alkane C-H bonds and only form σ-complexes with alkanes. In general, much more reactive metal centers are required to insert into C-H bonds of alkanes. The σ-complexes involve three-center two-electron bonds (M-HC or M-HSi) where there is still significant bonding between the hydrogen and carbon or silicon. By definition, the alkane case is an intermolecular agostic bond.[37] The extent of oxidative addition, *i.e.*, the extent of metal insertion into the C-H or Si-H bonds, is believed to be favored by electron-rich metal centers.[4] Many of the alkane adducts are unstable and difficult to handle. On the other hand, with silanes both insertion and σ-complexes have been isolated and extensively studied, and we anticipate the thermochemistry of metal-silane adducts will provide insight into the chemistry of the metal-alkane adduct. In any event, the metal-silane adducts are of interest in their own right. These adducts are important intermediates in hydrosilylation of olefins, and thermochemical data will provide mechanistic information about the process.

5.2 ENTHALPY OF TRIETHYLSILANE REACTION WITH METAL CARBONYLS

Our initial studies involved photolysis of various metal carbonyl complexes in neat triethylsilane (eq 13).[9] At the time, we were not set up for deconvolution analysis, and we

$$L_nM(CO) + HSiEt_3 \xrightarrow{h\nu} L_nM(HSiEt_3) + CO \qquad (13)$$

$$L_nM(CO) + RH \xrightarrow{h\nu} L_nM(RH) + CO \qquad (14)$$

needed to be sure that the reaction with the Si-H bond was faster than the transducer response. From ps flash photolysis studies we knew that the silane, as the solvent, would react with $Cr(CO)_5$ either at Si-H or an ethyl group in less than 100 ps.[13] Similarly studies of the rearrangement of $Cr(CO)_5$-C_2H_5X to $Cr(CO)_5$-XC_2H_5 (X = OH, Br, CN) established that when $Cr(CO)_5$ encounters an ethyl group before encountering the substituent the metal-ethyl complex will rearrange in less than 100 ps by an intramolecular process to the metal-substituent complex.[38] It can be concluded reaction 13 will be much faster than the frequency response of our 1 Mhz transducer.

It is also important to establish that the silane adduct is long lived compared to the response time of the transducer. In practice this would be on the order of a ms. The lifetime of the $Cr(CO)_5(HSiEt_3)$ adduct was observed to be long in preliminary flash photolysis experiments carried out in Dobson's laboratory.[39] This prompted us to examine the NMR spectra of $Cr(CO)_6$ in neat $HSiEt_3$ after UV photolysis. While the solution changed from colorless to yellow we had no sucess in observing a hydride peak. Dobson suggested the IR spectrum of the silane adduct might be observed, and indeed a peak at 1951 cm⁻¹ immediately appears upon photolysis and decays with a half-life of 10 to 30 minutes depending on the conditions.[9] The peak at 1951 cm⁻¹ did not occur in an alkane solvent in the absence of silane or in silane in the absence of $Cr(CO)_6$. The assignment of the peak to $Cr(CO)_5(HSiEt_3)$ is consistent with these results. This was a surprising result since it had been previously reported that silane adducts with $Cr(CO)_5$ could not be formed.[40] These results established that the decay of the silane adduct was much longer than the response time of the transducer and that subsequent reactions of the adduct would not contribute to the observed PAC signal. It can be concluded that a PAC signal would correspond to reaction 13.

Depending on the silane, the η^5-CpMn(CO)$_2$ fragment can insert into the Si-H bond or form a three-center two-electron bond. Various techniques indicate that the adduct with HSiEt₃ will form a three-center two-electron bond.[36,41] The other metal complexes we examined are less reactive than or have reactivity comparable to η^5-CpMn(CO)$_2$ and should not insert into the Si-H bond of HSiEt$_3$. Therefore three-center two-electron bonds are expected in each case. In any event, knowledge of the nature of the metal-silane interaction is not necessary for the PAC determination.

The results for the compounds photolyzed in HSiEt$_3$ are summarized in TABLE 3. For comparison, the corresponding values for alkane solvents (eq 14) have been included. It is immediately clear that the metal-silane interaction is stronger than the metal-alkane interaction since ΔH_{13} is more exothermic than ΔH_{14}. The difference between ΔH_{14} and ΔH_{13} corresponds to the enthalpy of alkane displacement of silane which will be equal to the difference in the metal-silane (BE$_{M-SiH}$) and metal-alkane (BE$_{M-RH}$) bond energies given that solvation of reactants and products are nearly the same. The difference for each of the metal centers is fairly constant, between 10 and 16 kcal/mol. This suggests that the metal-silane and metal-alkane bonding responds to changes in the metal center in a parallel manner and that the nature of the interactions are similar. Studies of these metal carbonyls in alkane solvents indicate that insertion into C-H bonds does not occur and labile σ-complexes are formed.[4,12,13,21] Therefore we predict that HSiEt$_3$ forms σ-complexes with these metal centers. This prediction is supported by the results of recent photoelectron spectroscopy studies of η^5-CpMn-silane adducts.[42]

TABLE 3. Enthalpy of Metal Carbonyl
Reactions with Triethylsilane and Alkanes[a]

L$_n$M(CO)	ΔH_{13}[b]	ΔH_{14}[c]
Cr(CO)$_6$	14.5(10)	25.3(26)
Mo(CO)$_6$	13.3(4)	23.2(13)
W(CO)$_6$	16.6(5)	31.3(6)
(η^6-C$_6$H$_6$)Cr(CO)$_3$	17.1(25)	32.8(3)
η^5-CpMn(CO)$_3$	30.9(10)	44.7(14)

[a]energy in kcal/mol [b]Φ assumed to be same as in heptane[18] [b]solvent was heptane except for η^5-CpMn(CO)$_3$, cyclohexane was used

All the reactions with silane and alkanes are endothermic because the M-CO bond that is broken is stronger than those formed with the silane or alkane. The ΔH_{13} and ΔH_{14} will be equal to the differences in the enthalpies of bonds broken and formed, so the enthalpy of the metal-silane (BE$_{M-SiH}$) and metal-alkane (BE$_{M-RH}$) bonds can be calculated from the corresponding metal-CO bond energies (BE$_{M-CO}$). These bond energies are summarized in TABLE 4.

TABLE 4. Enthalpy of Metal- Ligand Interactions[a]

L$_n$M(CO)	BE$_{M-SiH}$	BE$_{M-RH}$	BE$_{M-CO}$
Cr(CO)$_6$	22.3(22)	11.5(33)	36.8(20)
Mo(CO)$_6$	27.2(20)	17.3(27)	40.5(20)
W(CO)$_6$	29.4(21)	14.7(21)	46.0(20)
(η^6-C$_6$H$_6$)Cr(CO)$_3$	28	12	45[b]
η^5-CpMn(CO)$_3$	24	10	55[b]

[a]energy in kcal/mol [b]estimated values[9]

5.3 RELEVANCE TO MECHANISTIC STUDIES

Literature values are scarce but some comparisons are instructive. The kinetics of silane elimination from $Cr(CO)_5(HSiR_3)$ have recently been investigated, and an activation enthalpy of 20.0(5) kcal/mol for $HSiEt_3$ elimination in hexane was reported.[43] The result compares well with our value of 22.3 kcal/mol for BE_{Cr-SiH}. The lower value for the activation energy may be expected if there is residual metal-silane bonding in the transition state. Hart-Davis and Graham investigated the kinetics of reaction 15 which exhibited dissociative kinetics.[41] The rate determining step corresponds to the elimination of $HSiPh_3$, and they obtained an activation enthalpy of 29.2 (3) kcal/mol, a value substantially higher

$$\eta^5\text{-}CpMn(CO)_2(HSiPh_3) + PPh_3 \xrightarrow{\quad heptane \quad}$$

$$\eta^5\text{-}CpMn(CO)_2(PPh_3) + HSiPh_3 \qquad (15)$$

than our estimate of BE_{Mn-SiH} for $HSiEt_3$. This suggests that the Mn-silane interaction for $HSiPh_3$ is stronger than for $HSiEt_3$, which may explain why $\eta^5\text{-}CpMn(CO)_2 (HSiPh_3)$ is easily isolated and $\eta^5\text{-}CpMn(CO)_2(HSiEt_3)$ is not.[44]

Inspection of the data reveal other interesting trends. The variation in enthalpies of carbonyl displacement by silane (ΔH_{13}) and alkane (ΔH_{14}) is 18 and 22 kcal/mol, respectively, yet the corresponding variation in the metal-silane and metal-alkane bond energies is much smaller, 7 and 6 kcal/mol, respectively. The variation in the metal-CO bond energies is 18 kcal/mol and must be responsible for the large variation in the enthalpies of CO displacement by silane and alkane. Backbonding in metal-CO bonds is well established, and the greater variation of BE_{M-CO} relative to that of BE_{M-SiH} and BE_{M-RH} suggests that backbonding in the metal-silane and metal-alkanes bonds is not as important. This is entirely consistent with the formation of σ-complexes with silanes and alkanes since backbonding would lead to disruption of Si-H and C-H bonding, i. e., insertion.[4]

6. Photosubstitution in Perfluorocarbon Solvents

6.1 BACKGROUND

The PAC studies discussed so far have been generally restricted to those involving alkane solvents, primarily heptane. We have spent a great deal of effort on studies with perfluorocarbon solvents because of their general inertness. Matrix isolation studies demonstrate that the interaction of $Cr(CO)_5$ with CF_4 is very weak as indicated by the visible absorption spectrum of $Cr(CO)_5$ in CF_4. The spectrum is almost the same as in a neon matrix or the gas phase.[21] Kinetic studies also demonstrate a weak interaction between $Cr(CO)_5$ and perfluorocarbon solvent. The room temperature reaction of the $Cr(CO)_5$ fragment with CO in perfluoromethylcyclohexane is near or at the diffusion-controlled limit.[45] For comparison the same reaction in cyclohexane is three-orders of magnitude slower.[46] The difference suggests that the perfluorocarbon is more easily displaced than the alkane, a result that can be attributed to a weaker chromium interaction with the perfluorocarbon.

So far experimental limitations have prevented us from obtaining a value for a metal-fluorocarbon interaction: a situation we expect to overcome shortly. Beyond this there are experimental reasons for studying reactions in perfluorocarbon solvents. For select reactions, we anticipate that the enthalpy data in perfluorocarbon solvents will be a good approximation to that in the gas-phase. Furthermore, the weak interaction with perfluorocarbon solvents insures faster reactions rates allowing low nucleophile concentrations to be used. Nevertheless, a major disadvantage of using perfluorocarbons is the low solubility of organometallic compounds in perfluorocarbon solvents, although very slow dissolution is often mistaken for low solubility.

6.2 QUANTUM YIELDS FOR PHOTOSUBSTITUTION OF $Cr(CO)_6$ IN PERFLUORODECALIN

Our first study using a fluorocarbon solvent was of the reaction of various metal carbonyls with silanes in perfluorodecalin (PFD, eq 13). To our surprise we found that the fractional observed heat (α) was greater in perfluorocarbon solvent than in an alkane solvent. Further investigations with $Cr(CO)_6$ revealed this was true for any ligand. The enthalpy of reaction should not appreciably change upon changing the solvent from a hydrocarbon to perfluorocarbon. Examination of eq 3 indicates that a change in the quantum yield must account for the change in the observed heat. If the quantum yield for a reaction is known in a reference solvent (Φ_{REF}), then determining the fractional observed heat for the same reaction in the reference solvent (α_{REF}) and a perfluorocarbon solvent (α_{PFC}) allows the calculation of the quantum yield for that reaction in the fluorocarbon solvent. Hence eq 16

$$\Phi_{PFC}/\Phi_{REF} = (1-\alpha_{PFC})/(1-\alpha_{REF}) \qquad (16)$$

may be derived from eq 3. An advantage of PAC over conventional actinometry in determining quantum yields is that only a small fraction of the light need be absorbed, so very low concentrations of reactants may be used. Another advantage is that the product need not be stable; with deconvolution quantum yields could be determined for formation of intermediates whose lifetimes may be as short as a few ns.

Conventional actinometry was used to determined the quantum yield for substitution of $Cr(CO)_6$ with piperidine in reference solvents (α_{REF}). Piperidine was chosen because it had been used previously in quantum yield determinations with $Cr(CO)_6$.[47] However, piperidine was not convenient to use in PAC measurements since the product hydrogen bonds to free piperidine complicating the analysis (See Section 4).

It was convenient to determine α for the formation of solvent coordinated $Cr(CO)_5$. In such a case, the reference solvent acts as the trapping nucleophile as well as a dispersing medium, providing α_{REF}. The reference solvent can then be dispersed as a nucleophile in a perfluorocarbon solvent, providing α_{PFC}. In the absence of piperidine or other nucleophiles, the reference solvent is coordinated to the $Cr(CO)_5$ fragment in less than a ns and persists on the ms timescale.[13,46] The quantum yield of this reaction will be the same as that for formation of $Cr(CO)_5(c\text{-}HNC_5H_{10})$ when piperidine is present to scavenge $Cr(CO)_5$. That is, all the solvent coordinated $Cr(CO)_5$ would eventually be scavenged by piperidine. In a perfluorocarbon solvent, at mM concentrations of dispersed nucleophile, the $Cr(CO)_5$ fragment will be scavenged by the nucleophile on a time scale much faster than a 1 Mhz transducer. This is true for even poor nucleophiles like alkanes.[45]

Cyclohexane, benzene, and 1,2-dichloroethane were used as reference solvents to determine the quantum yield of piperidine substitution. Fractional observed heats were determined for substitution by cyclohexane, benzene, and 1,2-dichloroethane as nucleophiles neat and dispersed in PFD (TABLE 5).[48] The calculations of the quantum yields in PFD using eq 16 are reported in TABLE 5. In each case we found the quantum yield decreased to 0.3 in PFD. The low quantum yields in PFD determined by PAC were confirmed using conventional actinometry. A value of 0.31(1) was obtained for piperidine substitution on $Cr(CO)_6$. A similar result in perfluorohexane has been recently obtained by Wieland and van Eldik.[49]

TABLE 5. Fractional Observed Heats and Quantum Yields at 337 nm for $Cr(CO)_6$ Photosubstitution

solvent	ligand[a]	α	Φ
cyclohexane	cyclohexane	0.810(6)	0.67(2)[b]
PFD	cyclohexane	0.919(5)	0.28(2)
benzene	benzene	0.828(3)	0.67(2)[b]
PFD	benzene	0.923(3)	0.30(2)
1,2-dichloroethane	1,2-dichloroethane	0.845(17)	0.62(5)[b]
PFD	1,2-dichloroethane	0.933(16)	0.27(8)

[a]concentrations in PFD were 0.01-0.03 **M** [b]determined by conventional actinometry

6.3 WEAK COORDINATION OF PERFLUOROCARBON SOLVENT

The low quantum yield in PFD can be attributed to the weak interaction between the $Cr(CO)_5$ fragment and PFD. After photolysis, CO is trapped in the solvent cage, and there is a competition between CO and solvent for $Cr(CO)_5$ (See Scheme 1). If the solvent is cyclohexane, benzene, and 1,2-dichloroethane, once the solvent is coordinated, it is not easily

SCHEME 1

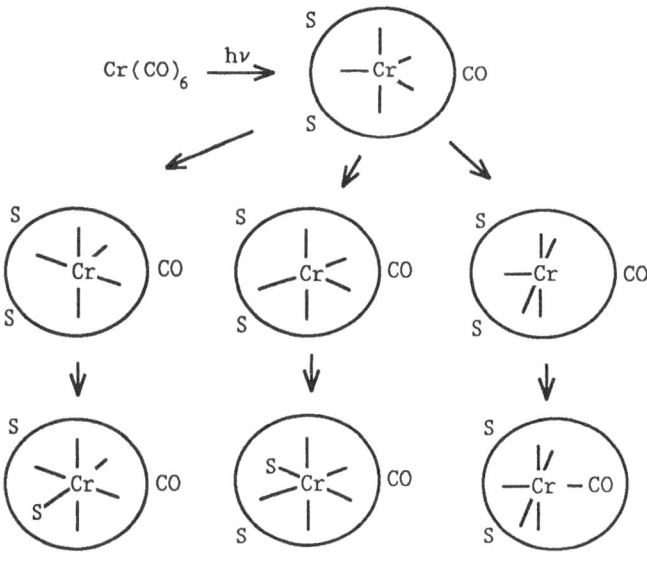

Figure reprinted with permission.[48] Copyright 1991, American Chemical Society.

displaced, and CO is forced to diffuse out of the solvent cage. The solvent coordinated fragment can eventually react with a dispersed nucleophile to form a stable complex. On the other hand, CO more easily displaces the PFD and more $Cr(CO)_6$ is reformed. The high quantum yields obtained in cyclohexane and other 'reactive' solvents occurs because the solvent 'holds' the $Cr(CO)_5$ long enough for the CO to escape. If the rate of CO escape is decreased, as it will be in highly viscous solvents, then the quantum yield should also decrease, and this is what is observed.[47] Interestingly, a quantum yield of unity is not obtained even in solvents that act as non-labile ligands.[18] A stereospecific rearrangement of the excited $Cr(CO)_5$ has been offered as an explanation of this result.[22,48] It has been hypothesized that the excited $Cr(CO)_5$ initially forms a trigonal bipyramid and then relaxes to a square pyramidal structure. The fraction of $Cr(CO)_5$ that opens toward CO with the square pyramidal structure immediately recombines with CO thus reforming $Cr(CO)_6$ and eliminating the opportunity for substitution and a unit quantum yield.

The lower quantum yield in PFD cannot simply be due to the greater viscosity of PFD (η = 6.02 cP) relative to other solvents like cyclohexane (η = 0.97 cP).[48] Greater viscosity would reduce the rate of CO escape from the solvent cage allowing more time for CO to recombine with $Cr(CO)_5$ in the solvent cage. The quantum yield in the more viscous paraffin I (η = 140 cP) is not less but significantly greater (Φ = 0.51) than in PFD.[47] We believe that the low quantum yield in PFD can be attributed to facile displacement of the fluorocarbon which leads to more efficient cage recombination with CO.

6.4 QUANTUM YIELDS FOR OTHER METAL CARBONYLS IN PERFLUORODECALIN

In preliminary studies, we have used PAC to determine the quantum yields for CO substitution of other metal carbonyls in PFD (Φ_{PFD}). The α value was determined for CO substitution by $HSiEt_3$ for each compound (eq 13) and used with eq 16 and the ΔH_{13} from Table 4 to calculate Φ_{PFD}. The results are reported in TABLE 6 along with available literature quantum yields in alkane solvents (Φ_{RH}). While the errors are somewhat large, the yields in PFD are definitely smaller than in alkane solvents.

TABLE 6. Quantum Yields of CO Substitution for Various Metal Carbonyls in Perfluorodecalin

$L_nM(CO)$	α	Φ_{PFD}	Φ_{RH}[a]
$Cr(CO)_6$	0.966(19)	0.20(11)	0.72(2)
$Mo(CO)_6$	0.913(15)	0.55(10)	0.93(1)
$W(CO)_6$	0.910(5)	0.46(3)	0.79(1)
$(\eta^6\text{-}C_6H_6)Cr(CO)_3$	0.901(4)	0.49(7)	0.72(7)
$\eta^5\text{-}CpMn(CO)_3$	0.823(25)	0.49(7)	0.65(15)

[a]references 18, 50, 51

We cannot rule out the possibility that lower quantum yields in PFD are due to a difference in the decay rates of the excited state species in PFD and the other solvents. However, if differences in decay rates were an important factor then we might expect a wavelength dependence on the quantum yield, but this is not the case.[47,48] Other evidence is consistent with the lower quantum yields resulting from weaker coordination by the perfluorocarbon solvent. Each of the compounds in TABLE 6 (except $(\eta^6\text{-}C_6H_6)Cr(CO)_3$) has been studied by flash photolysis, and the rates of reaction of the incipient intermediates with dispersed nucleophiles in alkane solvent are significantly less than the diffusion-controlled limit.[46,52] This has been attributed to each species having a significant interaction with the solvent. In addition we have evidence that each metal center interacts with alkanes in a manner similar to silanes and that these interactions are 10-17 kcal/mol (See Section 5). Thus the lower quantum yields in PFD relative to alkane solvents are consistent with greater cage recombination with CO due to easier displacement of PFD.

6.5 PAC MEASUREMENTS WITHOUT QUANTUM YIELDS

The technical problems of determining quantum yields by conventional means are sometimes prohibitive. If PAC is to be used to determine the enthalpy of a reaction then explicit knowledge of the quantum yield may not be necessary. If the enthalpy of a reference reaction (ΔH_r) has been independently determined and it is certain that the quantum yield of the reference reaction is the same as that for a sample reaction, then the enthalpy for the sample reaction (ΔH_s) can be calculated from eq 17. This equation is derived from eq 3 where α_s and α_r are the fractional observed heats for the sample and reference reactions.

$$\Delta H_s = (1 - \alpha_s)/(1 - \alpha_r)\Delta H_r \tag{17}$$

Ligand substitution reactions are an example where this equation may be applied. For example, if the enthalpy of CO substitution on a metal center by a reference nucleophile were known (ΔH_r) and the corresponding photosubstitution was independent of the nature of the nucleophile then ΔH_s could be calculated from eq 17 and the PAC determination of α_s and α_r. No quantum yield determination is necessary as long as there is a suitable reference reaction. Note that the same data provide a means to calculate the quantum yield.

7. Cr(CO)₅ Interaction with Weakly Coordinating Solvents: Comparison of Alkanes, Aromatics, and Chlorocarbons

7.1 Cr(CO)₅ BONDING TO ALKANES: PRIMARY VERSUS SECONDARY C-H

The iridium and rhodium complexes studied in Bergman's laboratory preferentially insert into alkane primary C-H bonds relative to secondary.[53] We were interested whether agostic type interactions (where the metal does not insert into the C-H bond) might show the same preference. Since the difference in the strength of interactions at primary and secondary carbons may be small it seemed prudent to compare PAC results for compounds that had purely secondary or primary C-H bonds.

Cycloalkanes are an obvious choice for the study of secondary C-H bonds, and cyclohexane can be used as a solvent and nucleophile. Examples of compounds having purely primary C-H bonds present a problem. The smallest such compounds would be methane, ethane, and 2,2-dimethylpropane; all are gases. The next homologue is hexamethylethane (2,2,3,3-tetramethylbutane), and it is a solid at room temperature, so it would have to be dissolved in a solvent. Alkanes and most other solvents would not be practical since they will have similar reactivity toward metal centers, and statistically the metal will prefer to react with the solvent. A perfluorocarbon solvent ought to be useful in such situations. Fortuitously hexamethylethane is soluble enough in PFD to make PAC measurements.

Preliminary results for the ligand substitution of hexamethylethane and other alkanes with Cr(CO)₆ are presented in TABLE 7. Within experimental error we observe no difference in the interaction of chromium with the primary C-H of hexamethylethane and the secondary C-H of cyclohexane. If there were a large preference for secondary C-H surely we would detect a difference in the enthalpies of reaction; however, if there were a preference for primary C-H, steric interactions with the bulky hexamethylethane may mask it. In pentane and heptane, the methyl groups are certainly less hindered than the methylene

TABLE 7. Enthalpies of CO Substitution on $Cr(CO)_6$ with Weakly Bound Ligands and Cr-Ligand Bonds[a]

ligand	Φ	ΔH_8	BE_{M-L}
hexamethylethane/PFD		25.6(23)	11.2(30)
pentane	0.72(2)[b]	26.0(25)	10.8(32)
heptane	0.72(2)	25.2(13)	11.6(26)
cyclohexane	0.67(2)	24.2(10)	12.6(22)
trifluoromethylbenzene	0.67[c]	22.7(8)	14.1(22)
benzene	0.67(2)	21.8(8)	15.0(22)
benzene/PFD		23.0(18)	13.8(27)
1,2-dichloroethane	0.62(5)	21.2(29)	15.6(35)
1,2-dichloroethane/PFD		20.0(50)	16.8(54)
chlorobenzene/PFD		17.6(75)	19.4(78)
tetrachloromethane	0.67[c]	17.3(5)	19.5(21)
1-hexene/heptane[d]	0.72(2)	14.0(16)	21.8(26)

[a]energy in kcal/mol, ligands are neat except in PFD where noted, ΔH in PFD are calculated using eq 17 and cyclohexane as the reference ligand [b]reference 18 [c]estimated [d]in heptane, reference 17.

groups and if there were a significant preference for binding to primary C-H, then the enthalpy of substitution for these compounds should be more exothermic than it is for cyclohexane. This is not the case, and within the errors reported we conclude there is no difference in the strengths of the primary and secondary C-H agostic bonds to $Cr(CO)_5$.

7.2 $Cr(CO)_5$ BONDING TO AROMATIC AND CHLOROCARBON COMPOUNDS

We are still refining data and determining quantum yields for aromatic and chlorocarbon solvents so only estimates are available in certain cases. Nevertheless, some useful trends are emerging. Consistent with the conclusions of flash photolysis studies we find that the Cr-L bond strength generally increases in the order of alkanes, aromatics, and chlorocarbons.[54] There again is little data for comparison. Zhang and Dobson have recently reported activation enthalpies for 1-hexene displacement of benzene and tetrachloromethane coordinated to $Cr(CO)_5$ (eq 18, S = benzene or tetrachloromethane, L = 1-hexene).[54] This reaction displays dissociative kinetics allowing the determination of the activation enthalpy for solvent dissociation, the first step in eq 18. The authors suggest that the activation enthalpies for solvent dissociation ought to be a good measure of the corresponding metal-solvent bond energies since the reverse reactions occur on the ps-fs

$$Cr(CO)_5S \xrightarrow{\longleftarrow} Cr(CO)_5 + S \xrightarrow{L} Cr(CO)_5L \qquad (18)$$

S = solvent

timescale and should have small activation enthalpies. They obtained activation enthalpies of 9.4(1) and 12.5(7) kcal/mol for benzene and tetrachloromethane and activation entropies of -2.4(3) and 7.0(24) eu, respectively. The activation enthalpies are significantly smaller than the BE_{M-L} values reported in TABLE 7. The activation entropies are not large perhaps indicating that there is residual Cr-S bonding in the transition state.

Residual Cr-S bond in the transition state brings up a potential contradiction. The kinetic results require that there is an intermediate formed following solvent dissociation that does not include solvent or 1-hexene. It seems inescapable that this intermediate is a truly coordinatively-unsaturated intermediate. This implies that the residual Cr-S bonding in the transition state is lost upon formation of the intermediate implying that the enthalpy of the intermediate is greater than that of the transition state. By definition the intermediate is lower in energy than the transition state leading to it, and so there is an apparent inconsistency. Yet it is the free energy of the transition state that must be lower for the intermediate not the enthalpy. Thus we conclude that the entropy must increase upon going from the transition state to the intermediate and that the free energy change due to the entropy increase is equal to or greater than that due to the enthalpy increase. Hence the free energy of the intermediate would be equal to or less than that of the transition state.

In the case of benzene, our average value for BE_{M-L} is 14.4 kcal/mol, 5 kcal/mol greater than the activation enthalpy. At 300 °K, a 5 kcal/mol contribution to free energy is equivalent to 17 eu, a reasonable value for the entropy change between the transition state and the intermediate considering the entropy change up to just the transition state is actually negative and the overall process is dissociative. A similar conclusion can be drawn for the tetrachloromethane data although the PAC result is based on an estimated quantum yield.

In conclusion, PAC provides energetic information that often is not available by kinetic and other techniques for studying reaction mechanisms and serves to provide a better understanding of organometallic transition states, intermediates, and stable complexes.

8. Acknowledgments

I would like to acknowledge the work of my coworkers S. Nayak, D. Robbins, G. Parker, J. Morse, D. Bobal and helpful discussions with G. Dobson, C. Harris, and G. Yang. This work was supported in part by Memphis State University Faculty Research Grants and the Donors of the Petroleum Research Fund, administered by the American Chemical Society.

9. References and Notes

1. (a) Rudzki, J.; Goodman, J. L.; Peters, K. S. *J. Am. Chem. Soc.* 1985, *107*, 7849. (b) Heihoff, K.; Braslavsky, S. E. *Chem. Phys. Lett.* 1986, *131*, 183. (c) Melton, L. A.; Ni, T.; Lu, Q. *Rev. Sci. Intrum.* 1989, *60*, 3217.

2. (a) Westric, J. A.; Goodman, J. L.; Peters, K. S. *Biochemistry* 1987, *26*, 8313. (b) Herman, M. S.; Goodman, J. L. *J. Am. Chem. Soc.* 1989, *111*, 1849.

3. (a) Nonell, S.; Aramendia, P. F.; Heihoff, K.;Negri, R. M.; Braslavsky, S. E. *J. Phys. Chem.* 1990, *94*, 5879. (b) Song, X.; Endicott, J. F. *Inorg. Chem.* 1991, *30*, 2214. (c) Patel, C. K. N.; Tam, A. C. *Rev. Mod. Phys.* 1981, *63*, 517.

4. Saillard, J.-Y.; Hoffmann, R. *J. Am. Chem. Soc.* 1984, *106*, 2006.

5. J. A. Davis, P. L. Watson, J. F. Liebman, and A. Greenberg (eds.) (1990) Selective Hydrocarbon Activation, VCH Publishers, New York.

6. (a) Brookhart, M.; Green, M. L. H. *J. Organometal. Chem.* 1983, *250*, 395. (b) Brookhart, M.; Green, M. L. H.; Wong, L.-L. *Prog. Inorg. Chem.* 1988, *36*, 2. *Chem. Rev.* 1990, *90*, 403.

7. (a) Peters, K. S.; Snyder, G. J. *Science* **1988**, *241*, 1053. (b) Rudzki, J. S.; Hutchings, J. J.; Small, E. W. *Proc. SPIE-Int. Soc. Opt. Eng.* **1989**, *1054*, 26. (c) Burkey, T. J.; Majewski, M.; Griller, D. *J. Am. Chem. Soc.* **1986**, *108*, 2218. (d) Bilmes, B. M.; Tocho, J. O.; Braslavsky, S. E. *Chem. Phys. Lett.* **1987**, *134*, 335. (e) Herman, M. S.; Goodman, J. L. *J. Am. Chem. Soc.* **1989** *111*, 9105. (f) Lynch, D.; Endicott, J. F. *Inorg. Chem.* **1988**, *27*, 2181. (g) Burkey, T. J.; Majewski, M.; Griller, D. *J. C. S. Chem. Comm.* **1985**, 1259. (h) Klassen, J. K.; Selke, M.; Sorensen, A. A.; Yang, G. K. *J. Am. Chem. Soc.* **1990**, *112*, 1267. (j) Caldwell , R. A.; Melton, L. A. *J. Am. Chem. Soc.* **1989**, *111*, 457. (k) Marr, K.; Peters, K. S. *Biochemistry* **1991**, *30*, 1254.

8. Burkey, T. J. *Polyhedron* **1989**, *8*, 2681.

9. Burkey, T. J. *J. Am. Chem. Soc.* **1990**, *112*, 8239.

10. (a) Hou, S.-Y.; Hetherington, W. M., III; Korenowski, G. M.; Eisenthal, K. B. *Chem. Phys. Lett.* **1979**, *68*, 282. (b) Maciejewski, A.; Jaworska-Augustyniak, A.; Szeluga, Z.; Wojtczak, J.; Karolczak, J. *Chem. Phys. Lett.* **1988**, *153*, 227

11. A possible exception may occur when reaction volumes are negative. See reference 2.

12. (a) Howell, J. A. S.; Burkinshaw, P. M. *Chem. Rev.* **1983**, *83*, 557. (b) Angelici, R. J. *Organomet. Chem. Rev.* 1968, *3*, 173.

13. (a) Simon, J. D.; Xie, X. *J. Phys. Chem.* **1989**, *93*, 291. (b) Joly, A. G.; Nelson *Chem. Phys.* **1991**, *152*, 69. (c) Joly, A. G.; Nelson, K. A. *J. Phys. Chem.* **1989**, *93*, 2876. (d) Simon, J. D.; Xie, X. *J. Phys. Chem.* **1987**, *91*, 5538. (e) Simon, J. D.; Xie, X. *J. Phys. Chem.* **1986**, *90*, 6751. (f) Lee, M.; Harris, C. B. *J. Am. Chem. Soc.* **1989**, *111*, 8963. (g) Yu, S.-C.; Xu, X.; Lingle, R., Jr.; Hopkins, J. B. *J. Am. Chem. Soc.* **1990**, *112*, 3668. (h) Joly, A. G.; Nelson, K. A. *Chem. Phys.* **1991**, *152*, 69.

14. Wang, L.; Zhu, X.; Spears, K. G. *J. Am. Chem. Soc.* **1988**, *110*, 8695.

15. (a) Bryndza, H. E.; Fong, L. K.; Paciello, R. A.; Tam, W.; Bercaw, J. E. *J. Am. Chem. Soc.* **1987**, *109*, 1444. (b) Bruno, J. W.; Marks, T. J.; Morss, L. R. *J. Am. Chem. Soc.* **1983**, *105*, 6824. (c) Gonzales, A. A; Zang, K.; Nolan, S. P.; de la Vega, R. L.; Mukerjee, S. L.; Hoff, C. D.; Kubas, G. J. *Organometallics* **1988**, *12*, 2429.

16. Morse, J. Jr.; Parker, G.; Burkey, T. J. *Organometallics* **1989**, *8*, 2471.

17. Yang, G. K.; Vaida, V.; Peters, K. S. *Polyhedron*, **1988**, *7*, 1619.

18. Wieland, S.; van Eldik, R. *J. Phys. Chem.* **1990**, *94*, 5865.

19. Errors for our data are one standard deviation and the values indicated in parentheses refer to the error in the last digit(s).

20. Lewis, K. E.; Golden, D. M.; Smith, G. *J. Am. Chem. Soc.* **1984**, *106*, 3905.

21. Perutz, R.; Turner, J. J. *J. Am. Chem. Soc.* **1975**, *97*, 4791.

22. (a) Burdett, J. K.; Grzybowski, J. M.; Perutz, R. N.; Poliakoff, M.; Turner, J. J.; Turner, R. F. *Inorg. Chem.* **1978**, *17*, 147. (b) Turner, J. J.; Poliakoff, M. *ACS Symp. Ser.* **1983**, *200*, 35.

23. Fletcher, T. R.; Rosenfeld, R. N. *J. Am. Chem. Soc.* **1988**, *110*, 2097.

24. Ganske, J. A.; Rosenfeld, R. N. *J. Phys. Chem.* **1990**, *94*, 4315.

25. Ishikawa, Y.-I.; Brown, C. E.; Hackett, P. A.; Rayner, D. M. *J. Phys. Chem.* **1990**, *94*, 2404.

26. Graham, J. R.; Angelici, R. J. *Inorg. Chem.* 1967, *6*, 2082.

27. Brown, C. E.; Ishikawa, Y.-I.; Hackett, P. A.; Rayner, D. M. *J. am. Chem. Soc.* **1990**, *112*, 2530.

28. Dobson, G. R.; Cate, C. D.; Cate, C. W. submitted to *Inorg. Chem.*

29. Dobson, G. R.; Spradling, M. D. *Inorg. Chem.* **1990**, *29*, 880.

30. Zhang, S.; Zang, V.; Bajaj, H. C.; Dobson, G. R.; van Eldik, R. *J. Organomet. Chem.* 1990, *397*, 279.

31. (a) Zhang, S.; Dobson, G. R.; Zang, V.;Bajaj, H. C.; van Eldik, R. *Inorg. Chem.* 1990, *29*, 3477. (b) Zhang, S.; Dobson, G. R. *Inorg. Chim. Acta* 1989, *165*, 11.

32. Dobson, G. R. *Accts. Chem. Res.* 1976, *9*, 300.

33. (a) Dennenberg, R. J.; Darensbourg, D. J. Inorg. Chem. 1972, *11*, 72. (b) Darensbourg, D. J.; Ewen, J. A. *Inorg. Chem.* 1981, 20, 4168.

34. Nakshima, M.; Adamson, A. W. *J. Phys. Chem.* 1982, *86*, 2905.

35. The literature gas-phase values are in general agreement for internal consistency we have chosen all the values from Reference 20.

36. (a) Jetz, W.; Graham, W. A. G. *Inorg. Chem.* 1971, *10*, 4. (b) Young, K. M.; Wrighton, M. S. *Organometallics* 1989, *8*, 1063. (c) Schubert, U.; Muller, J.; Alt, H. G. *Organometallics* 1987, *6*, 469. (c) Hill, R. H.; Wrighton, M. S. *Organometallics* 1987, *6*, 632. (d) Carre, F.; Colomer, E.; Corriu, R. J. P.; Vioux, A. *Organometallics* 1984, *3*, 1272.

37. (a) Brookhart, M.; Green, M. L. H. *J. Organometal. Chem.* 1983, *250*, 395. (b) Brookhart, M.; Green, M. L. H.; Wong, L.-L. *Prog. Inorg. Chem.* 1988, *36*, 2. *Chem. Rev.* 1990, *90*, 403.

38. (a) Xie, X.; Simon, J. D. *J. Am. Chem. Soc.* 1990, *112*, 1130. (b) O'Driscoll, E.: Simon, J. D. *J. Am. Chem. Soc.* 1990, *112*, 6580.

39. Private communication.

40. Geoffroy, G. L.; Wrighton, M. S. *Organometallic Photochemistry*; Academic Press: New York, 1979; Chapter 2.

41. Hart-Davis, A. J.; Graham, W. A. G. *J. Am. Chem. Soc.* 1971, *94*, 4288.

42. (a) Lichtenberger, D. L. ; Rai-Chaudhuri, A. *J. Am. Chem. Soc.* 1989, *111*, 3583. (b) Lichtenberger, D. L. ; Rai-Chaudhuri, A. *Inorg. Chem.* 1990, *29*, 975. (c) Lichtenberger, D. L. ; Rai-Chaudhuri, A. *J. Am. Chem. Soc.* 1990, *112*, 2492.

43. Zhang, S.; Dobson, G.: Brown, T. L. *J. Am. Chem. Soc.* 1991, *113*, 6908.

44. Jetz, W.; Graham, W. A. G. *Inorg. Chem.* 1971, *10*, 4.

45. Kelly, J. M.; Long, C.; Bonneau, R. *J. Phys. Chem.* 1983, *87*, 3344.

46. (a) Kelly, J. M.; Bent, D. V.; Hermann, H.; Schulte-Frohlinde, D.; Koerner von Gustorf, E. *J. Organometal. Chem.* 1974, *69*, 259. (b) Kelly, J. M.; Hermann, H.; Koerner von Gustorf, E. *J. C. S. Chem. Comm.* 1973, 105. (c) Hermann, H.; Grevels, F.-W.; Henne, A.; Schaffner, K. *J. Phys. Chem.* 1982, *86*, 5151. (d) Church, S. P.; Grevels, F.-W.; Hermann, H.; Schaffner, K. *Inorg. Chem.* 1985, *24*, 418.

47. (a) Nasielski, J.; Colas, A. *Inorg. Chem.* 1978, *17*, 237. (b) Nasielski, J.; Colas, A. *J. Organomet. Chem.* 1975, *101*, 215.

48. Nayak, S. K.; Burkey, T. J. *Organometallics* 1991, *10*, 3745.

49. Wieland, S.; van Eldik, R. *J. Phys. Chem.* 1990, *94*, 5865.

50. Wrighton, M. S.; Haverty, J. L. *Z. Naturforsch.* 1975, *30b*, 254.

51. Giordano, P. J.; Wrighton, M. S. *Inorg. Chem.* 1977, *16*, 160.

52. Creaven, B. S.; Dixon, A. J.; Kelly, J. M.; Long, C.; Poliakoff, M. *Organometallics* 1987, *6*, 2600.

53. (a) Periana, R. A.; Bergman, R. G. *J. Am. Chem. Soc.* 1986, *108*, 7332. (b) Buchanan, J. M.; Stryker, J. M.; Bergman, R. G. *J. Am. Chem. Soc.* 1986, *108*, 1537.

54. (a) Zhang, S.; Dobson, G. R. *J. Coord. Chem.* 1990, *21*, 155. (b) Zhang, S.; Dobson, G. R. *Inorg. Chim. Acta.* 1991, *181*, 103.

Electrochemistry of Some Organic and Organometallic Radicals and Their Application in Thermochemical Cycles[1]

Danial D.M. Wayner

Organic Reaction Dynamics Group
Steacie Institute for Molecular Sciences
National Research Council of Canada
Ottawa, Ontario, Canada K1A 0R6

ABSTRACT. A simple experiment is described (photomodulation voltammetry) that allows the electrochemical oxidation and reduction potentials of short lived intermediates to be measured directly. These data are used in thermochemical cycles to determine a number of radical cation σ-bond energies, $D^\circ(R\text{-}R^{+\bullet})$, in acetonitrile solution. By comparison with gas phase ionization potentials it is shown these bond energies are generally similar in the gas phase implying that the solvation energies of the radical cations and the product cations are the same. Only in those cases where a there is a significant change in the charge distribution of R^+ compared to $R\text{-}R^{+\bullet}$ are large differences between gas phase and solution observed. A comparison of hydrazine and diphosphine radical cation bond energies also is made. In contrast to hydrazines which form strong two-center three electron bonds upon removal of one electron, there appears to be very little three electron bonding interaction between the phosphorous atoms of tetraaryldiphosphine radical cations. This results in a modest weakening of the P-P bond relative to the neutral species.
    The standard potentials, E°, for the oxidation of the tributyl- and triphenylstannyl radicals were estimated from a combination of electrochemical and kinetic measurements. These data were combined with reduction potentials of nitroalkanes to estimate the thermochemistry for electron transfer between these species. The free energy change for the electron transfer reaction (outer sphere) must be > 12 kcal mol^{-1}. The rates of these reactions, measured by laser flash photolysis, were on the order of 10^8 M^{-1}s^{-1} showing, unequivocally, that mechanism of the reaction of stannyl radicals with nitroalkanes does not proceed via outer sphere electron transfer.

1. Introduction

There is growing interest in the application of thermochemical cycles to estimate thermodynamic properties that cannot be measured by direct experimentation. One of the first applications to organic chemistry was reported by Breslow[2] who combined the oxidation potentials of carbanions with homolytic bond dissociation energies (D°) in order to estimate the pKa of the corresponding carbon acids

J. A. Martinho Simões (ed.), Energetics of Organometallic Species, 95–108.
© 1992 *Kluwer Academic Publishers.*

A similar approach has been used by a number of groups to estimate homolytic bond energies of neutral molecules and radical ions[3-5] as well as heterolytic bond energies.[6] Most recently, Tilset and Parker have used thermochemical cycles to estimate $D°(M-H)$ values for series of organometallic species.[7]

We have pointed out that all of the thermochemical cycles describing the homolytic and heterolytic cleavage reactions of molecules and their radical ions can be conveniently represented in a thermochemical mnemonic (Figure 1)[8]. In this mnemonic the vertical arrows represent the loss of R' as a radical (i.e. the homolytic bond dissociation energy). The horizontal arrows represent the addition or removal of an electron while the diagonal

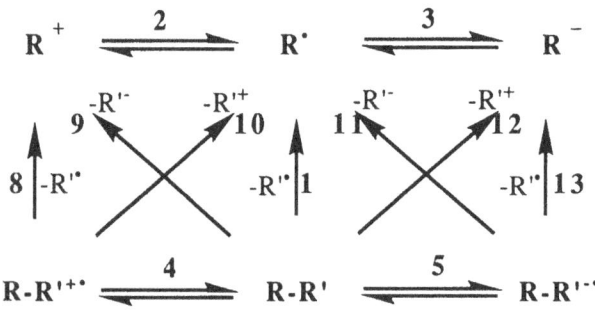

Figure 1. Thermochemical mnenomic describing all of the homolytic and heterolytic cleavage reactions of molecule R-R' and its corresponding radical ions.

arrows represent the loss of R'+ (up to the right) or R'- (up to the left). For the case where R'=H, the free energy change along the diagonals are related to more familiar thermodynamic parameters; reactions 9 and 11 are related to hydride transfer reactions while reactions 10 and 12 are related to proton transfer (pK_a's).

Only seven pieces of experimental data are required to complete the scheme for a given compound. Consequently, the thermochemistry for all of the possible homolytic and heterolytic cleavage reactions can be written simply as the sum of the homolytic bond dissociation energy (Equation 1) with the difference between two potentials (Equations 2-7). It is important to keep in mind that an entropy correction must be applied to the bond dissociation enthalpies that are available from data in the literature[9] since the redox potentials are actually *free energy* differences. For R'=H, this correction is ca. -8 kcal mol^{-1}.[8]   These free energy relationships are given in Equations 8-13 (F = 23.06 kcal mol^{-1}eV^{-1}).

$$R\text{-}R' \rightarrow R^{\bullet} + R'^{\bullet} \tag{1}$$

$$R^{\bullet} \rightarrow R^{+} \tag{2}$$

$$R^{-} \rightarrow R^{\bullet} \tag{3}$$

$$R\text{-}R' \rightarrow R\text{-}R'^{+\bullet} \tag{4}$$

$$R\text{-}R'^{-\bullet} \rightarrow R\text{-}R' \tag{5}$$

$$R'^{\bullet} \quad \rightarrow \quad R'^{+} \tag{6}$$
$$R'^{-} \quad \rightarrow \quad R'^{\bullet} \tag{7}$$

$$\Delta G^{\circ}_8 = \Delta G^{\circ}_1 + F(E^{\circ}_2 - E^{\circ}_4) \tag{8}$$
$$\Delta G^{\circ}_9 = \Delta G^{\circ}_1 + F(E^{\circ}_2 - E^{\circ}_7) \tag{9}$$
$$\Delta G^{\circ}_{10} = \Delta G^{\circ}_1 + F(E^{\circ}_6 - E^{\circ}_4) \tag{10}$$
$$\Delta G^{\circ}_{11} = \Delta G^{\circ}_1 + F(E^{\circ}_5 - E^{\circ}_7) \tag{11}$$
$$\Delta G^{\circ}_{12} = \Delta G^{\circ}_1 + F(E^{\circ}_6 - E^{\circ}_3) \tag{12}$$
$$\Delta G^{\circ}_{13} = \Delta G^{\circ}_1 + F(E^{\circ}_5 - E^{\circ}_3) \tag{13}$$

Getting reliable electrochemical data is not always straight forward. In many cases the necessary electrochemical measurements have been made in solutions of very long lived molecules or ions. These electrode reactions are often irreversible processes since the ionic or radical products tend to be highly reactive. As a result, the measured potentials are often corrupted by kinetic effects.[10] Furthermore, the number of ions that are amenable to this approach is limited. On the other hand, direct electrochemical measurements on solutions of radicals have been difficult since the radicals by their very nature are short-lived intermediates so only very low steady state concentrations can be achieved. The characterization of the electrochemical properties of these species has been made possible, only recently, by the development of ultramicroelectrode voltammetry[11] and photomodulation voltammetry (PMV).[12] Some recent results on the electrochemistry of carbon, phosphorus and tin centred radicals using the PMV method and application of these data in thermochemical cycles are described below.

2. Experimental Approach

We have shown that reliable electrochemical measurements can actually be made in solutions of transient radicals using the photomodulation voltammetry technique.[12,13] Two approaches have been used to generate radicals: photodecomposition of a dimer or symmetric ketone (Equation 14), and, photolysis of di-*tert*-butyl peroxide (DTBP) followed by hydrogen atom abstraction (Equations 15,16). Our experience is that the first approach (Equation 14) is preferred although the second approach (Equation 15,16) is far more general.

$$R\text{-}R \text{ \{or } R(CO)R\} \quad \xrightarrow{h\nu} \quad 2R^{\bullet} (+ CO) \tag{14}$$
$$DTBP \quad \xrightarrow{h\nu} \quad 2\,t\text{-BuO}^{\bullet} \tag{15}$$
$$t\text{-BuO}^{\bullet} + RH \quad \rightarrow \quad t\text{-BuOH} + R^{\bullet} \tag{16}$$

The electrochemical cell is a modified three electrode cell consisting of an optically transparent gold minigrid working electrode (400 wires/cm), a platinum coil counter electrode and a saturated calomel electrode (SCE) as a reference. A solution containing the

98

appropriate precursors is flowed slowly (ca. 5 ml/min) through the cell to prevent depletion of the precursors (the cell volume is ca. 500 μl).

The experiment is quite simple (Figure 2). The output of a photolysis lamp, focused through the back of the optically transparent working electrode, is modulated using a mechanical chopper with specially designed irises so that the intensity rises and falls as a sine wave (typically 20-200 Hz). This causes the radical concentration of modulate at the

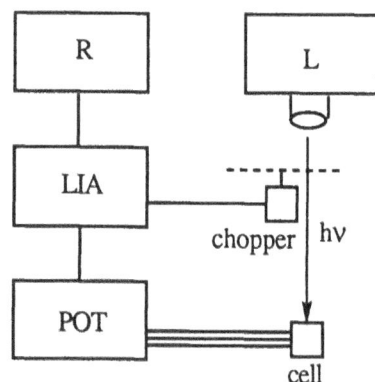

Figure 2. Block diagram of the PMV apparatus. L = lamp; LIA = lock-in amplifier; POT = potentiostat; R = recorder.

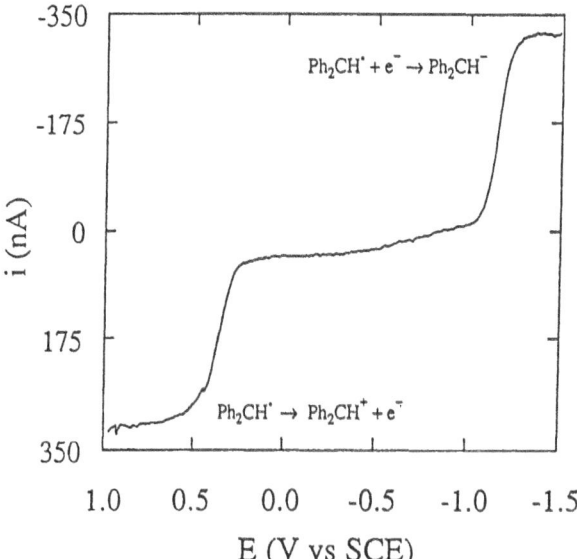

Figure 3. Voltammogram of $Ph_2CH^\bullet$ in acetonitrile with tetrabutylammonium perchlorate (0.1 M) as the electrolyte.

same frequency so the current that is detected by the potentiostat is made up of two components: a dc component and an ac component. It is possible to selectively amplify only those signals that are oscillating at the same frequency as the lamp with a lock-in amplifier using a reference signal from the chopper. A voltammogram of the radical can then be constructed by plotting the ac current versus the applied potential. An example is shown in Figure 3. In this case, the diphenylmethyl radical was generated by the photolysis of *sym*-tetraphenylacetone.

3. Bond Energies of Radical Cations

3.1 BACKGROUND

The addition or removal of an electron can have a profound effect on the strength of σ bonds. This phenomenon is well known in the gas phase where the removal of an electron from a molecule results in a characteristic mass spectrometric fragmentation pattern. Careful measurement of the energy required to fragment σ bonds of radical cations has been the basis for the estimation of a large number of thermochemical data in the literature (i.e. the appearance energy method).[9] Some bond energies for the cleavage of some simple first and second row dimers in the gas phase are shown in Table 1.

Table 1. Effect of Oxidation State on the σ-Bond Energies, $D°(R-R)$, in the Gas Phase (kcal mol^{-1}).[a]

R	$D°(R-R)$	$D°(R-R^{+\bullet})$[b]
CH_3	89.7	50.6 (0.56)
NH_2	67.4	137.5 (2.04)
OH	51.2	107.9 (2.11)
SiH_3	78.0	34.6 (0.44)
PH_2	50.6	74.1 (1.46)
SH	62.6	87.3 (1.39)

[a]Data from reference 9. [b]Values in parentheses are the relative σ bond energy of the radical cation compared to the neutral, $D°(R-R^{+\bullet})/D°(R-R)$.

It is clear from Table 1 that a decrease in $D°$ is observed only for carbon and silicon (Group IVB). In all of the other examples (Group VB and Group VIB), an increase in $D°$ in the radical cation, relative to the neutral species, is observed. These observations are easily understood when the nature of the highest occupied molecular orbital (HOMO), from which the electron is removed, is considered. For carbon and silicon, the HOMO is a σ orbital which is largely bonding between the two heavy atoms. Removal of an electron from this bonding orbital results in a reduction in D° to about half the original value.

On the other hand, for the Group VB and VIB dimers in Table 1, the HOMO is predominantly associated with the non-bonding lone pairs. Removal of one of these electrons has quite a different effect. While the interaction between two lone pairs in the neutral species is repulsive, in the radical cation a net bonding interaction between a lone pair and and the singly occupied molecular orbital (SOMO) is possible. This bonding interaction is referred to as a two-center three electron bond (Figure 4). The strength of the three electron bond will depend on the overlap between the interacting orbitals and their energy difference; greater overlap and smaller energy differences result in stronger interactions. The extent of these interactions are expected to decrease down a period (i.e. N > P and O >S).

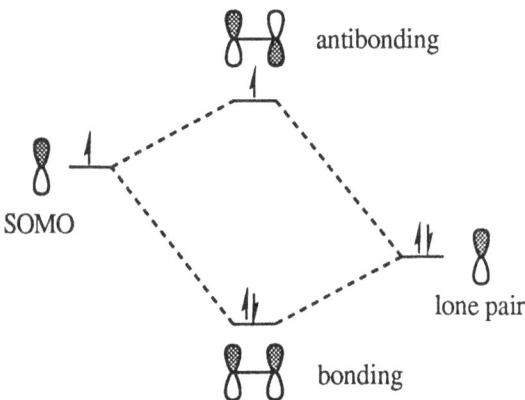

Figure 4. Three electron bonding interaction between the SOMO and lone pair.

3.2   SOLUTION BOND ENERGIES

While the effects in the gas phase are well understood, it is not obvious that even the trends will be the same in solution where differential solvation of the reactants or products may be important. The photomodulation voltammetry technique has provided some of the electrochemical data that are necessary to determine some radical cation bond energies that pertain to acetonitrile solution using Equation 8. Some of the electrochemical and bond energy data for carbon, tin and phosphorous centered systems are given in Table 2. Whenever possible, the solution data have been compared to data in the gas phase (values in parentheses in Table 2).

With only two exceptions, there is reasonably good correlation between the gas phase and solution bond energy differences. The correspondence between these values implies that the solvation energy of the radical cation and cation are similar (presumably because the extent of charge localization in the radical cation and cation product are about the same). As a result, the solvation terms approximately cancel. In most cases, the change in the σ-bond energy represents decrease of ca. 50 percent relative to the neutral species. This is somewhat surprising since the SOMO for the aryl radical cations is expected to be localized predominantly in the aromatic π system with little contribution from the σ bond.

Table 2. Oxidation Potentials (vs SCE) of Radicals and Bond Energy Differences for Molecules and Radical Ions in Acetonitrile/0.1M TBAP.

$$R_1\text{-}R_2^{+\bullet} \rightarrow R_1^+ + R_2^{\bullet}$$

R_1	R_2	$E_{1/2}^{ox}(R_1^{\bullet})^a$	$E_{1/2}^{ox}(R_1\text{-}R_2)^a$	$\Delta G^{\circ}_8\text{-}\Delta G^{\circ}_1{}^b$
		V vs SCE	V vs SCE	kcal mol^{-1}
PhCH$_2$	PhCH$_2$	0.73	2.4	-38 (-41)
Ph$_2$CH	Ph$_2$CH	0.35	2.04	-37
1-NapCH$_2$	1-NapCH$_2$	0.47	1.59	-26
Me$_2$NCH$_2$	Me$_2$NCH$_2$	-1.03	0.70	-40 (-36)
Me$_2$NCH$_2$	H	-1.03	1.15	-50 (-49)
Me$_2$NCH$_2$	PhCH$_2$	-1.03	1.15	-50 (-46)
PhCH$_2$	Me$_2$NCH$_2$	0.73	1.15	-10 (-11)
CH$_3$OCH$_2$	PhCH$_2$	-0.24	2.45	-62 (-43)
PhCH$_2$	CH$_3$OCH$_2$	0.73	2.45	-40 (-37)
Bu$_3$Sn	Bu$_3$Sn	<0.0^c	<1.2	<-27 (-16)^d
Bu$_3$Sn	Bu	<0.0^c	1.75	-40 (-41)^d
Ph$_2$P	Ph$_2$P	0.58	1.15	-11
Ar$_2$P^e	Ar$_2$P^e	0.25	0.57	-5
Ar'$_2$P^f	Ar'$_2$P^f	0.33	0.53	-7

[a]Measured by photomodulation voltammetry. [b]$\Delta G^{\circ}_8\text{-}\Delta G^{\circ}_1=\text{-}F[E_{1/2}^{ox}(R_1\text{-}R_2) - E_{1/2}^{ox}(R_1^{\bullet})]$. Values in parentheses are those calculted for the gas phase using data in reference 9. [c]Upper limit; see section 4. [d]Gas phase data for hexamethylditin. [e]Ar = 4-methoxy-2,6-dimethylphenyl. [f]Ar' = 2,4,6-triethylphenyl.

The fragmentation of PhCH$_2$CH$_2$OCH$_3$ is particularly interesting in this regard. The energetics for the fragmentation to give the benzyl cation are the same both in acetonitrile and in the gas phase. However, the fragmentation to give CH$_3$OCH$_2^+$ is much more favourable in acetonitrile than in the gas phase. *This is an example where solvation of the product provides an additional driving force.* The β-phenylether radical cation and benzyl cation are of similar size and extent of charge delocalization and hence have similar solvation energies. The methoxymethyl cation, on the other hand, is smaller and the charge is more localized than in the radical cation of the ether precursor and, consequently, has a solvation energy that is about 15 kcal mol^{-1} more exoergonic than that of the radical cation.

A decrease in the σ bond energies upon oxidation also is observed for the tin centered species. This reduction in bond energy is ca. 50 percent for hexabutyldistannane and 65 percent for tetrabutylstannane compared to the respective neutral species. The gas phase data that have been included are actually for hexamethyldistannane and the trimethylstannyl radical. Attempts to measure the oxidation potential of the trimethylstannyl radical (by photolysis of hexamethyldistannane) were inconclusive.[14] For this system, in acetonitrile, some modification of the electrode surface occured (i.e. the measured half-wave potential depended on the history of the electrode) so the measured potentials were not considered to be reliable. These surface effects were not apparent for the tributylstannyl system in this solvent.

The radical cations of Group V B dimers are of particular interest. It is well known that hydrazine radical cations are planar (or nearly so) and form strong three electron bonds.[15] This implies is that in solution, as well as in the gas phase, the oxidation potential (or IP) of the aminyl radical, $R_2N^\bullet$, will be higher than that of the hydrazine. This observation presents a practical problem in that all of the aminyl radical precursors will be more easily oxidized than the aminyl radical itself. Under these conditions, the photomodulation technique cannot be used since the precursor for the radical would be consumed in the diffusion layer before the oxidation potential of the radical was reached.

In contrast to hydrazine radical cations, both experimental[16a] and theoretical[16b,c] results have shown that the tricoordinate phosphorous atoms in diphosphine radical cations are highly pyramidalized (Figure 5). Nevertheless, in the gas phase, there is a 50 percent increase in $D°(P-P)$ in the radical cation of P_2H_4 relative to the neutral molecule (see Table 1) indicating that there must be some additional interaction between the phosphorous atoms; presumably three electon bonding. Surprisingly, in solution, we have found that the diarylphosphinyl radicals are more easily oxidized than the corresponding diphosphines which translates to an overall decrease in the σ-bond energy upon oxidation (Table 2).[17] This is in marked contrast to tetraphenylhydrazine radical cation which forms a strong three electron bond.[18] ESR evidence has shown that the spin is equally distributed between the two phosphorous atoms in the diphosphine radical cations.[16] This result suggests that the radical cation has an electronically delocalized structure with C_2 symmetry and the same

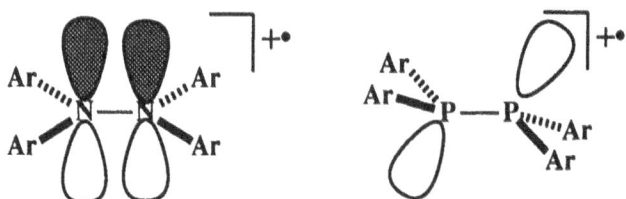

Figure 5. Structures of hydrazine and diphosphine radical cations.

degee of bending at the phosphorous, or, that the electron exchange between the phosphorous atoms is too rapid on the ESR time scale. The fact that $D°(P-P)$ appears to decrease on oxidation implies that there is actually very little π–interaction between the phosphorous atoms in these species and that delocalization between the phosphorous atoms must occur through the σ framework.

4. Thermochemistry and Kinetics for the Reaction Bu₃Sn˙ with Nitroalkanes

4.1 BACKGROUND

The role of electron transfer in the reactions of tin centered radicals in radical chain reductions has been questioned.[14] For example, two possible mechanisms for the reduction of nitroalkanes by tributyltin hydride have been proposed. Initially, the tin hydride reaction mechanism was thought to procede by an electron transfer reaction between the stannyl radical and the nitroalkane (Equations 17-19).[19a,b] An alternate

$$Bu_3Sn^\bullet + R\text{-}NO_2 \rightarrow Bu_3Sn^+ + R\text{-}NO_2^{-\bullet} \tag{17}$$

$$R\text{-}NO_2^{-\bullet} \rightarrow R^\bullet + NO_2^- \tag{18}$$

$$R^\bullet + Bu_3SnH \rightarrow R\text{-}H + Bu_3Sn^\bullet \tag{19}$$

mechanism was subsequently suggested[19c] in which the tin radical adds to the oxygen of the nitro group to give a nitroxyl radical which, in turn, fragments to give an alkyl radical (Equations 20,21).

$$Bu_3Sn^\bullet + R\text{-}NO_2 \rightarrow Bu_3SnO\overset{\overset{\displaystyle O^\bullet}{|}}{-}NR \tag{20}$$

$$Bu_3SnO\overset{\overset{\displaystyle O^\bullet}{|}}{-}NR \rightarrow Bu_3SnO\text{-}N{=}O + R^\bullet \tag{21}$$

Electron spin resonance studies of these reductions gave spectra which were equally consistent with both possibilities.[20] Since spectroscopic and product analysis failed to differentiate between these two possibilities, we have attempted to assess the thermochemistry associated with the key step (Equation 17). In addition, we have measured the kinetics for the reaction of tributylstannyl radical with a number of nitroalkanes. Analysis of the thermochemistry, in the context of the kinetics should provide a more reliable assessment of the reaction mechanism.

4.2 THERMOCHEMISTRY OF THE ELECTRON TRANSFER REACTION

In principle, it is possible to calculate the free energy change associated with reaction 17 from the difference between the oxidation potential of Bu₃Sn˙ and the reduction potential of RNO_2. The reduction potentials for α-nitrocumene, nitrocyclohexane and 2-methyl-2-nitropropane in acetonitrile (0.1 M TBAP) all were quasireversible on the cyclic voltammetric time scale (scan rate > 20 V/s) with half-wave potentials near -0.9 V vs SCE. In THF, quasireversible voltammograms near -1.1 V vs SCE were obtained.

The half-wave potential for the oxidation of the tributylstannyl radical represents an irreversible electrode process so the thermodynamic significance of this potential should be examined more closely before the thermochemistry for the reaction with nitroalkanes is discussed further. In principle, it is possible to bracket the value of $E°$ by first measuring the half-wave potential for oxidation of the radical and then, in a different experiment, measure the half-wave for reduction of the cation. Dessy and his coworkers[21] reported the preparation of Ph_3Sn^+ by the reaction of $AgClO_4$ with the stannyl chloride (Equation 22) or the distannane (Equation 23). Voltammetry on solutions of triphenyl- and tributylstannyl

$$AgClO_4 + Ph_3SnCl \rightarrow Ph_3Sn^+ ClO_4^- + AgCl \tag{22}$$

$$2AgClO_4 + Ph_6Sn_2 \rightarrow 2Ph_3Sn^+ ClO_4^- + Ag° \tag{23}$$

cations prepared in this way was complex and depended not only on the scan rate but also on the history of the electrode (i.e. number of scans, method of cleaning). Overall, the results indicated that surface films were forming on the electrode surface and that these films were corrupting the measurements of the redox potentials.

Since direct electrochemistry of the cations were complex, we attempted to estimate $E°(R_3Sn^+/R_3Sn^•)$ from the rate of homogeneous electron transfer between the stannyl cation and other electron donors in a stopped flow absorption spectrometer. The donors chosen were ferrocene (Fc, $E°=0.50$ V), dimethylferrocene (DiMFc, $E°=0.36$ V), dimethylphenozine (DMP, $E°=0.31$ V), N,N',N'',N'''-tetramethylphenylenediamine (TMPD, $E°=0.12$ V) and decamethylferrocene (DMFc, $E°=0.10$ V). The growth of the absorption due to the product cations were fit to second order kinetics to give the k_{obs} values shown in Table 3. In principle, $E°(R_3Sn^+/R_3Sn^•)$ can be estimated from the kinetic data using Marcus theory.[22] However, in this case the value of the solvent reorganization energy, λ, is not well defined so this approach is not really viable.

Nevertheless, it is possible to define a lower limit for $E°$ (Equation 24-26). Since $k_{obs} \leq k_{24}$ and, since k_{-24} cannot be greater than the diffusion controlled limit (10^{10} M^{-1} s^{-1}), then it follows that $K_{24} \geq 10^{-10}k_{obs}$. From this, a lower limit for $E°(R_3Sn^+/R_3Sn^•)$ can be determined (Table 3).

$$R_3Sn^+ + D \leftrightarrows R_3Sn^• + D^{+•} \tag{24}$$

$$2 R_3Sn^• \rightarrow R_6Sn_2 \tag{25}$$

$$\ln(K_{24}) = \ln(k_{24}/k_{-24}) = F[E°(R_3Sn^+/R_3Sn^•) - E°(D^{+•}/D)]/RT \tag{26}$$

Table 3. Electron Transfer Reactions of R_3Sn^+ in THF.[a]

R	Donor	k_{obs} (M s^{-1})	$E°(R_3Sn^+/R_3Sn^•)$[b] (V vs SCE)
Bu_3Sn^+	Fc	n.r.	
	DiMFc	n.r.	
	DMP	0.027	>-0.38
	TMPD	3.36	>-0.44
	DMFc	2.92	>-0.47
Ph_3Sn^+	Fc	n.r.	
	DiMFc	n.r.	
	DMP	0.86	>-0.28
	TMPD	7.32	>-0.42
	DMFc	19.0	>-0.42

[a]Measured in THF as solvent by stopped flow absorption spectrophotometry. n.r. = no reaction. [b]Lower limit estimated using Equation 26.

For comparison, photomodulation voltammograms of the radicals in THF as solvent were measured. These experiments gave half-wave potentials of +0.3 V and +0.12 V for the tributyl and triphenylstannyl radicals, respectively. Although there is a large variation between the two methods, these data, when combined with the reduction potential of the nitroalkanes lead to the conclusion that equation 17 must be endoergonic by at least 0.7 eV (ca. 15 kcal mol^{-1}).

4.3 KINETICS OF THE REACTION BETWEEN $Bu_3Sn^•$ AND RNO_2

The laser flash photolysis technique was used to measure rate constants for the reactions of $Bu_3Sn^•$ with a number of nitroalkanes. The stannyl radical was generated by photolysis of mixtures containing DTBP (25% v/v) and tributyltin hydride (0.3 M) in benzene as solvent. The observed pseudo first order rate constants, k'_{obs} for the decay of the radical were measured as a function of the nitroalkane concentration. The data were plotted according to Equation 27 where k_0 was the inverse of the lifetime of the stannyl radical in the absence of nitroalkane, and gave excellent straight lines from which the values of k_r were obtained (Table 4).

$$k'_{obs} = k_0 + k_r[RNO_2] \tag{27}$$

Attempts were made to meaure values of k_r in tetrahydrofuran (THF) and acetonitrile as solvents. However, in THF, reaction of the *tert*-butoxyl radicals with solvent diminished the yield of the stannyl radicals to the point where their absorptions could no longer be monitored. However, if the electron transfer mechanism were operative, the observed rate (as measured by laser flash photolysis) would almost certainly be greater in these solvents than in benzene. Thus, the data data in Table represent lower limits for k_r in the more polar solvents.

Table 4. Rate Contants for the Reaction of
$Bu_3Sn^\bullet$ with Nitroalkanes.[a]

Compound	$10^{-7}k_r$ $(M^{-1}s^{-1})$
CH_3NO_2	7.3
$(CH_3)_2CHNO_2$	9.5
$(CH_3)_3CNO_2$	9.3
$c\text{-}C_6H_{11}NO_2$	9.0

[a]Measured by laser flash photolysis in benzene.

The data in Table 3 demonstrate that the reactions between $Bu_3Sn^\bullet$ and nitroalkanes proceed with rate constants that are close to the diffusion limit. These results are not consistent with the electron transfer mechanism which was demonstrated to be endoergonic by ca. 12 kcal mo^{-1}. We can conclude, therefore, with some certainty that the addition/fragmentation mechanism (Equations 20,21) best describes these reactions since it is consistent with all of the available spectroscopic, kinetic and electrochemical data.

References
1. Issued as NRCC publication No. 33249.

2. (a) Breslow, R. and Balasubramanian, K. (1969) *J. Am. Chem. Soc.*, **91**, 5182. (b) Breslow, R. and Chu, W. (1970) *J. Am. Chem. Soc.*, **92**, 2165. (c) Breslow, R. and Goodin, R. (1976) *J. Am. Chem. Soc.*, **98**, 6067. (d) Wasielewski, M.R. and Breslow, R. (1976) *J. Am. Chem. Soc.*, **98**, 4222. (e) Juan, B., Schwartz, J. and Breslow, R (1980) *J. Am. Chem. Soc.*, **102**, 5741.

3. (a) Popielarz, R. and Arnold, D.R. (1990) *J. Am. Chem. Soc.*, **112**, 3068. (b) Nicholas, A.M. deP. and Arnold, D.R. (1982) *Can. J. Chem.*, **60**, 2165.

4. (a) Bordwell, F.G., Zhang, X. and Cheng, J.P. (1991) *J. Org. Chem.*, **56**, 3216. (b) Bordwell, F.G. and Cheng, J.P. (1991) *J. Am. Chem. Soc.*, **113**, 1736. (c) Bordwell, F.G. and Harrelson Jr., J.A. (1990) *Can. J. Chem.*, **68**, 1714. (d) Bordwell, F.G. and Cheng, J.P. (1989) *J. Am. Chem. Soc.*, **111**, 1792. (e) Bordwell, F.G. and Bausch, M.J. (1986) *J. Am. Chem. Soc.*, **108**, 1979.

5. (a) Maslak, P., Kula, J. and Chateauneuf, J.E. (1991) *J. Am. Chem. Soc.*, **113**, 2304. (b) Maslak, P. and Chapman Jr., W.H. (1990) *Tetrahedron*, **46**, 2715. (c) Maslak, P. and Narvaez, J.N. (1990) *Angew. Chem. Int. Ed.*, **29**, 283.

6. (a) Arnett, E.M., Amarnath, K., Harvey, N.G. and Cheng, J.P. (1990) *Sience*, **247**, 423. (b) Arnett, E.M., Armarnath, K., Harvey, N.G. and Cheng, J.P. (1990) *J. Am. Chem. Soc.*, **112**, 344. (c) Arnett, E.M., Amarnath, K., Harvey, N.G. and Venimadhavan, S. (1990) *J. Am. Chem. Soc.*, **112**, 7346.

7. (a) Tilset, M. and Parker, V.D. (1989) *J. Am. Chem. Soc.*, **111**, 6711. (b) Ryan, O., Tilset, M. and Parker, V.D. (1990) *J. Am. Chem. Soc.*, **112**, 2618.

8. Griller, D., Martinho Simões, J.A., Mulder, P., Sim, B.A. and Wayner, D.D.M. (1989) *J. Am. Chem. Soc.*, **111**, 7872.

9. Lias, S.G., Bartmess, J.E., Leibmann, J.F., Holmes, J.L., Levin, R.D. and Mallard, W.G. (1988) *J. Phys. Chem. Ref. Data*, **17**, supplement 1.

10. Bard, A.J. and Faulkner, L.R. (1980) Electrochemical Methods: Fundamentals and Applications, John Wiley and Sons, Inc., New York.

11. For a recent review see: Wightman, R.M. and Wipf, D.O. (1990) *Acc. Chem. Res.*, **23**, 64.

12. Wayner, D.D.M., McPhee, D.J. and Griller, D. (1988) *J. Am. Chem. Soc.*, **110**, 132.

13. (a) Wayner, D.D.M. and Griller, D. (1985) *J. Am. Chem. Soc.*, **107**, 7764. (b) Wayner, D.D.M., Dannenberg, J.J. and Griller D. (1986) *Chem. Phys. Lett.*, **131**, 189. (c) Sim, B.A., Griller, D. and Wayner, D.D.M. (1989) *J. Am. Chem. Soc.*, **111**, 754. (d) Nagaoka, T., Griller, D. and Wayner D.D.M. (1991) *J. Phys Chem.*, **95**, 6264.

14. Tanner, D.D., Harrison, J.D., Chen, J., Kharrat, A., Wayner, D.D.M., Griller, D. and McPhee, D.J. (1990) *J. Org. Chem.*, **55**, 3321.

15. (a) Nelsen, S.F. (1981) *Acc. Chem. Res.*, **14**, 131. (b) Nelsen, S.F., Cunkle, G.T. and Evans, D.H. (1983) *J. Am. Chem. Soc.*, **105**, 5928.

16. (a) Culcasi, M., Gronchi, G. and Tordo, P. (1985) *J. Am. Chem. Soc.*, **107**, 7191. (b) Feller, D., Davidson, E.R. and Borden, W.T., (1985) *J. Am. Chem. Soc.*, **107**, 2596. (c) Clark, T., (1985) *J. Am. Chem. Soc.*, **107**, 2597.

17. Wayner, D.D.M. and Tordo, P. *unpublished results*.

18. Cauquis, G., Delhomme, H. and Serve, D. (1975) *Electrochim. Acta*, **12**, 1019.

19. (a) Tanner, D.D., Blackburn, E.V. and Diaz, G.E. (1981) *J. Am. Chem. Soc.*, **103**, 1557. (b) Ono, N., Miyake, H., Tamura, R. and Kaji, A. (1981) *Tetrahedron Lett.*, **22**, 1705. (c) Kaminura, A. and Ono, N. (1988) *Bull. Chem. Soc. Jpn.*, **61**, 3629.

20. (a) Korth, H.G., Sustmann, R., Dupuis, J. and Geise, B. (1987) *Chem. Ber.*, **120**, 1197. (b) Davies, A.G., Hawari, J.A., Gaffney, C. and Harrison, P.G. (1982) *J. Chem. Soc. Perkin Trans. II*, 631.

21. Dessy, R.E., Kitching, W. and Chivers, T. (1966) *J. Am. Chem. Soc.*, **88**, 453.

22. Eberson, L. In: *Electron Transfer Reactions in Organic Chemistry*; Springer Verlag, Berlin, 1987.

Derivative Cyclic Voltammetry: Applications in the Investigation of the Energetics of Organometallic Electrode Reactions

Mats Tilset

Department of Chemistry

University of Oslo

P. O. Box 1033 Blindern

N-0315 Oslo 3, Norway

Abstract

Derivative cyclic voltammetry (DCV) can offer significant advantages in the investigation of electrode reactions when compared with conventional cyclic voltammetry (CV). Redox potentials can be accurately determined for incorporation in thermochemical cycles that allow for the determination of thermodynamic quantities that are of interest to organometallic chemists. Applications towards the determination of solution bond dissociation energies and Brønsted acidities for metal hydrides and metal hydride cation radicals are presented. However, perhaps the greatest advantages of DCV arise from the absence of double-layer charging currents in the derivative voltammograms. This effect makes the technique especially suitable for kinetic and mechanistic studies of the reactions of electrode-generated reactive intermediates. The use of DCV for this purpose is highlighted with case studies from our laboratories. This includes the substitution of one- and two-electron donors in 17-electron cation radicals by two-electron donor nucleophiles. Kinetic parameters show that both types of processes appear to involve 19-electron intermediates or transition states. Also included are some rather unusual reactions that proceed via initial reversible cation radical dimerization.

1   Introduction

During the past decade, the chemistry of 17- and 19-electron organotransition-metal species has been pursued with increasing intensity.[1] Considerable evidence now has established that 17/19-electron cycles may be as important in the reactions of odd-electron species as the 16/18-electron[2] cycles are in the chemistry of even-electron species.

The oxidation and reduction of coordinately saturated, 18-electron complexes represent one of the most direct methods for the generation of 17-electron and 19-electron systems. The electron-transfer reactions are commonly examined by a variety of electrochemical methods among which cyclic voltammetry (CV) appears to be employed most frequently. The measurement of electrode potentials by CV gives access to

J. A. Martinho Simões (ed.), Energetics of Organometallic Species, 109–129.
© 1992 Kluwer Academic Publishers.

thermochemically significant information. This potential scan reversal technique also allows for the use of CV in the investigation of the kinetics of homogeneous reactions succeeding the electron transfer.[3]

Parker and coworkers have promoted the use of derivative cyclic voltammetry[4] (DCV) as a valuable and efficient tool for the investigation of electron-transfer-induced reactions in organic chemistry.[5] Recently, the technique has found applications in the exploration of organometallic reactivity,[6] providing the basis for this contribution.

We will discuss briefly the basics of DCV, including the instrumentation setup as found in our laboratories. Some applications of DCV in the kinetics and thermodynamics of organometallic chemistry are then described.

2   Methods

2.1   INSTRUMENTATION

The electrochemical instrumentation (Figure 1), cells, and data handling procedures used in our group are, except for some minor modifications, the same as those described elsewhere.[4a,7] The potentiostat is an EG&G Princeton Applied Research Model 273 Potentiostat/Galvanostat. Positive feedback IR compensation is always applied during the electrochemical experiments. For easy access to high voltage sweep rates, the potentiostat is driven by a Hewlett Packard 3314A Function Generator, interfaced to an IBM PS/2 Model 50 computer via a GPIB-IEEE interface card and by the use of homemade programs. The output from the potentiostat is differentiated by passing the signal through an EG&G Princeton Applied Research Model 189 Selective Amplifier operated in the bandpass mode. The data are digitized by a Nicolet 310 Digital Oscilloscope interfaced to the computer. The 0.5 μs/point resolution allows for acceptable signal resolution at voltage sweep rates up to 2000 V/s or more. The data are read from the oscilloscope by the computer, and further manipulation of the data may be carried out with homemade or commercially available software.

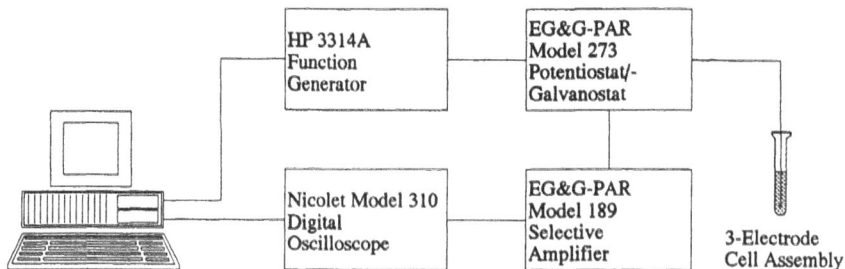

Figure 1. Schematic presentation of the electrochemical workstation used for DCV measurements.

2.2   DERIVATIVE CYCLIC VOLTAMMETRY

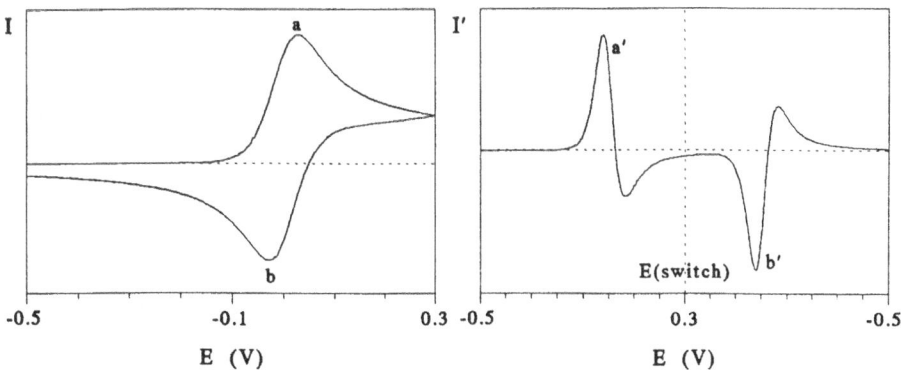

Figure 2. Simulated cyclic voltammogram (left) and derivative cyclic voltammogram (right) for a Nernstian one-electron transfer reaction.

Figure 2 displays a simulated cyclic voltammogram (left) and the corresponding derivative cyclic voltammogram (right) for a Nernstian one-electron transfer process.[8a] The CV peak current potentials (labeled a, b) correspond to the points where the rapidly descending (ascending) DCV curve crosses the base line after the derivative peak currents (a', b'). Clearly, the differentiation of the signal makes possible an accurate determination of the CV peak potentials. The differential response of the DCV technique frequently allows for the resolution of fine details that may not be readily discernible in normal cyclic voltammograms.

When conventional CV is employed as a tool for the investigation of electrode processes or of the reactions of electrode-generated species, the double-layer charging current causes a distortion of the base line.[3] This poses a problem for kinetic applications of CV,

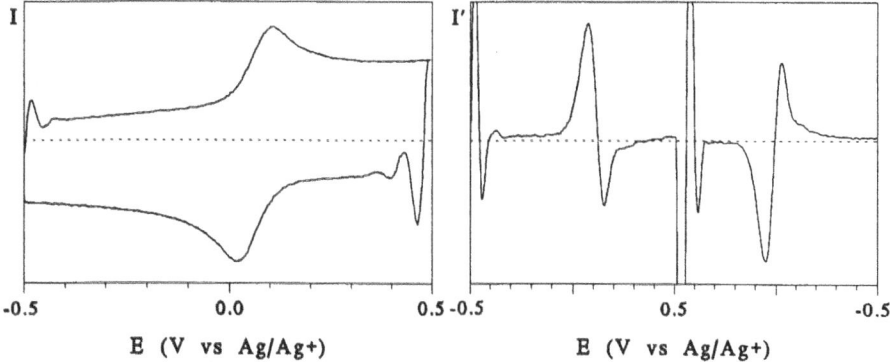

Figure 3. Cyclic voltammogram (left) and derivative cyclic voltammogram (right) for the oxidation of ferrocene (0.25 mM substrate in acetonitrile/0.1 M $Bu_4N^+PF_6^-$, 20 °C, 0.1 mm Pt disk electrode, sweep rate υ = 500 V/s).

especially at high voltage sweep rates and/or low substrate concentrations where the charging current may be quite dominant. This is illustrated with the cyclic voltammogram for the oxidation of ferrocene (0.25 mM) at a voltage sweep rate $\upsilon = 500$ V/s (acetonitrile/0.1 M $Bu_4N^+PF_6^-$, 20 °C, 0.1 mm Pt disk electrode), displayed in Figure 3 (left). The problem may be resolved by a number of different techniques, including subtraction of the background current recorded in the absence of electroactive species, or the application of microelectrode techniques.[9] The use of DCV also circumvents the double-layer charging problem in a convenient manner: because the charging current contribution to the overall current is constant at a given voltage sweep rate,[3] it is effectively eliminated by the differentiation of the signal. This effect is clearly demonstrated by comparison of the two voltammograms in Figure 3. The only observable effects of the double-layer charging current in the DCV trace appear at the start and switching potentials of the cyclic sweep, where abrupt charging current changes occur.

Figure 4 shows simulated voltammograms for a Nernstian electron transfer accompanied by a moderately fast homogeneous follow-up reaction. The DCV reaction-order relationship[10] provides a simple and elegant approach to the investigation of the kinetics and mechanisms of such reactions. In a typical DCV reaction-order analysis series, the cyclic voltammetry starting potential (E_{start}) is fine-tuned so that the switching potential (E_{switch}) is held at a constant value relative to the reversible potential (E_{rev}) for the electrode reaction throughout the experiment. The experimental parameter that is most conveniently used for the kinetic analysis is R_I', defined as the ratio of the derivative current peak heights for the reverse (I'_r) and forward (I'_f) scans as measured from the base line of the DCV trace (Figure 4). Since a wide range of values may be chosen for the experimental parameters υ, R_I', and ($E_{switch} - E_{rev}$), the method facilitates the investigation of reactions with rate constants spanning many orders of magnitude.

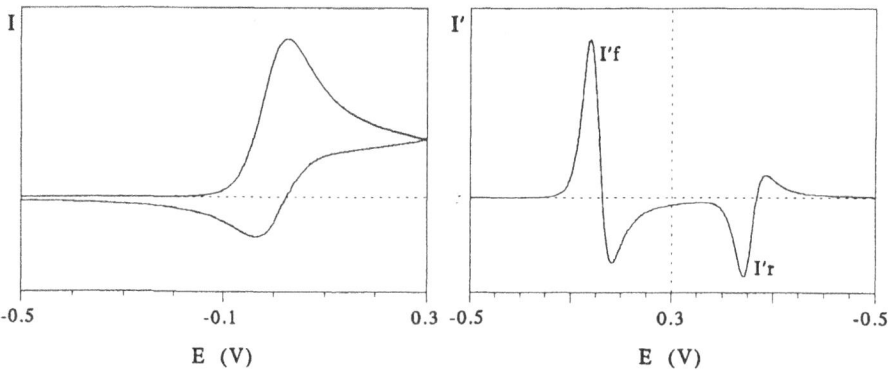

Figure 4. Simulated cyclic (left) and derivative cyclic (right) voltammogram for a Nernstian one-electron oxidation followed by a homogeneous reaction consuming the electrode-generated intermediate.

$$A \quad \rightleftharpoons \quad B + e^-$$

$$B + X \xrightarrow{\quad k \quad} \text{Products}$$

$$\text{rate} = -dC_B/dt = k\, C_A{}^a\, C_B{}^b\, C_X{}^x \tag{1}$$

$$R_{A/B} = a + b = 1 + d(\log v_c)/d(\log C_A) \tag{2}$$

$$R_X = x = d(\log v_c)/d(\log C_X) \tag{3}$$

$$k_H/k_D = v_c(H)/v_c(D) \tag{4}$$

If an electrode-generated species B undergoes a decomposition reaction obeying rate law (1), then the composite reaction order in terms of B and its precursor A, $R_{A/B}$, may be obtained from eq (2). The reaction order with respect to any added electro-inactive reagents X is given by eq (3). In eq (2) and (3), v_c is defined as the voltage sweep rate that causes the derivative peak current ratio R_I' to be equal to the constant c. Thus, for a series of experiments in which C_A and C_X are varied while ($E_{switch} - E_{rev}$) is maintained at a constant value, $R_{A/B}$ may be found as 1 plus the slope of a plot of log v_c vs log C_A. R_X on the other hand equals the slope of a plot of log v_c vs log C_X. Furthermore, H/D kinetic isotope effects are readily available from eq (4), where $v_c(H)$ and $v_c(D)$ represent observed values for normal and deuterated substrates, respectively. Finally, activation energies for electron-transfer induced reactions are available from variable-temperature DCV data - without prior knowledge of the reaction mechanism ! Plots of $\ln(v_c/T)$ vs $1/T$ provide Arrhenius-type relationships from which the activation energies may be calculated.[4b]

Deconvolution of the reaction orders with respect to A and B may be achieved by linear sweep voltammetry (LSV).[4b] The slopes of plots of E_p vs log v and of E_p vs log C_A (where E_p represents the cyclic voltammetry peak potential) are simple functions of the coefficients a, b, and x in the rate law. Here, E_p has to be measured with a high degree of accuracy and this is easily achieved by DCV. Once the rate law has been established, and provided that the overall stoichiometry of the electrode reaction is known, the DCV data may be compared with theoretical data obtained by digital simulation in order to arrive at the actual rate constants.

In the following, examples from our research in organometallic electron transfer chemistry will be used to highlight the use of DCV.

3   Applications of DCV in Organometallic Chemistry

3.1   KINETIC AND MECHANISTIC APPLICATIONS OF DCV

3.1.1. Ligand Substitution in Manganese Complexes. Ligand substitution reactions constitute one of the most thoroughly studied reaction types of 17-electron systems. The special lability of 17-electron metal centers with respect to substitution is striking when compared to that of 18-electron analogues. The reactions generally appear to proceed by associative mechanisms involving 19-electron intermediates or transition states.[11,12] In a classic study, Kochi and coworkers[12a] investigated the oxidatively induced, electrocatalytic ligand substitution reactions of Cp'Mn(CO)$_2$(NCMe) (Cp' = η^5-C$_5$H$_4$Me) and other complexes. Highly efficient electrocatalytic processes ensued when (as was commonly observed) the products Cp'Mn(CO)$_2$L$^{\cdot+}$ possessed a greater oxidizing power than the substrate cation radical. The overall mechanism, exemplified by the electrocatalytic substitution of PPh$_3$ for acetonitrile in Cp'Mn(CO)$_2$(NCMe)$^{\cdot+}$, is summarized in Scheme I.

Scheme I

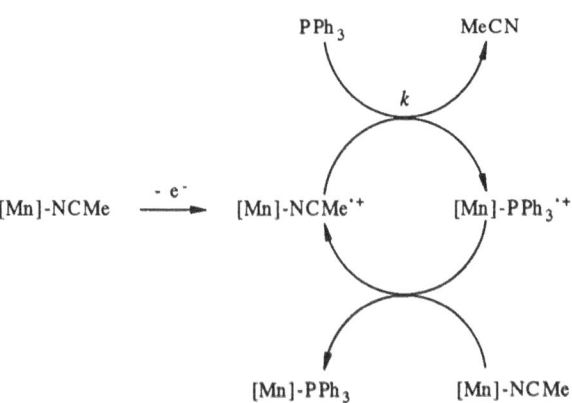

DCV has not before been applied to electrocatalytic reactions of this kind, and we decided to put the method to a test. The kinetics of this well-studied reaction was explored by DCV in acetonitrile so that the results could be compared with Kochi's data. The reaction-order analysis was in full agreement with an overall second-order reaction, first-order in cation radical and first-order in PPh$_3$. Variable-temperature DCV data are shown in Table I. An Arrhenius plot of $\ln(\upsilon_{0.5}/T)$ vs $1/T$ gives E_a = 4.2 ± 0.2 kcal/mol. Comparison of the DCV data with digital simulation data[8a] yielded k(25 °C) = 16.8 x 10^3 M^{-1}s^{-1}, and the Eyring equation finally gave $\Delta H^{\#}$ = 3.6 ± 0.2 kcal/mol and $\Delta S^{\#}$ = -27.0 ± 0.4 cal/(K·mol). These data agree quite well with the reported values (k(25 °C) = 13 x 10^3 M^{-1}s^{-1}, $\Delta H^{\#}$ = 4.4 kcal/mol and $\Delta S^{\#}$ = -25 cal/(K·mol)).[12a] It is noteworthy that these two approaches to investigate the

Table I. Variable Temperature DCV Data for the Reaction between $Cp'Mn(CO)_2(NCMe)^{\cdot+}$ and PPh_3

$T, ^\circ C$	$\upsilon_{0.5}, V/s^a$	$10^{-3} \times k, M^{-1}s^{-1}$
-25.0	9.0	4.1
-9.5	15.0	6.5
0.0	21.6	9.0
11.0	29.5	11.8
18.0	38.0	14.8
31.0	50.5	18.9

[a] Conditions: 1 mM substrate and 10 mM PPh_3 in acetonitrile/0.1 M $Bu_4N^+PF_6^-$, 0.4 mm Pt disk electrode, $E_{switch} - E_{rev} = 200$ mV.

kinetics operate under quite different conditions as far as the time scale is concerned. For example, at ca. 1:10 substrate to PPh_3 ratio, Kochi's measurements were carried out at relatively slow sweep rates. Consequently, the experimental time scale allowed for high catalytic turnover numbers, and most of the substrate conversion was due to the cross electron transfer in the catalytic cycle. Our measurements were performed at sweep rates about 100 times faster, with insufficient time for high catalytic turnover numbers. The magnitude of the simulated peak current amounted to about 90 % of that of the current for a reversible (no reaction) process, verifying that most of the substrate consumption under these conditions was due to direct oxidation at the electrode. Considering these differences, it is particularly pleasing to note that the results from the two methods agree well.

As a further check of the DCV method, digital simulations were carried out[8a] in order to obtain a best fit, as judged by visual inspection, of experimental and simulated derivative cyclic voltammograms under a given set of conditions (see caption to Figure 5). Disregarding

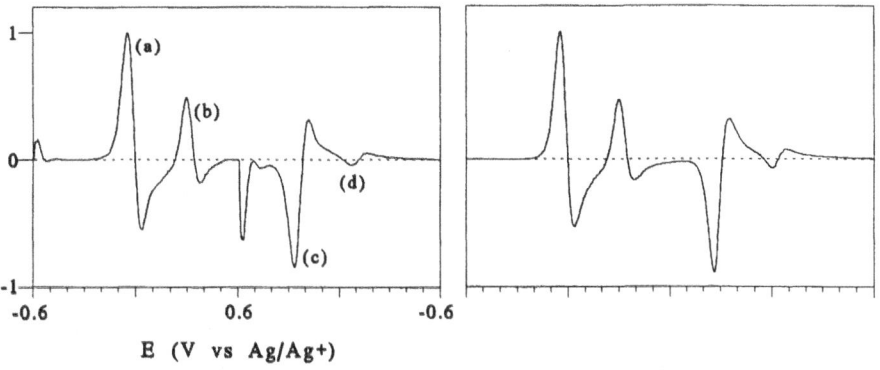

Figure 5. Experimental (left) and simulated[8a] (right) derivative cyclic voltammograms for the oxidation of $Cp'Mn(CO)_2(NCMe)$ (1 mM) in the presence of PPh_3 (10 mM). Experimental parameters: 18 $^\circ C$, 0.1 M $Bu_4N^+PF_6^-$, 0.4 mm Pt disk electrode, sweep rate $\upsilon = 35$ V/s. Simulated rate constant, $k = 12.2 \times 10^3$ $M^{-1}s^{-1}$.

Table II. Kinetic Parameters for Ligand Substitution Reactions of Manganese complexes

Substrate	Nucleophile	$10^{-3} \times k$ (25 °C), $M^{-1}s^{-1}$ [a]	$\Delta H^{\#}$, kcal/mol	$\Delta S^{\#}$, cal/(K·mol)
$Cp'Mn(CO)_2(NCMe)$[b]	PPh_3	16.8 (13)	3.7 ± 0.2	-26.7 ± 0.5
$Cp'Mn(CO)_2(NCMe)$[c]	PPh_3	12.0 (3.3)	4.2 ± 0.3	-25.6 ± 0.9
$Cp'Mn(CO)_2(NCMe)$[b]	$P(OMe)_3$	2.52	4.7 ± 0.6	-27.1 ± 2.0
$Cp'Mn(CO)_2(pyr)$[c]	PEt_3	1.18 (1.3)	2.8 ± 0.2	-35.2 ± 0.6
$Cp'Mn(CO)_2(pyr)$[c]	PBu_3	0.85	2.8 ± 0.2	-36.2 ± 0.6

[a] From the Eyring plot of the rate constants obtained by comparison of experimental DCV $\upsilon_{0.5}$ data (E_{switch} - E_{rev} = 200 mV) with digital simulation data. Literature values[12a] in parentheses. [b]Acetonitrile/0.1 M $Bu_4N^+PF_6^-$. [c]Acetone/0.1 M $Bu_4N^+PF_6^-$.

the discontinuity arising from reversal of the double-layer charging current at the switching potential, Figure 5 shows good agreement between experimental (left) and simulated (right) voltammograms. For this experiment, the conditions are different from those used for the kinetic DCV measurements in that the greater sweep width encompasses not only the peaks observed for the substrate electrode processes (peaks a, d), but also those for the product (peaks b, c). The best-fit simulation rate constant (12.2×10^3 $M^{-1}s^{-1}$) agrees well with the results from the DCV data (14.2×10^3 $M^{-1}s^{-1}$ at 18 °C by interpolation). It is particularly satisfying that both of these methods for extracting kinetic information from the DCV data give rate constants that agree well with those that were obtained previously by CV.

Several other electrocatalytic ligand substitution reactions between $Cp'Mn(CO)_2(NCMe)$ or $Cp'Mn(CO)_2(pyr)$ and phoshine or phosphite ligands were also investigated by DCV. The appropriate kinetic parameters for these reactions are summarized in Table II. Excellent agreement between our data and Kochi's is again noted for the reaction of $Cp'Mn(CO)_2(pyr)$ with PEt_3 in acetone. For $Cp'Mn(CO)_2(NCMe)$ and PPh_3 in acetone, for some reason the agreement is somewhat less satisfying. In all cases, the associative nature of these substitution reactions is clearly reflected in the highly negative entropies of activation which are near or within the range that has been commonly observed for associative substitution reactions at 17-electron metal centers.[11]

3.1.2. Oxidatively Induced Ligand Substitution in Cyclopentadienylruthenium Complexes. We have recently reported that oxidation of ruthenium complexes $CpRu(CO)(PR_3)CH_3$[6f] (R = Ph, Cy (cyclohexyl)) and $Cp^*Ru(CO)(PPh_3)CH_3$[6h] (Cp^* = η^5-C_5Me_5) in acetonitrile commences by competing $CH_3^{\cdot}$ and CO displacement by the acetonitrile as the primary reactions. Methane was obtained in near quantitative yields from the first two substrates. The substrate cation radicals underwent clean pseudo-first-order reactions in acetonitrile ($CpRu(CO)(PPh_3)CH_3$ showed second-order behavior at higher concentrations; vide infra).

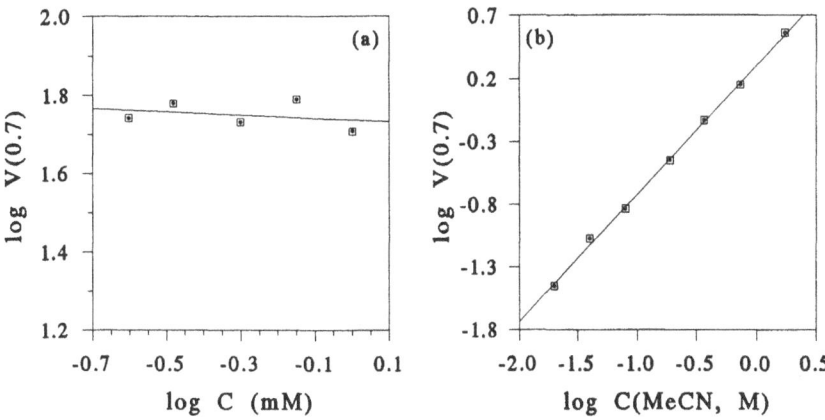

Figure 6. Reaction-order plots for the reaction of $Cp^*Ru(CO)(PPh_3)CH_3\cdot^+$ based on DCV data. (a) Varying substrate concentrations in acetonitrile/0.1 M $Bu_4N^+PF_6^-$ (20 °C, 0.4 mm Pt disk electrode, $E_{switch} - E_{rev}$ = 300 mV). (b) Varying acetonitrile concentrations in dichloromethane/0.2 M $Bu_4N^+PF_6^-$ (1.0 mM substrate, 20 °C, 0.6 mm Pt disk electrode, $E_{switch} - E_{rev}$ = 400 mV).

Figure 6 shows the results from a DCV reaction-order analysis of the reaction of $Cp^*Ru(CO)(PPh_3)CH_3$. Curve (a) depicts a reaction-order plot of log $v_{0.7}$ vs log $C_{substrate}$ in acetonitrile/0.1 M $Bu_4N^+PF_6^-$ at 20 °C with $E_{switch} - E_{rev}$ = 300 mV. The near-zero slope indicates a first-order reaction with respect to the cation radical. In curve (b), the dependence of the rate of reaction on the acetonitrile concentration in dichloromethane containing 0.02 - 1.74 M acetonitrile is shown. The unity slope of log $v_{0.7}$ vs log C_{MeCN} indicates a clear first-order dependence on the acetonitrile concentration.

The kinetic parameters for these reactions, obtained from Arrhenius plots of DCV kinetic data and calculation of the rate constants by comparison with theoretical data, are included in Table III. The negative entropies of activation suggest that associative reactions are involved, in agreement with the first-order dependence on the acetonitrile concentration. This conclusion was further corroborated by the observation of large solvent effects on the reaction rates. Minimum values for the relative rates in acetonitrile vs dichloromethane, $k(MeCN)/k(CH_2Cl_2)$, were 50 $(CpRu(CO)(PCy_3)CH_3)$ and 2000 $(Cp^*Ru(CO)(PPh_3)CH_3)$.

Table III. Kinetic Data for the Reactions of Cyclopentadienylruthenium Methyl Complexes in Acetonitrile[a]

Entry	Compound	$k(20\ °C), s^{-1}$ [b]	$\Delta H^{\#}$, kcal/mol	$\Delta S^{\#}$, cal/(K·mol)
1	$CpRu(CO)(PPh_3)CH_3$	29 ± 2	8.2 ± 0.6	-23 ± 2
2	$CpRu(CO)(PCy_3)CH_3$	0.26 ± 0.02	10.6 ± 0.3	-25 ± 1
3	$Cp^*Ru(CO)(PPh_3)CH_3$	50 ± 3	8.4 ± 0.2	-22 ± 1

[a] DCV measurements in acetonitrile/0.1 M $Bu_4N^+PF_6^-$. For entry 2, 10 vol-% dichloromethane was added as a cosolvent. [b] First-order rate constant for reaction of substrate cation radical.

118

Scheme II

$$CpRu(CO)(PR_3)CH_3{}^{\cdot+}$$
$$(17\ e)$$

$- CH_3{}^{\cdot}$ ⎯⎯ $CpRu(CO)(PR_3)(NCMe)CH_3{}^{\cdot+}$ ⎯⎯ $- CO$
$$(19\ e)$$

$$CpRu(CO)(PR_3)(NCMe)^+ \qquad\qquad CpRu(PR_3)(NCMe)CH_3{}^{\cdot+}$$
$$(18\ e) \qquad\qquad\qquad\qquad\qquad (17\ e)$$

$$\downarrow + MeCN$$

$$CpRu(PR_3)(NCMe)_2CH_3{}^{\cdot+}$$
$$(19\ e)$$

$$\downarrow - CH_3{}^{\cdot}$$

$$CpRu(PR_3)(NCMe)_2{}^+$$
$$(18\ e)$$

These and other data[6f,h] support the conclusion that not only the substitution of one two-electron donor for another, but also that of a two-electron donor for a one-electron donor, can proceed via 19-electron intermediates or transition states. This is indicated in the proposed mechanism in Scheme II.

3.1.3. Reversible Dimerization of Cation Radicals in Complex Mechanisms. Dimerization[1b,13-15] and atom abstraction[16] reactions are typical reactions of 17-electron metal centers that reflect their radical nature. Whereas 17-electron neutral species commonly dimerize via metal-metal bond formation,[13] the cation radicals often undergo coupling at the ligands, presumably due to coulombic repulsive forces between the charged metal centers.[1b,14] However, in some cases cation radicals have provided stable, dicationic metal-metal bonded dimers.[15] The electrostatic repulsion that must exist between the positively charged metal centers clearly may be overcome by strong metal-metal bonding.

The interesting reaction shown in Scheme III was communicated by us a while ago.[6c] The one-electron oxidation of the rhenium complex CpRe(NO)(PPh₃)CH₃ caused a quanti-tative disproportionation of the methyl group, yielding methane and metal-bound methylene.

Scheme III

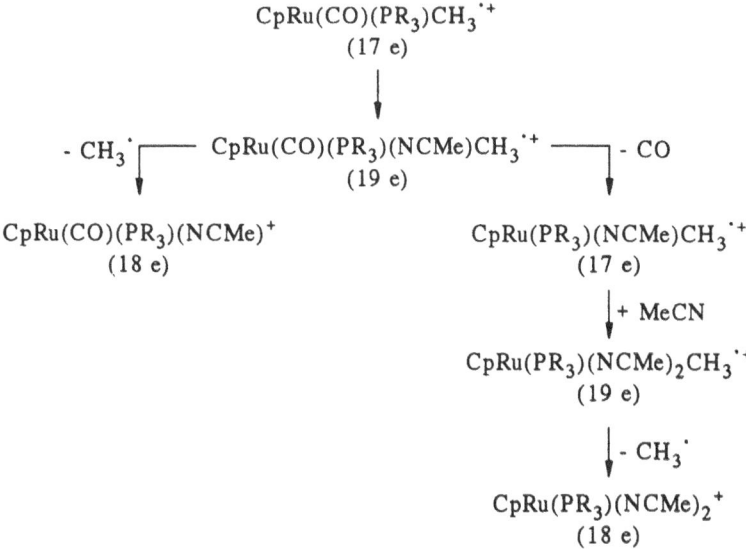

Table IV. Kinetic parameters for the reactions of $CpRe(NO)(PR_3)CH_3 \cdot^+$ in Acetonitrile as Determined by DCV

R =	Ph^a	Me^b
k (0 °C), $M^{-1}s^{-1}$	1130	25000
$\Delta H^{\#}$, kcal/mol	0.1	-1.0
$\Delta S^{\#}$, cal/(K·mol)	-46	-42
$k_H/k_D{}^c$	3.0	2.7
$k_+/k_{rac}{}^d$	1.6	n.a.

[a] From ref 6c. [b] From ref 17. [c] $CpRe(NO)(PR_3)CH_3$ vs $CpRe(NO)(PR_3)CD_3$. [d] Relative rates of enantiomerically pure (+)-(S)-CpRe-(NO)(PPh$_3$)CH$_3$ vs the racemate.

The oxidation of $CpRe(NO)(PR_3)CH_3$ (R = Ph,[6c] Me[17]) was probed in detail by DCV. The reactions of the cation radicals were found to be second-order processes with respect to the cation radicals. Representative kinetic data are listed in Table IV. Near-zero, or even negative, activation enthalpies along with highly negative activation entropies are often associated with multi-step reactions involving a reversible, exothermic dimerization prior to the rate-determining step.[18] In view of the aforementioned precedence for cation radical dimerization reactions, it seems reasonable to invoke a dinuclear, metal-metal bonded intermediate which eliminates methane in a rate-determining step that gives rise to the kinetic k_H/k_D isotope effects (Scheme IV). In the dinuclear intermediate suggested in this scheme, the largest ligands are oriented such as to relieve steric interactions most efficiently, with a ligand arrangement that is reminiscent of that found in the X-ray structure of $\{CpCr(CO)_2(P(OMe)_3)\}_2$.[19] The difference in steric effects between the methyl and nitrosyl ligands may not be significant, and this may allow for reversible formation of the (RR,SS) enantiomeric pair as well as the (RS) meso dimer. The chiral self-recognition effect must then be attributed to preferred elimination of methane from the (RR,SS) pair. This appears reasonable, since the two methyl groups are located on the same side of the Re-Re bond axis, conveniently set up for methane elimination by an unknown mechanism. In the (R,S) dimer, the methyl groups will be located on opposite sides of the Re-Re bond axis.

Scheme IV

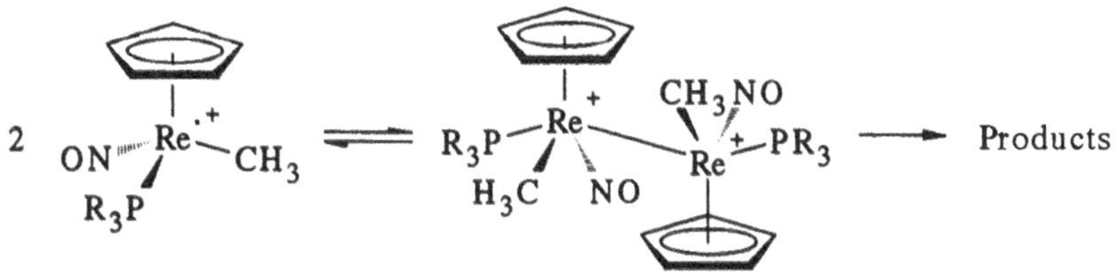

The ruthenium complexes $CpRu(CO)(PR_3)CH_3$ are isoelectronic with the methylrhenium systems discussed above. Ru has proved to have a rich cluster chemistry, indicative of good metal-metal bonding properties, and thence we suspected that some of these complexes might also exhibit second-order reactions subject to their oxidation. The DCV reaction-order analysis indeed proved this to be true for $CpRu(CO)(PMe_3)CH_3$, and for $CpRu(CO)(PPh_3)CH_3$ at substrate concentrations higher than ca. 1 mM. The overall reaction, observed by ^1H NMR spectroscopy at -40 °C, is depicted in Scheme V and involves the migration of a methyl group from one metal center to another.

Scheme V

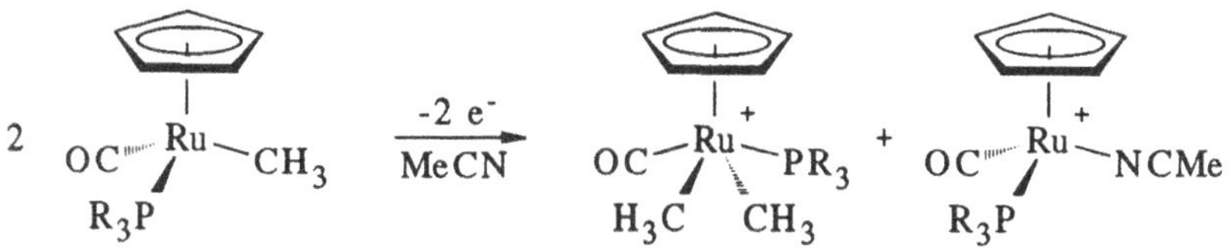

Kinetic parameters for the reactions, obtained by DCV, are shown in Table V. Again, near-zero, slightly negative activation enthalpies and highly negative activation entropies hint at the likely intervention of reversible dimerizations. Double labelling experiments involving $(\eta^5\text{-}C_5H_5)Ru(CO)(PPh_3)CH_3$ and $(\eta^5\text{-}C_5D_5)Ru(CO)(PPh_3)CD_3$ with incomplete substrate oxidation and analysis of unconsumed substrate established that the methyl transfer process was irreversible, so the methyl transfer presumably takes place within the reversibly formed dimer. The detailed DCV study certainly has unravelled mechanistic details that could not be as readily obtained by other methods.

Table V. Kinetic parameters for the reactions of $CpRu(CO)(PR_3)CH_3$·$^+$ in Acetonitrile as Determined by DCV

R =	Ph[a]	Me[b]
k (20 °C), $M^{-1}s^{-1}$	120000	2400000
$\Delta H^{\#}$, kcal/mol	-0.7	-2.6
$\Delta S^{\#}$, cal/(K·mol)	-38	-40
$k_H/k_D{}^c$	0.87	n.a.

[a] From ref 6f. [b] Unpublished results. [c] CpRu-$(CO)(PPh_3)CH_3$ vs $CpRu(CO)(PPh_3)CD_3$.

3.2   APPLICATIONS OF DCV IN THERMOCHEMICAL CYCLES

Following the pioneering work of Breslow and coworkers,[20] there has been a recent surge in the use of thermochemical cycles that incorporate electrode potentials as a means for obtaining estimates of thermodynamic quantities that are difficult to obtain directly. Such methodology has been used to extract bond dissociation energy (BDE) and thermodynamic acidity (pK_a) data for organic molecules, radicals, ions, and ion radicals.[21-25] Recently, we proposed to use such cycles to estimate metal-hydride BDE[23d,26] and metal-hydride cation radical pK_a values.[6e,g] In this section, some of these extremely useful thermochemical cycles, adapted for use in organometallic systems, will be presented, along with pertinent data related to the properties of metal hydrides and metal hydride cation radicals.

3.2.1. Metal-Hydride Bond Dissociation Energies and pK_a Data from Electrode Potentials.
The great influence that metal-ligand bond dissociation energies exert on organometallic reactivity has long been recognized.[27] Tabulations of metal-ligand BDEs and related data, as well as a discussion of the numerous methods that are available for their determination, have recently been presented in a comprehensive review.[27f]

Metal hydrides constitute a large class of compounds that is of fundamental importance in organometallic chemistry.[28] Metal hydrides have been recognized as crucial components in a variety of catalytic and stoichiometric organometallic reactions. The thermochemical properties of metal hydrides have therefore received considerable attention. A fair body of M-H pK_a data, important when considering heterolytic cleavage of metal-hydrogen bonds, exists in the literature.[29] Homolytic cleavage of M-H bonds is also commonly observed,[30] but in spite of this, homolytic bond dissociation energy data have been more scarce. This is in part due to the lack of generally suitable methods for their determination. Often, calorimetric methodology has been applied, but in many instances M-H BDE values depend on prior knowledge about sometimes controversial metal-metal BDE estimates.[26]

The thermochemical cycle shown in Scheme VI, based on a method devised for the investigation of organic systems by Arnold[21a] and later by Bordwell[22b,h] has recently been proposed by us for estimating metal hydride BDEs.[23d,26] The cycle requires knowledge about the reversible oxidation potential (E^{o}_{ox}(M$^-$,sol)) for the metal hydride conjugate base (anion) as well as the metal hydride pK_a for determination of the metal hydride BDE. As is apparent from eq (11), the application of the cycle also requires that E^{o}_{ox}(M$^-$,sol) is referred to the hydrogen electrode in the solvent of choice, and that the free energy of solvation of the hydrogen atom, ΔG^{o}_{sol}(H$^{\cdot}$, sol) is known. The latter is generally not available but has been assumed to equal the free energy of solvation of the hydrogen molecule. Furthermore, for practical purposes the hydrogen electrode is not used and the use of a different reference

Scheme VI

$$\Delta G^{\circ}$$

M-H(sol)	$\rightleftharpoons$	M$^-$(sol) + H$^+$(sol)	$2.301RTpK_a$	(5)
M$^-$(sol)	$\rightleftharpoons$	M$^{\cdot}$(sol) + e$^-$	FE°_{ox}(M$^-$,sol)	(6)
H$^+$(sol)	$\rightleftharpoons$	0.5 H$_2$(g)	-FE°(H$^+$/H$_2$,sol)	(7)
0.5 H$_2$(g)	$\rightleftharpoons$	H$^{\cdot}$(g)	ΔG°_f(H$^{\cdot}$,g)	(8)
H$^{\cdot}$(g)	$\rightleftharpoons$	H$^{\cdot}$(sol)	ΔG°_{sol}(H$^{\cdot}$,sol)	(9)
M-H(sol)	$\rightleftharpoons$	M$^{\cdot}$(sol) + H$^{\cdot}$(sol)	BDE	(10)

$$BDE = 2.301RTpK_a + F[E^{\circ}_{ox}(M^-,sol) - E^{\circ}(H^+/H_2,sol)] + \Delta G^{\circ}_f(H^{\cdot},g) + \Delta G^{\circ}_{sol}(H^{\cdot},sol) \quad (11)$$

$$BDE = 2.301RTpK_a + FE^{\circ}_{ox}(M^-, sol) + C \quad (12)$$

introduces an electrode potential conversion term, the magnitude of which depends on the chosen reference, into eq (11). Finally, eq (11) provides *free energy* BDEs rather than the more commonly used enthalpy-based BDEs which are normally obtained by calorimetry. Conversion of the free energy BDE to the enthalpy based value requires addition of $T\Delta S$ for the homolysis, approximated[24c] as the entropy of solvation of the hydrogen atom which again has been commonly taken to equal the entropy of solvation of the hydrogen molecule. All of these terms (H$^{\cdot}$ solvation, reference electrode correction, and entropy compensation) may be conveniently combined into one constant C as indicated in eq (12).

We have suggested that the ferrocene/ferricinium (Fc) couple should be used as the reference system. Ferrocene may be used as an internal or external (measurements on a separate solution) reference, and its use precludes the uncertainties and errors that frequently appear to arise from problems associated with the conversion of redox potentials from one reference electrode system to another. By the use of "best literature values" for the contributions to the constant (see ref 23d for details), we have arrived at a calculated value of 59.4 for the constant C when potentials are measured in acetonitrile at 25 °C and referred to the ferrocene/ferricinium couple.

Hoff and coworkers[31a] recently presented the first direct, calorimetric measurements of the M-H BDEs in a series of chromium carbonyl hydrides. The beauty of their experiments is that the monomeric nature of the corresponding chromium radicals in solution eliminated the need for prior knowledge about the related M-M bond strengths. The fact that these radicals do not dimerize in solution also permits the direct measurement of their *reversible* reduction potentials[31b] with no need to apply any kinetic potential shift corrections. Norton and coworkers have estimated the pK_a of CpCr(CO)$_3$H to be 13.3 in acetonitrile.[29a] Using their method, we recently estimated the pK_a of CpCr(CO)$_2$(PPh$_3$)H to be 21.8.[23d] These two compounds now represent two exceptional cases for which reliable BDE data, pK_a data, and

Table VI. Anion Oxidation Potentials, Acid Dissociation Constants, and M-H BDE of Metal Hydrides in Acetonitrile

Entry	Metal hydride	$E_{ox}(M^-)^a$	$pK_a(MH)^b$	BDE $(MH)^c$
1	CpCr(CO)$_3$H	-0.688^d	13.3	61.5^e
2	CpCr(CO)$_2$(PPh$_3$)H	$-1.289^{d,f}$	21.8^f	59.8^e
3	CpMo(CO)$_3$H	-0.501	13.9	69.2
4	Cp*Mo(CO)$_3$H	-0.709	17.1	68.5
5	CpW(CO)$_3$H	-0.491	16.1	72.3
6	CpW(CO)$_2$(PMe$_3$)H	-1.225	26.6	69.6
7	CpW(CO)$_2$(PMe$_3$)H$_2^+$	$+0.16^{g,h}$	$> 9^h$	> 75.5
8	Mn(CO)$_5$H	-0.555	14.1	68.0
9	Mn(CO)$_4$(PPh$_3$)H	-0.870	20.4	68.4
10	Re(CO)$_5$H	-0.690	21.1	74.7
11	CpFe(CO)$_2$H	-1.352	19.4	57.1
12	CpRu(CO)$_2$H	-1.057	20.2	64.9
13	Fe(CO)$_4$H$_2$	-0.403	11.4	67.6
14	Co(CO)$_4$H	-0.271	8.3	66.4
15	Co(CO)$_3$(PPh$_3$)H	-0.723	15.4	65.0
16	Co(CO)$_3$(P(OPh)$_3$)H	-0.489	11.3	65.2

[a] Electrode potentials vs Fc, from ref 23d unless otherwise noted, measured by DCV at 1.0 V/s. Not corrected for kinetic potential shifts. [b] pK_a data from Norton and coworkers[29a,b] unless otherwise noted. [c] Calculated from eq (12) using $C = 59.5$ kcal/mol, from ref 23d unless otherwise noted. [d] Reversible electrode potentials. [e] From ref 31. [f] From ref 23d. [g] Measured at 100 V/s; partial chemical reversibility. [h] From ref 6e.

reversible conjugate base oxidation potentials are available. The two systems allow for the independent experimental determination of the constant C in eq (12), yielding values of 59.2 and 59.8 kcal/mol, respectively. These are in excellent agreement with the value arrived at by the alternative method discussed in the previous paragraph. We suggest[23d] that the average value, 59.5 kcal/mol, should be used for C in eq (12) when electrode potentials are referred to the ferrocene/ferricinium couple in acetonitrile at 25 °C.

Now, eq (12) may be applied in order to estimate the values of BDE(M-H), provided that pK_a(M-H) and $E^o_{ox}(M^-)$ are available, and of pK_a(M-H), provided that $E^o_{ox}(M^-)$ and BDE(M-H) are available. In Table VI, a compilation of $E^o_{ox}(M^-)$, pK_a(M-H), and BDE(M-H) data for a wide range of metal hydrides is presented. Entries 1 and 2 form the basis for the choice of C equal to 59.5 kcal/mol. The rest provide estimates of BDE values based on pK_a(M-H) and $E^o_{ox}(M^-)$ measurements. Before the application of eq (12), kinetic potential shift corrections were applied to the irreversible $E^o_{ox}(M^-)$ potentials as described elsewhere.[23d]

For many of the entries in Table VI, the BDE values agree well with data obtained by other methods. However, it has been pointed out that for several entries there is a discrepancy between our data and literature values.[29e,30b] For example, for $Co(CO)_4H$, $Mn(CO)_5H$, $CpMo(CO)_3H$, and $CpW(CO)_3H$, the cycle yields BDE values that are consistently 3-4 kcal/mol higher than upper limits for BDEs based on kinetic measurements. Possible sources for these discrepancies has been discussed[29e,30b] but the question still remains unresolved.

3.2.2. Metal Hydride Cation Radical Acidity Data from Electrode Potentials. The thermochemical cycle presented in Scheme VII, which involves the measurement of oxidation potentials for the metal hydride and its conjugate base (anion) may be used to calculate the pK_a values of metal hydride cation radicals in solution.[6e] The cation radical acidity is estimated from eq (17) provided that pK_a data for the neutral precursors are available. In those instances where pK_a data are not available, *differences* in acid strengths between neutral and cationic hydrides may still be calculated from the two electrode potentials.

Table VII lists electrode potential data, obtained by DCV, for the oxidation of several metal anions and metal hydrides along with the relevant cation radical acidities as determined by the use of eq (17). Irreversible anion oxidation potentials (entries 3-7) have been corrected for the kinetic potential shifts due to radical dimerization as discussed elsewhere.[23d] All metal hydride oxidation waves were chemically irreversible due to the rapid deprotonation of the cation radicals. The rates of the deprotonation reactions are not known and for the sake of internal consistency no corrections have been made for the kinetic potential shifts. This introduces an uncertainty of unknown magnitude into the $pK_a(M-H^{\bullet+})$ data that are reported. However, the errors will at least in part cancel when comparisons are to be made between different metal hydrides.

Table VII demonstrates that the metal hydrides undergo significant activation with respect to heterolytic M-H bond cleavage resulting from their oxidation, corresponding to 19-24 pK_a units or about 26-33 kcal/mol. This is in agreement with recent reports that metal hydride cation radicals may act as strong Brønsted acids.[6e,g,j,k,32] For example, the cation radicals of entries 1, 3, 4, and 5 in Table VII undergoes spontaneous deprotonation to the

Scheme VII

			ΔG^o	
M-H	$\rightleftharpoons$	$M^- + H^+$	$2.301RT\,pK_a(M-H)$	(13)
M^-	$\rightleftharpoons$	$M^{\bullet} + e^-$	$FE^o_{ox}(M^-)$	(14)
$M-H^{\bullet+} + e^-$	$\rightleftharpoons$	M-H	$-FE^o_{ox}(M-H)$	(15)
$M-H^{\bullet+}$	$\rightleftharpoons$	$M^{\bullet} + H^+$	$2.301RT\,pK_a(M-H^{\bullet+})$	(16)

$$pK_a(M-H^{\bullet+}) = pK_a(M-H) + F/2.301RT\,[E^o_{ox}(M^-) - E^o_{ox}(M-H)] \qquad (17)$$

Table VII. Metal Anion and Hydride Oxidation Potentials, Metal Hydride Neutral and Cation Radical Acidities in Acetonitrile at 25 °C

Entry	Metal hydride	$E_{ox}(M^-)^a$	$E_{ox}(MH)^b$	$pK_a(MH)^c$	$pK_a(MH^{\cdot+})^{d,e}$	ΔpK_a
1	$CpCr(CO)_3H$	-0.688^f	0.668	13.3	-9.5	22.8
2	$CpCr(CO)_2(PPh_3)H$	$-1.289^{f,g}$	0.123^h	21.8^g	-2.1^h	23.9
3	$CpMo(CO)_3H$	-0.501	0.800	13.9	-6.0	19.9
4	$Cp^*Mo(CO)_3H$	-0.709	0.561	17.1	-2.5	19.6
5	$CpW(CO)_3H$	-0.491	0.758	16.1	-3.0	19.1
6	$CpW(CO)_2(PMe_3)H$	-1.225	0.195	26.6	+5.1	21.5
7	$CpRu(CO)(PPh_3)H$	-1.05^i	0.39^i	$27-28^i$	$6-7^i$	23

^{a} Electrode potentials vs Fc, from ref 23d unless otherwise noted. Not corrected for kinetic potential shifts. ^{b} Electrode potentials vs Fc, from ref 6e unless otherwise noted. Measured by DCV at 1.0 V/s. ^{c} pK_a data from Norton and coworkers[29a,b] unless otherwise noted. ^{d} From ref 6e unless otherwise noted. ^{e} Calculated from eq 17. ^{f} Reversible potentials. ^{g} From ref 23d. ^{h} This work. ^{i} From ref 6g.

medium (acetonitrile and residual water) subject to their oxidation.[6e] On the other hand, the less acidic cations of entries 6 and 7 transfer their proton to the neutral parent hydride, yielding the cationic dihydride[6e] $CpW(CO)_2(PMe_3)H_2^+$ and dihydrogen[6g] $CpRu(CO)(PPh_3)(\eta^2-H_2)^+$ complexes, respectively. Similar reactions of $CpRu(CO)-(PMe_3)H^{6j}$ and $CpRu(PPh_3)_2H,^{6k}$ for which pK_a data are not available, have also been interpreted in terms of proton transfer from metal hydride cation radical to neutral metal hydride. This behavior is in agreement with the significant activation of the M-H bond towards heterolytic cleavage that takes place following the oxidation.

4   Acknowledgment

We gratefully acknowledge support from Statoil under the VISTA program, administered by the Norwegian Academy of Science and Letters, and from the Norwegian Council for Science and the Humanities, NAVF. Professor Jack Norton is kindly thanked for providing preprints of references 29e and 30b.

5   References

(1)   (a) *Organometallic Radical Processes*; Trogler, W. C., Ed.; Elsevier: Amsterdam, 1990.   (b) Connelly, N. G. *Chem. Soc. Rev.* **1989**, *18*, 153.   (c) *Paramagnetic Organometallic Species in Activation/Selectivity, Catalysis*; Chanon, M., Julliard, M., Poite, J. C., Eds.; Kluwer Academic: Dordrecht, 1989. (d) Astruc, D. *Angew. Chem., Int. Ed. Engl.* **1988**, 27, 643. (e) Astruc, D. *Chem. Rev.* **1988**, 88, 1189. (f) Baird, M.

126

C. *Chem. Rev.* **1988**, *88*, 1217. (f) Kochi, J. K. *Organometallic Mechanisms and Catalysis*; Academic: New York, 1978.

(2)   (a) Tolman, C. A. *Chem. Soc. Rev.* **1972**, *1*, 337. (b) Collman, J. P.; Hegedus, L. S.; Norton, J. R.; Finke, R. G. *Principles and Applications of Organotransition Metal Chemistry*; University Science: Mill Valley, CA, 1987.

(3)   Bard, A. J.; Faulkner, L. R. *Electrochemical Methods: Fundamentals and Applications*; Wiley: New York, 1980.

(4)   (a) Ahlberg, E.; Parker, V. D. *J. Electroanal. Chem. Interfacial Electrochem.* **1981**, *121*, 73. (b) Parker, V. D. *Electroanal. Chem.* **1986**, *14*, 1.

(5)   See for example (a) Parker, V. D.; Tilset, M. *J. Am. Chem. Soc.* **1986**, *108*, 6371. (b) Reitstöen, B.; Norrsell, F.; Parker, V. D. *J. Am. Chem. Soc.* **1989**, *111*, 8463. (c) Parker, V. D.; Reitstöen, B.; Tilset, M. *J. Phys. Org. Chem.* **1989**, 2, 580. (d) Reitstöen, B.; Parker, V.D. *J. Am. Chem. Soc.* **1990**, *112*, 4968. (e) Parker, V. D.; Chao, Y.; Reitstöen, B. *J. Am. Chem. Soc.* **1991**, *113*, 2336. (f) Reitstöen, B.; Parker, V. D. *J. Am. Chem. Soc.* **1991**, *113*, 6954.

(6)   (a) Bodner, G. S.; Gladysz, J. A.; Nielsen, M. F.; Parker, V. D. *Organometallics* **1987**, 6, 1628. (b) Bodner, G. S.; Gladysz, J. A.; Nielsen, M. F.; Parker, V. D. *J. Am. Chem. Soc.* **1987**, *109*, 1757. (c) Tilset, M.; Bodner, G. S.; Senn, D. R.; Gladysz, J. A.; Parker, V. D. *J. Am. Chem. Soc.* **1987**, *109*, 7551. (d) Aase, T.; Parker, V. D.; Tilset, M. *Organometallics* **1989**, 8, 1558. (e) Ryan, O. B.; Tilset, M.; Parker, V. D. *J. Am. Chem. Soc.* **1990**, *112*, 2618. (f) Aase, T.; Tilset, M.; Parker, V. D. *J. Am. Chem. Soc.* **1990**, *112*, 4975. (g) Ryan, O. B.; Tilset, M.; Parker, V. D. *Organometallics* **1991**, *10*, 298. (h) Tilset, M.; Aase, T. In *Natural Gas Conversion*; Holmen, A., Eds.; Elsevier: Amsterdam, 1991; p. 197. (i) Skagestad, V.; Tilset, M. *Organometallics* **1991**, *10*, 2110. (j) Ryan, O. B.; Tilset, M. *J. Am. Chem. Soc.*, in print. (k) Ryan, O. B.; Smith, K.-T.; Tilset, M. *J. Organomet. Chem.*, in print.

(7)   (a) Ahlberg, E.; Parker, V. D. *J. Electroanal. Chem., Interfacial Electrochem.* **1981**, *121*, 57. (b) Ahlberg, E.; Parker, V. D. *Acta Chem. Scand., Ser. B* **1980**, *B34*, 97.

(8)   (a) Digital simulations were carried out as described by Feldberg.[8b] (b) Feldberg, S. W. *Electroanal. Chem.* **1969**, *3*, 199.

(9)   *Microelectrodes: Theory and Applications*; Montenegro, M. I., Queirós, M. A., Daschbach, J. L., Eds.; Kluwer Academic: Dordrecht, 1991.

(10)   Parker, V. D. *Acta Chem. Scand., Ser. B* **1984**, *B38*, 165.

(11)   For a recent review with a comprehensive list of pertinent references, see: Trogler, W. C., ref. 1a, p. 306.

(12)   For some leading references, see (a) Hershberger, J. W.; Klingler, R. J.; Kochi, J. K. *J. Am. Chem. Soc.* **1983**, *105*, 61. (b) Tyler, D. R., ref. 1a, p. 338. (c) Hepp, A. F.; Wrighton, M. S. *J. Am. Chem. Soc.* **1983**, *105*, 5935. (d) Doxsee, K. M.; Grubbs, R. H.; Anson, F. C. *J. Am. Chem. Soc.* **1984**, *106*, 7819. (e) Therien, M. J.; Trogler, W. C. *J. Am. Chem. Soc.* **1988**, *110*, 4942.

(13)   Meyer, T. J.; Caspar, J. V. *Chem. Rev.* **1985**, *85*, 187 and references cited.

(14)   (a) Geiger, W. E.; Gennett, T.; Lane, G. A.; Salzer, A.; Rheingold, A. L. *Organometallics* **1986**, *5*, 1352. (b) Iyer, R. S.; Selegue, J. P. *J. Am. Chem. Soc.* **1987**, *109*, 910. (c) Beddoes, R. L.; Bitcon, C.; Ricalton, A.; Whiteley, M. W. *J. Organomet. Chem.* **1989**, *367*, C21.

(15)   (a) DeLaet, D. L.; Powell, D. R.; Kubiak, C. P. *Organometallics* **1985**, *4*, 954. (b) Droege, M. W.; Harman, W. D.; Taube, H. *Inorg. Chem.* **1987**, *26*, 1309. (c) Fonseca, E.; Geiger, W. E.; Bitterwolf, T. E.; Rheingold, A. L. *Organometallics* **1988**, *7*, 567. (d) Pufahl, D.; Geiger, W. E.; Connelly, N. G. *Organometallics* **1989**, *8*, 412. (e) Moinet, C.; Le Bozec, H.; Dixneuf, P. H. *Organometallics* **1989**, *8*, 1493. (f) Bond, A. M.; Calton, R.; Mann, D. R. *Inorg. Chem.* **1989**, *28*, 54. (g) Einstein, F. W. B.; Jones, R. H.; Zhang, X.; Yan, X.; Nagelkerke, R.; Sutton, D. *J. Chem. Soc., Chem. Commun.* **1989**, 1424. (h) Herrmann, W. A.; Fischer, R. A.; Amslinger, W.; Herdtweck, E. *J. Organomet. Chem.* **1989**, *362*, 333. (i) Bitterwolf, T. E.; Spink, W. C.; Rausch, M. D. *J. Organomet. Chem.* **1989**, *363*, 189.

(16)   (a) Brown, T. L., ref. 1a, p. 67. (b) Brown, T. L.; Sullivan, R. J., ref. 1c, p. 187. (c) Goulin, C. A.; Huber, T. A.; Nelson, J. M.; Macartney, D. H.; Baird, M. C.*J. Chem. Soc., Chem. Commun.* **1991**, 798.

(17)   Tilset, M.; Bodner, G. S.; Senn, D. R.; Gladysz, J. A.; Parker, V. D. Unpublished results.

(18)   See ref 5a and references cited therein.

(19)   Goh, L.-Y.; D'Aniello, M. J.; Slater, S.; Muetterties, E. L.; Tavanaiepour, I.; Chang, M. I.; Fredrich, M. F.; Day, V. W. *Inorg. Chem.* **1979**, *18*, 192.

(20)   (a) Breslow, R.; Balasubramanian, K. *J. Am. Chem. Soc.* **1969**, *91*, 5182. (b) Breslow, R.; Chu, W. *J. Am. Chem. Soc.* **1970**, *92*, 2165. (c) Breslow, R.; Chu, W. *J. Am. Chem. Soc.* **1973**, *95*, 411. (d) Breslow, R.; Mazur, S. *J. Am. Chem. Soc.* **1973**, *95*, 584. (e) Wasielewski, M. R.; Breslow, R. *J. Am. Chem. Soc.* **1976**, *98*, 4222. (f) Breslow, R.; Goodin, R. *J. Am. Chem. Soc.* **1976**, *98*, 6076. (g) Breslow, R.; Grant, J. L. *J. Am. Chem. Soc.* **1977**, *99*, 7745. (h) Jaun, B.; Schwarz, J.; Breslow, R. *J. Am. Chem. Soc.* **1980**, *102*, 5741.

(21)   (a) Nicholas, A. M. de P.; Arnold, D. R. *Can. J. Chem.* **1982**, *60*, 2165. (b) Okamoto, A.; Snow, M. S.; Arnold, D. R. *Tetrahedron* **1986**, *22*, 6175. (c) Popielarz, R.; Arnold, D. R. *J. Am. Chem. Soc.* **1990**, *112*, 3068.

(22)   (a) Bordwell, F. G.; Bausch, M. *J. Am. Chem. Soc.* **1986**, *108*, 2473. (b) Bordwell, F. G.; Cheng, J.-P.; Harrelson, J. A. *J. Am. Chem. Soc.* **1988**, *110*, 1229. (c) Bordwell, F. G.; Cheng, J.-P.; Bausch, M. J. *J. Am. Chem. Soc.* **1988**, *110*, 2867. (d) Bordwell, F. G.; Cheng, J.-P.; Bausch, M. J. *J. Am. Chem. Soc.* **1988**, *110*, 2872. (e) Bordwell, F. G.; Cheng, J.-P.; Bausch, M. J.; Bares, J. E. *J. Phys. Org. Chem.* **1988**, *1*, 209. (f) Bordwell, F. G.; Bausch, M. J.; Branca, J. C.; Harrelson, J. A. *J. Phys. Org. Chem.* **1988**, *1*, 225. (g) Bordwell, F. G.; Cheng, J.-P. *J. Am. Chem. Soc.* **1989**, *111*, 1792. (h) Bordwell, F. G.; Harrelson, J. A.; Satish, A.V. *J. Org. Chem.* **1989**, *54*, 3101. (i) Bordwell, F. G.; Cheng, J.-P. *J. Am. Chem. Soc.* **1991**, *113*, 1736.

(23)   (a) Parker, V. D.; Tilset, M.; Hammerich, O. *J. Am. Chem. Soc.* **1987**, *109*, 7905. (b) Parker, V. D.; Tilset, M. *J. Am. Chem. Soc.* **1988**, *110*, 1649. (c) Parker, V. D.; Chao, Y.; Reitstöen, B. *J. Am. Chem. Soc.* **1991**, *113*, 2336. (d) Parker, V. D.; Handoo, K.; Roness, F.; Tilset, M. *J. Am. Chem. Soc.* **1991**, *113*, 7493.

(24)   (a) Wayner, D. D. M.; McPhee, D. J.; Griller, D. *J. Am. Chem. Soc.* **1988**, *110*, 132. (b) Sim, B. A.; Griller, D.; Wayner, D. D. M. *J. Am. Chem. Soc.* **1989**, *111*, 754. (c) Kanabus-Kaminska, J. M.; Gilbert, B. C.; Griller, D. *J. Am. Chem. Soc.* **1989**, *111*, 3311. (d) Griller, D.; Simões, J. A. M.; Mulder, P.; Sim, B. A.; Wayner, D. D. M. *J. Am. Chem. Soc.* **1989**, *111*, 7872. (e) Griller, D.; Wayner, D. D. M. *Pure Appl. Chem.* **1989**, *61*, 717.

(25)   (a) Arnett, E. M.; Harvey, N. G.; Amarnath, K.; Cheng, J.-P. *J. Am. Chem. Soc.* **1989**, *111*, 4143. (b) Arnett, E. M.; Amarnath, K.; Harvey, N. G.; Venimadhavan, S. *J. Am. Chem. Soc.* **1990**, *112*, 7346.

(26)   Tilset, M.; Parker, V. D. *J. Am. Chem. Soc.* **1989**, *111*, 6711; **1990**, *112*, 2843.

(27) (a) Skinner, H. A. *Adv. Organomet. Chem.* **1964**, *2*, 49. (b) Connor, J. A. *Top. Curr. Chem.* **1977**, *71*, 71. (c) Halpern, J. *Acc. Chem. Res.* **1982**, *15*, 238. (d) Skinner, H. A.; Connor, J. A. *Pure Appl. Chem.* **1985**, *57*, 79. (e) Marks, T. J.; Gagne, M. R.; Nolan, S. P.; Schock, L. E.; Seyam, A. M.; Stern, D. *Pure Appl. Chem.* **1989**, *61*, 1665. (f) Simões, J. A. M.; Beauchamp, J. L. *Chem. Rev.* **1990**, *90*, 629. (g) *Bonding Energetics in Organometallic Compounds*; Marks, T. J. Ed.; American Chemical Society Symposium Series, Vol. 428; Washington DC, 1990.

(28) (a) Moore, D. S.; Robinson, S. D. *Chem. Soc. Rev.* **1983**, *12*, 415. (b) Hlatky, G. G.; Crabtree, R. H. *Coord. Chem. Rev.* **1985**, *65*, 1. (c) Deutsch, P. P.; Eisenberg, R. *Chem. Rev.* **1988**, *88*, 1147. (d) Carneiro, T. M. G.; Matt, D.; Braunstein, P. *Coord. Chem. Rev.* **1989**, *96*, 49. (e) *Transition Metal Hydrides: Recent Advances in Theory and Experiment*; Dedieu, A., Ed.; VCH Publishers, New York, 1991.

(29) (a) Jordan, R. F.; Norton, J. R. *J. Am. Chem. Soc.* **1982**, *104*, 1255. (b) Moore, E.- J.; Sullivan, J. M.; Norton, J. R. *J. Am. Chem. Soc.* **1986**, *108*, 2257. (c) Kristjánsdóttir, S. S.; Moody, A. E.; Weberg, R. T.; Norton, J. R. *Organometallics* **1988**, *7*, 1983. (d) Weberg, R. T.; Norton, J. R. *J. Am. Chem. Soc.* **1990**, *112*, 1105. (e) Kristjánsdóttir, S. S.; Norton, J. R. In ref 28e.

(30) (a) For a recent review of hydrogen atom transfer reactions, see ref 30b. (b) Eisenberg, D. C.; Norton, J. R. *Isr. J. Chem.*, in print. (c) Eisenberg, D. C.; Lawrie, J. C.; Moody, A. E.; Norton, J. R. *J. Am. Chem. Soc.* **1991**, *113*, 4888.

(31) (a) Kiss, G.; Zhang, K.; Mukerjee, S. L.; Hoff, C. D. *J. Am. Chem. Soc.* **1990**, *112*, 5657. (b) O'Callaghan, K. A. E.; Brown, S. J.; Page, J. A.; Baird, M. C.; Richards, T. C.; Geiger, W. E. *Organometallics* **1991**, *10*, 3119.

(32) Westerberg, D. E.; Rhodes, L. F.; Edwin, J.; Geiger, W. E.; Caulton, K. G. *Inorg. Chem.* **1991**, *30*, 1107.

The Use of Calorimetric and Sublimation techniques to study Bond Properties in Organometallic Compounds.

A. S. Carson
School of Chemistry
The University of Leeds
Leeds LS2 9JT U.K.

Abstract
Rotating, aneroid combustion calorimetry combined with mass-loss, torsion Knudsen effusion is used to study the energetics of some organic compounds of Hg, Sn and Ge. Structural effects in some Cr, Mo and Cu compounds are also considered.

Introduction

Most of the information available on the strengths of bonds between metals and ligands has come from calorimetric measurements of the enthalpy changes which accompany reactions of the compounds concerned. These measurements yield the enthalpy of formation of the compound in the state at which it is stable at room temperature; this is usually solid or liquid and additional information on the enthalpies of sublimation or vaporization is required in order to transfer the enthalpy of formation to the gas phase. This eliminates the effects of the cohesive forces between molecules, which, in general, have no connections with the bonds at all.

Bond enthalpies have been used in two quite different ways and these must be carefully defined if confusion is to be avoided.
(1) Bond dissociation enthalpy (D). This is the enthalpy change accompanying the breaking of a particular bond in a molecule while the remainder of the molecule rearranges itself to make the best use of the available electrons. That is, D contains the reorganization energy of the radical or radicals produced by the dissociation.

J. A. Martinho Simões (ed.), Energetics of Organometallic Species, 131–158.
© 1992 *Kluwer Academic Publishers.*

To consider a simple example

$$CH_4 \quad \text{------>} \quad CH_3 + H$$
(tetrahedral)   (planar)

Planar CH_3 is stabilized by about 165 kJ mol-1, due to the increased s character of the bonds, compared to a CH_3, which remains in the tetrahedral configuration. To calculate the dissociation enthalpy, $D(CH_3–H)$, we require not only the standard enthalpy of formation of the gaseous molecule but also those of the fragments produced when the bond is broken.

$$D(CH_3 - H) = \Delta_f H^\circ(CH_3) + \Delta_f H^\circ(H) - \Delta_f H^\circ(CH_4)$$

(2) A bond enthalpy (E), may be defined as the mean contribution from each bond of a given type to the standard enthalpy of formation of the gaseous molecule from its constituent atoms. A quantity of energy equal to the sum of all the bond enthalpy terms present, would be sufficient to expand the molecule to infinite size while at the same time, maintaining its original shape. For example, the $C - H$ bond enthalpy term in CH_4, $E(C - H)$ is equal to the standard enthalpy of formation of the gaseous molecule from its constituent atoms, divided by four. If an imaginary reaction is visualised in which a quantity of energy, equal to the bond enthalpy term, is supplied to a CH_4 molecule, the result would be an H atom and a CH_3 radical in the tetrahedral, not the planar configuration.

Two things are clear from these definitions, E is not equal to D and the only connection between E and D is that the sum of the E's will equal the sum of the D's for the stepwise dissociation of the molecule into atoms. Also, the value of E is unambiguously defined only when one type of bond is present. E, like the dipole moment, the force constant and the normal modes of vibration of a bond, is associated with the undisturbed ground state of the molecule, whereas D depends on the state of the fragments produced at infinite separation so that only small changes in E may be accompanied by large variations in D.

The standard enthalpy of formation from atoms contains (1) the chemical binding energy, (2) the zero point energy, (3) the thermal energy, (4) steric effects and (5) the delocalization of the bonding electrons. The enthalpy of formation of the gaseous compound from its constituent atoms is equal to the sum of the bond enthalpies plus any stabilizing energy due to

Bond enthalpy relationships in the compound C_xH_yM (R/M)

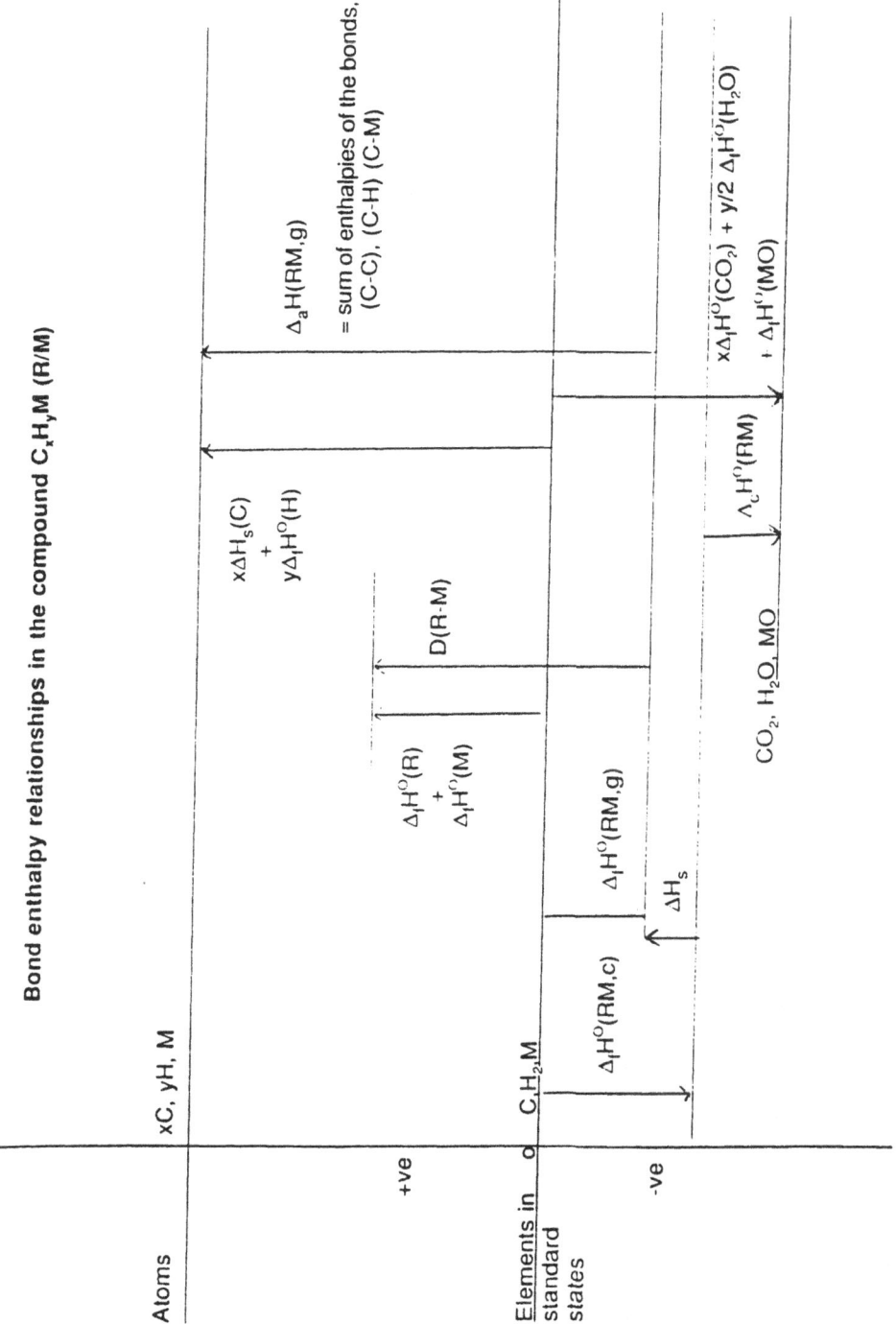

delocalization (5), less the strain energy due to steric effects (4). Because of (4) and (5) it is not possible to attach to a bond a single, transferable enthalpy term which is independent of the molecular environment. Bond enthalpy schemes[1] contain terms which allow for interactions of various types; often the corrections are entirely empirical, others are semi-empirical in that they were guided by theoretical calculations and predictions.

There is an apparent anomaly between the values usually quoted for $D(CH_3 - H)$ of 435 kJ mol^{-1} and $E(C - H)_{CH4}$ of 414 kJ mol^{-1}, since the former is stabilized by the rearrangement of the CH_3 radical. The anomaly disappears, however, when it is recalled that D is an experimental quantity, whereas E is defined and the value depends on the basis of the definition. If the tetrahedral valence state of carbon is used, instead of the ground state, to which the molecule dissociates, the value becomes about 586 kJ mol^{-1}. The ground state definition, however, is the preferred one and the matter is unimportant so long as there is general agreement on the definition to be used. The accompanying diagram shows the relationship between these various quantities for a hypothetical organometallic compound, $C_x H_y M$.

Apparatus

Differential scanning calorimetry[2] has been used to study enthalpies of decomposition and a wide range of reactions in solution has also been examined using a variety of calorimeters.[3,4,5,6]

In this chapter, however, the use of combustion calorimetry and sublimation measurements in studying the energetics of organometallic-metallic compounds will be discussed. The combustion calorimeter used in Leeds is a vacuum-jacketed aneroid type which can be rotated, so that with a suitable solvent present it acts both as a combustion and a solution calorimeter. Some organometallic-metallic compounds may be studied by combustion without rotation, for example, mercury compounds and tin compounds; the mercury remains as the free metal with only a trace of oxide present which may be easily estimated, and in the case of tin, the metal is all converted to the oxide. However, most of the reliable combustion results have been obtained from rotating instruments.

The Leeds calorimeter[7,8,] jacket and the method of rotation are shown in figures 1 and 2. The combustion bomb, J, which also acts as the calorimeter, was machined from high purity copper and is platinum-lined. A

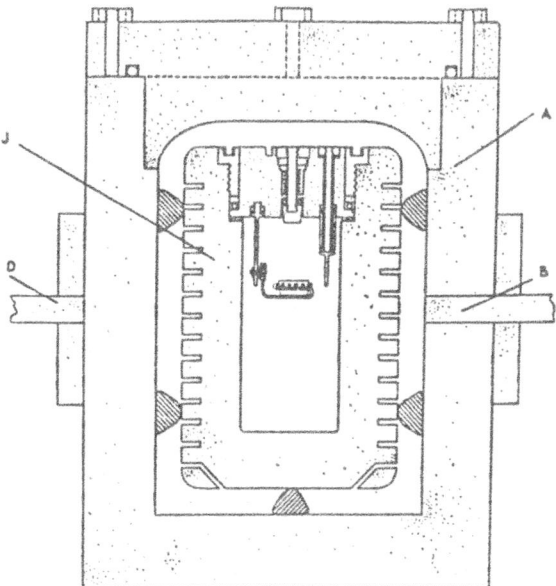

FIG.1

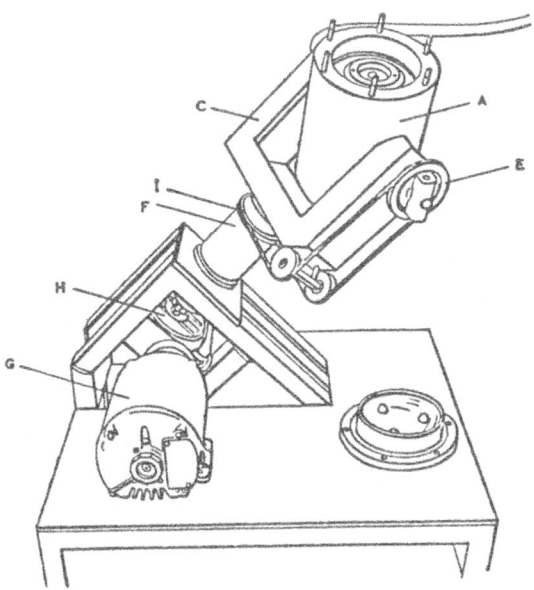

FIG. 2

long length of copper wire, wound non-inductively into grooves cut in the body of the calorimeter, acts as the resistance thermometer. The jacket, A, was also machined from high purity copper and it and the calorimeter were chromium plated and polished to reduce radiation effects. One of the supporting axles, B, was hollow and extended through the jacket wall. The electrical leads from the calorimeter were taken out through this shaft, which was also used as the connection to the vacuum line. The cradle, C, which carried the jacket could be rotated by a shaft which was driven by a motor, G, and which passed through a bearing, F. This bearing carried a fixed pulley, I, and during rotation the belt moved round the pulley, so that the movable pulley, E, attached to the solid axle, D, rotated also. Thus the jacket rotated simultaneously in two planes and so long as the fixed and movable pulleys are of the same diameter, the rotations cancelled one another so far as causing twist in the electrical leads and in the tubing connecting the jacket to the vacuum line. In a more recent modification, the belt has been replaced by shafts and gears but a diagram of the original system has been shown because it makes the method of rotation more
readily understandable.

An aneroid system has the advantage that, in the absence of a water jacket, the energy equivalent is only about one quarter of that of the conventional stirred liquid type. The same temperature rise can be obtained by the combustion of a smaller mass of material and so detonations become less likely.

A comprehensive account of the measurement of enthalpies of vaporization has been given by Majer and Svoboda[9]. Enthalpies of sublimation were derived in Leeds from measurements of vapour pressure as a function of temperature[10]. The vapour pressures were measured using a combined mass -loss and torsion Knudsen effusion apparatus, a diagram of which is shown in figure 3. The effusion cell, A, was suspended in a metal vacuum chamber, B, by a tungsten wire (diameter 0.0025 cm) from an automatic vacuum microbalance, C, which continuously recorded the mass loss from the cell throughout the experiment. There were two effusion holes on opposite sides of the cell equidistant from the points of suspension and the effusion of molecules through the holes caused the suspension to twist. The deflection of the cell was followed using a beam of light reflected from a mirror, D, attached to the suspension and the deflection was continuously recorded by a light spot follower. The effusion cell was heated in a thermostatically controlled furnace, E, and the temperature was found by

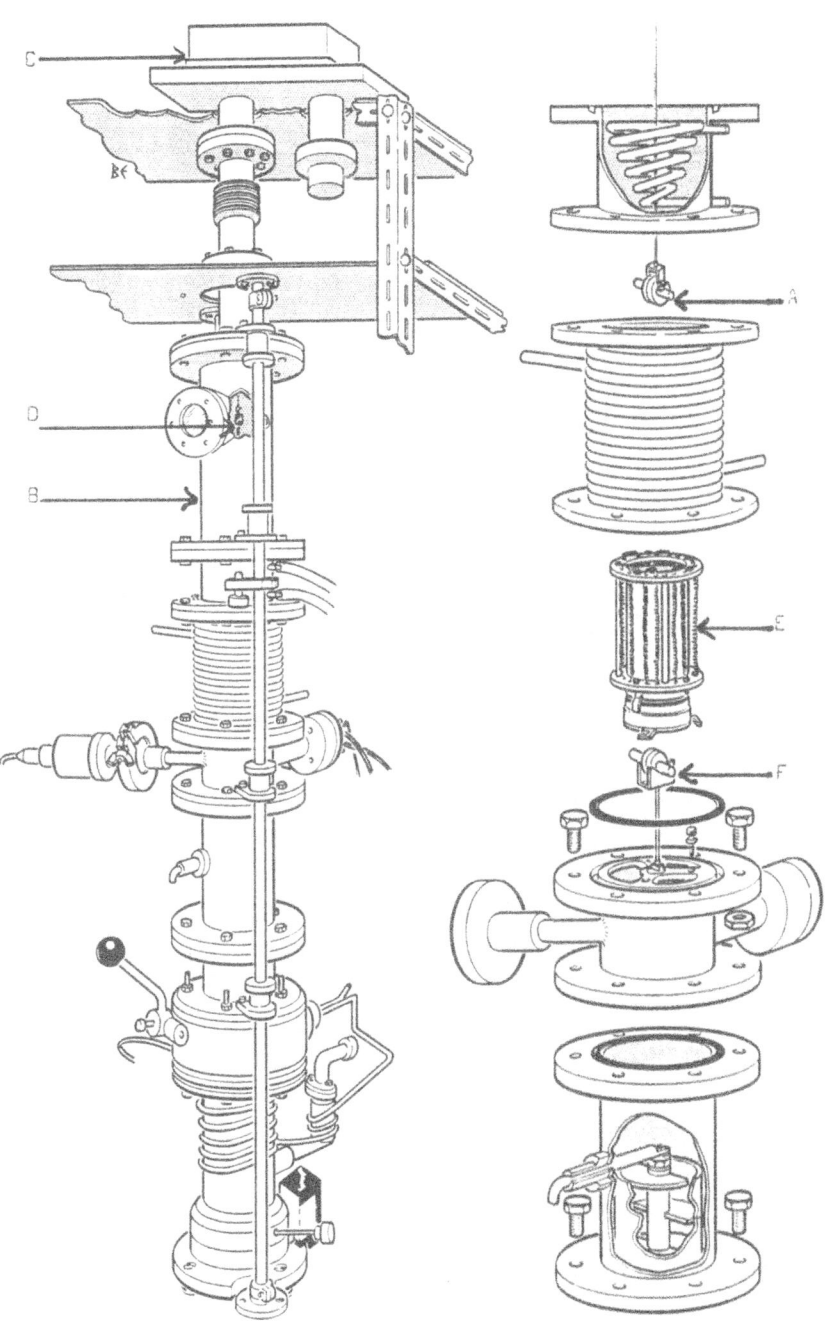

Fig 3

measuring the temperature of a similar cell, F, also mounted within the furnace and which contained a Pt resistance thermometer. The pressure is obtained from the Knudsen equation:

$$p = \left(\frac{dW}{dt}\right)\left(\frac{1}{K_c A}\right)\left(\frac{2\pi RT}{M}\right)^{\frac{1}{2}}$$

where dW/dt is the rate of mass loss, K_c is the Clausing factor, A is the hole area, M is the molar mass of the effusing species. Pressures may also be derived from the angular deflection without assuming a value for M :

$$p = \frac{D\varphi}{flA}$$

where D is the torsional constant of the wire, φ is the angle of deflection, f is the recoil force correction, A is the hole area and l is the distance of the holes from the suspension point. Agreement between the vapour pressures measured by both methods will confirm that the assumed value of M was correct.

Several important conditions must be satisfied for the Knudsen equation to be valid. The effusion hole is not infinitely thin and a correction must be applied for non-specular collisions with the walls of the hole which may return molecules to the cell; factors which allow for this, K_c and f are not the same; f is somewhat larger because the recoil force depends on the angular distribution of the effusing molecules as well as on their number. For a cylindrical tube Kc may be abbreviated for most purposes as:

$$K_c \sim \frac{1}{1 + \frac{L}{2r}}$$

where L is the length and r the radius of the hole. An equivalent value for f may be obtained from :

$$\frac{1}{f} = 0.0147\left(\frac{L}{r}\right) + 0.3490\left(\frac{L}{r}\right) - 0.9982$$

In general, the method may be used with confidence even when the hole size makes the steady state pressure less than the saturated one due to effusion loss. However, it is always worth while to use holes of different sizes, because other concealed sources of error may be revealed: there may be self cooling of the sample, the hole diameter-mean free path ratio may not be satisfactory, and there may be deviations from the assumed absolute vacuum on the low pressure side of the hole. In practice, a pressure of less than 10^{-6} torr is satisfactory. Effusion under ideal Knudsen conditions assumes that there is free molecular flow in which viscosity plays no part. This would hold when the ratio d/λ tends to zero, where d is the diameter of the hole and λ is the mean free path of the gas molecules. However, the assumption appears valid for $l/d = 10$ but there is evidence[11] that corrections should be applied if it lies in the range $1 - 5$. For values much greater than this a correction appears unnecessary as is shown by the good agreement between the enthalpies of sublimation of benzoic acid and 4-nitroaniline[12] obtained from direct calorimetry and Knudsen effusion.

Enthalpies of sublimation are obtained from the temperature variation of the vapour pressure by applying the Clausius-Clapeyron equation:

$$\frac{d\ln p}{d\left(\frac{1}{T}\right)} = -\frac{\Delta_{sub}H^o_m}{R}$$

to the linear plot of $\ln p$ against $1/T$.

Values may be corrected to 298K by using the difference in the heat capacities $\{C_p, g - C_p, cr\}$ when these are available, or by assuming that the difference between the two terms is $-2R$.

Experimental Results

During the past thirty years there has been a great upsurge of interest in organometallic chemistry, particularly in the field of transition metal complexes, which are important in analysis, biochemistry and catalysis. A very large number of compounds have been prepared, using about two hundred different ligands. The nature and the energies of the bonds in these compounds are important both for an understanding of their behaviour and as a link with theoretical predictions. Some examples of how calorimetry and sublimation have assisted in this process will now be considered.

The tetra-acetato complexes of chromium and molybdenum, $Cr_2(O_2CCH_3)_4$, $Mo_2(O_2CCH_3)_4$ and the fluorinated derivative, $Mo_2(O_2CCF_3)_4$ were studied calorimetrically in Manchester[13] and the enthalpies of sublimation were measured in Leeds[14]. The values are 299.6, 170.5 and 113.6 kJ mol^{-1} respectively and that for the chromium compound seems curiously high when compared with those for the molybdenum compounds. The lattice is formed through intermolecular metal---oxygen links between monomers. The chromium compound is more closely packed than the molybdenum one, since[15] the Cr---O distance between neighbours is only 234 pm compared with the Mo---O length of 264 pm. Mo---O bonds tend to be about 10 pm longer than the corresponding Cr---O ones so that the Cr---O length is very short compared with the Mo---O distance. The differences between the sublimation values are thus in accord with the crystal structures. In the trifluoroacetate crystal the Mo---O length has increased to 272 pm and the lower sublimation enthalpy reflects this.

The enthalpy of formation of the anhydrous and hydrated forms of the chromium compound have been measured by solution calorimetry[13] and this enables a standard enthalpy of hydration to be calculated,

$$Cr_2(O_2CCH_3)_4 2H_2O, cr \rightarrow Cr_2(O_2CCH_3)_4, cr + 2H_2O, g,$$

The enthalpy change is 94.3 ± 9.4 kJ mol^{-1}.

The vapour pressure work which was carried out on these compounds confirmed the correctness if this value. The pressure of water vapour above the crystal was measured at various temperatures,[14] in the temperature range at which previous experiments with the anhydrous compound had shown that there was no detectable mass loss. The temperature variation of the equilibrium constant expressed as: $\{p(H_2O/p°)\}^2$ leads to an enthalpy change of 96.3 ± 8.2 kJ mol^{-1}. Each water molecule is H-bonded to acetato oxygen atoms in adjacent molecules[16] and the enthalpy change corresponds to the rupture of these bonds so that two Cr–OH$_2$ links are replaced by two weaker Cr---O ones. The Cr–Cr bond is strengthened and shortened by the removal of the water molecules.

Another example is given by recent work on copper(II) complexes.[12] For a complex ML_n, the mean metal-ligand bond dissociation enthalpy is given by $\Delta H/n$, where the enthalpy change refers to the reaction :

MLn, g $\rightarrow$ M, g +nL, g.

If D(H − L) corresponds to HL, g $\rightarrow$ H, g +L, g then

$$D(M-L) - D(H-L) = 1/n[\Delta_f H°(M,g) - \Delta_f H°(ML_n,g)] - \Delta_f H°(H,g)$$
$$+ \Delta_f H°(H-L) \quad (1)$$

For most ligands D(H − L) values are not available, but there is experimental evidence (β-diketonate complexes of a wide variety of metals)[17] , that any change in the mean ligand-metal dissociation enthalpy caused by changing the ligand is counterbalanced by a corresponding change in D(H − L). The decision must be made as to whether the variations in D(H − L) represent changes in the binding energy, or are due to changes in the radical reorganizations. The reorganization variations may be eliminated by comparing the M-L dissociation enthalpy with that of H − L, but for most ligands this is an unknown quantity and an assumed value must be used. That this is a satisfactory procedure has been shown in several cases, for example, in a wide variety of diketonates a selected value for the O − H bond was assumed to be about the same in the various ligands and for nine copper complexes and five iron ones the differences
D(M − L) − D(H − L) were substantially the same. Recent work[5] in this field concerns $Cu_2(acetate)_4$; $Cu(dmg)_2$, (the ligand is dimethyl glyoxime) and $Cu(PhNacac)_2$, (the ligand is 4-phenylamino-3-penten-2-one). Reaction calorimetry in solution was used to obtain the standard enthalpies of formation of the crystalline compounds and the enthalpies of sublimation were derived from measurements of the vapour pressures as functions of temperature. (See, for example fig. 4). The enthalpy values are shown in Table 1.
D(H–L) values are not available but the standard enthalpies of formation of the gaseous HL compounds are known so that D(Cu − L) − D(H − L) may be found from the previously derived equation (1). These values are a measure of the enthalpy of transfer of the ligand from Cu to H. The crystal structure of the dmg complex, when compared with that of the $(accac)_2$ compound, shows the presence of two intramolecular H-bonds between the two dmg ligands attached to Cu, also the ligand is bound to Cu through N atoms, which form stronger links than the Cu − O bonds; in the free ligand the hydrogen is bound to oxygen. The prediction,therefore, is that the

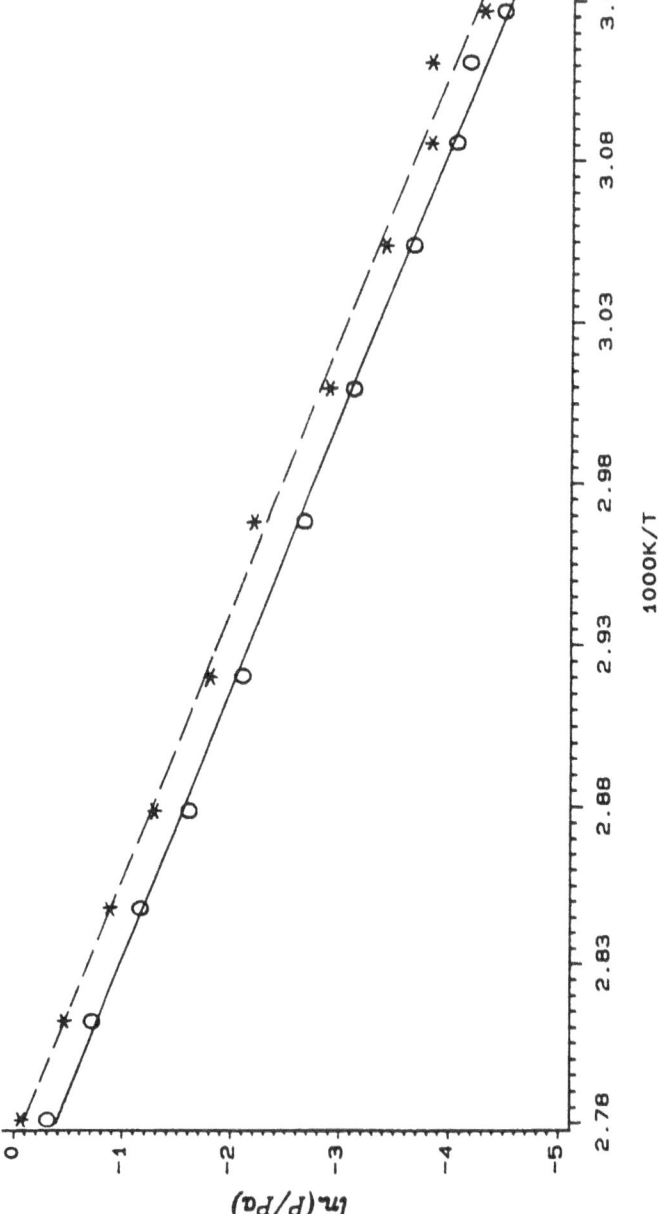

Fig. 4. Vapour pressure of cupric acetate: ○, mass loss; *, torsion.

Table 1

Complex	$\Delta_f H_m^o(cr)$	$\Delta_{sub} H_m^o$	$\Delta_f H_m^o(g)$	$\bar{D}(Cu-L)-D(H-L)$
Cu$_2$(acetate)$_4$	-1765.8 ± 2.7	106.1 ± 0.9	-1659.7 ± 2.8	-67.1 ± 1.8
Cu(dmg)$_2$	171.5 ± 2.3	93.1 ± 0.8	-78.4 ± 2.4	-63.7 ± 1.9
Cu(PhNacac)$_2$	-174.3 ± 2.7	128.1 ± 0.8	-46.2 ± 2.8	-92.3 ± 3.4
Cu(acac)$_2$	-809.9 ± 1.3	109.9 ± 3.4	-700 ± 3.8	-83.6 ± 2.4

binding in Cu(dmg)$_2$ is stronger than in Cu(acac)$_2$ and this is in accordance with the experimental results. Steric hindrance between the large attached ligands probably accounts for the weakened binding in Cu(PhNacac)$_2$. When the ligand is bound through oxygen atoms, Cu$_2$(acetate)$_4$ shows the greatest increase in the binding enthalpy when the ligand is attached to Cu. The existence of a Cu – Cu bond in this compound is controversial and thermochemistry cannot settle the question unambiguously.

The constancy of D(M – L) – D(H – L) is also illustrated by a wide range of copper and iron β-diketones, where the mean differences are – 84.2 and – 62.6 kJ mol^{-1} respectively.

A further example of combined calorimetry and vapour pressure work is given by the cyclopentadienyl molybdenum and tungsten hydrides, MCp$_2$H$_2$. These compounds are difficult to study because of their sensitivity to oxygen. The calorimetry was carried out in Lisbon and the vapour pressure measurements in Lisbon and Leeds[18].

$$\Delta_f H°[MoCp_2H_2, g] = 292.3 \pm 6.0 \text{ kJ mol}^{-1} \text{ and}$$

$$\Delta_f H°[WCp_2H_2, g] = 306.8 \pm 9.0 \text{ kJ mol}^{-1.}$$

The need to know $\Delta_f H°(MCp_2)$ is eliminated by considering the reaction,

$$MCp_2H_2, g + 2Cl, g \rightarrow MCp_2Cl_2, g + 2H, g$$

so that the enthalpy change of the reaction is,

$$\Delta H = 2D(M - H) - 2D(M - Cl).$$

No information on D(M – L) is available, but if it is assumed that the mean dissociation enthalpies D(Mo – Cl) and D(W– Cl) are the same as in MoCl$_6$ and WCl$_6$, this leads to the mean values,

$$D(Mo-H) = 256 \pm 8 \text{ kJ mol-1}$$
$$D(W-H) = 311 \pm 4 \text{ kJ mol-1}$$

Even though the absolute values may not be correct because of the above assumption, the difference between them should be sound.

Another series of compounds of considerable interest are the alkyls

Table 2

R	D(Hg – R)	D(Hg – R) – D(H – R)
Me	129.9 ± 3.0	-308.8 ± 3.0
Et	102.7 ± 3.2	-307.3 ± 3.2
n-Pr	107.8 ± 4.0	-309.9 ± 4.0
i-Pr	94.5 ± 4.5	-311.5 ± 4.5
Bz	91.7 ± 5.0	-308.3 ± 5.0
Ph	162.3 ± 2.5	-298.2 ± 2.5
(PhC_2)	312	-220

and aryls of the Group II element, mercury[3,19]. Mercury has a complete outer shell of s^2 electrons and so can only form two bonds of (s-p) type at 180° by promoting an electron. This means that the first dissociation enthalpy $D_1(RHg - R)$ is greater than the second, $D_2(Hg - R)$. In the case of $HgMe_2$, D1 is about 213 and D2 about 47 kJ mol^{-1}. Table 2 contains values for the mean dissociation enthalpies $D(Hg - R)$ and the differences between these and $D(H - R)$. It is difficult to attach an uncertainty to the value 312 kJ mol^{-1} for the radical $(PhC_2)_2$, since it depends on $\Delta_f H°(PhC \equiv CH, l)$, a calculated enthalpy of vaporization and the assumption that the C – H dissociation in ethynyl benzene is the same as in ethyne. Any error, however, will be eliminated in the comparison with $D(H - R)$. The comparisons with $D(H - R)$ enable a decision to be made as to whether the changes in the mean dissociation enthalpies are due to some intrinsic character of the Hg – C bond or to the inherent stability of the radical concerned. The constancy of the differences, $D(Hg - R) - D(H - R)$, show that with the exceptions of the phenyl and phenylethynyl radicals, the changes are due to the radical. However, when the metal is linked to an aromatic ring or an acetylenic group there is the possibility of partial π-bonding, using unoccupied orbitals of the metal. The figures indicate that, particularly in the ethynyl case, there is considerable strengthening of the Hg – C bond.

Similar effects have been found in the equivalent tin compounds[3,20] and the results are shown in Tables 3 and 4. The enthalpy of formation of the Ph_3Sn radical may be derived [21] from electron impact studies [22]. Relative appearance potentials give

$D(Ph_3Sn - Ph) - D(Ph_3Sn - SnPh_3) = 84.1 \pm 13.4$ kJ mol^{-1}.

Combining this with the standard enthalpies of formation of Ph_4Sn, g, Ph_6Sn_2, g and Ph, g yields the value

$$\Delta_f H°(Ph_3Sn) = 518.8 \pm 21.0 \text{ kJ mol}^{-1}.$$

The uncertainty in the dissociation enthalpies which involve this radical is not important from the point of view of comparison since all the listed compounds contain this radical.

The bonds of Sn-C(sp^2) type are very similar but in the ethynyl case the increase is much larger than the normally expected one in going from an sp^2 configuration to an essentially sp one. This indicates that, as in the mercury case, there is some interaction with the π-electrons of the bond.

Table 3

	$\overline{D}(Sn\text{-}R)$	$\overline{D}(Sn\text{-}R)\text{-}D(H\text{-}R)$
Me_4Sn	226.4 ± 1.1	-212.3 ± 1.1
Et_4Sn	194.7 ± 4.4	-215.3 ± 4.4
$n\text{-}Pr_4Sn$	206.4 ± 6.0	-211.3 ± 6.0
$i\text{-}Pr_4Sn$	190.1 ± 5.5	-215.9 ± 5.5
$n\text{-}Bu_4Sn$	201.8 ± 6.0	-212.9 ± 6.0
$(Ph_3Sn)_2$	252.1 ± 12.0	-208.4 ± 12.0
Ph_4Sn	257.2 ± 5.4	-203.3 ± 5.4

Table 4

$\Delta_f H°$, g, kJ mol-1		D(Ph$_3$Sn-R)	D-D(R-H)
Ph$_3$SnCH=CH$_2$ (525.5 ± 7.1)	→	D(Ph$_3$Sn-C$_2$H$_3$) (282)	-153.5
Ph$_3$Sn ≡ CPh (733.8± 8.1)	→	D(Ph$_3$Sn-C$_2$Ph) (427)	-105
Ph$_4$Sn (571.1 ± 5.4	→	D(Ph$_3$Sn-Ph) (274)	-186
Ph$_3$SnOH (178.7± 8.0	→	D(Ph$_3$Sn-OH) (379.1)	-123
(Ph$_3$Sn)$_2$O (612.5 ± 12.0)	→	D(Ph$_3$Sn-O) (338.9)	-121
(Ph$_3$Sn)$_2$ (848.5± 8.0	→	D(Ph$_3$Sn-SnPh$_3$) (189.1)	

Table 5

	D(Ge − R)	D(Ge − R) − D(H − R)
Me₄Ge	266.2 ± 8.3	− 172.5 ± 8.3
Et₄Ge	243.9 ± 6.7	− 166.1 ± 6.7
n-Pr₄Ge	246.7 ± 5.0	− 171.0 ± 5.0
Ph₄Ge	307.9 ± 8.0	− 152.6 ± 8.0
Bz₄Ge	228.6 ± 10.0	− 170.5 ± 10.0
(Ph₃Ge)₂	296.9 ± 8.0	− 163.6 ± 8.0

Table 6

$\Delta_f H°$, g kJ mol^{-1}		D(Ph$_3$Ge – R)
Ph$_3$GeC$_2$H$_3$ (362.6 ±7.5)	→	D(Ph$_3$Ge – CH=CH$_2$) (385)
Ph$_3$GeC$_2$Ph (579.0 ± 8.4)	→	D(Ph$_3$Ge – C ≡ CPh) (523)
Ph$_4$Ge (424.7 –12.9)	→	D(Ph$_3$Ge – Ph) (360)
(Ph$_3$Ge)$_2$ (662.9 ±12.8)	→	D(Ph$_3$Ge – GePh$_3$) (257)
(Ph$_3$Ge)$_2$O (259.1 ±11.6)	→	D(Ph$_3$Ge – O) (455)

Dissociation enthalpies (o) and bond enthalpies (•)
for some germanium and tin – carbon and – oxygen bonds

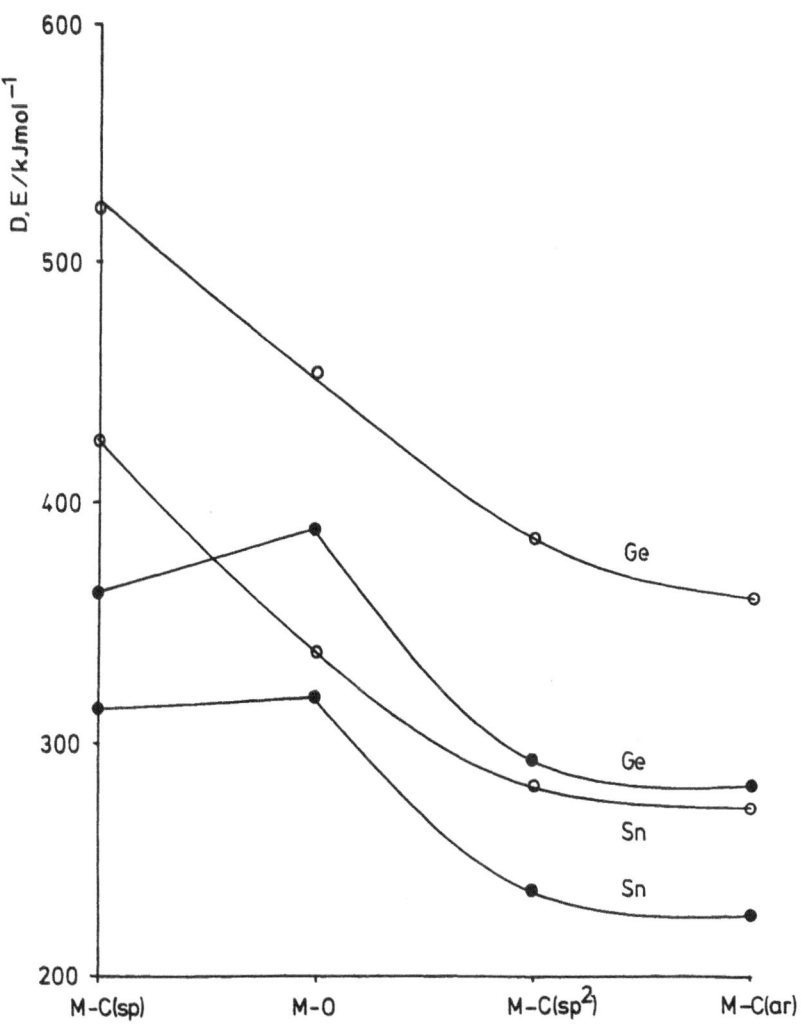

Figure 5

The dissociation enthalpy $D(Me_3Sn - Me) = 253$ kJ mol^{-1} would lead to a similar conclusion, if it is assumed that the bond strengthening in going from an sp^3 to an sp configuration is about 100 kJ mol^{-1}. There are similar results in the case of germanium compounds[3,23], An assumed value for $\Delta_fH°(GePh_3)$, based on $\Delta_fH°(SnPh_3)$, $D(Sn - Ph)$ and $D(Ge - Ph)$ may be used to obtain individual dissociation enthalpies. As with the tin compounds any error in this assumption is unimportant from the point of view of comparison. The results are shown in tables 5 and 6.

There has been controversy about the existence of a π-bonding effect in compounds where germanium is linked to an olefinic or an acetylenic group and it has been suggested that if it is present it will be to a lesser extent than in tin compounds. However, in going from an essentially sp^2 (Vi, Ph) configuration to an sp (alkynyl) one, the change is about 60 kJ mol^{-1} greater than the expected value in both the tin and germanium compounds. In both the tin and germanium compounds the dissociation enthalpies in going from M - C (sp) through $M - O$, $M - C$ (sp^2) to $M - C(ar)$ lie on smoothly descending curves. [23] Figure 5.

Bond enthalpy terms in organometallic compounds may be obtained from the enthalpies of formation of the gaseous compounds from their constituent atoms. These terms may be used to estimate enthalpies of formation for which no experimental data are available and to make predictions about environmental effects which may lead to strain in the molecule. Self-consistent bond enthalpy term schemes have been produced and have been widely used for organic compounds, but so far, have had only a limited application in the organometallic field. The values used in making predictions must be selected with care, making due allowance for any changes in the environment of the bond concerned. This may be illustrated briefly by considering some tin and germanium compounds. For example, Me$_4$Sn, g has an enthalpy of atomization[3], $- \Delta_aH°$, g of 5803.1 kJ mol^{-1}. This is equal to four times the tin-carbon and twelve times the carbon-hydrogen bond enthalpy terms. This leads to a value of 217.0 kJ mol^{-1} for the $[Sn - C (sp^3)]$ enthalpy term. This may be incorporated with the enthalpy of formation of gaseous Me$_3$SnPh to yield the tin-phenyl bond enthalpy term $E[Sn - C_b]$.

$$- \Delta_aH° [Ph] + E[Sn - C_b] + 3E[Sn - C(sp^3)] + 9E[H - C(sp^3)]$$
$$= 5103.7 + E[Sn - C_b] + 651.0 + 3701.3$$

153

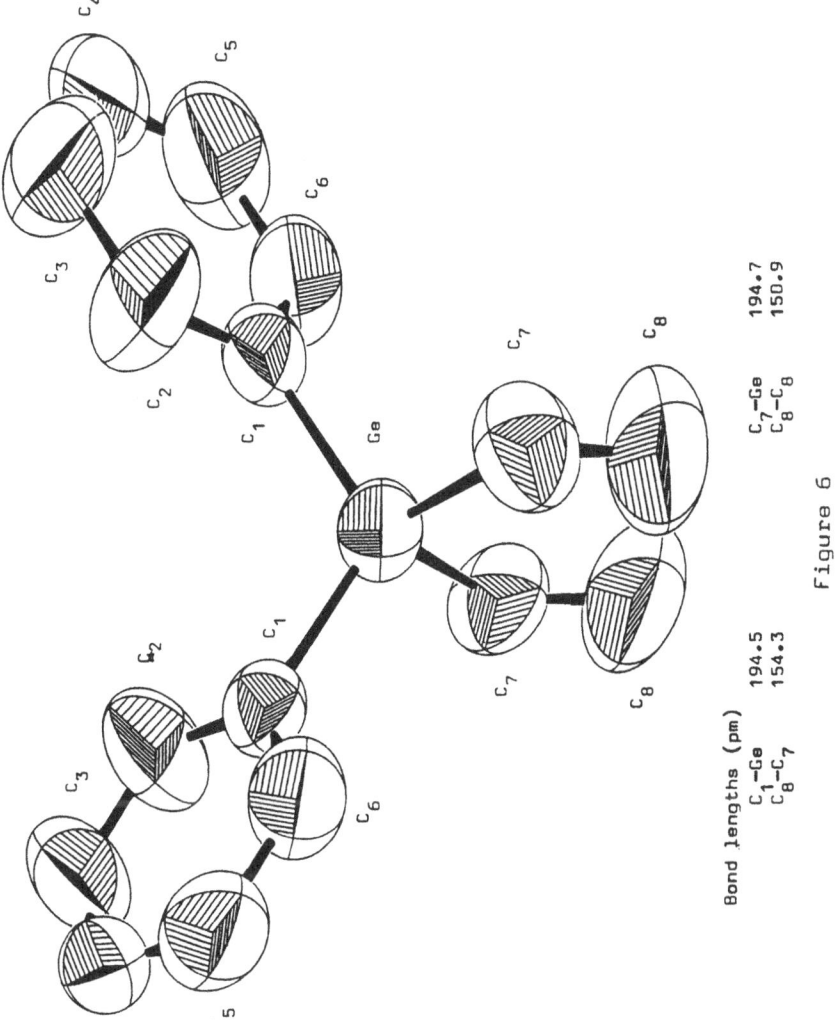

Bond lengths (pm)

$C_1–Ge$ 194.5 $C_7–Ge$ 194.7
$C_8–C_7$ 154.3 $C_8–C_8$ 150.9

Figure 6

so that $E[Sn\ C_b] = 233.8$ kJ mol^{-1}.

A value for this quantity may also be obtained from[25]

$$\Delta_f H°[Ph_4 Sn, c] = 410.0 \text{ kJ mol}^{-1} \text{ and}$$
$$\Delta_{sub} H°[Ph_4 Sn] = 161.1 \text{ kJ mol}^{-1}$$

so that $\Delta_f H°[Ph_4 Sn, g] = 571.1$ kJ mol^{-1}.

(Some values differ from those in ref. 25 because of subsequent changes in some of thermochemical quantities involved).

This leads to an enthalpy of atomization of $Ph_4 Sn$ of 21288.4 kJ mol^{-1} which may be equated to:

$-\Delta_a H°[Ph] + 4E[Sn - C_b]$ so that $E[Sn - C_b] = 218.4$ kJ mol^{-1}. The difference between the two bond enthalpy terms,

$E[Sn - C_b][Me_3 SnPh - Ph_4 Sn]$ of 15.5 kJ mol^{-1} may perhaps be explained by the greater steric hindrance in $Ph_4 Sn$.

It is of interest to use these bond enthalpy values to examine the question of strain in compounds in which the metal is a part of the ring, for example, 1,1diphenylstannocyclopentane $Ph_2 Sn(CH_2)_4$ and 1,1-diphenylstannocyclohexane $Ph_2 Sn(CH_2)_5$.

1,1diphenylgermanocyclopentane is also considered and its structure is shown in fig. 6. The standard enthalpy of formation of the gaseous tin cyclopentane is

$$\Delta_f H° [Ph_2 Sn(CH_2)_4, g] = 301.8 \pm 7.6 \text{ kJ mol}^{-1}$$

which leads to an enthalpy of atomization of 15390.1 kJ mol^{-1}. A predicted value based on bond enthalpy terms would be given by the contributions -

$- 2\Delta_a H°(Ph) + 2E\{ Sn - C_b\} + 2E\{ Sn - C(sp^3)\} +$
$3E\{ C - C(sp^3)\} + 8E\{ H - C(sp^3)\}$

$= 10207.4 + 467.6 + 434.0 + 1075.5 + 3259.2$

$= 15443.7$ kJ mol^{-1}

A comparison of the experimental and predicted values indicates that this compound is strained to the extent of 53.6 kJ mol^{-1}. Within the limits of experimental error, this strain energy may be explained by including the strain energy of 26.4 kJ mol^{-1}, which is the value usually accepted for five-membered carbon rings, and by assuming that the enthalpy term

E{Sn – C_b} has been reduced to 218 kJ mol^{-1} due to steric hindrance between the phenyl groups. A similar calculation for the six-membered ring system, based on

$$\Delta_f H°[Ph_2Sn(CH_2)_5, g] = 289.0 \pm 6.3 \text{ kJ mol}^{-1}$$

give an enthalpy of atomization of 16555.6 kJ mol^{-1}. A predicted value, based on bond enthalpy terms is 16617.0 kJ mol^{-1} which indicates that a strain energy of 61.4 kJ mol^{-1} is present. Because of the experimental uncertainty, therefore, there is no significant difference between the two strain energies. This is unexpected in tin cyclohexane since it is usually accepted that there is no strain energy in six-membered carbon rings. It may be that the larger tin atom and the much longer Sn – C(sp^3) links have distorted the ring system. Uncertainties in the bond enthalpy terms may be eliminated in comparing the two strain energies by considering the hypothetical reaction;

$$Ph_2Sn(CH_2)_4, g + C_6H_{12}, g \rightarrow Ph_2Sn(CH_2)_5, g + C_5H_{10}, g$$

The enthalpies of formation of gaseous cyclopentane and gaseous cyclohexane are – 76.4 $\pm$ 0.8 and – 123.4 $\pm$ 0.8 kJ mol^{-1} respectively, so that the enthalpy change for the reaction is 34.2 kJ mol^{-1}. This may be equated to the difference between the relative strain energies of the tin compounds and those of the hydrocarbons. That is, if Δ_s represents the strain energy,

$$34.2 = \Delta_s[Ph_2Sn(CH_2)_5] - \Delta_s[Ph_2Sn(CH_2)_4] + \Delta_s[C_5H_{10}] - \Delta_s[C_6H_{12}]$$

$\Delta_s[C_5H_{10}] = 26.4$ kJ mol^{-1}; $\Delta_s[C_6H_{12}]$ is zero, so that the difference between the strain energies of the six-membered and five-membered tin compounds is about 8 kJ mol^{-1} with an experimental uncertainty of $\pm$ 10 kJ mol^{-1}.

The bond enthalpy terms for the corresponding germanium compound, $Ph_2Ge(CH_2)_4$ are less secure and this is due both to experimental uncertainties in the enthalpy of formation of the compounds concerned and to a large uncertainty of $\pm$ 13 kJ mol^{-1} in the enthalpy of atomization of germanium. A recent determination[26] of the enthalpy of formation of Me_4Ge, g gives E{Ge – C(sp^3)} = 256.8 $\pm$ 15 kJ mol^{-1} and a value for E{Ge – C_b} of 280 $\pm$ 20 kJ mol^{-1} may be found from Ph_4Ge, g. The enthalpy of formation[27] of $Ph_2(CH_2)_4$, g is 138.9 $\pm$ 10.5 kJ mol^{-1}

which leads to an enthalpy of atomization of 15628 ± 16.0 kJ mol^{-1}. The value obtained from a summation of the bond enthalpy terms is 15615.6 ± 25 kJ mol^{-1} and it is clear, therefore, that because of the associated uncertainties, nothing of value can be said about the possibility of strain in the compound.

These examples involving tin and germanium compounds have been included to illustrate the point which was mentioned earlier, that bond enthalpy terms must be based on precise experimental data and used with caution.

References

1. Cox, J. D. and Pilcher, G. (1970) Thermochemistry of Organometallic Compounds, Academic Press, London.
2. Mortimer, C. T. (1984) 'Differential Scanning Calorimetry' in Ribeiro da Silva, M. A. V. (ed), Thermochemistry and its Applications to Chemical and Biochemical Systems, NATO ASI Series C, Vol. 119, Reidel, Dordrecht.
3. Pilcher, G. and Skinner, H. A. (1982) 'Thermochemistry of Organometallic Compounds' in Hartley, F. R. and Pataí, S. (eds) The Chemistry of the Metal-Carbon Bond, John Wiley and Sons, Ltd., London.
4. Marks, Tobin J. (1990) 'Importance of Metal-Ligand Energies', in Marks, Tobin J. (ed) Organometallic Chemistry, J. Amer. Chem. Soc. Symposium Series 428, Washington.
5. Burkinshaw, P. M. and Mortimer, C. T. (1983) Coordination Chemistry Review 48, 10
6. Mortimer, C. T. (1984) Rev. Organic Chemistry, 6, 233.
7. Adams, G. P., Carson, A. S. and Laye, P. G. (1969), Trans. Faraday Soc. 65, 113
8. Carson, A. S., Carson, E. M., Laye, P. G., Spencer, J. A. and Steele, W. V. (1970) Trans. Faraday Soc. 66, 2459.
9. Majer, V. and Svoboda, V. (1985) 'Enthalpies of Vaporization of Organic Compounds' in IUPAC Chemical Data Series No. 32, Blackwell Scientific Publications, Oxford.
10. Carson, A. S. (1984) 'The Measurement of Vapour Pressures', in Ribeiro da Silva, M. A. V. (ed) Thermochemistry and its Applications to Chemical and Biochemical Systems, NATO ASI Series C, Vol. 119, Reidel, Dordrecht.
11. Edwards, J. W. and Kington, G. L. (1962) Trans. Faraday Soc. 58, 1323
12. Ribeiro da Silva, M. A. V., Ribeiro da Silva, Maria D. M. C., Rangel, Maria C. S. S., PIlcher, G. Akello, Margaret J., Carson, A. S. and Jamea, E. H. (1990) Thermochim Acta 160, 267
13. Cavell, K. J., Garner, C. D., Pilcher, G. and Parkes, S. (1979) J. Chem. Soc. Dalton, 1714

158

14. Carson, A. S. (1984) J. Chem. Thermodynamics, 16, 855.
15. Cotton. F. A., Rice, C. E. and Rice, G. W. (1977) J. Am. Chem. Soc. 99, 4704.
16. Cotton, F. A., Deboer, B. G., Laprade, M. D., Pipal, J. R. and Ucko, D. A. (1971) Acta Cryst. B27, 1664.
17. Pilcher, G. (1989) Pure and Appl. Chem. 61, 855.
18. Dias, A. R., Diogo, H. P., Minas da Piedade, M.E., Martinho Simões, J. A., Carson, A. S. and Jamea, E. H. (1990) J. Organometallic Chem. 391, 361.
19. Carson, A. S. and Spencer, J. A. (1984) J. Chem. Thermodynamics, 16, 423.
20. Carson, A. S., Laye, P. G. and Spencer, J. A. (1985) J. Chem. Thermodynamics, 17, 277.
21. Carson, A. S., Franklin, J., Laye, P. G. and Morris, H. (1975) J. Chem. Thermodynamics, 7, 763.
22. Chambers, D. B. and Glockling, F. J. (1968) J. Chem. Soc.(A), 735.
23. Carson, A. S., Jamea, E. H., Laye, P. G. and Spencer, J. A. (1988) J. Chem. Thermodynamics, 20, 1223.
24. Carson, A. S., Laye, P.G., Spencer, J. A. and Steele, W. V. (1970) J. Chem. Thermodynamics, 2, 659.
25. Adams, G. P., Carson, A. S. and Laye, P. G. (1969) J. Chem. Thermodynamics, 1, 393.
26. Long, L. H. and Pulford, C. I. (1986) J. Chem. Soc. Faraday Trans.2, 82, 567.
27. Carson, A. S., Dyson, J., Laye, P. G. and Spencer, J. A. (1988) J. Chem. Thermodynamics, 20,1423.

Estimating Enthalpies of Sublimation of Hydrocarbons

JAMES S. CHICKOS AND DONALD G. HESSE
Department of Chemistry
University of Missouri-St. Louis
St. Louis Missouri 63121

JOEL F. LIEBMAN
Department of Chemistry and Biochemistry
University of Maryland-Baltimore County
Baltimore Maryland 21228

ABSTRACT. A general technique for the estimation of sublimation enthalpies is described and applied to hydrocarbons. The vaporization enthalpy of a hydrocarbon is obtained by using a simple relationship based on the type and number of carbons present. Fusion enthalpies are estimated using a group additivity approach. Addition of vaporization and fusion enthalpies affords a sublimation enthalpy in good agreement with experiment. The standard error of this estimation technique is 2.6 kcal/mole (10.9 kJ/mole) and compares with a standard experimental error of 1.74 kcal/mole (7.28 kJ/mole) obtained from a statistical analysis of literature data.

Fusion, vaporization and sublimation enthalpies are important physical properties of the condensed phase. Studies referencing the gas phase as a standard state require accurate values for these quantities[1,2]. The divergence in numbers between the many new organic compounds and the relatively few thermochemical measurements reported annually, has directed our interest toward the development of simple empirical relationships that can be used to estimate these enthalpies.

There are very few general techniques reported for estimating sublimation enthalpies although several different ones for homologous series[3] or for closely related structures[4] have been proposed. One of the earliest was application of Walden's Rule[5] to evaluate fusion enthalpies (equation 2) and Trouton's Rule[6] to obtain vaporization enthalpies (equation 3). The terms T_m and T_b in these equations refer to the melting and boiling points, respectively. Equation 1 is a thermodynamic equality for sublimation enthalpy if fusion and vaporization enthalpies are measured at or corrected to the same temperature. A good approximation of ΔH_s is obtained when using values measured at

$$\Delta H_s(298) \sim \Delta H_{fus}(T_m) + \Delta H_v(T_b) \tag{1}$$

$$\text{Walden's Rule: } \Delta H_{fus}(T_m)/T_m \sim 13 \text{ cal/(K·mole); } 54.4 \text{ J/(K·mole)} \tag{2}$$

$$\text{Trouton's Rule: } \Delta H_v(T_b)/T_b \sim 21 \text{ cal/(K·mole); } 87.9 \text{ J/(K·mole)} \tag{3}$$

159

J. A. Martinho Simões (ed.), Energetics of Organometallic Species, 159–169.
© 1992 Kluwer Academic Publishers.

different temperatures. The application of Walden's and Trouton's Rule to equation 1 provides a remarkably good approximation of ΔH_s if one considers that the estimation is based on only two "adjustable" parameters.

Another general technique for the estimation of sublimation enthalpies was developed by Bondi[7]. This method yields the sublimation enthalpy directly by using the method of group additivity. This method is quite useful for rough estimates of ΔH_s and has been used in cases where molecular instability is involved[8]. The major limitations of applying group additivity methods to obtain sublimation enthalpies directly are detailed below.

Equation 1 can be a useful equation if vaporization and fusion enthalpies at 298 K can be reliably estimated. A variety of general techniques for accurately estimating vaporization enthalpies of hydrocarbons and their derivatives are available[9-12]. The techniques based on group additivity relationships only require molecular structure information and in general are easy to use. Far fewer methods for estimating fusion enthalpies have been developed.

Sublimation Enthalpies-Semi-empirical Additivity of Latent Enthalpies

As an illustration of the usefulness of equation 1 in obtaining sublimation enthalpies, we have calculated the sublimation enthalpies of 44 solid hydrocarbons by combining the vaporization enthalpies of these solids to the experimental fusion enthalpy of each respective compound[13]. Vaporization enthalpies were estimated using equation 4, previously demonstrated to provide values for hydrocarbons within 5%[12]. The terms n_Q and $\tilde{n}_c$ in this equation refer to the number of quaternary and non-

$$\Delta H_v(298) = 1.12\, \tilde{n}_c + 0.31\, n_Q + 0.71 \qquad (4)$$

quaternary centers, respectively. The results of this semi-empirical approach of estimating sublimation energies are illustrated in Figure 1. The standard error associated

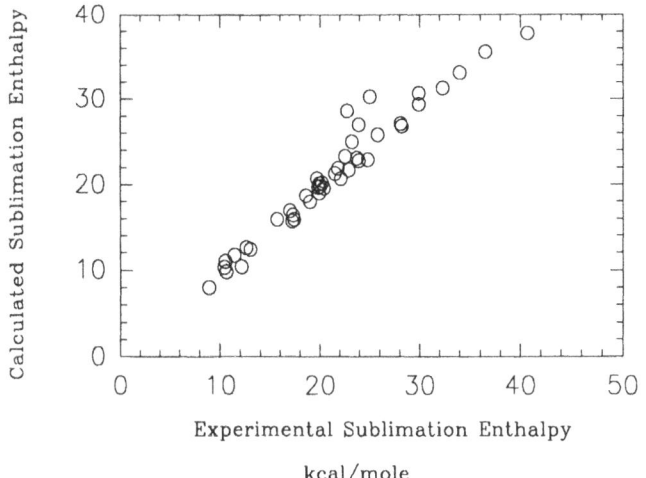

Figure 1. Experimental and calculated sublimation enthalpies of hydrocarbons using experimental fusion enthalpies and calculated vaporization enthalpies.

with this correlation is 1.66 kcal/mole (6.94 kJ/mole). The line obtained by a least squares fit has a slope of 0.986, an intercept of 0.15 and is characterized by a correlation coefficient of 0.9742.

Estimation of Fusion Enthalpies of Hydrocarbons

As noted above, very few methods of estimating fusion enthalpies have been reported. Recently, Domalski and Hearing[14] have reported a group additivity approach for calculating heats of formation of hydrocarbons in the condensed phases. Fusion enthalpies can be obtained by taking the difference of these two estimates. This method works well but the number of group types presently available limits the general applicability of this technique.

Two difficulties are anticipated in deriving a group method for estimating fusion enthalpies (and sublimation enthalpies) directly. Many organic solids undergo a variety of phase transitions below their melting point. These phase transitions are generally unpredictable and in some cases can dominate the magnitude of the total phase change enthalpy[15] observed. Examination of the thermochemical data of 191 hydrocarbons, identified 44 compounds (23 %) which exhibit phase transitions below their fusion points.

A second difficulty in deriving a group method for estimating fusion enthalpies (and sublimation enthalpies) was anticipated by the work of Leclercq, Jacques and Collet on chiral systems[16]. These workers investigated the fusion enthalpies of chiral molecules and their racemic modifications. They found that fusion enthalpies could vary by as much as 1-4 kcal/mole (4-16 kJ/mole) between the optically active and racemic form. Fusion enthalpies of common organic compounds average around 5-10 kcal/mole (20-40 kJ/mole). For molecules possessing identical molecular structures, this variation is substantial. This suggested that estimations of fusion enthalpies by group additivity methods would only enjoy limited success. The development of group parameters to estimate fusion entropies appears to be more promising. This optimism is based on the qualitative correlation observed between magnitude of the fusion enthalpy and the melting point of many isomeric compounds, including chiral systems.

Parameterization of Phase Change Entropies

The parameterization of fusion entropies is based on the assumption that the contribution of a particular atom or a group of atoms to the total entropy change associated in going from a rigid anisotropic solid to an isotropic liquid is fundamentally constant. The entropy associated with all phase transitions occurring in the solid state have been included in the parameterization of fusion entropies. All these transitions are associated with an increase in randomness and can be considered as partial transitions to the liquid state. The technique described below estimates the total phase change entropy associated with changes in the solid at 0 K to the isotropic liquid at the melting point (ΔS_{tpce}). **For most compounds the total phase change entropy (ΔS_{tpce}) is equivalent to the fusion entropy (ΔS_{fus}).**

The general philosophical guidelines followed in this work was to provide the best possible correlation with the fewest number of parameters. We have interpreted "best possible correlation" to mean statistical agreement within about two standard deviations of the experimental error of the measurements. The data used in developing these correlations were generally taken from the compilation of Domalski, Evans and Hearing[17-18]. The terms primary, secondary, tertiary and quaternary used below are

Table 1 Group Contributions to Hydrocarbon Fusion Entropies[18b]

	Entropy Number e.u.[20]	Number of Entries
Acyclic Portions of Hydrocarbons		
primary sp^3 carbon atom	4.38	22
secondary sp^3 carbon atom	2.25	22
tertiary sp^3 carbon atom	-3.87	17
quaternary sp^3 carbon atom	-9.25	14
Olefinic and Acetylenic Portions of Acyclic Hydrocarbons		
secondary sp^2 carbon	3.48	14
tertiary sp^2 carbon	1.16	18
quaternary sp^2 carbon	-2.72	7
tertiary sp carbon	[2.6]	1
quaternary sp carbon	[0.52]	6
Aromatic Hydrocarbons		
tertiary sp^2 carbon	1.54	44
quaternary sp^2 carbon adjacent to an sp^3 carbon	-2.47	37
peripheral quaternary sp^2 carbon adjacent to sp^2 carbon	-1.02	18
internal quaternary sp^2 carbon adjacent to sp^2 carbon	[0.1]	3
quaternary sp^2 carbon adjacent to sp carbon	[-0.6]	1
Cyclic Hydrocarbons		
contribution of the ring:		
ring size : n atoms; $\Delta S = 8.41 + 1.025[n-3]$		7
cyclic tertiary sp^3 carbon	-3.82	31
cyclic quaternary sp^3 carbon	[-7.88]	6
cyclic tertiary sp^2 carbon	-1.04	9
cyclic quaternary sp^2 carbon	-2.8	7
cyclic quaternary sp carbon	[-1.28]	1

Polycyclic Hydrocarbons
total number of ring atoms: R
number of rings: N; $\Delta S = [8.41]N + 1.025[R-3N]$

values in brackets are tentative assignments.

assigned on the number of hydrogens remaining on the carbon, 3,2,1,0, respectively, rather than on the usual carbon based convention.

An examination of the total phase change entropy of the n-alkanes versus the number of methylene groups in the alkane gives the linear correlation shown in Figure 2. The contribution of a methyl group to the total phase change entropy was obtained from the intercept and the contribution of a methylene group was obtained from the slope. These values were used in all subsequent correlations when needed to generate the group values listed in Table 1. The following summarizes the protocol used. All molecules were grouped according to the structural types listed in Table 1 to provide statistics for as many groups as possible. Once a group value was assigned, it was used as needed in subsequent correlations. Generally, the contributions of groups whose value was known were subtracted from the experimental fusion entropy. The remainder was averaged for all entries containing the structural parameter under evaluation. This value was then varied until the error was minimized by the method of least squares[19]. The number of entries used in the assignment of each group value in Table 1 is included in column 3. Values reported in brackets are based on limited experimental data and should be considered as tentative assignments. In some cases the experimental data was fit to more than one parameter by the least squares treatment.

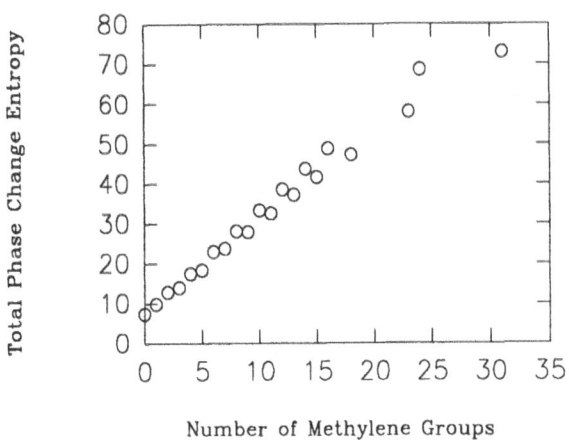

Figure 2. The total phase change entropy of the n-alkanes as a function of the number of methylene groups present.

The group values for cyclic hydrocarbons were obtained by considering the total phase change entropy of the cycloalkanes as a function of the number of carbons in the ring. This is shown in Figure 3 for cycloalkanes whose total phase change entropy was available (cyclopropane through to cyclotetradecane). The total phase change entropy for the cycloalkane can be obtained from the following equation: $\Delta S_{tpce} = 8.41 + 1.025\,[n\text{-}3]$ e.u.[20]; where n refers to the number of carbon atoms in the ring. In order to apply this equation to substituted cycloalkanes, a correction for the atom(s) modified by the substitution must be applied. These corrections are listed under Cyclic Hydrocarbons in Table 1 according to the substitution and hybridization of the carbon. Group values were obtained by the same procedure as described above. Group values used for the acyclic portion of the cycloalkane were the same as those used for acyclic systems.

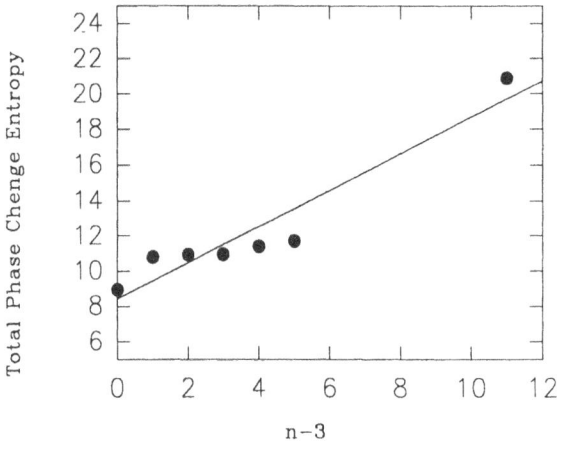

n = number of methylene groups

Figure 3. The total phase change entropy of the cycloalkanes as a function of ring size, n-3 (n = the number of ring atoms).

For polycyclic molecules, the equation for the cycloalkanes must be modified so that carbon atoms common to more than one ring are not counted more than once. The equation for polycyclic systems was derived from the cycloalkane equation and is given at the bottom of Table 1.

Most group types listed in Table 1 are easy to identify. Several entries under Aromatic Hydrocarbons are listed as quaternary sp^2 carbon atoms. To avoid any ambiguity in meaning, the molecule illustrated below contains an example of each structural type in this subgroup. The peripheral and internal quaternary sp^2 carbon atoms are entries most likely to cause confusion. The internal quaternary sp^2 carbon atom refers to an aromatic atom such as found in graphite whereas a peripheral quaternary sp^2 carbon atom refers to the type of atom frequently encountered in fused aromatic systems[21] such as naphthalene and anthracene.

internal quaternary sp^2 carbon
adjacent to sp^2 carbon

peripheral quaternary
sp^2 carbon adjacent to sp^2
carbon

—C≡CH

CH₃

quaternary sp^2 carbon
adjacent to sp carbon

quaternary sp^2 carbon
adjacent to sp^3 carbon

Estimation of Total Phase Change Entropies of Hydrocarbons-Applications

The application of the group values in Table 1 to estimate fusion entropies is relatively straightforward. The examples in Table 2 serve to illustrate the use of the group values to estimate phase change entropies. The first step in the estimation is to subdivide the target molecule into cyclic, and aromatic and acyclic components in order of decreasing priority. 1,1-Dimethylcyclopentane serves as a useful example. The

Table 2 Estimation of Fusion Entropies By Group Additivity

C_5H_8 2,3-pentadiene

2 primary sp^3 carbon atoms	8.76
quaternary sp carbon:	0.52
2 tertiary sp^2 carbon:	2.32
ΔS_{fus}(calcd):	11.6
ΔS_{fus}(expt):	10.7

C_7H_{14} 1,1-dimethylcyclopentane

cyclopentane:	[8.41]+1.025[5-3]
cyclic quaternary sp^3 carbon:	-7.88
2 primary sp^3 carbons:	8.76
ΔS_{fus}(calcd):	11.4
cII/cI (146.8 K):	10.6
cI/liq (203.6 K):	1.3
ΔS_{tpce}(expt):	11.9

C_9H_{12} isopropylbenzene

2 primary sp^3 carbons:	8.76
tertiary sp^3 carbon:	-3.87
5 tertiary aromatic sp^2 carbons:	7.70
quaternary aromatic sp^2 carbon adjacent to sp^3 carbon:	-2.47
ΔS_{fus}(calcd):	10.1
ΔS_{fus}(expt):	9.9

C_9H_{16} *trans* hexahydroindane

bicyclo[4.3.0]nonane:	[8.41]2 + 1.025[9-6]
2 cyclic tertiary sp^3 carbons:	-7.64
ΔS_{fus}(calcd):	12.3
ΔS_{fus}(expt):	12.2

$C_{10}H_8$ azulene

8 aromatic sp^2 carbons:	12.32
2 peripheral quaternary sp^2 carbons adjacent to sp^2 carbon:	-2.04
ΔS_{fus}(calcd):	10.3
ΔS_{fus}(expt):	11.5

$C_{20}H_{14}$ tryptycene

bicyclo[2.2.2]octane:	[8.41]2+1.025[8-6]
6 quaternary sp^2 carbons	16.8
2 tertiary cyclic sp^3 carbons:	-7.64
12 tertiary aromatic sp^2 carbons:	18.48
ΔS_{fus}(calcd):	12.9
ΔS_{fus}(expt):	13.7

$C_{16}H_{10}$ fluoranthene

12 tertiary aromatic sp^2 carbons:	15.4
5 peripheral quaternary sp^2 carbon adjacent to sp^2 carbon:	-5.1
internal quaternary sp^2 carbon adjacent to sp^2 carbon:	0.1
ΔS_{fus}(calcd):	10.4
ΔS_{fus}(expt):	11.7

$C_{11}H_{24}$ 2-methyldecane

3 primary sp^3 carbons:	13.14
tertiary sp^3 carbon:	-3.87
7 secondary sp^3 carbons:	15.75
ΔS_{fus}(calcd):	25.0
ΔS_{fus}(expt):	26.7

C_5H_8 spiropentane

spiro[2.2]pentane	[8.41]2 + 1.025[5-6]
cyclic quaternary sp^3 carbon:	-7.88
ΔS_{fus}(calcd):	7.92
ΔS_{fus}(expt):	9.25

$C_{14}H_{20}$ diamantane

pentacyclic ring system:	[8.41]5+1.025[14-15]
8 cyclic tertiary sp^3 carbons:	-30.6
ΔS_{fus}(calcd):	10.5
cIII/cII (407):	2.61
cII/cI (440):	4.86
cI/liq (517.9):	4.0
ΔS_{tpce}(expt):	11.5

molecule contains a cyclopentane ring and two methyl groups connected by a cyclic quaternary sp³ carbon atom. The total phase change entropy is estimated by addition of the contribution of the cyclopentane ring, the correction for the cyclic quaternary carbon which modifies the unsubstituted five membered ring and the two methyl groups. Estimation of the phase change entropy of tryptycene (**1**) illustrates the use of the ring equation for polycyclic molecules. The number of rings in a molecule is determined by the fewest number of carbon-carbon bonds to be broken in converting the polycyclic molecule to a completely acyclic structure. Aromatic rings are treated separately and are therefore ignored by this equation. Terms for the 6 cyclic sp² bridging carbons, 12 aromatic CH groups, and the two tertiary sp³ hybridized bridgehead carbons complete the estimation. Although the 6 bridging carbons are aromatic, they are also part of a non-

1   **2**

aromatic ring. The classification of these carbons as cyclic quaternary sp² carbons, the closest structural analog in the Cyclic Hydrocarbon category, takes priority. Estimation of the fusion entropy of fluoranthene (**2**) illustrates an example where the five membered ring is completely conjugated. This system is estimated as a completely aromatic compound[21]. In addition to the 10 tertiary aromatic carbons, the molecule contains 5 peripheral quaternary aromatic carbons and one internal quaternary aromatic carbon.

The quality of the correlation achieved by the process just outlined is shown in Figure 4. The calculated total phase change entropy is plotted against the experimental values for the 192 hydrocarbons that make up the data base. The equation of the best line through the data is characterized with a slope of 1.02, an intercept of -0.41 and a correlation coefficient of 0.9741. When the line is forced to pass through the origin, the standard error is 2.6 e.u. It will be noted that a significant portion of the database is clustered around 13 e.u. These are the molecules whose fusion entropy would be estimated quite well by Walden's Rule. From a structural perspective, most of these molecules are highly branched. The molecules that seem to deviate most significantly from Walden's Rule are linear molecules or molecules that contain long linear side chains.

Estimation of Total Phase Change Enthalpies of Hydrocarbons

Multiplying the total phase change entropy calculated by using the group values in Table 2 by the experimental melting point of each compound, $\Delta H_{tpce} = T_{fus} \cdot \Delta S_{tpce}(\text{calcd})$, gives the results shown in Figure 5. This figure compares the calculated total phase change enthalpy to the total experimental phase change enthalpy. The equation for the best straight line through these points is characterized by a slope of 0.994, an intercept of 0.063 and a correlation coefficient of 0.9733. The standard error in

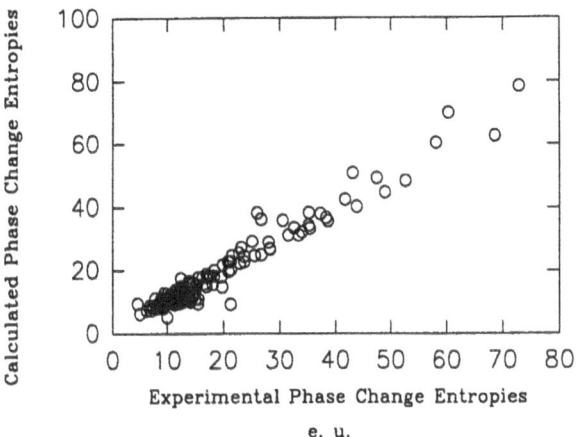

Figure 4. Experimental and calculated phase change entropy in the solid state (1 e.u. equals 1 cal/(mole K); 4.184 J/(mole K)). For most compounds this is equivalent to the fusion entropy.

the calculated total phase change enthalpy when the line is forced to pass through the origin is 0.92 kcal/mole (3.8 kJ/mole). This error can be compared to errors in experimental phase change enthalpies calculated from the Domalski, Evans and Hearing compendium[17a]. A standard error of 0.66 kcal/mole (2.8 kJ/mole) was obtained from the numerical agreement observed for each compound with multiple experimental citations.

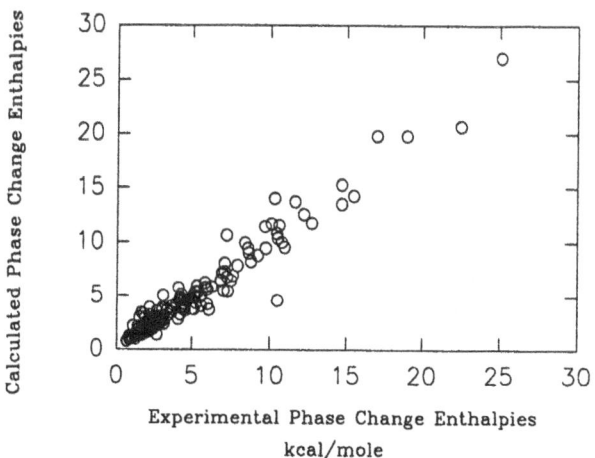

Figure 5. Experimental and calculated phase change enthalpy of hydrocarbons in the solid state. For most compounds this is equivalent to the fusion enthalpy.

Estimation of Sublimation Enthalpies of Hydrocarbons

Estimation of the enthalpy of sublimation of hydrocarbons using equation 1 is now relatively straightforward. Using equation 4 to estimate vaporization enthalpies, the

168

group values in Table 2 to estimate total phase change entropies, and the experimental melting point, the sublimation enthalpies of most hydrocarbons can now be estimated. Estimations by equation 1 are compared to 137 experimental sublimation enthalpies in Figure 6. The equation of the best straight line passing through these points is characterized by a slope of 0.994, an intercept of 0.0 and a correlation coefficient of 0.963. The standard error is 2.6 kcal/mole (10.9 kJ/mole). This error can be compared to a standard error of 1.74 kcal/mole (7.28 kJ/mole) calculated from the experimental agreement observed from 191 independent experimental measurements reported on 44 different hydrocarbons[22]. The standard error in using equation 1 to estimate sublimation enthalpies of hydrocarbons (2.6 kcal/mole, 10.9 kJ/mole) falls within two standard deviations of the experimental error.

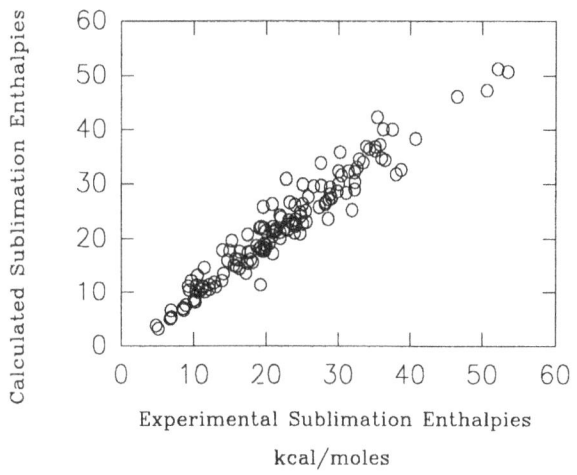

Figure 6. Calculated and experimental sublimation enthalpies of hydrocarbons calculated using equation 4 to obtain the vaporization enthalpy and the group values of Table 1 to obtain fusion enthalpies.

Estimation of Sublimation Enthalpies of Hydrocarbon Derivatives

The estimation of sublimation enthalpies of hydrocarbon derivatives according to equation 1 requires methods of estimation of vaporization and fusion enthalpies of these derivatives. Methods for the estimation of vaporization enthalpies of these compounds are known[11,12] and recently, group parameters for the estimation of fusion enthalpies (total phase change enthalpies) of hydrocarbon derivatives have been reported[23]. The method is similar to the method described above. For additional details the reader is directed to the literature.

Acknowledgment. We are grateful to the Weldon Spring Fund of the University of Missouri, the Petroleum Research Fund, administered by the American Chemical Society and the Thermodynamics and Kinetics Division of the National Institutes of Standards and Technology for support of portions of this work.

References

1. See, for example: Benson, S. W. *Thermochemical Kinetics: Methods for the Estimation of Thermochemical Data and Rate Parameters,* 2nd ed.; Wiley: New York, 1976.
2. Greenberg, A. and Liebman, J. F. *Strained Organic Molecules;* Academic Press: New York, 1978.
3. Davies, M. *J. Chem. Educ.* **1971**, *48*, 591; Gardner, P. J. and Hussain, K. S. *J. Chem. Thermodyn.* **1972**, *4*, 819.
4. Morawetz, E. *J. Chem. Thermodyn.* **1972** *4*, 455, 461.
5. Walden, P. Z. *Elektrochem.* **1908**, *14*, 713.
6. Trouton, F. *Phil Mag.* **1884**, *18*, 54.
7. Bondi A. *J. Chem. Eng. Data* **1963**, *8*, 371.
8. Domalski, E. S. *Environ. Sci. Technol.* **1971**, *5*, 443.
9. Laidler, K. *Canadian J. Chem.* **1956**, *34*, 626.
10. Guthrie, J. P. and Taylor, K. F. *Canadian J. Chem.* **1983**, *61*, 602.
11. Ducros, M.; Greison, J. F., and Sannier, H. *Thermochim. Acta* **1980**, *36*, 39; Ducros, M., Greison, J. F., Sannier, H., and Velasco, I. *Thermochim. Acta* **1981**, *44*, 134;. Ducros, M. and Sannier, H. *Thermochim. Acta* **1984**, *75*, 329.
12. a) Chickos, J. S., Hyman, A. S., Ladon, L. H., and Liebman, J. F. *J. Org. Chem.* **1981**, *46*, 4295; b) Chickos, J. S., Hesse, D. G., Liebman, J. F., and Panshin, S. Y. *J. Org. Chem.* **1988**, *53*, 3424; c) Chickos, J. S., Hesse, D. G., and Liebman, J. F. *J. Org. Chem.* **1989**, *54*, 5250.
13. Chickos, J. S., Annunziata, R., Ladon, L. H., Hyman, A. S., and Liebman, J. F. *J. Org. Chem.* **1986**, *51*, 4311;
14. Domalski, E. S. and Hearing, E. D. *J. Phys. Chem. Ref. Data* **1988**, *17*, 1637.
15. The total phase change entropy or enthalpy refers to the total entropy or enthalpy change associated with any discontinuity in the heat capacity curve.
16. Leclercq, M., Collet, A., aand Jacques, J. *Tetrahedron* **1976**, *32*, 821; Jacques, J., Collet, A., and Wilen, S. H. *Enantiomers, Racemates and Resolutions,* Wiley and Sons: New York, N.Y., 1981, Chapter 2.
17. a) Domalski, E. S., Evans, W. H., and Hearing, E. D. *J. Phys Chem. Ref. Data* **1984**, *13*, Suppl.1; b) Domalski, E. S. and Hearing, E. D. *J. Phys Chem. Ref. Data* **1990**, *19*, *(4)* 881.
18. Some additional data was obtained from the following: a) Kratt, G., Beckhaus, H. -D., Bernlohr, W., and Ruchardt, C. *Thermochimica Acta* **1983**, 62, 279; b) Chickos, J. S., Hesse, D. G., Liebman, J. F. *J. Org. Chem.* **1990**, *55*, 3833.
19. The parameter that was minimized by the method of least squares was the fractional error in the total phase change entropy, $[\Delta S_{tpce}(\text{expt}) - \Delta S_{tpce}(\text{calcd})]/\Delta S_{tpce}(\text{expt})$[18b].
20. 1 e.u. equals 1 cal/K·mole; 4.184 J/K·mole.
21. The term aromatic is used in a general sense to include all fully conjugated planar non-benzenoid aromatic systems as well. This includes molecules such as azulene and acenapthylene.
22. Chickos J. S. "Heats of Sublimation" Chapter 2, in J.F. Liebman and A. Greenberg (Eds.), *Molecular Structure and Energetics, Physical Measurements,* VCH Publishers, Inc., New York, 1987, Vol. 2.
23. Chickos, J. S., Braton, C. M., Hesse, D. G., and Liebman, J. F. *J. Org. Chem.* **1991**, *56*, 927.

DERIVATION OF THERMODYNAMIC QUANTITIES FROM KINETIC MEASUREMENTS IN GAS PHASE SILANE CHEMISTRY

R.WALSH
Department of Chemistry
University of Reading
Whiteknights, P.O.Box 224
Reading RG6 2AD
U.K.

ABSTRACT. This paper describes recent kinetic measurements of elementary reactions involving reactive silicon species such as SiH_3 and SiH_2, which have led to improvements in our knowledge of fundamental thermodynamic quantities. In particular a new value for $DH^o(SiH_3\text{-}H)$ and a refined value for $\Delta H_f^o(SiH_2)$ are derived. It is shown that these improvements have come about through a combination of careful measurement and the elimination of previous assumptions used in estimation of these quantities. The article also explores the wider implications of the new numbers as well as some of the reasons for previous (erroneous) estimates. The continuing need for reliable (ancillary) enthalpies of formation is emphasised.

1. Introduction

The traditional methods of determining thermodynamic quantities are *calorimetry* and the study of *equilibria* [1,2]. These methods are appropriate for pure stable compounds and clean reactions but are not generally possible for reactive species of lower (or higher) valence states than those commonly encountered. For such species an approach which has been frequently adopted is that based on a variant of the equilibrium law, viz. the principle of microscopic reversibility (see, e.g. [3]). In a reversible process such as

$$A + B \underset{r}{\overset{f}{\rightleftharpoons}} C + D$$

the equilibrium constant K is obtained as the ratio of forward and reverse rate constants k_f/k_r. This procedure is valid provided the elementary processes can be clearly identified and their rate constants measured under identical conditions (of temperature and pressure, in order to maintain the same, normally Boltzmann, molecular distribution over internal energies states for all species). Measurements of rate constants (and their temperature dependences) can therefore be used to obtain useful thermodynamic information on a variety of interesting reactive species.

This article describes the results of selected studies, mainly for reactions of silicon-containing molecules. These studies appear to have reached the state of reliability where, in the most favourable cases, enthalpy changes (for example bond dissociation enthalpies) are determined with uncertainties of no more than ± 2 kJ mol^{-1}. This article is also used as a

J. A. Martinho Simões (ed.), Energetics of Organometallic Species, 171–187.

TABLE 1. Landmark values of DH^o (SiH$_3$-H)

Authors	Ref	DH^o (SiH$_3$-H) / kJ mol^{-1}
Saalfeld and Svec (1966)	[4]	398 ± 8
Doncaster and Walsh (1981)	[5]	378 ± 5
Boo and Armentrout (1987)	[6]	387 ± 7
Seetula et al (1991)	[7]	384 ± 2

vehicle to try to illustrate how accepted values for some quantities have evolved. Changes in generally accepted values come about, not so much through remeasurement of an existing quantity but rather through elimination, via new measurement, of an assumption. It has almost invariably been necessary to make assumptions in the chain of reasoning between measurement and evaluation of a desired quantity. Progress and change come when these assumptions are directly tested.

These points may be illustrated by the data in tables 1 and 2. Table 1 shows a series of landmark measurements of DH^o(SiH$_3$-H), the Si-H bond dissociation enthalpy (BDE) in silane itself. These values show a range of 20 kJ mol^{-1}. This might not appear to be a large uncertainty, set beside the assumption often made prior to 1960 and still persisting in some quarters, of equating DH^o(SiH$_3$-H) to the average bond enthalpy of 322 kJ mol^{-1}. Close examination of the table shows a kind of oscillation in magnitude but refinement in values. Sequential values do not quite overlap within the claimed error limits. Boo and Armentrout [6] using guided Si$^+$ ion beams (of known kinetic energy) to measure the threshold energy for SiH$^+$ + SiH$_3$ appearance from SiH$_4$, was essentially a more controlled mass-spectrometric based method than the old electron impact studies of Saalfeld and Svec [4]. The studies of Seetula et al [7] of the kinetics of the SiH$_3$ + HBr/HI reactions, essentially eliminated one of the assumptions of Doncaster and Walsh [5] in their analysis of the kinetics of the I + SiH$_4$ reaction.

Table 2 shows data published in an earlier review [8] by the present author of BDE values for Si-Si bonds in disilane and hexamethyldisilane. The higher value for the latter led me to suggest that methyl groups can act as bond-strengthening substituents in silicon chemistry. This seductive idea, contrasting with organic chemistry, appeared to be reinforced by known Pauling electronegativities for carbon (2.6) and hydrogen (2.2). It will be seen later how this idea led to erroneous conclusions.

TABLE 2. 1981 values [8] of DH^o(Si-Si) in disilane and hexamethyldisilane

Bond	DH^o / kJ mol^{-1}	Authors and reference
H$_3$Si-SiH$_3$	308	Doncaster and Walsh [5]
Me$_3$Si-SiMe$_3$	337	Davidson and Howard [9]

2. Basic Principles of the Kinetic Method

The basic principles of the kinetic method are embodied in equations (1) - (3) :

$$K_{eq} = k_f / k_r \tag{1}$$

$$\Delta H^o = E_f - E_r \tag{2}$$

$$\Delta S^o = R \ln(A_f / A_r) \tag{3}$$

where ΔH^o and ΔS^o are the standard enthalpy and entropy changes resp., E_f and E_r are the forward and reverse activation energies, resp., and A_f and A_r the forward and reverse Arrhenius A-factors. These relationships assume an Arrhenius form for the temperature dependence of the rate constants. When applying these equations, since the standard state pressure is 1 bar ($= 10^5$ Nm^{-1}), strict convention would suggest the use of rate constants with these units. This is not common practice. (It is not even common practice to use 1 atm ($= 1.01 \times 10^5$ Nm^{-2}), the older standard pressure, which is the basis for most of the thermodynamic data quoted and derived in this article). More commonly concentration units are used (e.g. cm^3 molecule^{-1} s^{-1} or dm^3 mol^{-1} s^{-1} for second order rate constants) and equations (1) - (3) are employed in slightly modified form (see ref [10] for fuller details). Further complications will arise if non-Arrhenius forms such as $k = AT^n \exp(-E/RT)$ are used for rate constants (as is increasingly the practice for many rate processes). These complications can be by-passed by use of a version of the so-called "third-law" method when the main object is to obtain ΔH^o. In this case equation (1) is used to calculate K_{eq} (converted to standard pressure units) and then ΔG^o and ΔH^o are calculated via:

$$\Delta G^o = -RT \ln K_{eq} \tag{4}$$

$$\text{and } \Delta H^o = \Delta G^o + T \Delta S^o \tag{5}$$

This requires a knowledge of ΔS^o for reaction, but this is often known or can be calculated with reasonable precision.

One of the advantages of the equilibrium method, and therefore of the use of rate constant ratios, is that even quite rough values for K_{eq} can lead to reasonably precise values for ΔG^o. For instance, uncertainties of 10% in k_f and k_r, leading to $\pm 14\%$ in K_{eq}, correspond to only ± 0.33 kJ mol^{-1} in ΔG^o at 298 K. Even factors of 2 errors in the rate constants lead to an error of only 2.6 kJ mol^{-1} in ΔG^o. Even at higher temperatures of ca 600K, where these rate constants are often measured the uncertainties are only doubled. At its most pessimistic, this uncertainty is usually beyond the capability of calorimetry, or threshold energy measurements.

Two difficulties of this method, which are the cause of most of the uncertainty in derived enthalpies of formation, are the following. First it is not easy to study reversible elementary processes *under the same conditions*. This is most easily illustrated for a dissociation reaction such as

$$C_2H_6 \underset{r}{\overset{f}{\rightleftharpoons}} 2CH_3$$

The forward reaction requires high temperatures ($T > 700$ K whereas the reverse methyl radical recombination reaction is most easily studied (using flash photolysis techniques) at room temperature. However these days higher temperature flash photolysis experiments are routinely carried out, the only limitations being those of the temperature stability of the radical-precursor molecules. Another problem with such reactions is that of pressure dependences. For relatively small molecules and radicals, the rate constants for unimolecular dissociations and radical associations are subject to *pressure dependencies* [11]. Moreover the extent of pressure dependence is itself a function of both temperature and reaction mixture diluent gas. Thus a proper comparative measurement of k_f and k_r in these systems requires not only the same temperature and pressure but also the same diluent bath gas. This can be avoided if high enough pressures are used to reach the so-called high-pressure limiting rate constant (or if extrapolation from lower pressures can be carried out). A review of recent rate data in several reaction systems of this type has been carried out by Hughes and Pilling [12].

The second general difficulty is that of temperature extrapolation down to 298 K of enthalpy changes derived from rate data at high temperature. In principle the procedure is straightforward and is carried out via equation (6):

$$\Delta H^\circ(298 \ K) = \Delta H^\circ(T) - \int_{298 \ K}^{T} \Delta C_p dT \qquad (6)$$

The problem is that ΔC_p, the specific heat change, may not be known with precision, especially for a reaction involving large radicals with several internal rotational degrees of freedom. Griller and colleagues [13] have noted that different values for alkyl radical enthalpies of formation have been obtained by different groups [14,15] using similar experimental rate data but different temperature extrapolations. O'Neal & Ring et al [16] have suggested the safest procedure is to extrapolate k_f to 298 K (using Transition State Theory, TST) before combination with k_r to avoid these problems. The trouble with this procedure is that the transition state structure is sensitive to errors in the measured A factor, and additionally the structure may not remain fixed over a temperature range of several hundreds of Kelvins. For a dissociation reaction with a typical activation energy of *ca* 210 kJ mol^{-1}, k_f will decrease by a factor of *ca* 10^{18} between 600 K and 298 K. A small deviation in the TST form (or a small error in E_a) can easily change this by a factor of 10. Thus the problem is not really avoided by extrapolating rate data rather than equilibrium data.

In the following sections, two reversible reaction systems are put under the microscope, viz. the iodination reaction devised by Benson and colleagues to obtain bond dissociation enthalpies [17] and the thermal dissociation reactions of disilanes which are used as the basis of silylene thermochemistry [16]. It should be emphasised that this article is intended to illuminate progress rather than be a comprehensive review of data.

3. Bond Dissociation Enthalpies via the Iodination method

3.1. ABSOLUTE VALUES

Until just over 10 years ago, it appeared that the prototype C-H BDEs in alkanes were well established [18]. The basis of their determination was the reaction

$$I + RH \underset{r}{\overset{f}{\rightleftharpoons}} R + HI$$

For this reaction we may write the exact enthalpic equation:

$$DH^o(\text{R-H}) = DH^o(\text{H-I}) + E_f - E_r \qquad (7)$$

The value of $DH^o(\text{H-I})$ is very precisely known [17] and in the experiment E_f is measured usually with a precision of $\pm$ 2 kJ mol^{-1}. Details of the experimental procedure and data evaluation have been extensively discussed [8,17]. They are not thought to be contentious and are not therefore included in the present article. The reverse reaction was thought to be fast but direct measurements of k_r were not available. In their absence it was assumed (based on plausible argument) that E_r values were small and lay in the range 4-8 kJ mol^{-1}.

Recently Gutman's group have carried out direct time-resolved measurements of rate constants for reactions of alkyl radicals with HI, using photo-ionisation mass spectrometry to monitor the decay of alkyl radicals produced by pulsed laser photodissociation of suitable precursors in an excess of HI [19]. The results of these measurements carried out over a typical temperature range of 300-650 K are shown in table 3. It seems clear from the data shown that

TABLE 3. Measured activation energies for reactions of alkyl radicals with HI [19]

Radical	E_a / kJ mol^{-1}
CH_3	- 1.2 $\pm$ 0.6
C_2H_5	- 3.2 $\pm$ 0.6
i-C_3H_7	- 5.1 $\pm$ 0.7
t-C_4H_9	- 6.3 $\pm$ 0.8

the assumption of small positive activation energies for these reactions has to be replaced. It should be added that while a mechanistic explanation for these negative activation energies has been offered [19,20], there are still disagreements over the data in some cases [21,22]. Nevertheless at the present time, the weight of evidence (analogous bromination kinetic studies [23,24]) supports these new results. The revised BDE values calculated from the new data are compared with the previously estimated values in table 4. These data appear to resolve in large measure long-standing discrepancies in hydrocarbon fragmentation processes [14].

In the past year Gutman's group in collaboration with that of Pilling [7] have extended their studies to include the reaction of SiH_3 with HI (and with HBr). The reaction, studied between 297 and 550 K, had an activation energy of -2.0 kJ mol^{-1}. When combined with the measured E_f for I + SiH_4 [5] this yielded $DH^o(\text{SiH}_3\text{-H}) = 384 \pm 2$ kJ mol^{-1}, an increase of 6 kJ mol^{-1} relative to our earlier value [5,8] based on an *estimate* of E_r. The question now arises as to whether *all* Si-H bond dissociation enthalpies are too low by similar amounts. In view of the apparent insensitivity of earlier DH^o values for Si-H in methylsilanes [8] to methyl substituents it would seem reasonable to expect this. We recommend that, until the appropriate R + HI studies are carried out, a value of 384 kJ mol^{-1} represents the best estimate for $DH^o(\text{Me}_n\text{SiH}_{3-n}\text{-H})$ for all values of n (0-3).

TABLE 4. Bond dissociation enthalpies
determined by iodination kinetics

Bond	DH^o / kJ mol^{-1}	
	Old [18]	New [19]
CH_3 - H	440	438
C_2H_5 - H	411	421
i-C_3H_7 - H	398	411
t-C_4H_9 - H	390	401

3.2. RELATIVE VALUES

The assumption of a constant value for E_r , based on the now-known value for the reaction of SiH_3 + HI, would appear reasonable. Thus the relativities between all DH^o values for Si-H previously determined by the iodination method [8] should be unaltered. The substituent effects already discussed [8] are therefore unaffected by the changes brought about by the Gutman determination. However, our knowledge of substituent effects has been extended by the application of photoacoustic calorimetry (PAC) to the measurement of Si-H bond dissociation enthalpies [25,26]. Table 5 shows a more extended list of these effects than available previously

TABLE 5. Substituent effects on Si-H bond dissociation enthalpies.

Substituent(s)	ΔDH^o / kJ mol^{-1} [a]	Reference
Me,Me$_2$,Me$_3$	0	This review
Ph	- 9	[8]
Ph$_2$	- 18	[26]
Ph$_3$	- 26	[26]
SiH$_3$	- 17	[8]
(SiMe$_3$)$_3$	- 46	[25]
Cl$_3$	+ 4	[8]
F$_3$	+41	[8]

[a]   Relative to DH^o (SiH_3-H)

[8,27]. In this table the relativities amongst values determined by PAC are based on comparison with DH^o(Et$_3$Si-H) which is assumed to have the same value as DH^o(Me$_3$Si-H) even though the original value (377 kJ mol^{-1}) [25] was lower than that proposed above.

The figures in table 5 show that silyl and phenyl substituents are bond weakeners while

Cl and F are bond strengtheners. These effects have been discussed elsewhere [8,26] and it is not the purpose of this article to pursue that discussion further here.

3.3 DERIVED VALUES

The availability of Si-H bond dissociation enthalpies means that values for Si-C and Si-Si bonds (and others) may be derived. This is simply done by use of Hess's law cycles, employing equations such as (8) and (9)

$$\Delta H_f^o(R) = DH^o(R-H) + \Delta H_f^o(RH) - \Delta H_f^o(H) \tag{8}$$

$$DH^o(R-X) = \Delta H_f^o(R) + \Delta H_f^o(X) - \Delta H_f^o(RX) \tag{9}$$

If $\Delta H_f^o(RH)$ represents the enthalpy of formation of a methylsilane then the Si-C bond dissociation enthalpies shown in table 6 may be derived via $\Delta H_f^o(R)$ values (which are not shown). $\Delta H_f^o(RH)$ values are reasonably reliably known [28] and have not altered their values since 1981. Although the Si-C bonds apparently show small increases with methyl substitution, the variations are scarcely beyond experimental error. They are approximately constant.

When X in equation (9) represents a silyl group then the data may be used to derive Si-Si bond dissociation enthalpies. This requires a knowledge of $\Delta H_f^o(RX)$, which represent, in this case, enthalpies of formation of disilanes. While $\Delta H_f^o(Si_2H_6)$ appears to be reliably known [29], values for other (methyl-substituted) disilanes have been subject to considerable dispute. These uncertainties have been largely resolved by a recent, solution calorimetric, determination of ΔH_f^o (Si_2Me_6) by Pilcher's group [30]. Reliable calorimetric data for organosilanes is still much in need and this recent determination illustrates this point. The consequential Si-Si bond dissociation enthalpies are shown in table 7. The striking feature of this data is the reversal in trends of methyl substitution resulting from the new data. Methyl groups now appear to be bond weakening rather than strengthening although it should be said that there is probably enough uncertainty that $DH^o(Si-Si)$ are almost the same in Si_2H_6 and Si_2Me_6. The data of tables 6 and 7 show that methyl groups, whether bond-weakening or strengthening, have rather small effects. Incidentally the older value for $DH^o(Me_3Si-SiMe_3)$ is based on an activation energy measurement [9] in the pyrolysis of Si_2Me_6 rather than a Hess's Law cycle. We are currently reinvestigating this pyrolysis [31] to try to resolve the discrepancy in $DH^o(Me_3Si-SiMe_3)$.

TABLE 6. Derived values for Si-C bond dissociation enthalpies.

| Bond | DH^o / kJ mol^{-1} | |
	Old [8]	New (see text)
H_3Si-CH_3	369	376
$MeSiH_2-CH_3$	369	378
$Me_2SiH-CH_3$	369	381
Me_3Si-CH_3	374	382

TABLE 7. Derived values for Si-Si bond dissociation enthalpies.

Bond	DH^o / kJ mol^{-1}	
	Old [8]	New (see text)
$H_3Si-SiH_3$	308	321
$Me_3Si-SiMe_3$	337	310

4. Silylene studies

4.1. GENERAL REMARKS

Silylenes are ubiquitous intermediates in thermal and photochemical decompositions of silanes and organosilanes. This comes about, in part, because of their special stability, arising from the low energy of the in-plane lone pair orbital, with its high s character. We have defined the term divalent state stabilisation energy (DSSE) to try to quantify this effect [32]. Operationally the DSSE of a silylene, SiX_2, is the difference between the first and second dissociation enthalpies in any SiX_4 species. Thus,

$$DSSE\ (SiX_2) = DH^o\ (SiX_3\text{-}X) - DH^o\ (SiX_2\text{-}X) \tag{10}$$

If the definitions of DH^o are incorporated, equation (10) can be transformed into (11)

$$DSSE\ (SiX_2) = 2\ \Delta H_f^o\ (SiX_3) - \Delta H_f^o\ (SiX_4) - \Delta H_f^o\ (SiX_2) \tag{11}$$

Thus it is clear that for any discussion of the stability of silylenes, a knowledge of $\Delta H_f^o\ (SiX_2)$ is important. This section is devoted to its measurement for SiH_2 and $SiMe_2$. Recently Grev [33] has extended the idea of DSSE to carbenes and germylenes.

4.2. ENTHALPY OF FORMATION OF SiH_2

Until ca 1985 the most commonly quoted value for $\Delta H_f^o\ (SiH_2)$ was 242 kJ mol^{-1} [34]. This was based on rate constant data for the reactions

$$SiH_4 \rightleftarrows SiH_2 + H_2$$

$$Si_2H_6 \rightleftarrows SiH_2 + SiH_4$$

In fact it is these two reactions which provide the basis up to the present time for estimates of $\Delta H_f^o\ (SiH_2)$. Progress has come in a steady refinement of knowledge of the rate constants in these systems. Before 1985, the forward unimolecular decomposition steps were reasonably well studied and their rate constants fairly reliably measured. However the reverse reactions had not been studied at all except relative to one another and some other insertion processes. The John and Purnell value [34] for $\Delta H_f^o\ (SiH_2)$ was based on $estimates$ of their rate constants. The situation changed dramatically in 1985/6, with two papers, the first by Inoue and Suzuki [35] and the second by Jasinski [36], who reported for the first time direct time-resolved studies of SiH_2 in which the key rate constants were measured at room temperature. The values obtained were *ca* 10^4 greater than the earlier estimates and in the case of $SiH_2 + SiH_4$, the reaction occurs at practically every collision.

Jasinski [36] pointed out that his measurements implied an upward revision of ΔH_f^o (SiH_2) towards a new figure of 285 kJ mol^{-1} obtained by theoretical calculation [37]. We also pointed out by analysis of both reactions [32, 38] that a value of 273 $\pm$ 6 kJ mol^{-1} was implied by the new measurements. In these estimates it was assumed that the reactions had zero activation energy (rate constants were only measured at 298 K). This has subsequently been borne out by measurements in our laboratory for the reaction of SiH_2 with D_2 [39].

In this article we refine ΔH_f^o (SiH_2) further by employing new rate constants for $SiH_2 + SiH_4$ measured as a function of both temperature and pressure [40]. These rate constants have been obtained by laser flash photolysis. SiH_2 is created by 193 nm excimer laser photolysis of phenylsilane and detected and monitored via CW single-mode dye laser absorption of the $^RQ_{0.5}$ rovibrational transition in the $^1B_1(0,2,0) \leftarrow {}^1A_1(0,0,0)$ visible vibronic spectrum. The rate constants have been obtained in the pressure range 1-100 torr in the presence of both Ar and SF_6 bath gases and show the expected pressure dependences associated with a third body association process. Selected low pressure and also high pressure limiting rate constants (measured directly at lower temperatures and obtained by extrapolation at higher temperatures) are shown in table 8. The high pressure limiting rate constants fit an Arrhenius equation

$$\log (k/\text{cm}^3 \text{ molecule}^{-1} \text{ s}^{-1}) = -9.98 \pm 0.07 + (3.7 \pm 0.6 \text{ kJ mol}^{-1}) / RT \ln 10.$$

TABLE 8. Rate constants for reaction of SiH_2 with SiH_4

T/K	$k / 10^{-10}$ cm^3 molecule^{-1} s^{-1}		
	10 Torr (Ar)	10 Torr (SF$_6$)	∞
296	3.3	4.2	4.2
363	2.2	3.3	3.6
432	1.9	2.9	3.2
488	1.1	1.9	2.8
576	0.55	1.3	2.2
653	0.29	0.72	1.8

This equation with its negative activation energy emphasises the point that for a correct thermodynamic evaluation for this reaction the room temperature rate constant is not adequate. For convenience the calculation of ΔH_f^o (SiH_2) was based on rate constants measured at 552 K. The details are shown in table 9. The data have been processed via the so-called "third law" method in which the entropy change is calculated independently, and used to convert ΔG^o to ΔH^o. In this reaction entropy and specific heat changes are precisely obtained by combining tabulated values [41, 42] for SiH_4 and SiH_2 with values calculated from spectroscopic data [43] for Si_2H_6. The derived ΔH^o(298 K) value is turned into $\Delta H_f^o(SiH_2)$ using the known experimental values for ΔH_f^o (SiH_4) and ΔH_f^o (Si_2H_6) [29, 41].

TABLE 9. Calculation of ΔH_f^o (SiH_2) from experimental rate constants for the reaction:

$$Si_2H_6 \underset{r}{\overset{f}{\rightleftarrows}} SiH_4 + SiH_2$$

$T = 551.6$ K	k_f^∞	$=$	1.28×10^{-5} s^{-1}	(ref [44])
	k_r^∞	$=$	$(2.4 \pm 0.4) \times 10^{-10}$ cm^3 molecule^{-1} s^{-1}	(ref [40])
	K_p	$=$	$(4.0 \pm 0.7) \times 10^{-15}$ atm	
	ΔG^o	$=$	151.9 ± 0.8 kJ mol^{-1}	
	ΔH^o	$=$	225.9 ± 0.8 kJ mol^{-1} (via ΔS^o)	
$T = 298$ K	ΔH^o	$=$	227.2 ± 0.8 kJ mol^{-1} (via $\overline{\Delta C_p^o}$)	
	ΔH_f^o (SiH_2)	$=$	272.8 ± 2.1 kJ mol^{-1}	

This result is now quite precise and the uncertainties are experimental, rather than any built-in assumptions. It is sobering to compare it with the many values published since 1985 shown in table 10. Clearly the value derived here is in good agreement with several previous

TABLE 10. Recent values for ΔH_f^o (SiH_2) / kJ mol^{-1}

	Value	Authors	Year	Ref.
	265	Pople, Luke, Frisch, Binkley	1985	[45]
	285	Ho, Coltrin, Binkley, Melius	1985	[37]
*	273 ± 6	Frey, Walsh, Watts	1986	[38]
	273	Gordon, Gano, Binkley, Frisch	1986	[46]
	289 ± 13	Shin, Beauchamp	1986	[47]
*	273 ± 3	Berkowitz, Green, Cho, Rusic	1987	[48]
*	287 ± 6	Boo, Armentrout	1987	[6]
*	269 ± 1	Martin, Ring, O'Neal	1987	[44]
*	274 ± 7	Van Zoeren, Thomas, Steinfeld, Rainbird	1988	[49]
	266	Curtiss, Pople	1988	[50]
	272	Horowicz, Goddard	1988	[51]
*	267	O'Neal, Ring, Richardson, Licciardi	1989	[16]
	271 ± 9	Ho, Melius	1990	[52]
	274 ± 4	Moffatt, Jensen, Carr	1991	[53]
*	273 ± 2	Becerra, Frey, Mason, Walsh (unpublished)	1991	[40]

* Experimental values

determinations. Its merit however, lies in the fact that it is now the value with the highest precision. Experiment and theory are in tolerably good agreement over ΔH_f^o (SiH$_2$).

4.3 ENTHALPY OF FORMATION OF SiMe$_2$

In 1981 we proposed [8] ΔH_f^o (SiMe$_2$) = 109 kJ mol^{-1} based on a Me-for-H replacement enthalpy of -67 kJ mol^{-1}. Subsequent to this there have been various determinations of this quantity, the principal ones being shown in table 11. Clearly our own estimate, three times

TABLE 11. Recent values for ΔH_f^o (SiMe$_2$) / kJ mol^{-1}

Value	Authors	Reference
92 ± 8	Walsh	[54]
109 ± 8	Walsh	[55, 56]
155 ± 25	Shin, Irikura, Beauchamp, Goddard III	[57]
141 ± 8	O'Neal, Ring, Richardson, Licciardi	[16]
136 ± 13	Gordon, Boatz	[58]

reassessed since 1981, is out of line with the others. Once again the basis of this was kinetic measurements, this time based on the reaction

$$Me_3SiSiMe_2H \rightleftarrows Me_3SiH + SiMe_2$$

The assessment by ourselves [56] and by O'Neal et al [16] came up with values of 220 ± 6 kJ mol^{-1} and 218 kJ mol^{-1}, respectively, for ΔH^o. This is not, therefore, the major source of disagreement. Nevertheless these numbers may be refined since at the time of their publication [16, 56] the temperature dependence of the reverse reaction, the insertion of SiMe$_2$ into the Si-H bond of Me$_3$SiH had not been investigated; in other words, k_r had been estimated. We have subsequently carried out a full study of the kinetics of this reaction [59] over the temperature range 300-600 K.

These rate constants were again obtained by laser flash photolysis. SiMe$_2$ is formed by the 193 nm photolysis of pentamethyldisilane and detected and monitored via CW Argon Ion laser absorption in the broad band $^1B_1 \leftarrow {}^1A_1$ transition. The measured rate constants also show substantial decreases with increasing temperature, again revealing the complexity of the rate process and difficulty of inferring thermochemistry from a room temperature value alone.

For convenience the calculation of ΔH^o was based on rate constants measured at 603 K. The details are shown in table 12. Once again as for the Si$_2$H$_6$ dissociation reaction, the third law method was used. There is, however, greater uncertainty associated with the ΔS^o and $\overline{\Delta C_p}^o$ values in this case. They are obtained by additivity for Me$_3$SiSiMe$_2$H [61], from tabulated data for Me$_3$SiH [28] and by use of tabulated data on Me$_2$S [1]. The latter was used as a model for SiMe$_2$, based on the structural analogy principle [10], since entropy and specific heat data are not available for SiMe$_2$ itself. The derived ΔH^o is slightly lower than the earlier values [16, 56] but represents an improvement. Error limits for ΔH^o are larger for the pentamethyldisilane

TABLE 12. Calculation of ΔH^o, from experimental rate constants for the reaction:

$$Me_3SiSiMe_2H \underset{r}{\overset{f}{\rightleftarrows}} Me_3SiH + SiMe_2$$

$T = 603$ K	k_f^∞	=	$8.5 \times 10^{-5} \text{ s}^{-1}$	(ref [59])
	k_r^∞	=	$(4.5 \pm 1.0) \times 10^{-13} \text{ cm}^3 \text{ molecule}^{-1} \text{ s}^{-1}$	(ref [60])
	K_p	=	$(1.55 \pm 0.34) \times 10^{11}$ atm	
	ΔG^o	=	124.8 ± 1.3 kJ mol^{-1}	
	ΔH^o	=	204.9 ± 5.0 kJ mol^{-1} (via ΔS^o)	
$T = 298$ K	ΔH^o	=	209.6 ± 5.4 kJ mol^{-1} (via $\overline{\Delta C_p}^{\,o}$)	

decomposition than for disilane decomposition because of greater uncertainties in the entropies of Me_5Si_2H and $SiMe_2$ (uncertain barriers to internal rotations).

To determine $\Delta H_f^o(SiMe_2)$ requires a knowledge of $\Delta H_f^o(Si_2Me_5H)$ and $\Delta H_f^o(Me_3SiH)$. The latter is reliably known [28] at -163.4 ± 4.0 kJ mol^{-1}. But the former depends on estimates of which there have been several [16, 56]. For instance O'Neal et al [16] used a value of -241 kJ mol whilst we [56] preferred -276 kJ mol^{-1}, which was the principal cause of the difference (see table 11) in ΔH_f^o (SiMe$_2$). The reasons for these choices may be found in the two papers. The question of the value of ΔH_f^o (Si$_2$Me$_5$H), although not directly determined, is now much clearer as a result of the new measurement of ΔH_f^o (Si$_2$Me$_6$) [30]. This value anchors ΔH_f^o for all the methyl substituted disilanes, by use of the Allen bonding scheme. It yields a value of ΔH_f^o (Si$_2$Me$_5$H) of -238 ± 6 kJ mol^{-1} (thus demonstrating our earlier estimate [56] was erroneous). When combined with the figure in table 12 this now gives a value for ΔH_f^o (SiMe$_2$) of 135 ± 8 kJ mol^{-1}, which is in much better agreement with the other values in table 11.

4.4. DIVALENT STATE STABILISATION ENERGIES (DSSE) FOR SILYLENES

The derivation of more precise and reliable values for ΔH_f^o (SiH$_2$) and ΔH_f^o (SiMe$_2$) means that better values than before [32] may now be derived for the DSSE for these two silylenes. Using the data from this paper in conjunction with equation (10) the values for SiH$_2$ and SiMe$_2$ are derived. These are shown in table 13 along with earlier values [32] for the halosilylenes, for which there have been no recent experimental refinements. The values for SiH$_2$ and SiMe$_2$ are very close together, thus indicating that there is rather little effect of methyl-for-hydrogen substitution on the DSSE. This revises our earlier suggestion [56] that methyl groups in silylenes are exhibiting an electronegative substituent effect. If the latest DSSE values are plotted against unshielded core potential (UCP) for substituent X in the series SiX$_2$, figure 1 is obtained. UCP, has been recently proposed by Benson and Luo [62] as the most successful modern scale of electronegativity. Superficially the correlation line appears quite good, but it fails to fit the most reliable values, viz those for SiH$_2$ and SiMe$_2$ (it is not improved if Pauling electronegativities are used). This is a cautionary tale about correlations (one of chemistry's favourite devices). We were strengthened in our belief in our earlier preferences [56] for ΔH_f^o and DSSE for SiMe$_2$ by the fact of a better correlation than in figure 1. We have had to relearn that correlations of this kind, whilst often useful guides to trends, must not be pushed to their quantitative limit.

TABLE 13. Divalent State Stabilisation Energies (DSSE)

Species	$DSSE$ / kJ mol^{-1}	Reference
SiH$_2$	93 ± 3	This review
SiMe$_2$	104 ± 11	This review
SiF$_2$	205 ± 42	[32]
SiCl$_2$	159 ± 17	[32]
SiBr$_2$	142 ± 50	[32]
SiI$_2$	134 ± 54	[32]

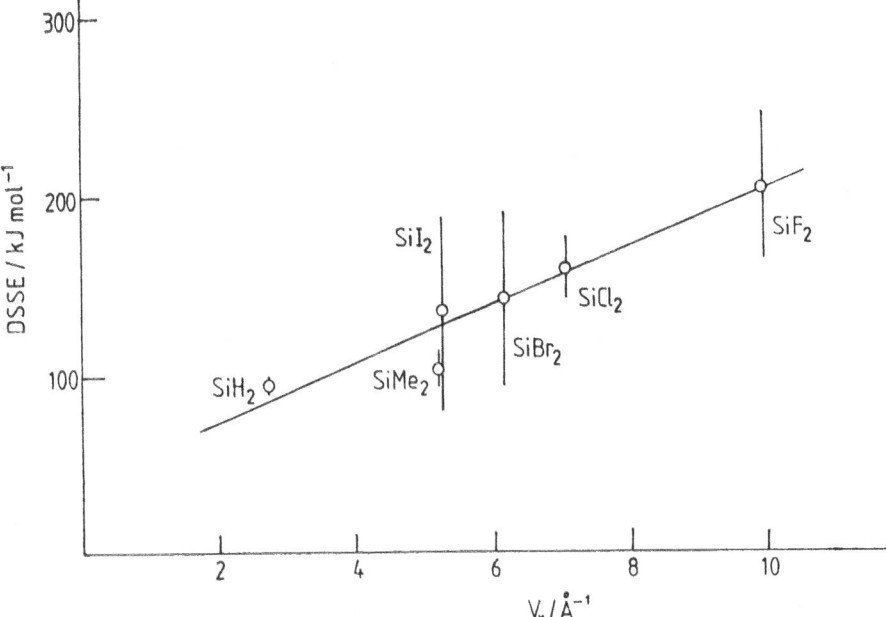

Figure 1. Correlation of DSSE for silylenes with unshielded core potential, V_x, for substituent.

5. Summary

This article has shown that carefully measured rate constants of forward and reverse elementary reactions *when measured under identical conditions* can lead to reliable thermodynamic data. This has been applied to two systems of general interest in silicon chemistry. In the first, the iodination reaction of silanes, we have shown how new measurements of rate constants for silyl radical with hydrogen iodide by Gutman and Pilling and their groups, have led to more precise value for $DH^o(SiH_3\text{-}H)$, the prototype Si-H bond dissociation enthalpy. In the second, the decomposition process of disilanes, we have shown how our own measurements of rate constants for silylene insertion reactions, have led to more precise values for ΔH_f^o (SiH_2) and ΔH_f^o ($SiMe_2$). These cases have also emphasised the continuing need for reliable enthalpies of ancillary compounds, since a new calorimetric value for ΔH_f^o (Si_2Me_6) by Pilcher's group has led to substantial changes in related derived quantities, such as DH^o ($Me_3Si\text{-}SiMe_3$) and $DSSE$ ($SiMe_2$).

6. Acknowledgements

The author would like to acknowledge support from the UK Science and Engineering Research Council which permitted the initiation of some of the studies described here. I would also like to thank members of the Reading Kinetics group, past and present, for their essential contributions to much of the work described here. Finally, I would acknowledge the stimulating and friendly interaction with the wider community of silicon chemists over many years.

7. References

[1]   Stull, D.R., Westrum, E.F.Jr., and Sinke, G.C. (1969) The Chemical Thermodynamics of Organic Compounds, Wiley, New York.

[2]   Cox, J.D., and Pilcher, G. (1970) Thermochemistry of Organic and Organometallic Compounds, Academic Press, London.

[3]   Moore, J.W., and Pearson, R.G. (1981) Kinetics and Mechanism, Wiley, New York, p.307.

[4]   Saalfeld, F.E., and Svec, H.J., J. Phys. Chem. (1966) **70**, 1753.

[5]   Doncaster, A.M., and Walsh, R., Int. J. Chem. Kinet. (1981) **13**, 503.

[6]   Boo, B.H., and Armentrout, P.B., J. Am. Chem. Soc. (1987) **109**, 3549.

[7]   Seetula, J.A., Feng, Y., Gutman, D., Seakins, P.W., and Pilling, M.J., J. Phys. Chem. (1991) **95**, 1658.

[8]   Walsh, R., Acc. Chem. Res. (1981) **14**, 246.

[9]   Davidson, I.M.T., and Howard, A.V., J. Chem. Soc., Faraday Trans. 1 (1975) **71**, 69.

[10]   Benson, S.W. (1976) Thermochemical Kinetics, 2nd ed., Wiley, New York.

[11]   Robinson, P.J., and Holbrook, K.A. (1972) Unimolecular Reactions, Wiley-Interscience, New York.

[12] Hughes, K.J., and Pilling, M.J., Annu. Rpts. Prog. Chem. Sect.C (1989) **85**, 91.

[13] Griller, D., Kanabus-Kaminski, J.M., and Maccoll, A., J. Mol. Str. (Theochem) (1988) **163**, 125.

[14] Tsang, W., J. Am. Chem. Soc. (1985) **107**, 2872.

[15] Baldwin, R.R., Drewery, G.R., and Walker, R.W., J. Chem. Soc., Faraday Trans. 1 (1984) **80**, 2827.

[16] O'Neal, H.E., Ring, M.A., Richardson, W.H., and Licciardi, G.F., Organometallics (1989) **8**, 1968.

[17] Golden, D.M., and Benson, S.W., Chem. Rev. (1969) **69**, 125.

[18] McMillen, D.F., and Golden, D.M., Annu. Rev. Phys. Chem. (1982) **33**, 493.

[19] Seetula, J.A., Russell, J.J., and Gutman, D., J. Am. Chem. Soc. (1990) **112**, 1347.

[20] McEwan, A.B., and Golden, D.M., J. Mol. Str. (Theochem) (1990) **224**, 357.

[21] Gutman, D., Acc. Chem. Res. (1990) **23**, 375.

[22] Muller-Markgraf, W., Rossi, M.J., and Golden, D.M., J. Am. Chem. Soc. (1989) **111**, 956.

[23] Russell, J.J., Seetula, J.A., Timonen, R.S., Gutman, D., and Nava, D.F., J. Am. Chem. Soc. (1988) **110**, 3084.

[24] Russell, J.J., Seetula, J.A., and Gutman, D., J. Am. Chem. Soc. (1988) **110**, 3092.

[25] Kanabus-Kaminska, J.M., Hawari, J.A., Griller, D., and Chatgilialoglu, C., J. Am. Chem. Soc. (1987) **109**, 5267.

[26] Griller, D., Kanabus-Kaminska, J.M., and Martinho Simões, J.A. (1990) Chap. 14 in ACS Symposium Series No.428, Bonding Energetics in Organometallic Compounds, Ed., Tobin Marks.

[27] Walsh, R., (1989) Thermochemistry, Chap. 5 in The Chemistry of Organic Silicon Compounds, Eds., S. Patai and Z. Rappoport, Wiley, New York.

[28] Doncaster, A.M., and Walsh, R., J. Chem. Soc., Faraday Trans. 1 (1986) **82**, 707.

[29] Pedley, J.B., and Iseard, B.S. (1972) Catch Tables for Silicon Compounds, University of Sussex.

[30] Pilcher, G., Leitão, M.L.P., Yang, M-Y., and Walsh, R., J. Chem. Soc., Faraday Trans. (1991) **87**, 841.

[31] Bullock, W., King, K.D., and Walsh, R., experiments in progress.

[32] Walsh, R., Pure & Appl. Chem. (1987) **59**, 69.

[33] Grev, R.S., Advances in Organometallic Chem. (1991) **33**, in press.

[34] John, P., and Purnell, J.H., J. Chem. Soc., Faraday Trans 1. (1973) **69**, 1455.

[35] Inoue, G., and Suzuki, M., Chem. Phys. Lett. (1985) **122**, 361.

[36] Jasinski, J.M., J. Phys. Chem. (1986) **90**, 555.

[37] Ho, P., Coltrin, M.E., Binkley, J.S. and Melius, C.F., J. Phys. Chem. (1985) **89**, 4647.

[38] Frey, H.M., Walsh, R., and Watts, I.M., J.C.S., Chem. Comm. (1986) 1189.

[39] Baggott, J.E., Frey, H.M., Lightfoot, P.D., Walsh, R., and Watts, I.M., J. Chem. Soc., Faraday Trans. (1990) **86**, 26.

[40] Becerra, R., Frey, H.M., Mason, B.P., and Walsh, R., unpublished results.

[41] JANAF Thermochemical Tables (1978 Supplement), J. Phys. Chem. Ref. Data, **7**, 793.

[42] Fredin, L., Hauge, R.H., Kafafi, Z.H., and Margrave, J.L., J. Chem. Phys. (1985) **82**, 3542.

[43] Fraioli, P., and Walsh, R., unpublished calculation.

[44] Martin, J.G., Ring, M.A., and O'Neal, H.E., Int. J. Chem. Kinet. (1987) **19**, 715.

[45] Pople, J.A., Luke, B.T., Frisch, M.J., and Binkley, J.S., J. Phys. Chem. (1985) **89**, 2198.

[46] Gordon, M.S., Gano, D.R., Binkley, J.S., and Frisch, M., J. Am. Chem. Soc. (1986) **108**, 2191.

[47] Shin, S.K., and Beauchamp, J.L., J. Phys. Chem. (1986) **90**, 1507.

[48] Berkowicz, J., Greene, J.P., Cho, H., and Ruscic, B., J. Phys. Chem. (1987) **86**, 1235.

[49] Van Zoeren, C.M., Thoman, J.W., Steinfeld, J.I., and Rainbird, N., J. Phys. Chem. (1988) **92**, 9.

[50] Curtiss, L.A., and Pople, J.A., Chem. Phys. Lett. (1988) **144**, 38.

[51] Horowicz, D.S., and Goddard lll, W.A., J. Mol. Str. (Theochem.) (1988) **163**, 207.

[52] Ho, P., and Melius, C.F., J. Phys. Chem. (1990) **94**, 5120.

[53] Moffat, H.K., Jensen, K.F., and Carr, R.W., J. Phys. Chem. (1991) **95**, 145.

[54] Walsh, R., J. Phys. Chem. (1986) **90**, 389.

[55] Walsh, R., Organometallics (1988) **7**, 75.

[56] Walsh, R., Organometallics (1989) **8**, 1973.

[57] Shin, S.K., Irikura, K.K., Beauchamp, J.L., and Goddard lll, W.A., J. Am. Chem. Soc. (1988) **110**, 24.

[58] Gordon, M.S., and Boatz, J.A., Organometallics (1989) **8**, 1978.

[59] Becerra, R., Bertram, J.S., Walsh, R., and Watts, I.M., J. Chem. Soc., Faraday Trans. 2 (1989) **85**, 1837.

[60] Baggott, J.E., Blitz, M.A., Frey, H.M., and Walsh, R., J. Am. Chem. Soc. (1990) **112**, 8337.

[61] O'Neal, H.E., and Ring, M.A., J. Organomet. Chem. (1981) **213**, 419.

[62] Luo, Y-R., and Benson, S.W., J. Phys. Chem. (1990) **94**, 914.

BOND ENTHALPY TRANSFERABILITY - IS IT ACHIEVABLE?

J.A. Connor
Chemical Laboratory
University of Kent
Canterbury
Kent, CT2 7NH, U.K.

1. Introduction

The provision of experimental data for compiling a table of values from which the heat of every possible chemical reaction can be calculated was declared as the aim of thermochemistry by Rossini,[1] many years ago.   Subsequently, this purpose was broadened[2] to include the derivation of the heats of formation of compounds from their constituent elements and the relationship of these results to the chemical binding energies in the molecules. The "classical" examples of chemical binding energies come close to the beginning of anyone's serious study of chemistry. Much understanding of deep significance has followed from these data: electronegativity, ideas about aromaticity, bond polarity and ionic character, are examples. Coherent with these has been the development of a relation between bond length and bond enthalpy (strength). This leads to the still widely held perception that a "long" bond is a "weak" bond. This, in turn, is related to bond order: an interatomic bond of high order is both shorter and stronger than a bond of lower order between the same two elements. There is a lot of reliable experimental evidence to support these principles when elements of the first short period (n = 2) are considered. In the second short period (n = 3; $\ell = 0,1$) the position is more confused. The lack of a substantial base of data for enthalpies of formation must be held responsible for this, in part at least.

Progressing to the first $\underline{d}$-transition series (n = 3, $\ell = 2$) presents the extent of the problem. Any prospective system of bond enthalpy transfer must, surely, take into account the influences of oxidation state (or d-electron configuration), of coordination number, of other ligands and their steric and electronic properties, and of structure - but how is this to be achieved? If it is accepted that, notwithstanding the efforts of several people during the past thirty years, there remains an almighty lack of thermochemical information about both molecular and ionic species containing transition metals, then it is necessary to make the best of the meagre information that is available.

J. A. Martinho Simões (ed.), Energetics of Organometallic Species, 189–196.
© 1992 *Kluwer Academic Publishers.*

2. Simple Metal Carbonyls

2.1 Metal-CO Bonds

The enthalpy of formation of the hexacarbonyl metal compounds have been determined many times by measuring the heat of combustion[3] by static bomb calorimetry and by measuring the heat of thermal decomposition[4] in a microcalorimeter or, on a macro scale, in a hot-zone reaction calorimeter. Taking $Mo(CO)_6$ as an example, the agreement between the results of these procedures is quite good. The problem of interpretation arises with the discovery[5] that the solid product of thermal decomposition, a bright, shiny film is neither the metal nor a metal carbide but a new oxocarbide, in this case $Mo(O)C$, so that the reaction is described by:

$$Mo(CO)_6 \rightarrow Mo(O)C + 5CO \tag{1}$$

The value of $\Delta H_f^{\circ}[Mo(O)C,\underline{c}]$ can be estimated to be - 436.6 kJ mol^{-1}.

The mean M-CO bond dissociation energies, $\bar{D}$, which result from the recommended[6] values of the standard enthalpy of formation of various simple metal carbonyls are compared with the values obtained[7] by laser pyrolysis or photodetachment experiments for the first dissociation energy, D_1 in Table 1, and with values of these quantities calculated[8] by density functional theory. The influence of spin multiplicity changes[9] in $Fe(CO)_5/Fe(CO)_4$ has been accounted for in a detailed[10] study of the $Mn(CO)_x^+$ (x = 1-6) system.

TABLE 1. Experimental and Calculated Values of Mean, $\bar{D}$, and First, D_1, Bond Dissociation Energies in some Metal Carbonyls. (values in kJ mol)

	$\Delta H_f^{\circ},g$	$\bar{D}$	D_1	$\bar{D}$(calc.)	D_1(calc.)
$Cr(CO)_6$	-979.7±2	107±1	154±13	110	147
$Mo(CO)_6$	-989.1±1.7	152±1	170±13	126	119
$W(CO)_6$	-960.2±2.9	178±1	192±13	156	142
$Fe(CO)_5$	-765.1±6.6	118±2	174±13	118	185
$Ni(CO)_4$	-626.4±4.3	147±2	104±13		106

The fact that the experimental value of the D_1 is rather greater than $\bar{D}$, particularly for Cr and Fe, must mean that some of the remaining M-CO bonds are very easily broken and there is some evidence for this.[7,11]

2.2 Metal-Metal Bonds

The problem of the value to be assigned to a particular metal-metal bond enthalpy contribution in polynuclear metal carbonyls and their derivatives appears to be becoming more complicated (and obscure, if a simple explanation is sought) as the information base increases. The case of $Mn_2(CO)_{10}$ is taken as a good example of this problem.

The enthalpy of formation of the compound was determined[12] long ago by bomb calorimetry. In this compound, the Mn-Mn distance (290 pm at 296K (ref. 13)) is longer than that[14] in manganese metal (274 pm). Manganese is a relatively volatile[6] (ΔH_f°(Mn,g)=279.1 kJ mol^{-1}) body centred cubic metal. Following the model based[15] on a straightforward geometric concept, the bond enthalpy contribution of the metal-metal bond in $Mn_2(CO)_{10}$ is calculated[16] as 35 kJ mol^{-1}. Mass spectrometric measurements have produced[17] higher values for D(Mn-Mn) between 96 and 104 kJ mol^{-1}. Kinetic studies of substitution reactions in solution have concluded[18] that a lower limit value of D(Mn-Mn) is 154 kJmol^{-1}. Photoacoustic calorimetry[19] has led to D(Mn-Mn)=(159$\pm$21) kJ mol^{-1} and, at the upper boundary (so far?), an ion cyclotron resonance and photoelectron spectroscopic study of [Mn(CO)$_5$(CH$_2$Ph)] implies[20] that D(Mn-Mn) ~ (171$\pm$38) kJ mol^{-1}, and a laser pyrolysis measurement[21] gives D(Mn-Mn) > 176 kJ mol^{-1}. The authoritative review[22] by Martinho Simoes and Beauchamp selects D(Mn-Mn)=(159$\pm$21) kJ mol^{-1}, which leads to the important ΔH_f°(Mn(CO)$_5$,g) = -(713$\pm$11) kJ mol^{-1}. It is obvious that the naive idea of an inverse relation between bond length and bond enthalpy contribution, so familiar from carbon chemistry, is untenable in this simple transition metal system. The implications of this finding for other polynuclear metal compounds are unpredictable.

We have made microcalorimetric measurements of the thermal decomposition and the iodination of the hetero-bimetallic carbonyls [MnRe(CO)$_{10}$], [CoMn(CO)$_9$] and [CoRe(CO)$_9$]. The values obtained for the standard enthalpies of formation of these and related homometallic carbonyls are collected in Table 2.

TABLE 2. Standard Enthalpy of Formation of Solid Metal Carbonyls and Enthalpy of Vaporisation and Disruption (all values in kJ mol^{-1})

	$\Delta H_f^{\circ},c$	ΔH_{sub}°	$\Delta H_f^{\circ},g$	ΔH_{dis}	reference
MnRe(CO)$_{10}$	-1634$\pm$24	86$\pm$4	-1548$\pm$25	1505	23
CoMn(CO)$_9$	-1441$\pm$28	71$\pm$4	-1370$\pm$30	1080	23
CoRe(CO)$_9$	-1443$\pm$36	83$\pm$6	-1360$\pm$40	1574	23
Mn$_2$(CO)$_{10}$	-1677$\pm$6	80$\pm$4	-1597$\pm$7	1050	12
Re$_2$(CO)$_{10}$	-1660$\pm$11	101$\pm$2	-1559$\pm$11	2019	24
Co$_2$(CO)$_8$	-1250$\pm$5	65$\pm$3	-1185$\pm$6	1151	25
Co$_4$(CO)$_{12}$	-1845$\pm$13	96$\pm$4	-1749$\pm$13	2123	26

From these data and using the following values of the enthalpy of formation of the component metal carbonyl radicals : ΔH_f°(Mn(CO)$_5$,g) = -713$\pm$11; ΔH_f°(Re(CO)$_5$,g) = -686$\pm$6; ΔH_f°(Co(CO)$_4$,g) = -561$\pm$8 kJ mol^{-1}, the values of the various metal-metal bond enthalpy contributions are calculated as D(Mn-Re) = (149$\pm$11), D(Co-Mn) = (96$\pm$12) and D(Co-Re) = (113$\pm$15) kJ mol^{-1}. The value of D(Mn-Re) is larger than that (110 kJ mol^{-1}) obtained[27] by photocalorimetry and smaller than that (157 kJ mol^{-1}) derived[18] from kinetic measurements. Whatever the disagreements may be, all agree that D(Mn-Re) is less than 0.5[D(Mn-Mn)+D(Re-Re)] (173 kJ mol^{-1}). This result is in contrast to the conclusion of a study[28] of Fe$_2$Ru(CO)$_{12}$ and FeRu$_2$(CO)$_{12}$ which indicated that D(Fe-Ru) is greater than 0.5 [D(Fe-Fe)+D(Ru-Ru)] - however, this presumed that the Fe-CO and Ru-CO bond enthalpy contributions are unchanged from their values in the homonuclear $M_3(CO)_{12}$. Other estimates of D(Mn-Re) have been made from mass spectrometric measurements[17,29] which give values >200 kJ mol^{-1} . The values of D(Co-Mn) and D(Co-Re) are also both smaller than the average of the homometallic bond enthalpy contributions.

3. Organo-metal carbonyl compounds

3.1 Allyl derivatives of Manganese Carbonyl.

Several years ago we made[30] microcalorimetric measurements on [Fe(η^3-allyl)(CO)$_3$I] and other olefin-iron complexes. The average bond enthalpy contribution per electron pair was fond to be fairly constant (94 $\pm$ 2 kJ mol^{-1}) and less than that of CO in these systems. We have now made microcalorimetric measurements[31] on [Mn(CO)$_5$(η^1-C$_3$H$_5$)] and [Mn(CO)$_4$(η^3-C$_5$H$_5$)], which are particularly interesting because the former is converted directly to the latter by the action of heat with elimination of one CO ligand. The results are shown in Table 3.

TABLE 3. Enthalpies of formation and disruption (kJ mol^{-1})

	$\Delta H^\circ_f,c$	ΔH°_{sub}	$\Delta H^\circ_f,g$	ΔH_{dis}
[Mn(CO)$_5$(η^1-C$_3$H$_5$)]	-771$\pm$8	95(estimate)	-676$\pm$9	569
[Mn(CO)$_4$(η^3-C$_3$H$_5$)]	-649$\pm$6	98$\pm$3	-551$\pm$8	555

Analysis of the mixture of hydrocarbon products (hexa-1,5-diene, propa-1,2-diene and propene) is a problem here. Using the accepted[22] value of ΔH°_f(Mn(CO)$_5$,g) = - (713$\pm$11) kJ mol^{-1} together with the most recent value[32] of ΔH°_f(C$_3$H$_5$,g)=(166.9$\pm$2) kJ mol^{-1} gives a value of D(η^1-C$_3$H$_5$-Mn) = (130$\pm$11) kJ mol^{-1}, which is consistent with the value[22,33] for D(PhCH$_2$-Mn)=(129$\pm$10) kJ mol^{-1}, a linkage which can also exhibit $\eta^1 \leftrightarrow \eta^3$ behaviour. The value of D(η^3-C$_3$H$_5$-Mn) = (203$\pm$10) kJ mol^{-1} may be compared with D(η^3-C$_3$H$_5$-Fe)=176 kJ mol^{-1}. If the transfer of D(Mn-CO) = 88 kJ mol^{-1} from Mn(CO)$_5$,g, is justified, then the enthalpy change in the $\eta^1 \leftrightarrow \eta^3$ process is seen to be strongly encouraged by the nucleophilic displacement of a gaseous product.

3.2 η^6-Arenetricarbonylmetal complexes.

A precise example of transferability is presented by the process:

$$[M(arene)_2,g] + M(CO)_6,g \rightarrow 2[M(CO)_3(arene),g] \qquad (2)$$

In earlier work[34], on chromium complexes of various methyl-substituted benzenes, some of the limitations of the procedure were explored. Studies of the thermochemistry of bis-arene complexes of molybdenum and tungsten[35] indicated that it would be sensible to examine the appropriate tricarbonylmetal complexes. We have therefore made microcalorimetric studies[23] of the thermal decomposition and iodination of $[Mo(CO)_3(C_6H_6)]$ and $[W(CO)_3(C_6H_5Me)]$ with the following results:

TABLE 4. Enthalpies of formation (kJ mol^{-1})

	$\Delta H_{f}^{\circ},c$	ΔH_{sub}°	$\Delta H_{f}^{\circ},g$
$[Mo(CO)_3(\eta^6\text{-}C_6H_6)]$	-(420±8)	105±5(est.)	-315±10
$[W(CO)_3(\eta^6\text{-}C_6H_5Me)]$	-(395±10)	108±6(est.)	-287±13
$[Cr(CO)_3(\eta^6\text{-}C_6H_6)]$	-(443±8)	91±4	-(352±9)

when combined with the values for $\Delta H_{f}^{\circ}(M(CO)_6,g)$ and $\Delta H_{f}^{\circ}(M(arene)_2,g)$, the net enthalpy change of the transferability process (equation 2) is -(29±20) (Cr), -(62±23) (Mo) and -(42±27) (W). Given the errors involved, one might conclude that transfer of bond enthalpy contributions is possible in these systems, but nagging doubts remain: the arenetricarbonylmetal complex is thermodynamically more stable than either of the components in every case.

3.3 Methylidene and Methylidyne Complexes.

The isolobal principle developed by Hoffmann[36] has had far-reaching significance for transition metal chemistry. It would be interesting to explore the isolobal relationships of CH_2 with $[Fe(CO)_4]$ and of CH with $[Co(CO)_3]$. This can be regarded as a natural extension of work done earlier on $[Fe(CO)_4(C_2H_4)]$ (ref. 37) and $Fe_3(CO)_{12}$ and $Co_4(CO)_{12}$ (ref. 26). We have made microcalorimetric measurements[31] of the thermal decomposition and of the halogenation (I_2, Br_2, CBr_4) of $[Fe_2(\mu\text{-}CH_2)(CO)_8]$, $[Co_3(\mu\text{-}CH)(CO)_9]$ and $[Co_2(\eta\text{-}C_2H_2)(CO)_6]$. These have not been convenient systems with which to work. The composition of the hydrocarbon product mixture is neither simple nor easily reproducible, so the errors are substantial. The results of our studies are summarised in Table 5.

TABLE 5. Enthalpies of formation, vaporisation and disruption (kJ mol^{-1})

	$\Delta H_{f}^{\circ},c$	ΔH_{sub}°	$\Delta H_{f}^{\circ},g$	ΔH_{dis}
$[Fe_2(\mu\text{-}CH_2)(CO)_8]$	-1133±50	110±6	-1023±56	1356
$[Co_3(\mu\text{-}CH)(CO)_9]$	-1164±60	125±4	-1039±64	1913
$[Co_2(\eta\text{-}C_2H_2)(CO)_6]$	-864±42	114±6	-750±48	2125

The values calculated for ΔH_{dis} are based on $\Delta H_{f}^{\circ}(CH_2,g) = 385$ kJ mol^{-1} and $\Delta H_{f}^{\circ}(CH,g) = 594$ kJ mol^{-1}. As usual, the interpretation of ΔH_{dis} has to proceed from assumptions about the component bond enthalpy contributions based on a simple description of the structures. Taking $[Fe_2(\mu\text{-}CH_2)(CO)_8]$ as an example, $\Delta H_{dis} = $ D(Fe-Fe) + 8T(Fe-CO) + 2D(Fe-C) and using D(Fe-Fe) = 67 kJmol^{-1}, and $\bar{T}$(Fe-CO) = 117 kJ mol^{-1} gives D(Fe-C) = 183 kJ mol^{-1}. In $[Co_3(\mu\text{-}CH)(CO)_9]$, D(Co-CH) = 142 kJ

mol^{-1}, but in $[Co_2(\eta-C_2H_2)(CO)_6]$, D(Co-CH) = 218 kJ mol^{-1}. The enthalpy of the reaction.

$$Co_2(CO)_8 + C_2H_2 \rightarrow [Co_2(\eta-C_2H_2)(CO)_6] + 2CO$$

is calculated to be -14 kJ mol^{-1}.

4.   Sulphide and Thiolato Derivatives of Iron Carbonyls.

An interesting group of simple complexes which has been known for many years and have been fully characterised by X-ray crystallography is presented by $[Fe(CO)_3(SMe)]_2$, $[Fe(CO)_3S]_2$ and $[Fe_3(CO)_9S_2]$. The thermal decomposition of these compounds has been studied by microcalorimetry. Careful analysis of the products of decomposition of the methanethiolate complex showed that the products include ethane and dimethylsulphide in addition to iron(II) sulphide, iron and CO. The derived enthalpies of formation are shown in Table 6.

TABLE 6.   Enthalpy of formation, vaporisation and disruption of iron carbonyl sulphide complexes 9kJ mol^{-1}).

	$\Delta H^{\circ}_{f},c$	ΔH°_{sub}	$\Delta H^{\circ}_{f},g$	ΔH_{dis}
$[Fe_2(CO)_6(SMe)_2]$	-1190 ± 18	110 ± 2	-1080 ± 20	1528
$[Fe_2(CO)_6S_2]$	-1068 ± 20	100(est)	-968 ± 20	1695
$[Fe_3(CO)_9S_2]$	-1530 ± 20	110(est)	-1420 ± 20	2232

Presuming that D(Fe-Fe) and T(Fe-CO) can be transferred from the binary iron carbonyls unchanged allows calculation of the bond enthalpy contributions D(Fe-SMe) = (188 ± 5) kJ mol^{-1} and D(Fe-S) = (168 ± 5) kJ mol^{-1} which can be compared with the dissociation energy $D^{\circ}(Fe=S) = (334 \pm 21)$ kJ mol^{-1} (ref. 39).

5.   Conclusion

The interpretation of the enthalpies of formation of the compounds considered in this paper points to the problems which have to be recognised in any bond enthalpy transfer scheme. The first and foremost presumption must be that the calorimetric measurements are reliable and that the products have been correctly identified and quantified - the latter is by no means obvious or straightforward in some of the "simple" systems we have studied. Having accepted the value of the enthalpy of formation of the gaseous species of interest, it is necessary to rely on the inadequate framework of existing information in order to make any progress in the direction of identifying the enthalpy contributions of individual bonds - whether this information comes from diatomic molecules or from the kinetics of reactions in solution or from whatever other useful, reliable source. It is clear that there are already many problems in the way : heterometallic complexes, molecules containing formal multiple bonds between metal atoms, and molecules containing fluxional/bridging ligands to mention only three of them. Bond enthalpy contributions may be transferable between organometallic species, just as they are between organic species. The conditions and restrictions to be taken into account in making such transfers are only slowly becoming apparent and understood.

I wish to thank the SERC for supporting our work, and Susan Addison, John Kinkaid and Dr. Andreas Göbel for their contributions to it.

References

1. F.D. Rossini, Chem. Rev., 1936, 18, 233.
2. J.D. Cox and G. Pilcher, Thermochemistry of Organic and Organometallic Compounds. Academic Press, London 1970.
3. F.A. Cotton, A.K. Fischer and G. Wilkinson, J. Amer. Chem. Soc., 1956, 78, 5168; D.S. Barnes, G. Pilcher, D.A. Pittam, H.A. Skinner and D. Todd, J. Less-Common Metals, 1974, 38, 53 and refs. therein.
4. D.S. Barnes, G. Pilcher, D.A. Pittam, H.A. Skinner, D. Todd and Y. Virmani, J. Less-Common Metals, 1974, 36, 177; J.A. Connor, H.A. Skinner and Y. Virmani, J. Chem. Soc. Faraday 1, 1972, 1754.
5. I.M. Watson, J.A. Connor and R. Whyman, Polyhedron, 1989, 8, 1794.
6. G. Pilcher and H.A. Skinner, in The Chemistry of the Metal-Carbon Bond (F.R. Hartley and S. Patai, eds.). Wiley. Chichester, 1982.
7. K.E. Lewis, D.M. Golden and G.P. Smith, J. Amer. Chem. Soc., 1984, 106, 3905; A.E. Stevens, C.S. Feigerle and W.C. Lineberger, J. Amer. Chem. Soc., 1982, 104, 4825.
8. T. Ziegler, V. Tschinke and C. Ursenbach, J. Amer. Chem. Soc., 1987, 109, 4825.
9. T.J. Barton, R. Grinter, A.J. Thomson, B. Davies and M. Poliakoff, J. Chem. Soc., Chem. Commun., 1977, 841.
10. D.V. Dearden, K. Hayashibara, J.L. Beauchamp, N.J. Kirchner, P.A.M. van Koppen and M.T. Bowers, J. Amer. Chem. Soc., 1989, 111, 2401.
11. M.R.A. Blomberg, U.B. Brandemark, P.E.M. Sieghahn, J. Wennerberg and C.W. Bauschlicher, J. Amer. Chem. Soc., 1988, 110, 6650.
12. W.D. Good, D.M. Fairbrother and G. Waddington, J. Phys. Chem., 1958, 62, 853.
13. M.R. Churchill, K.N. Amoh, and H. Wasserman, Inorg. Chem., 1981, 20, 1609.
14. J.A. Oberteufer and J.A. Ibers, Acta Crystallogr., Sect. B., 1970, 26, 1499.
15. C.E. Housecroft, K. Wade and B.C. Smith, J. Chem. Soc., Chem. Commun, 1978, 765.
16. J.A. Connor in Transition Metal Clusters (B.F.G. Johnson, ed.) Wiley. Chichester 1980, p. 345.
17. D.R. Bidinosti and N.S. McIntyre, Can. J. Chem., 1970, 48, 593; G.A. Junk and H.J. Svec, J. Chem. Soc. A., 1970, 2102.
18. A. Marcomini and A. J. Poë, J. Chem. Soc., Dalton Trans., 1984, 95.
19. J.L. Goodman, K.S. Peters and V. Vaida, Organometallics, 1986, 5, 815.

196

20.   J. Martinho Simoes, J.C. Schulz and J.L. Beauchamp, Organometallics, 1985, 4, 1238.
21.   G.P. Smith, Polyhedron, 1988, 7, 1605.
22.   J.A. Martinho Simoes and J.L. Beauchamp, Chem. Rev. 1990, 90, 629.
23.   S.J. Addison and J.A. Connor, unpublished results.
24.   D.L.S. Brown, J.A. Connor and H.A. Skinner, J. Organometallic Chem., 1974, 81, 403: G. Al-Takhin, J.A. Connor and H.A. Skinner, J. Organometallic Chem., 1983, 259, 313.
25.   P.J. Gardner, A. Cartner, R.G. Cunningham and B.H. Robinson, J. Chem. Soc., Dalton Trans., 1978, 2582.
26.   J.A. Connor, H.A. Skinner and Y. Virmani, Symp. Faraday Div., 1974, 3, 18.
27.   Y. Harel and A.W. Adamson, J. Phys. Chem., 1986, 90, 6693.
28.   A.K. Baev, J.A. Connor, N.I. El-Saied and H.A. Skinner, J. Organometallic Chem., 1981, 213, 151.
29.   W.K. Meckstroth and D.P. Ridge, J. Amer. Chem. Soc., 1985, 107, 2281; note A.L. Rheingold, W.K. Meckstroth and D.P. Ridge, Inorg. Chem., 1986, 25, 3706.
30.   J.A. Connor, C.P. Demain, H.A. Skinner and M.T. Zafarani-Moattar, J. Organometallic Chem., 1979, 170, 117.
31.   S.J. Addison, J.A. Kinkaid and J.A. Connor, unpublished work.
32.   W.R. Roth, F. Bauer, A. Beitat, T. Ebbrecht and M. Wüstefeld Chem. Ber., 1991, 124, 1453.
33.   J.A. Connor, M.T. Zafarani-Moattar, J. Bickerton, N.I. El-Saied, S. Suradi, R. Carson, G. Al Takhin and H.A. Skinner, Organometallics, 1982, 1, 1166.
34.   J.A. Connor, J.A. Martinho-Simoes, H.A. Skinner and M.T. Zafarani-Moattar, J. Organometallic Chem., 1979, 179, 331.
35.   J.A. Connor, N.I. El-Saied, J.A. Martinho-Simoes and H.A. Skinner, J. Organometallic Chem., 1981, 212, 405.
36.   R. Hoffmann, Angew, Chem. Internat. Ed., 1982, 21, 711.
37.   D.L.S. Brown, J.A. Connor M.L. Leung, M.I. Paz Andrade and H.A Skinner, J. Organometallic Chem., 1976, 110, 79.
38.   A. Göbel and J.A. Connor, unpublished work.
39.   J. Drowart, A. Pattoret and S. Smoes, Proc. Brit. Ceram. Soc., 1967, 67.

ESTIMATES OF THERMOCHEMICAL DATA FOR ORGANOMETALLIC COMPOUNDS

J. A. Martinho Simões
Departamento de Engenharia Química
Instituto Superior Técnico
1096 Lisboa Codex
Portugal

ABSTRACT. Literature methods that have been used to estimate the energetics of organometallic reactions are critically surveyed and illustrated with selected examples.

1. Introduction

Despite the limited size of our thermochemical data bank, estimates to within 10-20 kJ/mol of the energetics of many reactions, many more than the species for which experimental thermochemical information is available, can be made.

The most popular methods for predicting enthalpies of organic reactions, in particular those involving carbon, hydrogen, oxygen, and nitrogen species, are based on the additivity of bond enthalpy terms or group enthalpies [1,2]. These terms or group enthalpies generate standard enthalpies of formation in the gas phase, from which the enthalpy of a given reaction can be calculated [3]. Deriving the enthalpy of the same reaction in solution requires other types of data (enthalpies of vaporization, sublimation, and solution), but some of these are usually easier to measure or to estimate than standard enthalpies of formation.

Most of the tabulated bond enthalpy terms or group enthalpies rely on experimental values of standard enthalpies of formation [3,4]. The determination of these quantities is therefore of central

197

J. A. Martinho Simões (ed.), Energetics of Organometallic Species, 197–232.
© 1992 Kluwer Academic Publishers.

importance in organic thermochemistry. The large majority of data were obtained by static or rotating-bomb combustion calorimetry, one of the few "absolute" thermochemical methods available. Reactions 1 and 2 can

$$C_4H_4O_3(c) + 3.5O_2(g) \rightarrow 4CO_2(g) + 2H_2O(l) \tag{1}$$

$$C_4H_4O_3(c) + H_2O(l) \rightarrow C_4H_6O_4(c) \tag{2}$$

be used to clarify this designation. The combustion of succinic anhydride yields carbon dioxide and water, whereas its hydrolysis produces succinic acid. Both reactions can be (and have been [1]) used to determine the standard enthalpy of formation of the anhydride, but in the second case the standard enthalpy of formation of succinic acid must be available. The same condition applies, of course, to carbon dioxide and water, but the standard enthalpies of formation of these and other combustion products (such as nitric acid, sulphuric acid, hydrochloric acid, etc.) are known accurately. Combustion calorimetry is therefore an absolute method in the sense that the enthalpies of formation of the products are usually well established quantities. As it happens, the enthalpy of formation of succinic acid has also been determined accurately by combustion calorimetry.

As discussed in the chapter by Pilcher, the success of combustion calorimetry experiments for probing C,H,N,O compounds and some other heteroatom organic molecules is not observed for many main group and transition metal-organo compounds. It is thus not surprising that the number of enthalpies of formation available for these substances is relatively small and so are the tabulated bond enthalpy terms and group enthalpies [2,3,5]. Much more abundant are data for enthalpies of reactions other than combustion, measured e.g. by reaction-solution calorimetry, by equilibrium methods (van't Hoff plots), or derived from kinetic experiments. Most of these enthalpies of reaction, however, do not afford standard enthalpies of formation since these parameters are unknown for at least two of the species involved.

Even if a large number of standard enthalpies of formation of organometallic molecules were known, the virtually infinite variety of bonds and groups in this family of substances would hinder the application of the above traditional prediction schemes.

The question is, therefore, how to use the relatively vast collection of enthalpies of reaction to predict new values. The aim of this chapter is to survey the methods that have been suggested for that purpose, including those developed in our own laboratory.

2. Enthalpies of Reaction and Bond Dissociation Enthalpies

A reaction, other than combustion, usually involves the cleavage and the formation of only a few bonds. In this case, it is simple to relate the enthalpy of the reaction to the energetic balance of those bonds. Consider, for example, reaction 3. $\Delta H^o(3)$ reflects the

$$ML(g) + L'(g) \longrightarrow ML'(g) + L(g) \tag{3}$$

difference between M-L and M-L' bond dissociation enthalpies, D(M-L)-D(M-L'). As pointed out elsewhere [6], devising an estimation method for these quantities is apparently easier than for enthalpies of formation, since $\Delta H^o_f(ML,g)$ and $\Delta H^o_f(ML',g)$ would require energetic information on each chemical bond in the molecules, and it is also less demanding in terms of size of the experimental data bank.

Two difficulties are commonly faced when bond dissociation enthalpies are to be derived from enthalpies of reaction. One is that chemical reactions seldom involve the cleavage or the formation of only one bond, and thus the measured enthalpies yield differences between bond dissociation enthalpies. This problem, illustrated by reaction 3, is often tackled by estimating a value for one of those bond dissociation enthalpies, say for D(M-L'). In other words D(M-L) will by *anchored* on the value chosen for D(M-L'). While this method may lead to a rather inaccurate value for an individual M-L bond dissociation enthalpy, if applied to a series of ligands L it affords a reliable relative scale of D(M-L). Many examples of this procedure, together with methods used to estimate the *anchor*, are found in the literature [7]. The second difficulty concerns the fact that most enthalpies of reaction involving organometallic complexes are determined in solution and therefore have to be related to gas phase enthalpies before deriving information on the energetics of chemical bonds. This subject, which has been discussed in detail [7], is beyond the scope of the present paper. It is just noted that in many cases it seems reasonable to identify the energetics in solution with the energetics in the gas phase. There are, however, very few examples where this assumption can be checked, so it has to be taken with great caution. Another way of facing the problem is simply to define bond dissociation enthalpies in solution and assume, for example in the solution analog of reaction 3, that different solvation enthalpies for a series of L will have a negligible influence on the trend of D(M-L).

3. Metal-Ligand Bond Enthalpy Terms

The use of "bond enthalpy terms", as defined below, for estimating the energetics of organometallic compounds was suggested 10 years ago [8,9]. It all arose from a question posed by Henry Skinner to the author of the present paper during his Ph. D. exam. The question was whether the chromium-benzene "bond strength" in $Cr(CO)_3(C_6H_6)$ could be identified with $D[(CO)_3Cr-C_6H_6]$. The answer was a prompt, No (chances of correctness were 50%). In the discussion that followed it became clear that bond strengths should always be higher than the corresponding bond dissociation enthalpies since they do not include the (necessarily) exothermic rearrangement of the fragments. In other words, the fragments $Cr(CO)_3$ and C_6H_6 will relax from the geometry they had in the complex to their most stable individual geometry. Weeks later, these thoughts led to the possibility of estimating the relaxation or reorganization energies of the fragments produced by cleavage of M-L bonds in the complexes of the type $M(Cp)_2L_2$, where M=Ti, Mo, W and $Cp=\eta^5-C_5H_5$, by using standard quantum chemistry calculations. As the extended Hückel molecular orbital method was then the favorite theoretical approach to deal with complex molecules, it was decided to ask Roald Hoffmann to make an estimate of the reorganization energies of the fragments $M(Cp)_2$. Contrary to his preliminary expectations, the values were quite high: about -40 kJ/mol both for molybdenum and tungsten [10].

The bond enthalpy method has been described in detail elsewhere [7] and can be summarized by considering Scheme 1. Here, $E(M-L)$ represents the enthalpy change associated to the cleavage of M-L bonds, but where the fragments keep the same configuration as in the initial complex (starred fragments), $\bar{D}(M-L)$ is the mean bond dissociation enthalpy, and ER_1 and ER_L are the reorganization energies of the fragments $M(Cp)_2^*$ and L^*, respectively.

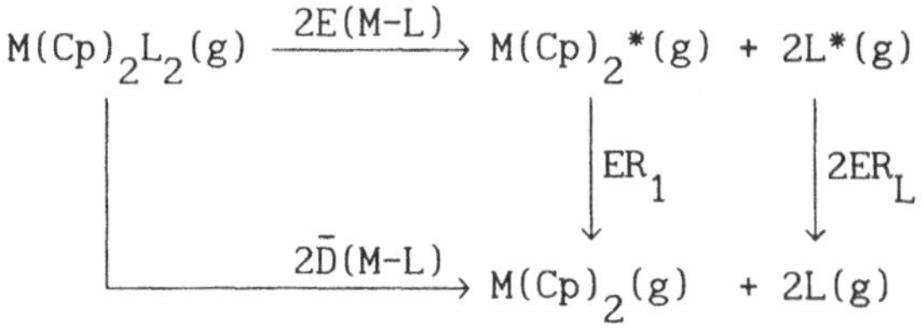

$$M(Cp)_2L_2(g) \xrightarrow{\ 2E(M-L)\ } M(Cp)_2^*(g) + 2L^*(g)$$

$$\downarrow ER_1 \qquad \downarrow 2ER_L$$

$$\xrightarrow{\ 2\bar{D}(M-L)\ } M(Cp)_2(g) + 2L(g)$$

Scheme 1

For the present discussion it is not important to know how $\bar{D}(M-L)$ was obtained. It suffices to say that in the case of the Ti, Mo, and W cyclopentadienyl systems, $\bar{D}(M-L)$ is derived from reaction-solution calorimetry data and is anchored on estimates of $E(M-Cl)$ [7]. As indicated above, ER_1 can be estimated by using the extended Hückel molecular orbital method. The values obtained have, however, only a semi-quantitative value and it is probably wise to leave them out from the calculation of $E(M-L)$. Moreover, for a series of ligands L, it is reasonable to assume similar values of ER_1, so that the influence of these values on the $E(M-L)$ trend will be small. We are therefore left with the calculation of ER_L.

L is commonly an organic radical, such as an alkyl, an aryl, an alkoxy, etc., so there are a number of theoretical methods which could be used to calculate ER_L. However, another approach was chosen. A perusal of literature indicates that the geometry of L^* is often

similar in the organometallic compound and in LH. In these cases,
Scheme 2 can be used to derive ER_L, provided that E(L-H) is known.

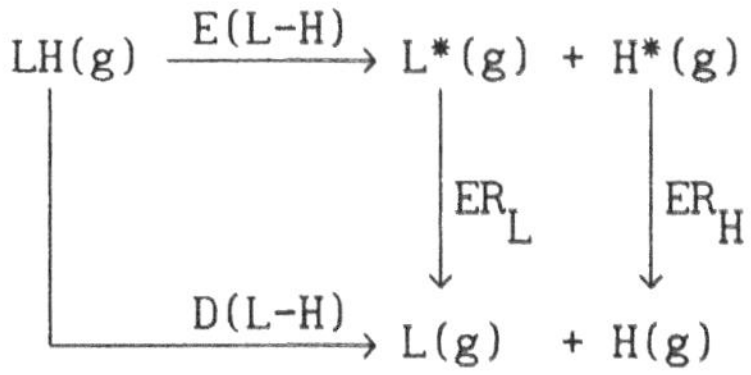

$$LH(g) \xrightarrow{E(L-H)} L^*(g) + H^*(g)$$

with vertical arrows labeled ER_L and ER_H, and

$$LH(g) \xrightarrow{D(L-H)} L(g) + H(g)$$

Scheme 2

Table 1   Laidler terms, E(L-H), bond dissociation enthalpies, D(L-H),
and reorganization energies, ER_L, of selected radicals.
Values in kJ/mol [1,6,7].

Radical, L	D(L-H)	E(L-H)	ER_L
Me	439.4	415.8	24
Et	421	410.8	10
Pr	418	410.8	7
t-Bu	396	403.9	-8
Ph	465	420.6	44
PhCH$_2$	368	410.8	-43
PhO	362	451.2	-89
PhCO$_2$	436	451.2	-15
EtS	376	360.3	16
PhS	346	360.3	-14

The reorganization energy of the hydrogen atom (or any other monoatomic
species) is taken as zero. Several methods are available to obtain
E(L-H), including empirical correlations between bond enthalpy terms
and bond lengths [9], and the Laidler scheme [1]. As this scheme is
one of the favorite methods to estimate the energetics of organic

compounds, Laidler parameters being tabulated for a wide variety of bonds, its use in Scheme 2 seemed the best choice. A look to the values in Table 1 shows, however, that the values of ER_L obtained for several organic radicals are positive, implying that they cannot be regarded as true reorganization energies. Having in mind this limitation, let us finally use the ER_L values in Table 1 to calculate a few metal-ligand bond enthalpy terms.

Table 2   Bond dissociation enthalpies, D(M-L), mean bond dissociation enthalpies, $\bar{D}$(M-L), and bond enthalpy terms, E(M-L), for several organometallic compounds. Values in kJ/mol [6,7,11].

Molecule	D or $\bar{D}$(M-L)	E(M-L)
$Mo(Cp)_2Me_2$	172±11	148
$Mo(Cp)_2Et_2$	156±9	146
$Mo(Cp)_2Bu_2$	154±12	147
$Mn(CO)_5Me$	187±4	163
$Mn(CO)_5Ph$	207±11	163
$Mn(CO)_5CH_2Ph$	129±10	172
$Ti(Cp)_2(OPh)_2$	373±11	462
$Ti(Cp)_2(O_2CPh)_2$	440±6	455
$Ti(Cp)_2(SEt)_2$	341±10	326
$Ti(Cp)_2(SPh)_2$	344±10	358

The striking feature of the numbers in Table 2 is the constancy of E(M-L) for some families of compounds, despite the large differences observed in the respective D or $\bar{D}$(M-L). For example, the manganese-benzyl bond dissociation enthalpy is some 60 kJ/mol lower than manganese-methyl bond dissociation enthalpy, but $E(Mn-CH_2Ph)$ differs from E(Mn-Me) by only 9 kJ/mol. This approximate constancy of E(M-L) for a given *family* of ligands has been observed in many cases

[7] and can be used to predict bond dissociation enthalpy data. A new value for, say, D(M-L'), is obtained by taking the average result for E(M-L) together with the reorganization energy of L', derived from D(L'-H) and the Laidler term E(L'-H).

One must of course recognize that the method just described is somewhat crude. First, it involves some degree of risk when assigning an average value to a new E(M-L'). This is observed in Table 2 for the thiolate complexes: the bond dissociation enthalpies are actually closer to each other than the bond enthalpy terms! Secondly, it usually does not allow any appreciation of small variations in bond dissociation enthalpies, since it deals with "ball park" numbers. Yet, we must not forget that most data for organometallic compounds are subject to considerably larger uncertainties than for organic substances and so even if it is just taken as a guideline, the bond enthalpy term method is useful.

4. Bond Enthalpy Terms vs. Bond Lengths

Short bonds are stronger than long bonds. As mentioned above, this relationship can be used to estimate bond enthalpy terms in organic molecules. Plots of bond enthalpy terms vs. bond lengths have been drawn for several types of bonds, including C-C, C-H, C-O, O-H, N-H, N-N, etc. [9]. Early versions of some of these curves were published by Skinner [12], almost 50 years ago.

Bond enthalpy term-bond length plots have more recently been used by Skinner and coworkers [13,14] to probe the energetics of metal-metal and metal-oxygen bonds in chromium and molybdenum complexes. An example of such a curve is given in Figure 1.

As can be verified by reading the original literature [13,14], it is not simple to obtain a plot like the one shown. Some points are almost pure estimates! Others rely on assumptions concerning the transferability of bond enthalpies. For example, the calculation of E(Mo-O) in $Mo_2(O_2CMe)_4$ required the value of the metal-metal bond enthalpy term, which was obtained from a similar type of plot, now

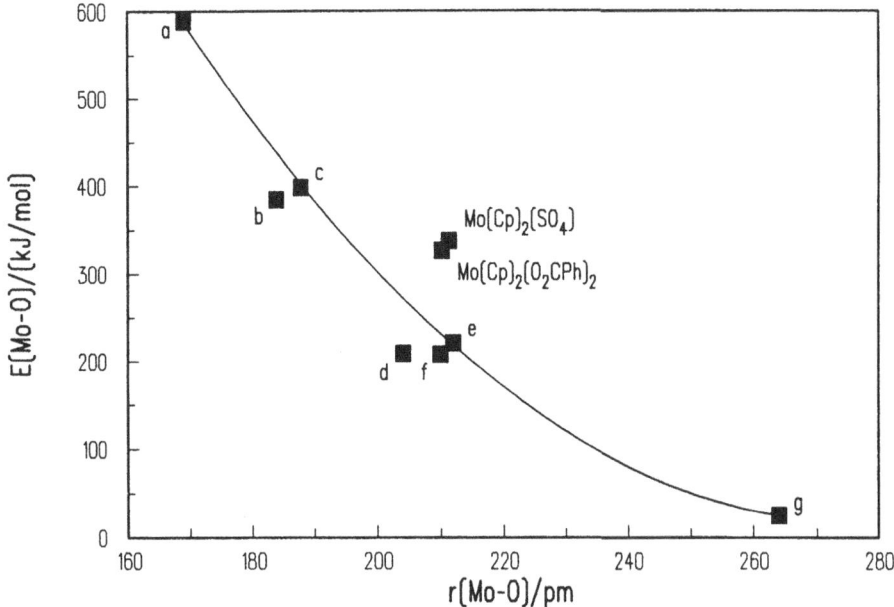

Figure 1 - Bond enthalpy term-bond length plot for Mo-O. a=MoO$_3$, b=Mo(OPr-i)$_4$, c=Mo$_2$(OPr-i)$_6$, d=Mo(acac)$_3$, e=Mo$_2$(O$_2$CMe)$_4$, f=Mo$_2$(acac)$_2$(O$_2$CMe)$_2$, g=Mo$_2$(O$_2$CMe)$_4$, intermolecular. Adapted from [13].

involving E(Mo-Mo) vs. r(Mo-Mo).

Bond enthalpy term-bond length curves are useful to tackle very specific problems, like those addressed by Skinner *et al.* [13,14], and yield "ball park" values. Both the lack of structural data and our present ignorance about the rules that dictate the transferability of bond enthalpies, hinder a wider application of the method. Incidentally, two values of E(Mo-O) reported for Mo(Cp)$_2$(O$_2$CPh)$_2$, 327±12 kJ/mol, and for Mo(Cp)$_2$(SO$_4$), 338±9 kJ/mol, (bond lengths 210.2 and 211.3 pm, respectively), lie clearly above the line of Figure 1. However, it must be stressed that these bond enthalpies are relative to an estimate for E(Mo-Cl) [15].

5. Bond Enthalpies and Stretching Frequencies

Infrared or Raman spectra are usually easier to obtain than
thermochemical data, hence it would be convenient if one could
correlate stretching frequencies with bond enthalpy data. In
principle, this idea looks quite sensible since the force constants of
harmonic oscillators – which should reflect the bond strengths – are
proportional to the squares of the frequencies. In reality, however,
early attempts by Cottrell [16] and by Gaydon [17] had very limited
success. This was not entirely unexpected since a 'force constant is a
measure of the resistance of the bond to small perturbations, and the
perturbations involved in chemical reactions are not small' [16].

 The bond enthalpy-frequency correlation has been reexamined by
McKean and applied to C-H bonds [18]. He found that the complications
in the vibrational spectra, e.g. due to the coupling between C-H bonds
in methyl and methylene groups, could be removed by replacing all the
hydrogens but one by deuterium. The measured C-H stretching
frequencies, called *isolated*, $\nu(CH)^{is}$, were plotted against C-H bond
dissociation enthalpies. An updated version of this correlation has
been published recently [19], leading to equation 4 (correlation
coefficient 0.991; D(C-H) in kJ/mol and $\nu(CH)^{is}$ in cm^{-1}).

$$D(C-H) = (0.375\pm0.014)\nu(CH)^{is} - (688.2\pm40.8) \qquad (4)$$

 It is important to note that McKean's linear correlation relies
on radicals which have negligible stabilization energies (such as
alkyl, fluoroalkyl, phenyl, etc.). For a given radical, the difference
between the predicted (from equation 4) and the experimental D(C-H)
values was identified with the stabilization energy of the radical.

 Measured $\nu(CH)^{is}$ values and equation 4 (or its earlier version)
have been used to estimate carbon-hydrogen bond dissociation enthalpies
in organometallic compounds. For instance, a slight increase in
D(C-H), ca. 7 kJ/mol, was predicted for the group 12 dimethyls, MMe_2,
when Zn is replaced by Hg, the value for Cd lying in-between [20]. An
opposite trend was found for $Mn(CO)_5Me$ and $Re(CO)_5Me$, where D(C-H) is

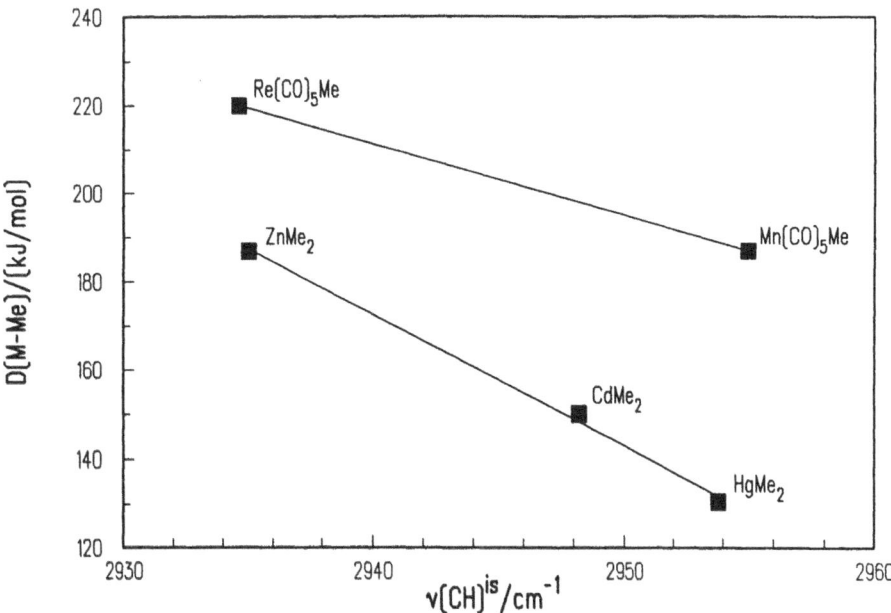

Figure 2 - Correlation between isolated C-H stretching frequencies
of methyl groups and M-Me bond dissociation enthalpies (for
$M(CO)_5Me$) or M-Me mean bond dissociation enthalpies (for
MMe_2).

lower for the third row metal complex (by 7 kJ/mol) [21]. On the other
hand, as shown in Figure 2, it was observed that the mean bond
dissociation enthalpies in the series MMe_2, $\bar{D}$(M-Me), and the
metal-methyl bond dissociation enthalpies in $M(CO)_5Me$ decrease with
increasing $\nu(CH)^{is}$. A similar reverse trend observed for the group 14
tetramethyls [20] led to the conclusion that differences of 1 cm^{-1} in
$\nu(CH)^{is}$ in structurally related molecules imply differences of 3-5
kJ/mol in metal-carbon bond dissociation enthalpies [21]. The
application of this rule to the complexes $M(Cp)_2Me_2$ (M=Ti, Zr, Hf) and
$M(Cp)(CO)_3Me$ (M=Cr, Mo, W), led to semi-quantitative trends of D(M-Me)
for each triad [22,23]. Gas phase $\nu(CH)^{is}$ were estimated by adding a
constant value of 10 cm^{-1} to the solution data.

A particularly interesting application of the previous method concerns the effect of replacing a carbonyl group by N_2 in the complexes $M(CO)_5Me$ (M=Mn, Re). IR spectra of the matrix-isolated species $M(CO)_4(N_2)Me$ have shown that $\nu(CH)^{is}$ decrease by 13-23 cm^{-1}, relative to the pentacarbonyl analogues [24]. This is in line with the poor electron acceptor character of N_2 as compared with CO. If the above rule is accepted, this shift implies that D(M-Me) increases by ca. 50-90 kJ/mol when CO is replaced by N_2 [24]!

McKeans' method needs, of course, to be subject to further testing in order to assess its quantitative value. Furthermore, there is at least one example where bond dissociation enthalpies do not correlate with isolated stretching frequencies: the scatter observed in the plot of D(Si-H) in substituted silanes against $\nu(SiH)^{is}$ led McKean *et al.* to use a different approach to probe the effect of methylation in silane [25]. This procedure relied on the assumption that the isolated species MH can be treated as a diatomic molecule and so the parameters ω_e and $\omega_e x_e$ of the Morse function could be evaluated from the measured $\nu(MH)^{is}$ and its first overtone, $2\nu(MH)^{is}$, and used to estimate the spectroscopic bond dissociation energy (0 K) from equation 5. Here, 0.87 is an empirical constant which accounts for the fact

$$D(M-H; \ 0 \ K) \approx 0,87 \ [\omega_e^2/(4\omega_e x_e) - \omega_e/2 + \omega_e x_e/4] \qquad (5)$$

that the potential energy curve of MH is not exactly represented by a Morse function. The results of this exercise were as follows:
$D(H_3Si-H) = D(MeSiH_2-H)+14 = D(Me_2SiH-H)+21 = D(Me_3Si-H)+28$ kJ/mol [25]. This large effect of methyl substitution conflicts with the currently accepted values of D(Si-H) in methylsilanes [26], which are identical within the assigned experimental uncertainties (average: 376±2 kJ/mol). The evidence that D(Si-H) are nearly independent of the extent of alkylation is also supported by a linear correlation between $\Delta H_f^o(Me_nSiH_{4-n}, g)$ and $\Delta H_f^o(Me_{n+1}SiH_{3-n}, g)$ [27].

Equation 5 (with the factor 0.87 replaced by 0.81) has also been used to study the alkylation effect of germanes and the conclusions are

similar, e.g. $D(H_3Ge-H)$ is about 29 kJ/mol higher than $D(Me_3Ge-H)$.
Again this is at variance with results by Walsh, namely $D(H_3Ge-H)=$
346±10 kJ/mol and $D(Me_3Ge-H)=340±10$ kJ/mol [28]. The latter is in
excellent agreement with the value obtained in a recent photoacoustic
calorimetry study, 341±2 kJ/mol [29].

It remains to be said that McKean *et al.* have concluded that the
direct source of the above discrepancies is the empirical factor in
equation 5, which varies with methyl substitution [30]. This, in turn,
seems to result from neglecting the cubic term $\omega_e y_e$ in the equation for
the vibrational energy [17].

Zavitsas has also explored a correlation between bond
dissociation enthalpies and uncoupled stretching frequencies in a
series of organic compounds, equation 6 [31]. Here, c_1 is a constant

$$D = c_1 + \nu^2/4908 \quad kJ/mol \tag{6}$$

characteristic of the two bonded atoms and does not depend on the
multiplicity of the bond (ν is in cm^{-1}). Several values of c_1 were
given (C-C 166.1, C-N 111.7, C-O in alcohols and carbonyls 141.4, C-O
in ethers and esters 101.3, C-F 215.1, C-Cl 177.8, and C-Br 169.9) and
it was claimed that an error of 10 cm^{-1} in ν leads to an error of about
4 kJ/mol in the bond dissociation enthalpy.

A word of caution about Zavitsas' method concerns the radical
bond enthalpy data base used to derive the parameters of equation 6.
The recent reevaluation of enthalpies of formation of many radicals may
lead to some changes of the parameters listed above. Finally, it is
noted that the application of Zavitsas' method to organometallic
substances has not been attempted yet. The same can be said about
another procedure, suggested by Nonhebel and Walton [32], which
involves a linear correlation between barriers to internal rotation and
bond dissociation enthalpies. An example of a successful prediction by
this correlation is the value of $D(MeCOCH_2-H)$, later experimentally
confirmed [33].

6. Enthalpies of Reaction vs. Electronic and Steric Parameters

The enthalpy of a metal-ligand bond depends on the electronic and steric features of the ligand. One of the first successful attempts to quantify this fact was due to C. A. Tolman, who, in a study focussing on the role of phosphorus ligands in organometallic chemistry and homogeneous catalysis, introduced the parameters χ and θ, known as the *electronic parameter* and the *ligand cone angle*, respectively [34]. Values of χ were obtained by comparing the A_1 carbonyl stretching frequencies in a series of $Ni(CO)_3L$ complexes. Basic ligands L induce an increase of the electron population in the anti-bonding orbital π^* of the carbonyl group, resulting in a decrease of the C-O vibration frequency. The cone angle θ of a given ligand, on the other hand, reflects its bulkiness and was originally obtained as the 'apex angle of a cylindrical cone centered 228 pm from the center of the P atom, which just touches the van der Waals radii of the outermost atoms of the ligand' [34]. Tolman's tabulated values χ and θ have been used by many authors to rationalize a variety of data, including enthalpies of reaction. One interesting example is given by $\Delta H^O(7)$, determined for

$$(\eta^3\text{-}MeC_3H_3Me)Ni \overset{Me}{\underset{Me}{\cdots\cdots}} Ni(\eta^3\text{-}MeC_3H_3Me) + 2L \longrightarrow 2\ (\eta^3\text{-}MeC_3H_3Me)Ni\overset{Me}{\underset{L}{\diagdown}} \quad (7)$$

a series of phosphines and phosphites. The results were fitted to equation 8, which indicates that the exothermicity of reaction 7 increases with acidity of L (basic ligands have small values of χ) and decreases for bulky ligands [35].

$$-\Delta H^O(7) = -1.65\chi + 0.56\theta + 172.0 \quad kJ/mol \qquad (8)$$

The σ-donation of a ligand L is often more conveniently expressed by the value of $pK_a(LH^+)$. This parameter has been used by Geno and Halpern to investigate the trends of cobalt-benzyl bond

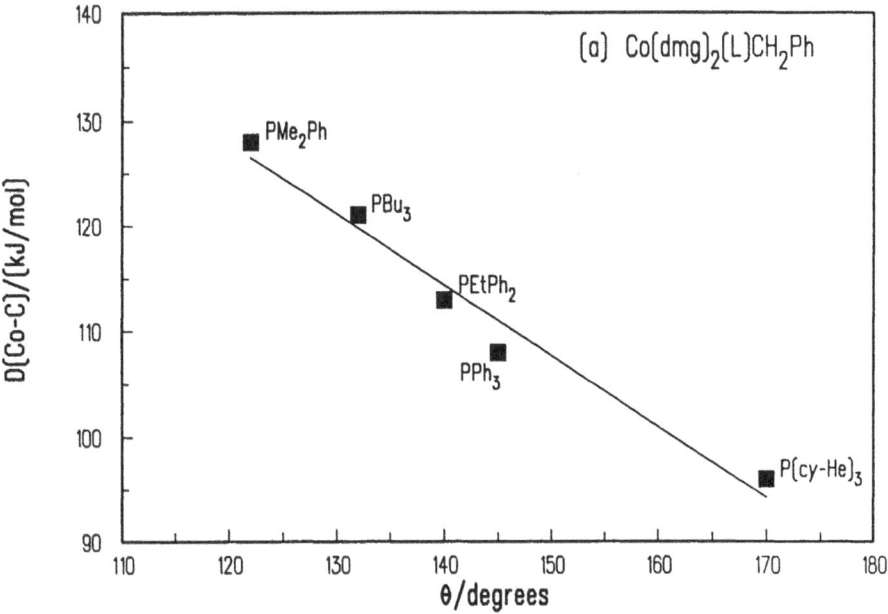

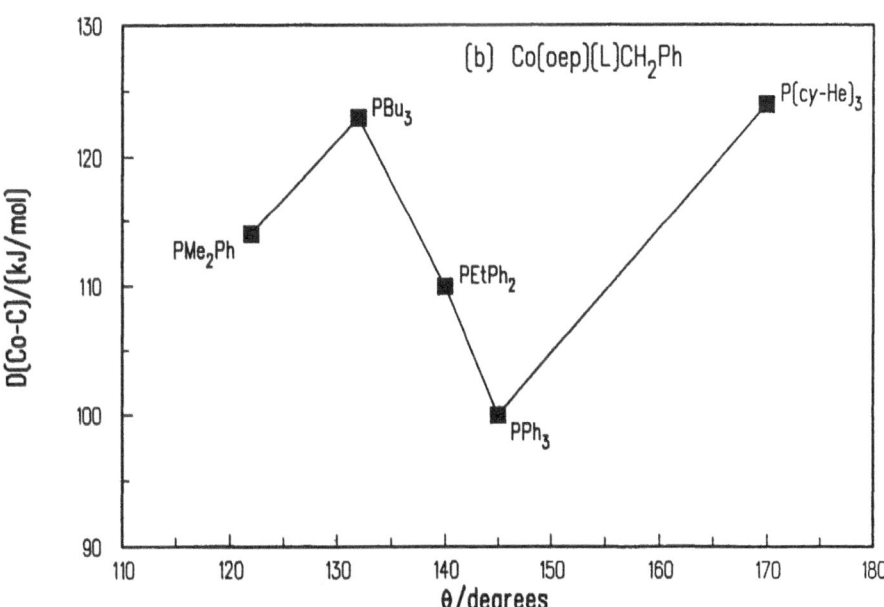

Figure 3

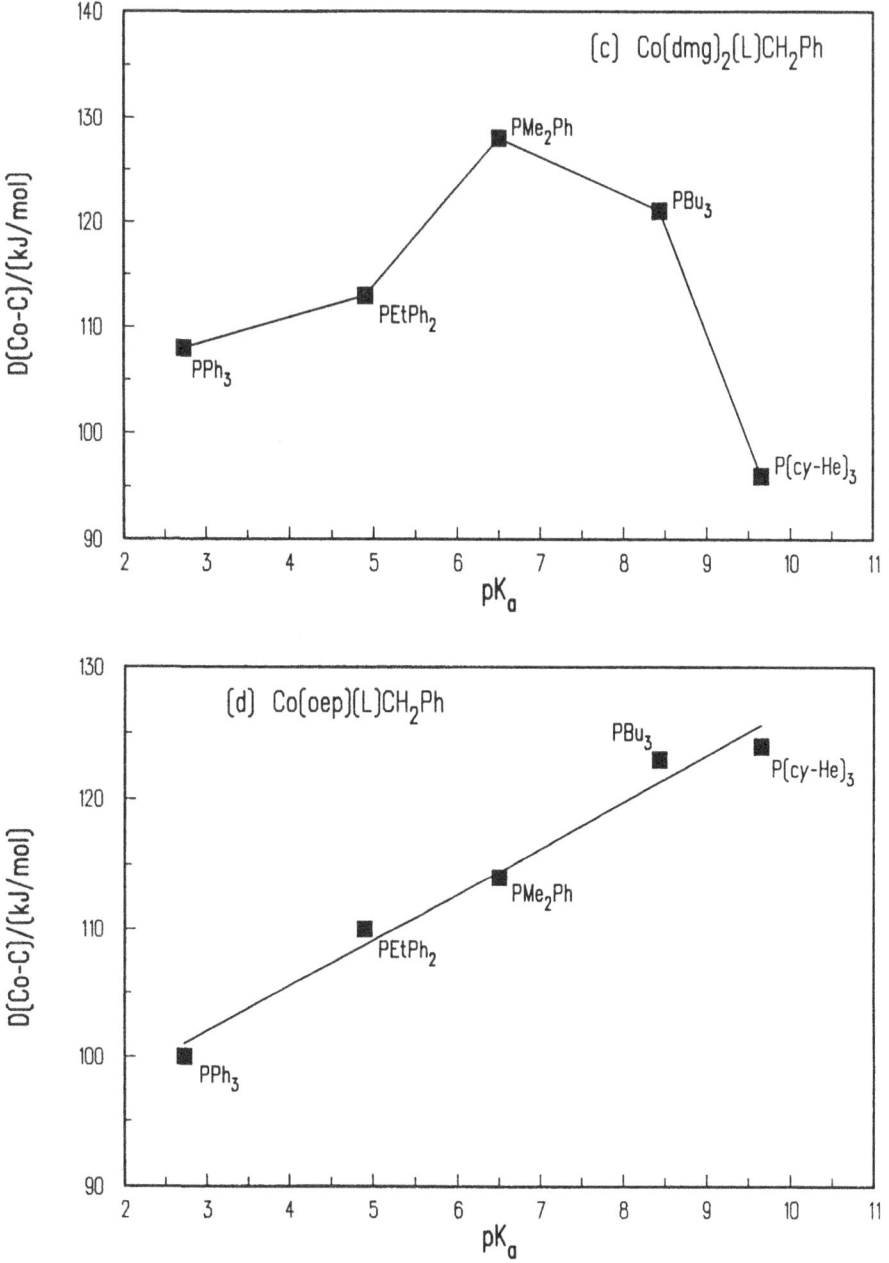

Figure 3 – Correlation of cobalt-benzyl bond dissociation enthalpies with cone angles (θ) and pK$_a$'s of the protonated ligands in the complexes Co(dmg)$_2$(L)CH$_2$Ph and Co(oep)(L)CH$_2$Ph. Adapted from [36].

dissociation enthalpies in the complexes $Co(dmg)_2(L)CH_2Ph$ and $Co(oep)(L)CH_2Ph$, where Hdmg stands for dimethylglyoxime, oep for octaethylporphyrin, and L is a phosphine [36]. Solution values of $D(Co-C)$ for these molecules are plotted against the pK_a's and the θ's of the phosphines in Figure 3. As indicated by the linear variation in Figure 3a, the dominating effect in the glyoxime complexes is the bulkiness of the ligand, whereas for the porphyrin series, Figure 3d, the electronic factors become the most important.

The effects of electronic and steric features of ligands on bond dissociation enthalpies (or enthalpies of reaction) and other types of data (e.g. rate constants) has been discussed in detail by Giering and coworkers in several recent papers [37]. A very brief description of their *quantitative analysis of ligand effects* (QALE) method is as follows. The enthalpy of a reaction involving the cleavage or the formation of a metal-phosphorus bond can be decomposed into four contributions, according to equation 9. Here, the superscript "st"

$$\Delta H = \Delta H_{\sigma} + \Delta H_{\sigma}^{st} + \Delta H_{\pi} + \Delta H_{\pi}^{st} \qquad (9)$$

indicates a steric contribution, and ΔH_{σ} and ΔH_{π} stand for the σ and π electronic contributions to ΔH. The basic assumption of QALE is the classification of phosphorus(III) ligands, PR_3, into three classes (I, σ-donor/π-donor; II, σ-donor; III, σ-donor/π-acceptor) according to their position in a plot of $pK_a(PR_3H^+)$ vs. the reversible oxidation potentials (E°) of a family of transition metal complexes containing those phosphines or phosphites. An inverse linear dependence was observed for class II ligands, while those of class I (π-bases) and class III (π-acids) appear below and above the line, respectively. The π-basicity $(E_{\pi b})$ or the π-acidity $(E_{\pi a})$ of a ligand of these classes is measured by the difference between the experimental E° and the value predicted from its pK_a and the straight line defined by class II ligands. These two parameters, $E_{\pi a}$ and $E_{\pi b}$, can then be used to analyze equation 9, since ΔH_{π} is assumed to be a linear function of $E_{\pi a}$ (or $E_{\pi b}$). ΔH_{σ}, on the other hand, is taken as a linear function of pK_a. The remaining two assumptions of the method are that enthalpies

214

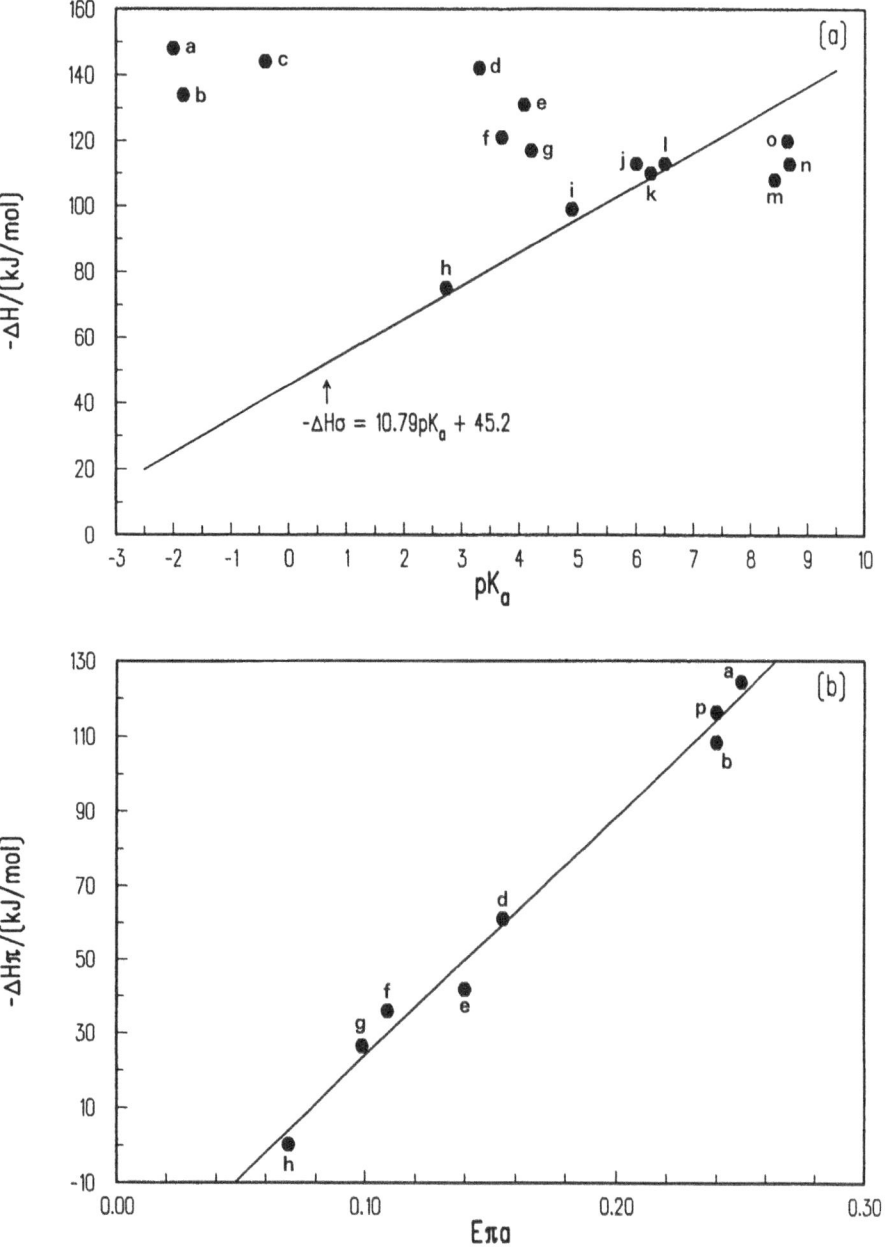

Figure 4 – (a) σ-electronic profile for reaction 7. (b) π-electronic
profile for class III ligands of the same reaction.

a=P(OPh)$_3$, b=P(O-o-MeC$_6$H$_4$)$_3$, c=PPh(PPh)$_2$, d=P(OEt)$_3$,
e=P(OPr-i)$_3$, f=P(OC$_{10}$H$_{19}$)$_3$, g=PPh(OC$_{10}$H$_{19}$)$_2$, h=PPh$_3$,
i=PEtPh$_2$, j=P(CH$_2$Ph)$_3$, k=PEt$_2$Ph, l=PMe$_2$Ph, m=PBu$_3$, n=PEt$_3$,
o=PMe$_3$, p=P(O-o-PhC$_6$H$_4$)$_3$. Adapted from [37b].

of reaction for the three classes of ligands have identical pK_a dependence and are linear functions of the ligand cone angle, pK_a and $E_{\pi a}$ (or $E_{\pi b}$) above the *steric threshold*, i.e. the minimum value of θ for which steric effects are apparent.

The application of QALE to reaction 7 above led to Figures 4a and 4b, which show the σ and the π electronic profiles of that process. ΔH_{π} represent the deviations of ΔH for class III ligands from the line in Figure 4a. Equation 10, obtained from these plots, is valid for

$$- \Delta H = - \Delta H_{\sigma} - \Delta H_{\pi} =$$

$$= 10.79 pK_a + \lambda(643.92E_{\pi a} - 40.0) + 45.2 \qquad (10)$$

ligands of classes II and III. λ is a switching function that turns on the π effect when $E_{\pi a}$ exceeds the so-called π-electronic threshold, i.e. the minimum value of $E_{\pi a}$ for which the π-acidity becomes operative (the $E_{\pi a}$ intercept in Figure 4b): $\lambda = 0$ for $E_{\pi a} < 0.062$ (class II ligands) and $\lambda = 1$ for $E_{\pi a} > 0.062$ (class III ligands).

As stressed by Giering and coworkers [37b], a conclusion of the QALE study leading to equation 10 is that steric effects are not needed to fit the experimental thermochemical data, in contrast to equation 8. These steric effects are, however, important when discussing the results for other, more crowded, systems, such as $Mo(CO)_3(PR_3)_3$ and $[Pt(PMe_2Ph)_2(Me)(PR_3)]PF_6$ [37b].

As the values of pK_a, $E_{\pi a}$, $E_{\pi b}$, and θ are tabulated for a variety of phosphines, its is possible to use QALE to predict enthalpies of reaction. The method is still undergoing further test and development, including its extension to other types of ligands. A revision of fundamental parameters may, however, be useful. For example, Tolman's values for the cone angles were listed almost fifteen years ago and the structural data now available are considerably larger. This point is illustrated by a recent study where the enthalpies of reaction 11 (C_7H_{11}=2,4-dimethylpentadienyl; L=phosphine

$$Ti(C_7H_{11})_2L(soln) \rightarrow Ti(C_7H_{11})_2(soln) + L(soln) \qquad (11)$$

or phosphite) were correlated with Tolman's cone angle [38]. A good linear trend was observed, except for P(OMe)$_3$ and P(OEt)$_3$. Ruling out steric effects as the source of this discrepancy, the authors came to the conclusion that updated values of those cone angles (128 and 134°, compared to the early 107 and 109°, respectively for P(OMe)$_3$ and P(OEt)$_3$), led to a much better correlation. It is worth pointing out that Tolman's cone angle, which can also be defined as the *fan angle*, FA=arctan(r/b), r being the ligand diameter (considering the van der Waals radii of the outmost atoms) and b the metal-ligand bond length, is less informative than more elaborate parameters like the *solid angle*, SA=2π(1-cosFA) or the *solid angle factor*, SAF=SA/4π. The latter is particularly useful, since it represents the fraction of a unit radius sphere, centered on the metal, "occupied" by a given ligand. More quantitative treatments will probably include these parameters.

Another method to assess electronic contributions to enthalpies of reaction has been devised by Drago and coworkers [39]. It relies on the empirical equation 12, where ΔH represents the enthalpy of reaction

$$-\Delta H = E_A E_B + C_A C_B \qquad (12)$$

between the acceptor A and the donor B, yielding the adduct AB, and E and C are parameters reflecting the tendency of A and B to undergo electrostatic and covalent bonding, respectively. These parameters have been obtained for a large number of acids and bases by fitting equation 12 to experimental values of enthalpies of reaction. They can also be estimated from Hammett substituent constants [40].

The previous relationship has been modified into equation 13 to

$$-\Delta H = -W + E_A E_B + C_A C_B \qquad (13)$$

account for reactions more complex than the simple association of A and B, for instance reactions where both the cleavage and the formation of bonds occur. W is a constant for a given donor reacting with a series of acceptors or for an acceptor reacting with several donors. The application of Drago's method to organometallic compounds is still at an early stage. One of the few examples concerns the recent application of equation 13 to thermochemical data for the reactions involving the fission of cobalt-carbon bonds in $Co(dmg)_2(L)CH(Me)Ph$ complexes (L is a substituted pyridine). Good agreement between experimental and calculated values of ΔH was obtained. It was also possible to predict the Co-C bond dissociation enthalpy, in the absence of the axial base L, as 46 kJ/mol [40].

7. Electronegativities and Bond Enthalpies

An account of this subject would justify at least one long chapter, such is the number of recent studies on electronegativity scales and their relation to the energetics of chemical bonds. Rather than attempting this comprehensive survey, only a few methods will be highlighted, and even these with brevity.

Sanderson's electronegativity scale has been extensively used to predict atomization enthalpies of organic and inorganic compounds [41]. Its application to organometallic compounds, particularly in the case of transition metals, is, however, problematic. Even for species like silane and germane, SiH_4 and GeH_4, the experimentally derived enthalpies of atomization are in substantial disagreement with the calculated values.

A concept closely related to electronegativity, χ, is the *electronic chemical potential*, μ, defined by equation 14, where E is

$$\mu = \left(\frac{\partial E}{\partial N}\right)_v = -\chi \tag{14}$$

the electronic energy, N, the number of electrons, and v the potential due to the nuclei. The electronic chemical potential is also related to the *chemical hardness*, η, through equation 15. Both equations are

$$\eta = \frac{1}{2}\left(\frac{\partial \mu}{\partial N}\right)_v \tag{15}$$

exact and have been derived by using the density functional theory.
Approximate values of χ and η can be calculated from the experimental
values of the vertical ionization energy, I, and the vertical electron
affinity, A [42]:

$$\chi = (I+A)/2 \tag{16}$$

$$\eta = (I-A)/2 \tag{17}$$

According to the principle of electronegativity equalization,
when two mono or polyatomic moieties, X and Y, form a chemical bond,
there will be a flow of electrons from the less electronegative to the
more electronegative of those moieties until $\chi_X = \chi_Y$. Equation 18 is

$$\Delta N = (\chi_X - \chi_Y)/[2(\eta_X + \eta_Y)] \tag{18}$$

an approximate relationship, obtained by Pearson [42], to calculate the
number of electrons (ΔN) transferred. Strong bonds should correspond
to large values of ΔN, so that it is not surprising that the above
method had been used to predict trends of bond dissociation enthalpies,
including those in organometallic compounds [42,43]. Take, for
example, the molecules M-Me and M-Cl, M being a transition metal-organo
moiety. How does the difference D(M-Cl) - D(M-Me) vary across the
first row transition elements? A naive use of equation 18, by taking
the values of χ and η for two transition metal atoms [44], say Ti and
Ni, together with the parameters for Cl and Me [43], leads to the
conclusion that the difference is relatively constant (a slight
decrease is predicted for Ni). A similar, though more pronounced trend
is observed when Cl is replaced by I: D(M-I) - D(M-Me) decreases for
late transition metals. An opposite trend is, however, observed for
the difference D(M-H) - D(M-Me), i.e. metal-methyl bond dissociation
enthalpies are closer to metal-hydrogen bond dissociation enthalpies
for early transition metals than for the more electronegative, softer,
late transition elements.

As far as our present experimental evidence can tell us, all the

three previous trends are correct. Actually, as pointed out by
Labinger and Bercaw, the D(M-H) - D(M-Me) trend is predicted from the
well known original Pauling relationship, equation 19 [45]. That

$$D(X-Y) = [D(X-X)+D(Y-Y)]/2 + 96.5(\chi_X - \chi_Y)^2 \qquad (19)$$

difference was calculated from equation 20, after introducing the

$$D(M-H) - D(M-Me) =$$
$$193.0(\chi_{Me} - \chi_H)\chi_M + [D(H-H)-D(Me-Me)]/2 + 96.5(\chi_H^2 - \chi_{Me}^2) =$$
$$19.3\chi_M - 14.3 \qquad (20)$$

values of the electronegativities [46] and the bond dissociation
enthalpies. When the same exercise is performed for D(M-Cl) - D(M-Me)

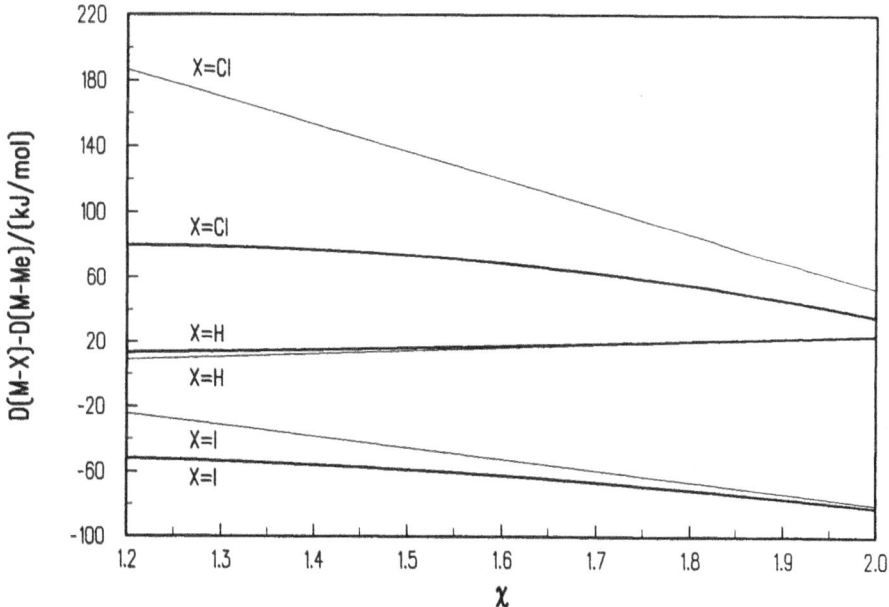

Figure 5 - D(M-X)-D(M-Me) vs. the electronegativity of the metal, as
predicted by Pauling's equation (thin lines) and Matcha's
equation (thick lines).

and for D(M-I) - D(M-Me), the conclusions are qualitatively correct, as can be observed in Figure 5. Yet, for instance D(M-I) is predicted to be always smaller than D(M-Me). This is not unexpected, given the limitations of the arithmetic mean in Pauling's equation .

More reliable trends of the above differences are probably obtained through equation 21, adapted from another derived by Matcha

$$D_{XY} = [D(X-X)+D(Y-Y)]/2 + K\{1-\exp[-96.5(\chi_X-\chi_Y)^2/K]\} \qquad (21)$$

[47], where K=439 kJ/mol is an empirical parameter. Again the arithmetic mean had to be used instead of the geometric mean, in order to obtain the plots shown in Figure 5 (the absolute values of D(M-X) - D(M-Me) in this figure are, therefore, unreliable). A comparison of D(M-I) - D(M-Me), calculated from equation 21, with experimental data has been been made by Schock and Marks [48].

Let us now leave all these qualitative or, at best, semi-quantitative trends, and address the application of Yuan's electronegativity scale, westernized by Luo and Benson [49]. Yuan's scale relies on the *unshielded core potential* of the atom X, V_X, given by the ratio between its number of valence electrons, n_X, and its covalent radius, r_X. Luo and Benson have found that many thermochemical data fit linear relationships illustrated by equation 22, where $\Delta\Delta H_f^o(RX/MeX)$ represents the difference between the standard

$$\Delta\Delta H_f^o(RX/MeX) = I_m + S_m V_X \qquad (22)$$

enthalpies of formation of gaseous RX and MeX, and I_m and S_m are empirical parameters. This correlation has been used to estimate enthalpies of formation of organic substances and of some main group organometallic molecules [49,50]. As it deals with gas phase values, it can also provide information on bond dissociation enthalpies. Let us look at a couple of simple examples, involving silanes. Figure 6 shows the plots of $\Delta\Delta H_f^o(H_3SiX/MeX)$ and $\Delta\Delta H_f^o(Me_3SiX/MeX)$ for X=F, Cl, Br, I. Data for the silanes are from Walsh's review [26] and for the halomethanes from [3]. Both correlations are fairly good and the

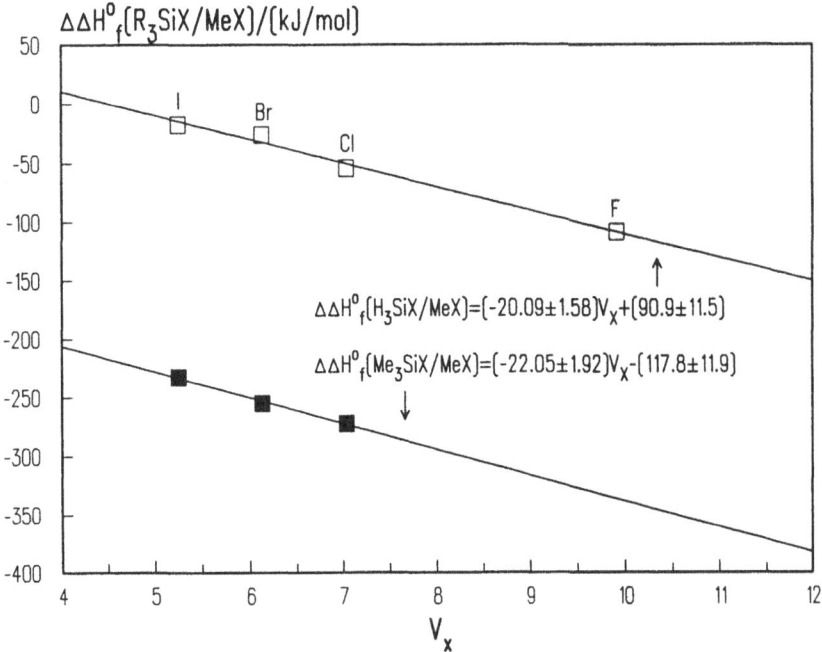

Figure 6 - $\Delta H_f^o(H_3SiX,g)-\Delta H_f^o(MeX,g)$ and $\Delta H_f^o(Me_3SiX,g)-\Delta H_f^o(MeX,g)$ as a function of the electronegativity of X (Yuan's scale). Adapted from [50c-d]

second leads to an estimated value of −583±22 kJ/mol for $\Delta H_f^o(Me_3SiF,g)$. This is in good agreement with Walsh's own estimate, −589±21 kJ/mol [26], but far from the value recommended in NIST publications (e.g. in [3]), −527 kJ/mol.

A comparison of Yuan's scale with sixteen other electronegativity scales has been made by Luo and Benson [51], showing that the unshielded core potential is the parameter which correlates best with thermochemical data. The same authors have also shown that V_X varies linearly with the ionization energies of the main group elements [52], e.g., for group 14, $V_X = (0.622\pm0.051)I(eV) - (1.78\pm0.44)$. This means, of course, that ionization energies should also yield good correlations with thermochemical data. Plots involving

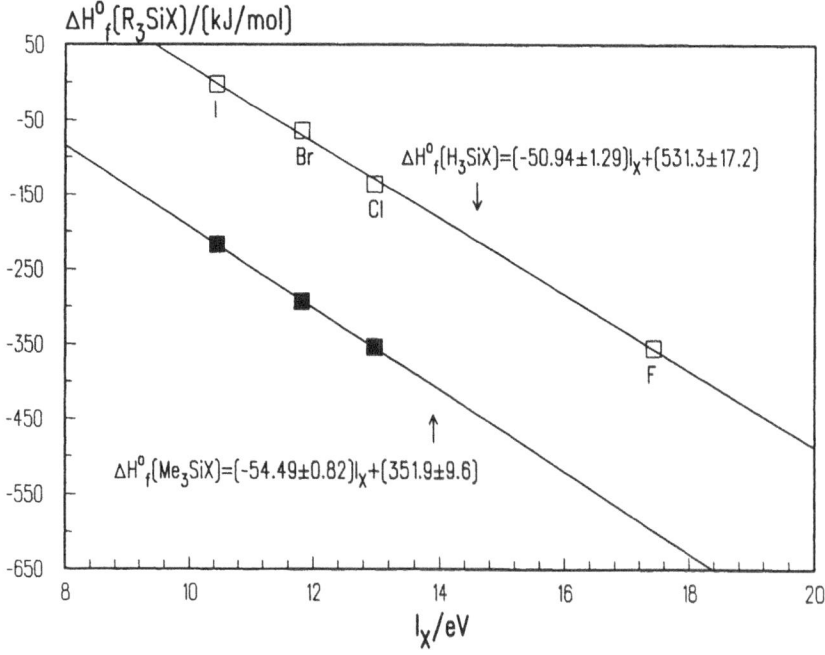

Figure 7 - $\Delta H_f^O(H_3SiX,g)$ and $\Delta H_f^O(Me_3SiX,g)$ as a function of the ionization energy of X. Adapted from [53].

enthalpies of formation and ionization energies are in fact the basis of the estimation method suggested by Screttas and Screttas [53]. To continue with our example of estimating the enthalpy of formation of Me_3SiF, plots of $\Delta H_f^O(H_3SiX,g)$ and $\Delta H_f^O(Me_3SiX,g)$ vs. I(X), X= F, Cl, Br, I are shown in Figure 7. The latter leads to $\Delta H_f^O(Me_3SiF,g)$ = -597±17 kJ/mol, in fair agreement with the value obtained by Luo and Benson's method.

An interesting point about Luo and Benson's method is related to the fact that equation 22 fits X groups which are as different as H, Me, NH_2, OH, SH, and halogens. If, for instance, $\Delta H_f^O(MeX,g)$ is plotted against V_X a considerable scatter will be observed, although the points of the four halogens define a perfect straight line. The "ligands" H, Me, NH_2 and OH belong to different *families* each of them defining a linear correlation with a given intercept. These sets of (probably parallel) lines yield a single line by using $\Delta\Delta H_f^o$(RX/MeX), indicating that the difference between the intercepts of two families is the same for RX and MeX.

A final note just to remind us that as the electronegativities of the transition metals are difficult to determine [54], the methods attempting to relate these parameters with thermochemical data are essentially restricted to main group elements.

8. Enthalpies of Formation vs. Enthalpies of Formation

As hinted before, the scarce use of the traditional schemes, developed for organic molecules, for the prediction of the energetics of organometallic compounds, is not due to a failure of the basic assumption of those methods, namely that the energetic content of a molecule reflects the energetic content of the moieties from which it is built. This idea justifies the method briefly described below, where the standard enthalpies of a series of compounds MX_nL_m yield good linear relationships when plotted against the enthalpies of formation of the ligands LH [55]. As these are usually available, the correlations can be used to derive unknown values of ΔH_f^o(MX_nL_m), to assess experimental data, or even to probe the contributions of steric factors or strain to the energetics of the molecule.

One of the most interesting features of this method is that the linear correlations hold for MX_nL_m and LH in their standard reference states (i.e. their stable physical states at 298.15 K and 1 bar). Consider Scheme 3, which represents an hypothetical reaction

$$MX_nL_m(rs) + mHCl(g) \xrightarrow{\Delta H^o(1)} MX_nCl_m(rs) + mLH(rs)$$

$$\left\downarrow \Delta H^o_{V1} \qquad \left\downarrow 0 \qquad\qquad \left\downarrow \Delta H^o_{V2} \qquad \left\downarrow m\Delta H^o_{V3}$$

$$MX_nL_m(g) + mHCl(g) \xrightarrow{\Delta H^o(2)} MX_nCl_m(g) + mLH(g) \qquad \underline{\text{Scheme } 3}$$

involving the compounds MX_nL_m. $\Delta H^o(1)$ and $\Delta H^o(2)$ are the enthalpies of reaction with MX_nL_m in the standard reference state (usually solid) and in the gas phase, respectively, and ΔH^o_V are vaporization or sublimation enthalpies. As implied by the empirical linear relationship, equation 23, $\Delta H^o(1)$ is constant for the series of compounds MX_nL_m. This

$$\Delta H^o_f(MX_nL_m, rs) = a\Delta H^o_f(LH, rs) + b \qquad (23)$$

enthalpy of reaction, on the other hand, can be expressed, by using Scheme 3, in terms of the bond dissociation enthalpies, equation 24.

$$\Delta H^o(1) = m[\bar{D}(M-L) - D(L-H)] + [\Delta H^o_{V1} - m\Delta H^o_{V3}]$$

$$+ [mD(H-Cl) - m\bar{D}(M-Cl) - \Delta H^o_{V2}] \qquad (24)$$

Here, it is seen that the third bracketed term is constant, which implies that $m[\bar{D}(M-L) - D(L-H)] + [\Delta H^o_{V1} - m\Delta H^o_{V3}]$ is also constant for the series of compounds that obey the linear correlation. In other words it is very likely that $\bar{D}(M-L)$ and $D(L-H)$ follow nearly parallel trends.

The method has been tested with a variety of organic, inorganic, and organometallic thermochemical data, which revealed some guidelines for its correct use and also some limitations [55,56]. Three examples are given here. One, Figure 8a, shows the correlation for compounds of the type $Mo(Cp)_2L_2$ (data from [6] and [11]). The values of the enthalpies of formation of several zirconium alkoxides, $Zr(Cp)_2(Cl)OR$, [57] are plotted in Figure 8b. Finally, Figure 8c displays values of the enthalpies of formation of $Th(Cp^*)_2R_2$ relative to $\Delta H^o_f[Th(Cp^*)_2(O-t-Bu)_2, c] - 2\Delta H^o_f(t-BuOH, l)$, calculated from data

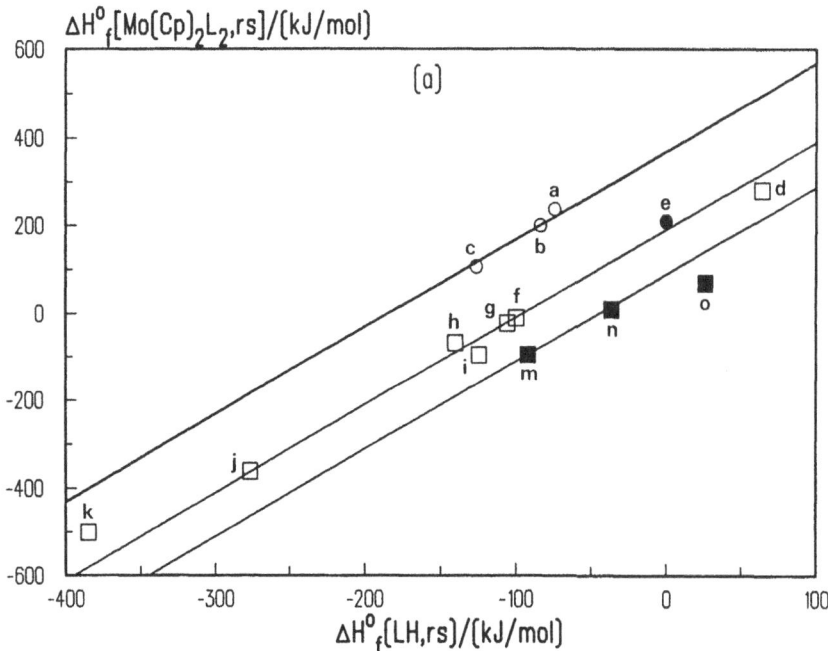

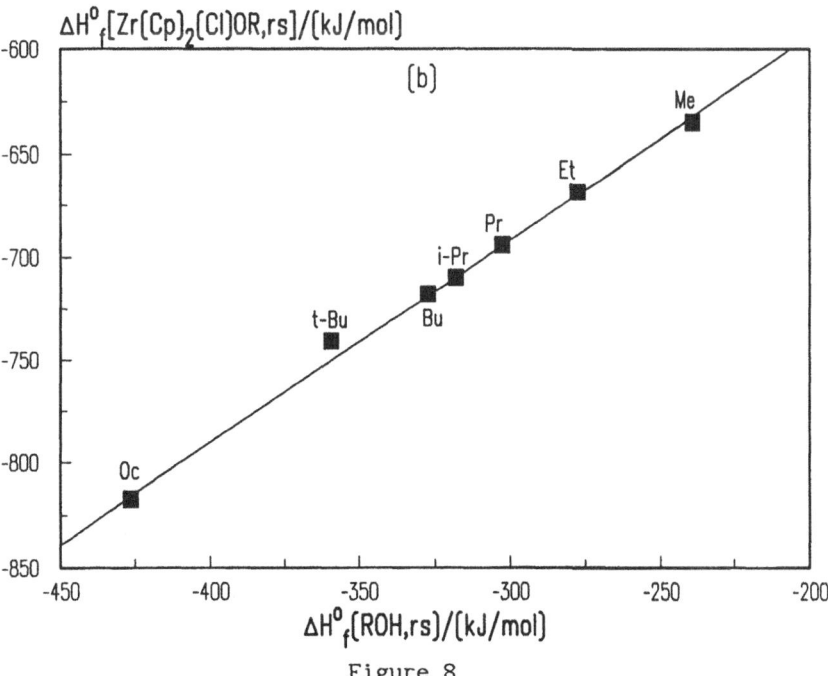

Figure 8

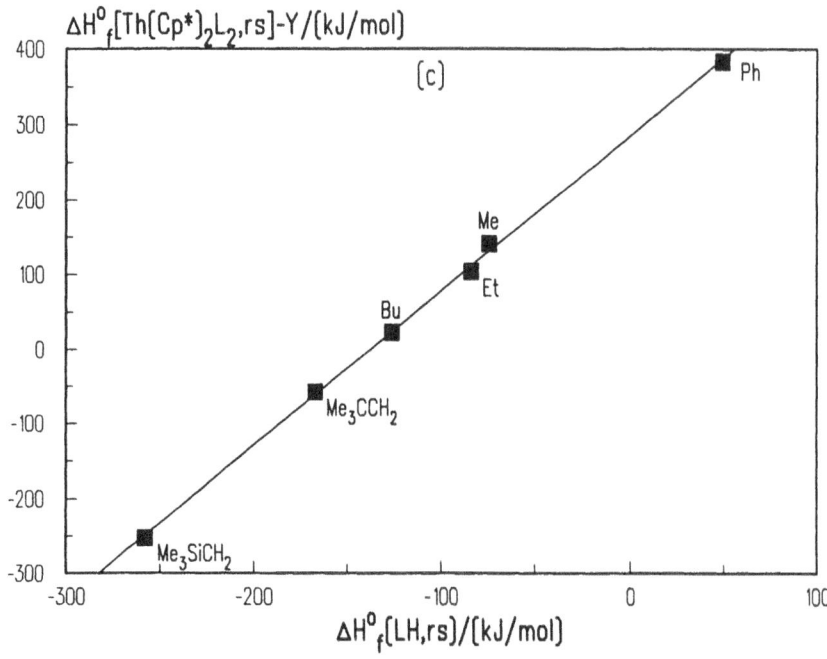

Figure 8 - Enthalpies of formation of solid complexes MX_nL_m vs. the enthalpies of formation of LH in their standard reference states. (a) $Mo(Cp)_2L_2$: a=Me, b=Et, c=Bu, d=PhS, e=H, f=PrS, g=i-PrS, h=t-BuS, i=BuS, j=$C_8H_{17}S$, k=PhCOO, l=CF_3COO (not shown), m=Cl, n=Br, o=I. (b) $Th(Cp^*)_2L_2$ - Y, where Y = $\Delta H_f^o[Th(Cp^*)_2(O-t-Bu)_2,c] - 2\Delta H_f^o(t$-BuOH,l). (c) $Zr(Cp)_2(Cl)(OR)$. (a) and (c) were adapted from [55b].

obtained by Marks and coworkers [58] ($Cp^* = \eta^5$-C_5Me_5).

Three "families" of compounds were considered in Figure 8a, one for L=alkyl, another involving oxygen and sulfur ligands, and the third defined by the halogen complexes. These correlations (slope 2) suggest a few comments. The one for the halogens indicates that the value recommended for $\Delta H_f^o[Mo(Cp)_2I_2,c]$ is a low limit. This is indeed a possibility, since the value relies on an estimate for the enthalpy of solution of HI in toluene. The oxygen and sulfur family, on the other hand, should probably be split into at least two families, in order to

improve the fit. More detailed information can be drawn from Figure
8b, where there is no doubt that a single family is defined. For
instance, it can be concluded that for R=t-Bu the difference between
the experimental enthalpy of formation and the value calculated from
the correlation (10 kJ/mol) is probably due to the steric effects
caused by the bulky t-butyl group. Finally, the alkyl correlation in
Figure 8a enables an estimate of the enthalpies formation of higher
alkyl complexes to be made. Interestingly, the point for methyl falls
slightly above the line. This is also observed in other similar
correlations, e.g. in Figure 8c.

As a final illustration of the method, it is noted that a plot
of $\Delta H_f^o[Me_3SiX,g]$ vs. $\Delta H_f^o(HX,g)$, X = Cl, Br, I, leads to $\Delta H_f^o(Me_3SiF,g)$ =
-564±10 kJ/mol. Again, this value is much higher than the one
recommended in [3], although it is 25 kJ/mol lower than Walsh's
estimate.

9. Bond Dissociation Enthalpies vs. Bond Dissociation Enthalpies

In the preceding paragraph it was stated that the linear correlations
would probably imply that $\bar{D}(M-L)$, or $D(M-L)$, and $D(L-H)$ follow parallel
trends. This was indeed suggested by Bryndza et al. [59], and
confirmed by plotting a large number of metal-ligand bond dissociation
enthalpies, measured in solution, against the respective gas phase
$D(L-H)$. Figure 9 illustrates the plot for complexes $Ru(Cp^*)(PMe_3)_2L$
and $Pt(dppe)(Me)X$ (dppe = 1,2-bis(diphenylphosphino)ethane). The data
are relative to $D(M-OH)=0$.

The large deviation observed for L=CN in Figure 9 was attributed
to ligand back-donation, while the case of L=SH was justified by an
improved overlap between Ru or Pt σ orbitals and the 3s/3p orbitals of
sulfur, as compared to the 2s/2p orbitals of oxygen [59a]. A parallel
line was thus suggested by Bryndza et al. for second row ligands.
Further examples of $D(M-L)/D(L-H)$ correlations have been discussed by
Marks and coworkers (see, e.g., [48], [60], and the chapter in this
volume). Here it is only stressed that despite the excellent

statistical behavior of many of those correlations, the slope is often different from unity.

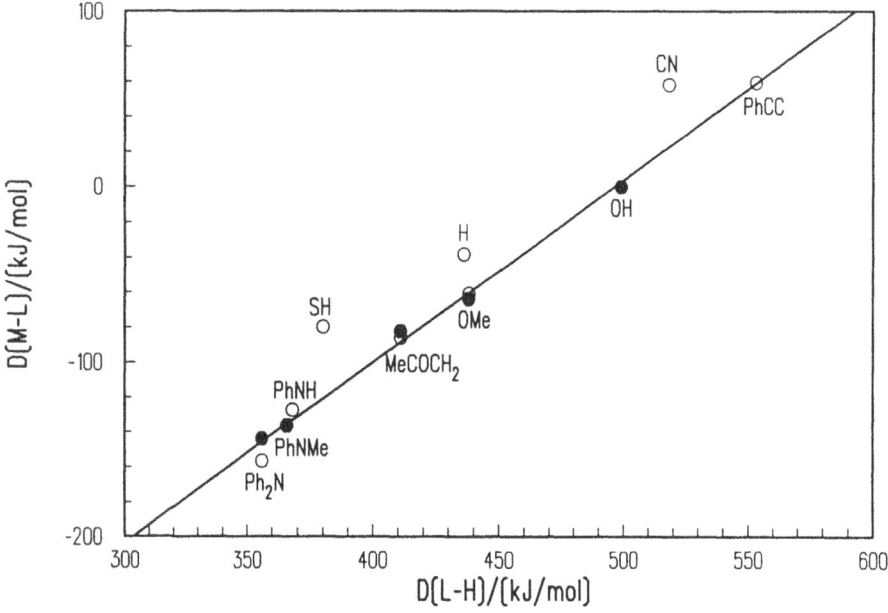

Figure 9 - Bond dissociation enthalpies D(Ru-L) and D(Pt-L) (filled circles), measured in solution, versus D(L-H). D(M-L) values are relative to D(M-OH) = 0. Adapted from [59a].

10. Conclusions

Estimating reliable thermochemical data for organometallic compounds is often a difficult exercise, in particular for transition metal complexes. Intuition still plays an important role, but as new experimental data become available a better understanding of the application of the above methods can be achieved. Undoubtedly, the growth of the thermochemical data bank will also foster the development of other prediction methods. One of the challenging issues, which was not addressed in this survey, is the relation between gas phase and solution data, namely the estimation of enthalpies of sublimation. Despite some recent efforts in this area [61], almost everything remains to be done!

11. Acknowledgement

Financial support from Junta Nacional de Investigação Científica e Tecnológica (project PMCT/C/CEN/42/90) is gratefully acknowledged.

12. References

1.   Cox, J. D. and Pilcher, G. (1970) Thermochemistry of Organic and Organometallic Compounds, Academic Press, London.
2.   Benson, S. W. (1976) Thermochemical Kinetics, Wiley, New York.
3.   Stein, S. E., Rukkers, J. M., and Brown, R. L. (1991) NIST Structures & Properties Database and Estimation Program, NIST Standard Reference Database 25, U.S. Department of Commerce, Gaithersburg, MD.
4.   Pedley, J. B., Naylor, R. D., and Kirby, S. P. (1986) Thermochemical Data of Organic Compounds, Chapman and Hall, London.
5.   Pilcher, G. and Skinner, H. A. (1982) in F. R. Hartley and S. Patai (Eds.), The Chemistry of Metal-Carbon Bond, Wiley, New York.
6.   Dias, A. R. and Martinho Simões, J. A. (1988) Polyhedron 7, 1531.
7.   Martinho Simões, J. A. and Beauchamp, J. L. (1990) Chem. Rev. 90, 629.
8.   Dias, A. R., Salema, M. S., and Martinho Simões, J. A. (1981) J. Organometal. Chem. 222, 69.
9.   (a) Dias, A. R. and Martinho Simões, J. A. (1982) Rev. Port. Quím. 24, 191. (b) Martinho Simões, J. A. unpublished data.
10. Hoffmann, R., Chester, N., and Tatsumi, K. (1981) private communication.
11. Calhorda, M. J., Carrondo, M. C. T., Dias, A. R., Galvão, A. M., Garcia, M. H., Martins, A. M., Minas da Piedade, M. E., Pinheiro, C. I., Romão, C. C., Martinho Simões, J. A., and Veiros, L. F. (1991) Organometallics 10, 483.
12. Skinner, H. A. (1945) Trans. Faraday Soc. 41, 645.
13. Cavell, K. J., Connor, J. A., Pilcher, G., Ribeiro da Silva, M. A. V., Ribeiro da Silva, M. D. M. C., Skinner, H. A., Virmani, Y, and Zafarani-Moattar, M. T. (1981) J. Chem. Soc., Faraday 1 77, 1585.

14. Cavell, K. J., Garner, C. D., Martinho Simões, J. A., Pilcher, G., Al-Samman, H., Skinner, H. A., Al-Tekhin, G., Walton, I. B., and Zafarani-Moattar, M. T. (1981) J. Chem. Soc., Faraday 1 77, 2927.

15. Calhorda, M. J., Carrondo, M. C. T., Dias, A. R., Domingos, A. M. T. S., Martinho Simões, J. A., and Teixeira, C. (1986) Organometallics 5, 660.

16. Cottrell, T. L. (1954) The Strengths of Chemical Bonds, Butterworths, London.

17. Gaydon, A. G. (1968) Dissociation Energies and Spectra of Diatomic Molecules, Chapman and Hall, London.

18. McKean, D. C. (1978) Chem. Soc. Rev. 7, 399.

19. McKean, D. C. (1989) Int. J. Chem. Kinet. 21, 445.

20. McKean, D. C., McQuillan, G. P., and Thompson, D. W. (1980) Spectrochim. Acta 36A, 1009.

21. Long, C., Morrisson, A. R., McKean, D. C., and McQuillan, G. P. (1984) J. Am. Chem. Soc. 106, 7418.

22. McQuillan, G. P., McKean, D. C., and Torto, I. (1986) J. Organometal. Chem. 312, 183.

23. McKean, D. C., McQuillan, G. P., Morrisson, A. R., and Torto, I. (1985) J. Chem. Soc., Dalton Trans. 1207.

24. Firth, S., Horton-Mastin, A., Poliakoff, M., Turner, J. J., McKean, D. C., McQuillan, G. P., and Robertson, J. (1989) Organometallics 8, 2876.

25. McKean, D. C., Torto, I., and Morrisson, A. R. (1982) J. Phys. Chem. 86, 307.

26. Walsh, R. (1988) in S.Patai and Z. Rappoport (Eds.), The Chemistry of Organic Silicon Compounds, Wiley, New York.

27. Dias, A. R., Diogo, H. P., Griller, D., Minas da Piedade, M. E., and Martinho Simões, J. A. (1990) in T. J. Marks (Ed.), Bonding Energetics in Organometallic Compounds, ACS Symposium Series, Washington, DC.

28. Noble, P. N. and Walsh, R. (1983) Int. J. Chem. Kinet. 15, 547.

29. Clark, K. B. and Griller, D. (1991) Organometallics 10, 746.

30. McKean, D. C., Morrisson, A. R., and Kelly, M. I. (1984) Chem. Phys. Lett. 109, 347.

31. Zavitsas, A. (1987) J. Phys. Chem. 91, 5573.

32. Nonhebel, D. C. and Walton, J. C. (1984) J. Chem Soc., Chem. Commun. 731.

33. Holmes, J. L. and Lossing, F. P. (1986) J. Am. Chem. Soc. 108, 1086.

34. Tolman, C. A. (1977) Chem. Rev. 77, 313.

35. Schenkluhn, H., Scheidt, W., Weimann, B., and Zähres, M. (1979) Angew. Chem. Int. Ed. Engl. 18, 401.

36. Geno, M. K. and Halpern, J. (1987) J. Am. Chem. Soc. 109, 1238.

37. See, for example: (a) Golovin, M. N., Rahman, M. M., Belmonte, J. E., and Giering, W. P. (1985) Organometallics 4, 1981. (b) Rahman, M. M., Liu, H.-Y., Prock, A., and Giering, W. P. (1987) Organometallics 6, 650. (c) Rahman, M. M., Liu, H.-Y., Prock, A., and Giering, W. P. (1989) Organometallics 8, 1. (d) Liu, H.-Y., Eriks, K., Prock, A., and Giering, W. P. (1990) Organometallics 9, 1758. (e) Tracey, A. A., Eriks, K., Prock, A., and Giering, W. P. (1990) Organometallics 9, 1399.

38. Stahl, L. and Ernst, R. D. (1987) J. Am. Chem. Soc. 107, 5673.

39. Drago, R. S., Vogel, G. C., and Needham, T. E. (1971) J. Am. Chem. Soc. 92, 6014.

40. Drago, R. S., Wong, N., Bilgrien, C., and Vogel, G. C. (1987) Inorg. Chem. 26, 9.

41. (a) Sanderson, R. T. (1976) Chemical Bonds and Bond Energy (2nd ed.), Academic Press, New York. (b) Sanderson, R. T. (1983) Polar Covalence, Academic Press, New York.

42. (a) Pearson, R. G. (1990) Coord. Chem. Rev. 100, 403. (b) Pearson, R. G. (1990) in T. J. Marks (Ed.), Bonding Energetics in Organometallic Compounds, ACS Symposium Series, Washington, DC.

43. Pearson, R. G. (1988) J. Am. Chem. Soc. 110, 7684.

44. Pearson, R. G. (1988) Inorg. Chem. 27, 734.

45. Labinger, J. A. and Bercaw, J. A. (1988) Organometallics 7, 926.

46. Huheey, J. E. (1975) Inorganic Chemistry, Harper & Row, New York.

47. Matcha, R. L. (1983) J. Am. Chem. Soc. 105, 4859.

48. Schock, L. E. and Marks, T. J. (1988) J. Am. Chem. Soc. 110,

7701.

49. Luo, Y.-R., and Benson, S. W. (1988) J. Phys. Chem. 92, 5255.

50. See, for example: (a) Luo, Y.-R. and Benson, S. W. (1989) J. Am. Chem. Soc. 111, 2480. (b) Luo, Y.-R. and Benson, S. W. (1989) J. Phys. Chem. 93, 3304. (c) Luo, Y.-R. and Benson, S. W. (1989) J. Phys. Chem. 93, 1674. (d) Luo, Y.-R. and Benson, S. W. (1989) J. Phys. Chem. 93, 4643. (e) Luo, Y.-R. and Benson, S. W. (1989) J. Phys. Chem. 93, 3791.

51. Luo, Y.-R. and Benson, S. W. (1990) J. Phys. Chem. 94, 914.

52. Luo, Y.-R. and Benson, S. W. (1989) J. Phys. Chem. 93, 7333.

53. Screttas, C. G. and Micha-Screttas, M. (1989) J. Org. Chem. 54, 5132.

54. Allen, L. C. (1989) J. Am. Chem. Soc. 111, 9003.

55. (a) Dias, A. R., Martinho Simões, J. A., Teixeira, C., Airoldi, C., and Chagas, A. P. (1987) J. Organometal. Chem. 335, 71.
(b) Dias, A. R., Martinho Simões, J. A., Teixeira, C., Airoldi, C., and Chagas, A. P. (1989) J. Organometal. Chem. 361, 319.
(c) Dias, A. R., Martinho Simões, J. A., Teixeira, C., Airoldi, C., and Chagas, A. P. (1991) Polyhedron 10, 1433.

56. Griller, D., Martinho Simões, J. A., and Wayner, D. D. M. (1991) in Sulfur-Centered Reactive Intermediates in Chemistry and Biology (C. Chatgilialoglu and K.-D. Asmus, Eds.), Plenum, New York.

57. Diogo, H. P., Simoni, J. A., and Martinho Simões, J. A. to be published.

58. Bruno, J. W., Marks, T. J., and Morss, L. R. (1983) J. Am. Chem. Soc. 105, 6824.

59. (a) Bryndza, H. E., Fong, L. K., Paciello, R. A., Tam, W., and Bercaw, J. E. (1987) J. Am. Chem. Soc. 109, 1444. (b) Bryndza, H. E., Domaille, P. J., Tam, W., Fong, L. K., Paciello, R. A., and Bercaw, J. E. (1987) Polyhedron 7, 1441.

60. Nolan, S. P., Stern, D., and Marks, T. J. (1989) J. Am. Chem. Soc. 111, 7844.

61. Chickos, J. S. (1987) in Molecular Structure and Energetics, Vol. 2 (J. F. Liebman and A. Greenberg, Eds.), VCH, New York; see also his chapter in this volume.

BONDING AND SOLVATION ENERGETICS FOR ORGANOMETALLIC GAS-PHASE MOLECULES AND IONS: EXPERIMENTAL APPROACHES AND THERMODYNAMIC INSIGHTS

D. E. Richardson
Department of Chemistry
University of Florida
Gainesville, Florida 32611

ABSTRACT. Experimental techniques for study of the gas-phase energetics of organometallic oxidation-reduction processes via Fourier transform ion cyclotron resonance mass spectrometry are described. In particular, sampling methods for typical compounds are mentioned along with approaches for determining the thermodynamics and kinetics of electron-transfer reactions. Results obtained recently for adiabatic ionization energies and electron attachment energies involving organometallic species are presented. Thermodynamic insights that arise from the experimental results are described, including applications to bond energetics, solvation, substituent effects, redox entropies, and thermodynamic effects in electron-transfer kinetics.

1. Introduction

Oxidation-reduction reactions have been extensively studied in organometallic chemistry, with the bulk of thermochemical information coming from electrochemical studies in the form of electrode potentials [1]. Such potentials relate to the free energy of adding or subtracting electrons from a compound in a solvent (when reversible processes are studied). Thus, one is obviously not studying the intrinsic properties of the compound but rather the properties of a strongly solvated redox couple, and the solvent can have profound effects on the ease or difficulty of a given redox process. In addition, reduction or oxidation of organometallic compounds at an electrode is often followed by chemical reactions that make direct measurement of standard electrode potentials difficult or impossible. Many of these interfering reactions involve the solvent or other components of the electrochemical cell.

To avoid the various problems associated with measurement and interpretation of the redox thermochemistry of organometallic compounds in solution, one might imagine an experiment in which no solvent or counterions are present and the thermodynamics of attaching electrons to or removing electrons from a compound are determined. We have employed a technique that closely approaches this ideal experiment and provides direct insight into the intrinsic tendency of organometallic compounds to be oxidized or reduced [2-5]. The method, Fourier transform ion cyclotron resonance mass spectrometry (FTICR-MS) [6], has provided data that lead to fundamental information on redox processes, bonding, and solvation energetics for organometallic molecules and ions.

In this article, I will summarize relevant experimental techniques and some of the results we have obtained recently for adiabatic ionization energies and electron attachment energies

J. A. Martinho Simões (ed.), Energetics of Organometallic Species, 233–251.
© 1992 *Kluwer Academic Publishers.*

involving organometallic species. In particular, sampling methods for typical compounds are described along with approaches for determining the thermodynamics and kinetics of electron-transfer reactions involving them. Some results of such studies are then summarized for negative and positive ions. Finally, thermodynamic insights that arise from the experimental results are described, including applications to bond energetics, solvation energetics, substituent effects, redox entropies, and thermodynamic effects in electron-transfer kinetics.

2. Experimental Approaches for Determining Ionization and Electron Attachment Energetics for Organometallic Molecules

2.1 SAMPLING TECHNIQUES FOR ION CYCLOTRON RESONANCE STUDIES OF ORGANOMETALLICS

The FTICR-MS technique has been extensively reviewed [6], and general features of its application to coordination compounds have been discussed [2]. Only brief descriptions of the instruments at the University of Florida are given here along with some pertinent details regarding sampling techniques for organometallic compounds.

The FTICR-MS instrument is based on a roughly cubic ion trap placed in a high vacuum system that has a background pressure in the 10^{-9} torr range. The trap is connected to a computer and electronics that provide for the programming of pulse sequences to control the production of ions in the trap and the subsequent detection of the mass spectrum of the ion population. The trap is located in a magnetic field that allows the trapping of ions in the cell via combined magnetron and cyclotron motions. At the University of Florida, 2 tesla and 3 tesla instruments are available that have a practical upper mass to charge (m/z) limit of ~3000 u.

For the studies described here, compounds are usually introduced as neutral vapor through a precision leak valve or by heating the sample on a solids probe located in the main vacuum chamber near the trap. Typical sample operating pressures are in the 10^{-8} to 10^{-6} torr range, so samples with relatively low vapor pressures are often easily studied. Parent ions are typically produced in the trap by electron impact ionization (positive ions) or low energy electron attachment (negative ions). In some cases, direct electron beam ionization or attachment leads to unacceptable fragmentation, so electron-transfer chemical ionization (ETCI or "charge-transfer CI" [7]) is employed. In the ETCI method, a secondary gas is introduced that is readily ionized by electron impact and then transfers its charge in an ion/molecule reaction with the neutral organometallic species. In this way, the energy released upon ionization is reduced and less excess energy is available for fragmentation. Thus, for example, the benzene cation ($C_6H_6^+$, IP benzene = 9.2 eV) cleanly oxidizes $Cr(CO)_6$ (IP ~ 8.4 eV) to $Cr(CO)_6^+$ with minimal CO loss [8]. Another advantage of ETCI, particularly in negative ion studies, is the increased yields of parent ions for neutrals with relatively low cross sections for direct electron beam ionization.

In principle, the FTICR trap can be combined with virtually any ion source to provide for ion/molecule studies involving ions not obtainable in the traditional manner described above. In this way, ions of low volatility or nonvolatile compounds can be introduced following production by alternative methods. For example, at the University of Florida, laser desorption of a solid sample using either a UV-visible-nearIR Nd-YAG laser or a CO_2 IR laser can be used to produce ions directly [9]. Ions produced in external sources (e.g., fast atom bombardment (FAB) or electrospray) can be introduced into ICR traps by using ion transfer techniques. With these various sampling methods, studies on a wide range of organometallic compounds should be possible.

2.2 EQUILIBRIUM AND BRACKETING METHODS

The goal of this work is to determine the thermodynamics of electron ionization or electron attachment for organometallic couples in which both members of the redox couple are thermally equilibrated. The energy required to remove an electron from a neutral molecule A or an anion A$^-$ can be obtained by photoelectron spectroscopy (PES) [10-12], but resulting spectral profiles are usually broadened in the case of organometallic samples and vibrational fine structure is typically not resolved. If the A(v=0)→A$^+$(v=0) or A$^-$(v=0)→A(v=0) transition is observed and assigned correctly, the energy of that transition is technically the enthalpy of electron loss at 0 K (i.e., the adiabatic IP for A → A$^+$ or the electron affinity, EA, for A$^-$ → A). To obtain the thermodynamics of electron loss at a temperature such as 298 K for use in thermochemical cycles, a statistical mechanical calculation must be done to estimate the changes in enthalpy and entropy between 0 K and the desired temperature. Given the paltry amount of spectroscopic data for organometallic molecules and their corresponding anions or cations, such an estimate is difficult at best.

Since it is not straightforward to measure directly the free energy changes for equilibria such as those in eqs 1 and 2,

$$A(g) \rightleftharpoons A^+(g) + e^- \tag{1}$$
$$A(g) + e^- \rightleftharpoons A^-(g) \tag{2}$$

the values of $\Delta G_i°$ (ionization, eq 1) and $\Delta G_a°$ (electron attachment, eq 2) are usually derived indirectly. By far most of these quantities for organic and inorganic compounds have been produced by electron transfer (or "charge-transfer") equilibrium (ETE) methods using either ion cyclotron resonance mass spectrometry [13,14] or pulsed high pressure mass spectrometry (PHPMS) [15-17]. The specific techniques have been described in detail elsewhere [2,13-17]. In essence, a reaction such as in eq 3 is allowed to come to equilibrium in the ion trap or high pressure source, and from the ratios of the ion intensities and the neutral pressures a value for $\Delta G_{et}°$ is obtained (eq 4) at the experimental temperature.

$$A(g) + B^+(g) \rightleftharpoons A^+(g) + B(g) \tag{3}$$

$$\Delta G_{et}° = -RT \ln \left(\frac{P_B I_{A^+}}{P_A I_{B^+}} \right) \tag{4}$$

If the value of $\Delta G_i°(B)$ is known, $\Delta G_i°(A)$ is immediately obtained from eq 5.

$$\Delta G_{et}° = \Delta G_i°(A) - \Delta G_i°(B) \tag{5}$$

From temperature dependence studies, corresponding values of $\Delta H_i°(A)$ and $\Delta S_i°(A)$ can be derived (assuming values for the B$^{0/+}$ couple are known). Ladders of $\Delta G_i°$, $\Delta S_i°$, $\Delta H_i°$, etc., derived by equilibrium methods must be anchored to a compound for which the thermodynamic quantities are well established (generally by spectroscopic methods and statistical mechanics). Completely analogous treatments are used to derive $\Delta G_a°$ and related thermodynamic functions from equilibria involving negative ions.

In the event that a neutral pressure ratio cannot be established that allows determination of $\Delta G_{et}°$ via eq 4, estimates of $\Delta G_i°$ or $\Delta G_a°$ can be made by using bracketing methods. For example, although a significant pressure of Cl(g) cannot be produced near room temperature for use in ETE experiments, rapid electron transfer from the ion Cl$^-$(g) to an acceptor B(g) suggests that the electron affinity of B exceeds that of Cl(g). However, this method can be inaccurate since measurable rates of electron transfer can be followed under some circumstances for reactions that are endoergic by up to 10 - 20 kJ mol^{-1}. Furthermore, the absence of observable electron transfer does not require that the process be endoergic since large kinetic barriers may make an exoergic process too slow to detect prior to ion loss from the trap ($k < \sim 10^8$ M^{-1}s^{-1} or $\sim 10^{-12}$ cm^3 s^{-1}). For these reasons, equilibrium results are generally more reliable.

Gas-phase ion/molecule equilibria are normally obtained under conditions that should insure thermalization of both ions. In the ICR trap with a total neutral pressure of $\sim 10^{-6}$ torr, ions undergo ~ 30 collisions per second. Approach to equilibrium is normally followed for ~ 10 - 20 s with apparent equilibrium typically established in a few seconds for most reactions we have observed. Thus, trapped ions will normally have hundreds of collisions with neutrals before the final equilibrium ratio is measured, and all species are assumed to be at the temperature of the ion trap. This assumption has been partially confirmed by studies using "thermometer" reactions [18], but nonthermal ions are always a concern in FTICR equilibrium studies.

2.3 TEMPERATURE DEPENDENCE OF EQUILIBRIA IN THE ICR TRAP

The vast majority of temperature dependent ETE data have been generated by PHPMS [15-17]. The FTICR-MS technique has rarely been applied in temperature dependence studies [13,14], and most ICR equilibrium data have been obtained at effective temperatures of ~ 350 K.

In order to assess the accuracy of enthalpy and entropy changes derived for electron transfer equilibria studied by FTICR-MS, we investigated the equilibrium constant for eq 6 as a function of the trap and vacuum system temperature (350 - 500 K).

$$CO^+(g) + Kr(g) \rightleftharpoons Kr^+(g) + CO(g) \qquad (6)$$

For this model reaction, the heat capacities and entropies of all species can be accurately calculated by using spectroscopic data and statistical mechanics. The experimentally derived parameters in the range 350 - 500 K ($\Delta H_{et}° = 1.2 \pm 3.3$ kJ mol^{-1}, $\Delta S_{et}° = 13 \pm 6$ J mol^{-1}K^{-1}) compare favorably to the estimates from statistical mechanics ($\Delta H_{et}° = -0.96 \pm 0.04$ kJ mol^{-1}, $\Delta S_{et}° = 5.9 \pm 0.3$ J mol^{-1}K^{-1}). The enthalpy change obtained is well within the estimated errors normally quoted for equilibrium ladders (± 0.1 eV or 10 kJ mol^{-1}). The error in the observed $\Delta S_{et}°$ value most probably arises from errors in determination of the relative pressures of the neutral gases (pressure measurement errors affect derived entropy values but derived enthalpies are unaffected). Generally, neutral pressures are determined by using an ion gauge calibrated for the reactant gases against a capacitance manometer. The pressure characteristics of the U. of Florida FTICR/MS vacuum systems have been carefully investigated by placement of ion gauges at trap locations when the magnetic field is off [19].

2.4 KINETICS OF GAS-PHASE ELECTRON TRANSFER REACTIONS

By acquiring spectra as a function of time after the ionization event, the rates of organometallic electron-transfer reactions can be derived for reversible or essentially irreversible reactions

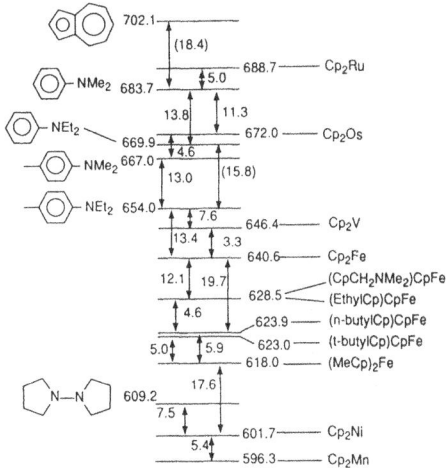

Figure 1. Free energy ladder for ionization of metallocenes (in kJ mol⁻¹). Reference compounds at left.

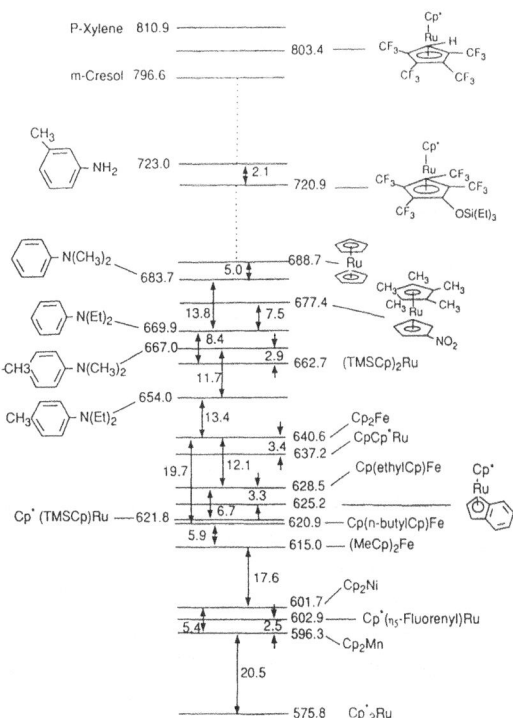

Figure 2. Free energy ladder for ionization of ruthenocene derivatives (in kJ mol⁻¹). Organic reference compounds at left, other metallocene referenced from Figure 1. Cp* = C₅Me₅.

[8,20-22]. The rates of self-exchange reactions (such as in eq 7) can be determined by initially ejecting one or more of the isotopes of the parent ion from the trap and following the reestablishment of the natural isotope distribution. Generally, the second order rate constants for gas-phase electron-transfer reactions are much greater than those observed for the same reactions in solution. For example, the reaction in eq 7

$$Cp_2Fe + Cp_2{}^*Fe^+ \rightleftharpoons Cp_2Fe^+ + Cp_2{}^*Fe \qquad (7)$$

is governed by a near-collisional rate constant in the gas-phase ($k_{gp} = 1.5 \times 10^{11}$ M^{-1}s^{-1} at 350 K), which is ~10^4 - 10^5 times greater than observed in polar solvents (e.g., $k_{MeOH} = 1.85 \times 10^7$ M^{-1}s^{-1} at 298 K). In other cases, however, gas-phase self-exchange reactions can be much slower than collisional (e.g., the SF$_6{}^{0/-}$ exchange is immeasurably slow and $k_{gp} < 3 \times 10^7$ M^{-1}s^{-1}).

3. Results from Fourier Transform ICR Studies

In this section, some recent results for electron-transfer equilibria involving metal complexes are summarized. As will be seen, studies of positive ions provide data that complement and extend PES results. On the other hand, the negative ion data provide some previously unknown information on the energetics of electron attachment processes for metal complexes in the gas phase. The analogous spectroscopic experiment, PES of gas-phase negative ions [23], is more difficult than PES of neutral molecules and has provided relatively little data for comparison to ICR results.

3.1. EQUILIBRIA INVOLVING POSITIVE METAL COMPLEX IONS

3.1.1. *Metallocenes.* The electrochemical properties of most known bis(cyclopentadienyl) complexes have been studied, and many of these show reversible behavior in the Cp$_2$M$^{+/0}$ couples [1]. Some of the metallocenes are irreversibly oxidized, however, so absolute comparisons of electrode potentials can only be made with confidence for a few couples. Photoelectron data on metallocenes are much more complete, and extensive analyses of the trends in ionization potentials (Cp$_2$M(g) $\rightarrow$ Cp$_2$M$^+$(g)) have appeared [12]. With this background, the metallocenes were the ideal choice for incorporation of organometallic compounds into electron-transfer equilibrium studies. An experimental free energy ladder (T = 350 K) is shown in Figure 1 for metallocenes with $\Delta G_i°$ values in the 6.2 - 7.2 eV (600 -700 kJ mol^{-1}) range.

Generally, the derived $\Delta G_i°$ values are on the low energy side of the vertical ionization energies determined by PES. This difference is easily understood to originate from the adiabatic nature of the thermal equilibrium, where both ion and neutral are produced in their Boltzmann equilibrated ground electronic state. Large differences between vertical and thermal ionization energies suggest relatively large structural rearrangements accompanying the ionization (i.e., substantial changes in bond lengths and/or angles).

Of the metallocenes in Figure 1, the smallest difference in equilibrium geometries of the neutral and ion occurs for vanadocene, which has an unusually narrow first ionization band in the PES. The close agreement between the first vertical IP determined by Lichtenberger and coworkers [24] and the $\Delta G_i°$ value here (both 6.75 eV) is noteworthy, but the $\Delta H_i°$ value needs to be determined for the equilibrium experiment for a better comparison to the PES data.

3.1.2. *Substituent Effects.* Electrochemical [1] and PES studies [12] on a limited number of metallocene derivatives have explored the effects of Cp substituents on the ionization energies. Where comparisons are possible, parallel trends are observed in the results of ETE studies on ferrocene and ruthenocene derivatives (Figures 1 and 2). The data for ruthenocene derivatives are of particular note; with the exception of decamethylruthenocene, the +/0 couples are electrochemically irreversible [25]. Therefore, ETE is the only method that has been applied for direct determination of the free energies of oxidation for most ruthenocene derivatives.

3.1.3. *Temperature Dependencies of Electron-Transfer Equilibria.* A reported PHPMS temperature dependence study of ferrocene [17] resulted in a ΔH_i° value of 657.3 kJ mol^{-1} (6.81 eV) based on equilibria with diethyltoluidine (DET) and dimethyltoluidine (DMT), for which ΔH_i° values had been previously determined [13,14]. Our FTICR temperature dependence study of the ferrocene/DET equilibrium (eq 8) agrees well with the PHPMS result for ΔH_{et}° (ICR: 0.4 ± 4.1 kJ mol^{-1}, PHPMS: -0.4 kJ mol^{-1}). Thus, the adiabatic IP of ferrocene is well established as 6.81 ± 0.10 eV taking the usual error limits for reference compounds into account.

$$DET^+ + Cp_2Fe \rightleftharpoons Cp_2Fe^+ + DET \qquad (8)$$

However, the ICR and PHPMS studies diverge in the ΔS_{et}° values for eq 8 (ICR: 39 ± 5 J mol^{-1}K^{-1}, PHPMS: 9.2 J mol^{-1}K^{-1} (no error limits reported)). A difference of this magnitude is explicable if the experimental pressure ratio used for eq 4 is in error by a factor of ~30 for one of the experiments. Since different methods are used to establish experimental pressures in ICR and PHPMS studies, it is conceivable that a systematic pressure error is introduced in one or both of the methods. Our pressure measurements are obtained directly in the vacuum system by using an ion gauge and are corrected by independent calibrations for each reactant gas. In addition, comparative studies of organic equilibria (e.g. DET/DMT) by using FTICR show excellent agreement with the PHPMS results, and the free energy ladder of Figure 1 shows good internal consistency between ferrocene, ferrocene derivatives and the organic reference compounds. The second order rate constants determined on our instruments for strongly exoergic gas-phase electron-transfer reactions [8,20-22] involving metallocenes are fully compatible with estimates for collision-limited rate constants. However, the PHPMS results [17] do not show any large inconsistencies in rate constants, either, and the smaller ΔS_{et}°(eq 8) value is actually more consistent with estimates based on statistical mechanics. The origin of the unexpectedly large FTICR value for ΔS_{et}°(eq 8) is under investigation.

3.2. EQUILIBRIA INVOLVING NEGATIVE IONS

3.2.1. *Complexes of Acetylacetonate and Related Ligands.* An ETE ladder for representative tris(β-diketonate) complexes is shown in Figure 3. In this study, extensive use was made of ETE ladders for organic compounds published primarily by Kebarle and co-workers [15]. The hexafluoroacetylacetonate (hfac) complexes were found to have ΔG_a°(350 K) values near the extreme high end of the organic ladder, and only estimates could be made in the cases of M = Mn, Fe, Ru, and Co, all of which exceed Cl(g) in electron affinity! Metal acetylacetonate complexes were more amenable to ETE studies since their electron affinities fall in the range for which numerous organic reference compounds are available.

3.2.2. *Metallocenes.* Nickelocene is unique among the first row metallocenes in its ready formation of the parent negative ion, and ETE studies set its ΔG_a°(350 K) value at -79 ± 7 kJ

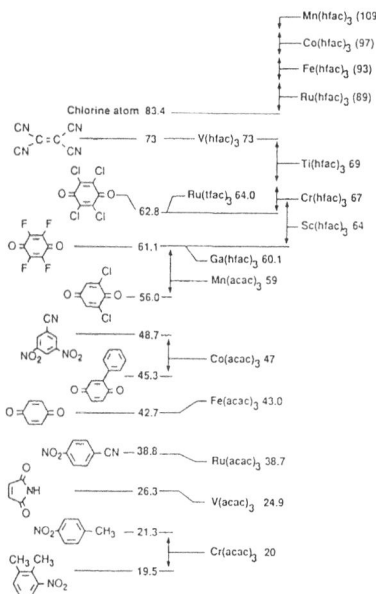

Figure 3. Free energy ladder for electron attachment to metal acac and hfac complexes (in kcal mol^{-1}). From P. Sharpe, Ph.D. Dissertation, University of Florida, 1990.

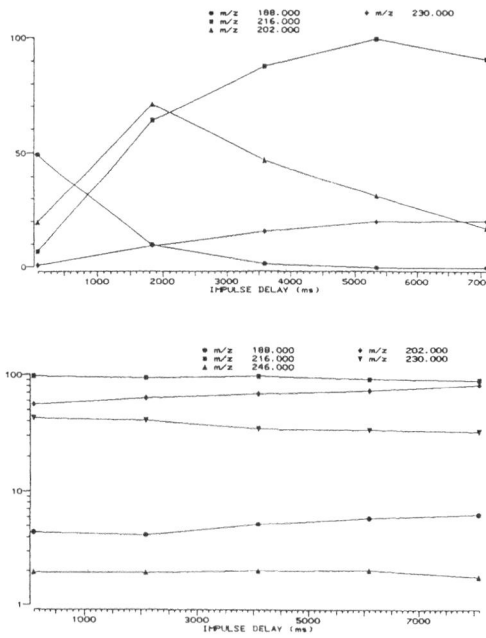

Figure 4. Time dependence of ion populations for a mixture of methylated nickelocenes. (Top) Time plot of ion intensities from positive ion spectrum. (Bottom) Time plot of ion intensities from negative ion spectrum. M/z values: 188, nickelocene; 202, monomethylnickelocene; 216, 1,1'-dimethylnickelocene; 230, trimethylnickelocene (several isomers).

mol^{-1}. Nickelocene is therefore an excellent choice for investigating the dependence of substituent effects on the nature of the process under consideration (here, reduction vs. oxidation). We have investigated the effect of alkylation on the electron affinity and ionization energy of nickelocene [26]. As noted for ferrocene and ruthenium derivatives, increasing alkylation of the Cp rings lowers the ionization energy, as can be seen from the time dependence of the charge transfer in the positive ion mode spectra of a mixture of alkylated nickelocenes (Figure 4). Remarkably, the corresponding negative ion spectra indicate that the electron affinities of the derivatives are very close since all the various ions are present at equilibrium (Figure 4). If one of the ions is ejected from the ion trap (e.g., the dialkyl ion), the original equilibrium quotients are rapidly reestablished via electron transfer from the remaining anions to the neutrals.

3.2.3. *Other Organometallic Anions.* Metal carbonylate anions of the types $M(CO)_n^-$ and $CpM(CO)_n^-$ are popular organometallic nucleophiles for the synthesis of new M-C bonds via group transfer reactions with RX substrates. The relative reactivities of various metal carbonylates have been studied in solution, and either direct associative displacement or electron transfer/radical coupling paths have been suggested. We established [27] that the reactivities observed in solution are paralleled by rates observed in gas-phase reactions with various RX substrates. With respect to the electron-transfer path, the electron affinities of the corresponding neutral metal radicals $M(CO)_n$ and $CpM(CO)_n$ are of interest and can be estimated by gas-phase bracketing methods. Unfortunately, since the neutral precursors are dimers rather than free radicals, ETE methods cannot be used. However, the bracketed $\Delta G_a°$ values follow the same trends observed for the electrode potentials observed for the anions in solution.

4. Thermodynamic Insights

4.1. METAL-LIGAND BOND ENERGETICS

A thermochemical cycle of the type shown in Figure 5 can be used to extract average metal-ligand bond disruption enthalpies ($\Delta H_{dis}°$(M-L) or BDE) for ML_n^z (z = +1, -1) complexes from ETE data [5,28]. Enthalpies of formation of the neutral member of the redox couple are often available from combustion or reaction calorimetry, so $\Delta H_a°$ or $\Delta H_i°$ values allow the estimation of $\Delta H_{dis}°$(M-L) values for the corresponding ions. Thus, it is possible to compare bond energetics for two or three members of redox couples of the general type $ML_n^{-/0/+}$ and assess the thermochemical origins of redox energetics in the absence of solvent. Appropriate caution must be taken in using the ETE results for free energies as enthalpies in such cycles, but the errors introduced are expected to be relatively small (typically ~5 kJ mol^{-1} or less per M-L bond) [4].

Table 1 summarizes bond disruption enthalpy values for several neutral compounds and their accessible gas-phase ions. For many of the ions, the BDE values have been estimated from ETE data; however, some additional estimates for cations based on PES data are provided. The latter values are included for comparison of complexes with the same metal or ligand but are not as reliable since the adiabatic IP values needed are difficult to determine by PES. For L = acac, alkyl, and Cp, the heterolytic disruption yields anionic ligands, whereas the heterolytic cleavage of M-CO or M-arene bonds yields neutral ligands.

In comparing BDE values for these various types of complexes, bond disruptions that yield the same metal product are generally used. In Figure 6, BDE values are compared for disruptions that produce the M^0 atom (from neutral complexes) and the M^+ ion (from cationic complexes).

242

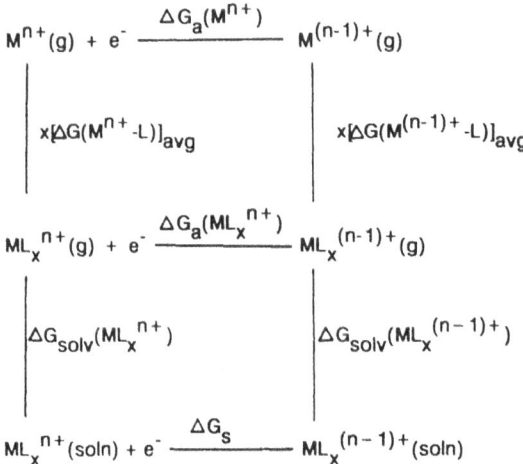

Figure 5. Traditional free energy cycle for thermochemistry of metal ions and their complexes in the gas phase and solution. If the ligand has a -1 charge, the charges on the complexes are (n - x)+ and (n - 1 - x)+. Completely analogous cycles can be constructed for enthalpies, entropies, etc. (Reprinted with permission from *Inorganic Chemistry*. Copyright 1990 American Chemical Society.)

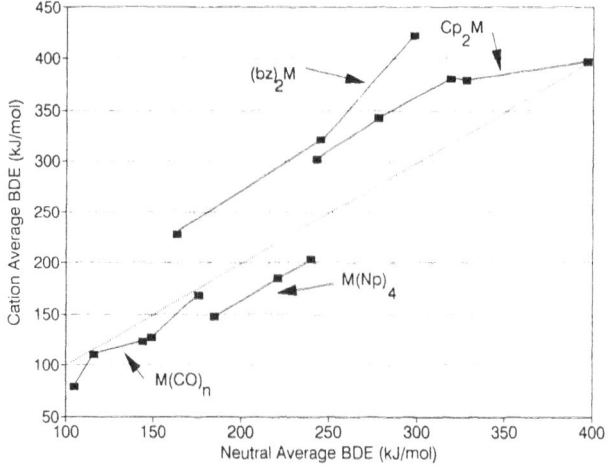

Figure 6. Plot of average BDE values for organometallic cations (ML_n^+) vs. the average BDE values for the corresponding neutral compound. Abbreviations: bz = benzene, Cp = cyclopentadienyl, Np = neopentyl. Data used are from Table 1. The BDE values in all cases are for production of metal atom and neutral ligands (neutral complexes) or metal monocation and neutral ligands (cation complexes).

Table 1. Selected Average Bond Disruption Enthalpies for Neutral and Ionic Gas-Phase Organometallics[a]

Complex	$\Delta\bar{H}°_{hom}(M\text{-}L)^{b,c}$/kJ mol^{-1}	$\Delta\bar{H}°_{het}(M\text{-}L)^{c,d}$/kJ mol^{-1}
$[Fe^{II}(acac)_3]^-$	360 ± 25	895 ± 67
$Fe^{III}(acac)_3$	301 ± 25	1820 ± 40
$[Fe^{IV}(acac)_3]^+$	$(290)^e$	$(3500)^e$
Cp_2Fe^{II}	328 ± 6	1330 ± 20
$[Cp_2Fe^{III}]^+$	379 ± 8	2480 ± 20
$Fe^0(CO)_5$	-	$(116)^f$
$[Fe^I(CO)_5]^+$	-	$(110)^e$
$[Mn^{II}(acac)_3]^-$	>360	862 ± 67
$Mn^{III}(acac)_3$	276 ± 25	1860 ± 40
Cp_2Mn^{II}	243 ± 8	1200 ± 20
$[Cp_2Mn^{III}]^+$	302 ± 10	2520 ± 20
$[Cp_2Ni^I]^-$	261 ± 9	527 ± 17
Cp_2Ni^{II}	278 ± 7	1360 ± 20
$[Cp_2Ni^{III}]^+$	343 ± 9	2760 ± 20
$Ni^0(CO)_4$	-	$(144)^f$
$[Ni^I(CO)_4]^+$	-	$(122)^e$
$[Cr^{II}(acac)_3]^-$	384 ± 25	904 ± 67
$Cr^{III}(acac)_3$	377 ± 25	1870 ± 40
Cp_2Cr^{II}	(319)	(1280)
$[Cp_2Cr^{III}]^+$	$(380)^e$	$(2680)^e$
Bz_2Cr^0	-	171 ± 5
$[Bz_2Cr^I]^+$	-	237 ± 7^e
$Cr^0(CO)_6$	-	(107)
$[Cr^I(CO)_6]^+$	-	$(85)^e$
Bz_2Mo^0	-	$(245)^f$
$[Bz_2Mo^I]^+$	-	$(321)^e$
$Mo^0(CO)_6$	-	$(149)^f$
$[Mo^I(CO)_6]^+$	-	$(127)^e$
Bz_2W^0	-	$(298)^f$
$[Bz_2W^I]^+$	-	$(422)^e$
$W^0(CO)_6$*	-	$(176)^f$

Complex	$\Delta\bar{H}^\circ_{hom}(M\text{-}L)^{b,c}/kJ\ mol^{-1}$	$\Delta\bar{H}^\circ_{het}(M\text{-}L)^{c,d}/kJ\ mol^{-1}$
$[W^I(CO)_6]^+$	-	$(168)^e$
$Ti(np)_4$	$(185)^f$	-
$[Ti(np)_4]^+$	$(148)^e$	-
$Zr(np)_4$	$(221)^f$	-
$[Zr(np)_4]^+$	$(185)^e$	-
$Hf(np)_4$	$(240)^f$	-
$[Hf(np)_4]^+$	$(203)^e$	-

[a]Derived in part from selected data in references 3, 4, 5, 26, 38. Additional BDE values for other neutrals and ions are available in the references. Except where noted, all values are derived from FTICR-MS ETE studies. Auxiliary thermochemical data taken from reference 40, except where noted. (Abbreviations: acac = acetylacetonate, Cp = cyclopentadienyl, Bz = benzene, np = neopentyl). [b]The average homolytic bond disruption enthalpy per ligand in kJ mol^{-1}. [c]Estimated errors are total bond disruption enthalpy errors divided by the number of ligands. [d]The average heterolytic bond disruption enthalpy per ligand in kJ mol^{-1}. [e]PES spectra used to estimate adiabatic ionization energy and BDE values. [f]BDE value quoted by Skinner and Connor (*Pure Appl. Chem.* **1985**, *57*, 79).

In the cases shown in Figure 6, the carbonyl and neopentyl complex cations have lower BDE values than the neutrals, and the arene and Cp neutral complexes have lower BDE values compared to the cations.

The general trends in Figure 6 have slopes of ~1 for all four types of complex, and oxidation of a neutral complex does not have a large effect on BDE values (the points cluster in the region around the 1:1 line shown). With reference to the thermochemical cycle used (Figure 5), this near equality arises from a similar ionization potentials for the metal atom and the complex (for $ML_n^{0/+}$ couples) or, in the case of neutral-anion comparisons in Table 1, similar electron affinities for the metal atom and the complex. In the cases showing large dependence of homolytic BDE on the central atom oxidation state, unusually low electron affinities (e.g., Mn(g)) or high ionization potentials (e.g., Fe(g)) are seen to be the origin of the observed difference. These observations serve to emphasize the danger of predicting changes in bond energetics solely on the basis of the change in the electronic structure accompanying oxidation or reduction of metal complexes. As always, one must be aware of the nature of the final states produced (M^z + nL).

The heterolytic BDE values in Table 1 for acac and Cp complexes follow relatively predictable trends. For L^- = acac$^-$ and Cp$^-$, the heterolytic BDE values for the M^{3+} complexes are ca. twice those of the corresponding M^{2+} complexes. A factor of two is also observed in comparing aquo complexes of M^{2+} and M^{3+} transition metals [29], suggesting that heterolytic BDE changes with oxidation state for $M^{2+/3+}$ complexes are rather insensitive to the metal, the nature of the ligand, and the overall charge of the complexes.

For the case of Cp_2Ni, the bond energy changes are tabulated for the Ni(I), Ni(II), and Ni(III) complexes. It is tempting to evaluate the dependence of the homolytic energies on oxidation state in terms of the electronic structure of the metallocene, but again the final states must be kept in mind.

The gas-phase $\Delta G_i°$ values also allow refinement of some M-H bond energies derived from proton affinity measurements for organometallic compounds [30]. These revised values are given in Table 2.

Table 2. Reevaluated Metal-Hydride Homolytic Bond Dissociation Energies

Compound	PA/kJ mol^{-1} [a]	$\Delta H_i°$/kJ mol^{-1} [b]	$\Delta H_{dis}°$(M-H)/kJ mol^{-1} [c]
Cp$_2$Fe	866 ± 5^d	657 ± 10	211 ± 11
Cp$_2$Ru	912 ± 12	692 ± 10	292 ± 16
Cp$_2$Ni	933 ± 12	605 ± 10	226 ± 16

[a]Proton affinity values from reference 40, except Cp$_2$Fe. [b]From FTICR-MS ETE studies (references 26, 38).[c]PA + $\Delta H_i°$(Cp$_2$M) - $\Delta H_i°$(H).[d]Reference 17.

4.2. SOLVATION ENERGETICS FOR ORGANOMETALLIC REDOX COUPLES

The thermodynamics of gas-phase and solution oxidation/reduction of metal complexes can be connected via a cycle of the type shown in Figure 5. Cycles of this type have long been used to estimate solvation energies of simple species such as metal ions (e.g., Na$^+$) and anions (e.g., Cl$^-$), but the necessary gas-phase data for metal complexes have not been available until recently. If one wishes to tabulate absolute solvation energies, the difficulty of defining an anchor point arises. However, accepting the recommended value [31] for the absolute potential of the standard hydrogen couple (4.44 V) leads immediately to estimates for $\Delta G_a°$(soln) and $\Delta G_i°$(soln) based on observed $E_{1/2}$ values. Thus, the differential solvation energies ($\Delta\Delta G_{solv}°$) for the various homogeneous redox couples can be derived from the cycle in Figure 5, and some representative values are given in Table 3.

Differential free energies correspond to processes such as those in eq 9 and 10, in which a charge is effectively transferred from a gas-phase complex

$$Cp_2M^+(g) + Cp_2M(soln) \rightleftharpoons Cp_2M(g) + Cp_2M^+(soln) \qquad (9)$$

$$M(acac)_3^-(g) + M(acac)_3(soln) \rightleftharpoons M(acac)_3(g) + M(acac)_3^-(soln) \qquad (10)$$

to the same entity in solution. This process is therefore comparable to the process described by the Born equation, which models electrostatic solvation free energies for a conducting sphere of radius r and charge q in a dielectric continuum of dielectric constant ε (eq 11) [32,33].

$$\Delta G_{el} = (-q^2/2r)(1 - 1/\varepsilon) \qquad (11)$$

In the case of the metallocenes, an effective r of 3.8 Å derived from crystal structures leads to a ΔG_{el} value of -180 kJ mol^{-1}, which is very close to the average values of $\Delta\Delta G_{solv}°$ given for the metallocenes in Table 3. Metallocene redox couple solvation energetics can apparently be modelled reasonably well by spheres, which is not so surprising given the compact structure of metallocenes and the absence of specific interactions with solvent (i.e., no H-bonding or strong acid-base interactions).

Table 3. Differential Solvation Free Energies for Transition Metal Complex Couples.

Couple	$E_{1/2}$(solvent)[a]	$\Delta\Delta G°_{solv}$/kJ mol^{-1} [b]
metallocenes		
$Cp_2V^{+/0}$	-0.55 (THF)[c]	252[g]
$Cp_2Fe^{+/0}$	0.31 (CH_3CN)[c]	166
$Cp_2Ni^{+/0}$	-0.09 (CH_3CN)[c]	162
$Cp_2Ru^{+/0}$	~0.8 (CH_2Cl_2)[d]	163
$Cp_2Ni^{0/-}$	-1.66 (CH_3CN)[c]	-209
$(Me_5C_5)_2Ru^{+/0}$	0.42(CH_2Cl_2)[d]	87
β-*diketonates*[e]		
$Cr(acac)_3{}^{0/-}$	-1.81 (CH_3CN)	-193
$Mn(acac)_3{}^{0/-}$	-0.09 (CH_3CN)	-196
$Fe(acac)_3{}^{0/-}$	-0.67 (CH_3CN)	-207
$Co(acac)_3{}^{0/-}$	-0.34 (CH_3CN)	-222
$Ru(acac)_3{}^{0/-}$	-0.70 (CH_3CN)	-222
	-0.73 (DMF)[f]	-219
$Ru(dpm)_3{}^{0/-}$	-1.038 (DMF)[f]	-173
$Ru(tfac)_3{}^{0/-}$	-0.016 (DMF)[f]	-185
$Ru(hfac)_3{}^{0/-}$	0.726 (CH_3CN)[f]	-150

[a]V vs. SCE reference electrode. [b]Derived by using ETE values for $\Delta G_i°$ and $\Delta G_a°$ (references 3, 4, 5, 26, 38). Estimated error in these values is ± 15 kJ mol^{-1}. [c]Geiger, W.; Holloway *J. Am. Chem. Soc.* **1978**, *101*, 2038. [d]Gassman, P.; Winter, C. *J. Am. Chem. Soc.* **1988**, *110*, 6130. [e]Original references to $E_{1/2}$ values tabulated in reference 4, except where noted. [f]Patterson, G.; Holm, R. H. *Inorg. Chem.* **1972**, *11*, 2285. [g]Oxidized form probably has inner-sphere solvent bound, leading to unusually high $\Delta\Delta G_{solv}°$.

In contrast to the metallocenes, $\Delta\Delta G_{solv}°$ values for $M(acac)_3{}^{0/-}$ couples are not readily predicted from the structure of the complex and eq 11. It is difficult to assign an effective radius for such chelate complexes from their structure, but the thermochemical radius for $Ru(acac)_3{}^{0/-}$ in water is estimated at 2.9 Å [4], substantially less than the extreme distance from the metal center to the ligand periphery. It is also noteworthy that fluorination of the acac at the methyl groups leads to lower values of $\Delta\Delta G_{solv}°$ in the ruthenium couples (Table 3) and higher thermochemical radii, and the intrinsic effect of the F substitution on the oxidation potential of the complex (Figure 3) is moderated somewhat by solvation (e.g., if the solvation energetics for the $Ru(hfac)_3{}^{0/-}$ couple were the same as for the $Ru(acac)_3{}^{0/-}$ couple, the former would have an electrode potential ~0.8 V higher than observed).

With the assessment of both bonding and solvation and solvation energetics for these prototypical complexes, a complete thermochemical picture emerges for the trends in redox potentials in metal complexes [28]. Electron attachment energetics are illustrated on a common energy scale in Figure 7 for a variety of metal complexes, including two discussed in detail here.

Gas-phase electron attachment energies for complexes with charges other than +1 or zero are estimates. Remarkably, the shifts in solution electrode potentials seem rather insignificant on the scale of Figure 7, but it is these relatively subtle variations that have occupied the attention of inorganic chemists. The obvious point is that the sum of the inner sphere bonding and solvation effects is a roughly constant quantity consisting of two large contributions, so a detailed analysis of $E°$ values and trends is difficult without the necessary gas-phase data of the type described here.

4.3. SUBSTITUENT EFFECTS IN REDOX THERMOCHEMISTRY

Alkylation of the cyclopentadienyl ligand reduces the $\Delta G_i°$ values of ferrocenes and ruthenocenes, indicating that the alkyl groups stabilize the ion relative to the neutral (Figures 1 and 2). Similar trends have been noted in the PES [12] and electrochemistry [1] results for ferrocenes. Furthermore, increasing the number of carbons in the substituents decreases the ionization energy in a predictable way.

The traditional attribution of "electron donating" character to alkyl groups appears to offer a suitable explanation for the alkyl effects on metallocene oxidation, but this "property" of alkyl substituents is clearly related to the nature of the process under study (here, the formation of a positive charge in the parent ferrocene unit). When electron attachment to form a negative metal complex *in the gas phase* is considered $(ML_n \rightarrow ML_n^-)$, the supposed "electron donating" character of alkyl groups can disappear. For example, Sharpe [34] determined the gas-phase electron affinities of Ru(acac)$_3$ (**1**) and Ru(dpm)$_3$ (**2**) and found that **2** has the higher EA value

by ~17 kJ mol^{-1}. In contrast, under the same solution conditions, **1** is the stronger oxidant of the two by 30 kJ mol^{-1}, a total inversion of 47 kJ mol^{-1} or ~0.5 V! Clearly, the t-Bu group would traditionally be considered the more "electron donating" relative to methyl group, but this characterization of alkyl effects obviously does not hold in general. The intrinsic order (observed in the gas phase) is reversed in solution, presumably because the t-Bu groups disrupt the solvation of the anion (relative to the neutral complex) to a greater degree than the methyl groups.

The data for nickelocene and its alkyl derivatives (Figure 4) clearly show the difference in the effects of alkyl groups on oxidation and reduction *of the same complex*. The effect on oxidation follows the usual order, with the more substituted nickelocenes exhibiting the lower ionization energies. In contrast, alkylation of nickelocene evidently has a rather small effect on the electron affinity, since the same mixture of alkylated nickelocenes comes to equilibrium in the trap in the negative ion ETE experiment.

Obviously, care must be taken in rationalizing substituent effects on redox thermochemistry of organometallic couples. If one must apply a simple model to these complex effects, we prefer to rationalize the alkyl group effects as being partially the result of polarization effects in the cation or anion. Of course, specific electronic effects must also be considered, but polarization effects may predominate in some cases. Larger alkyl substituents have the greater polarizabilities

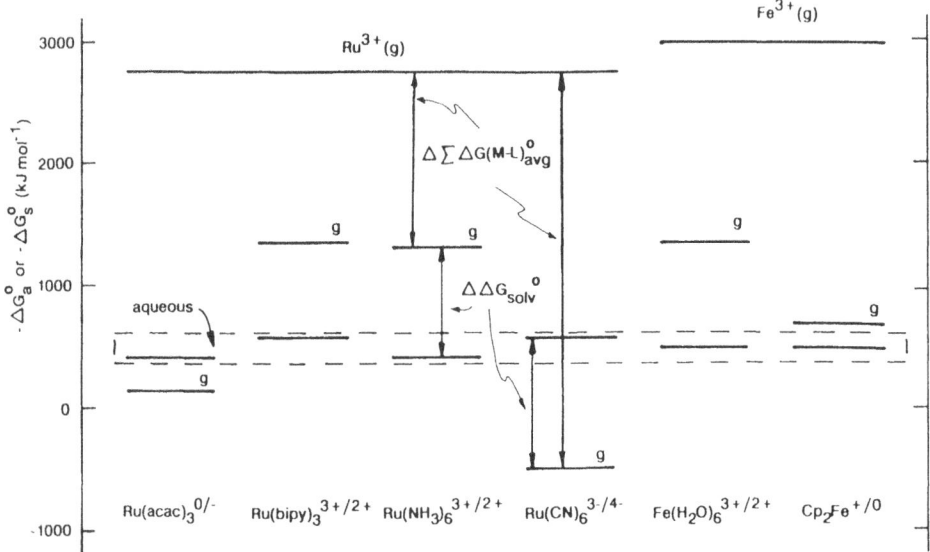

Figure 7. Diagram illustrating electron attachment free energies and single electrode free energies on one graph. The lines at the top are the $-\Delta G_a^\circ$ values for Ru^{3+} and Fe^{3+} ions. The horizontal lines indicate the values of the aqueous $-\Delta G_s^\circ$ values (in the dashed box) and the estimated or experimental gas-phase $-\Delta G_a^\circ$ values, indicated by "g", for the complexes shown at the bottom. $\Delta\Sigma\Delta G(M\text{-}L)^\circ$ is the difference in the sum of the average heterolytic bond free energies for the M(3+) and M(2+) complexes, while $\Delta\Delta G_{solv}^\circ$ is the differential solvation free energy for the oxidized and reduced complexes. These two quantities are indicated for the $[Ru(NH_3)_6]^{3+/2+}$ and $Ru(CN)_6^{3-/4-}$ couples. (Reprinted with permission from *Inorganic Chemistry*. Copyright 1990 American Chemical Society.)

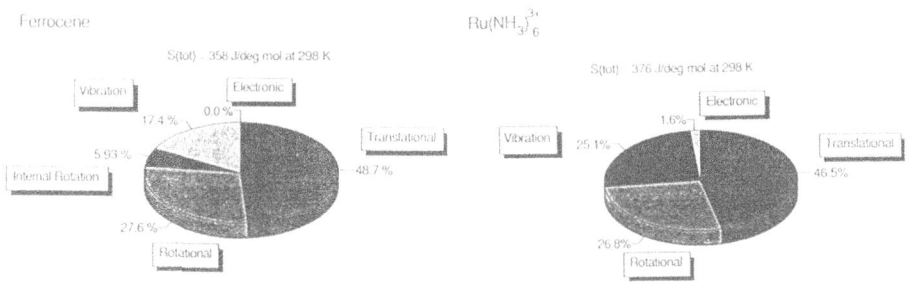

Figure 8. Illustration of intramolecular contributions to third law entropies for metal complexes ferrocene and hexaammineruthenium(III).

and thus contribute more to intramolecular ion-induced dipole stabilization in the charged complexes relative to the neutral molecules. Such an explanation is comparable to that used to rationalize the trends in the gas-phase electron affinities of RO· radicals (R = alkyl) [35].

From the above examples, it is clear that interpretation of solution electrochemical results in terms of the intrinsic effect of substituents can be misleading. Only the gas-phase results address the intrinsic effects; the solution energetics must include both the intrinsic effects and the differential solvation energies of the redox couple. These comments are quite analogous to interpretation of solution acidities and basicities, which can also show opposite trends from the gas-phase values.

4.4. ENTROPY EFFECTS IN ORGANOMETALLIC REDOX THERMOCHEMISTRY

In the study of ionization energies and electron affinities of organic molecules, it is usually found that temperature dependent ETE studies reveal low values of $\Delta S_{et}°$ for equilibria involving molecules of a similar type. Furthermore, the values of $\Delta S_i°$ and $\Delta S_a°$ derived for individual couples are usually small for most aromatic compounds (usually < 15 J mol^{-1} K^{-1}). This is easily understood since ionization of or electron attachment to an aromatic compound usually produces a radical cation or anion that is quite similar in structure and bonding to the parent neutral [13,14]. Equilibria involving metal complexes might initially be expected to follow the same pattern given the usually small changes in structure that occur upon 1-e$^-$ oxidation or reduction.

However, transition element coordination complexes and organometallics differ from aromatic organic compounds in two important respects: (1) they have an abundance of low frequency skeletal vibrations (15 in an ML_6 unit, typically below 600 cm^{-1}) [36] and (2) they are prone to undergo significant changes in multiplicity when the metal oxidation state changes (i.e., high spin-low spin conversions).

The consequences on these two factors on entropy changes can be dramatic [37]. A significant percentage of the third law entropy for metal complexes is associated with S_{vib} (Figure 8). In ferrocene, ~15% of S_{total} arises from vibrational degrees of freedom [38], while in Ru(NH$_3$)$_6^{3+}$ ion the fraction is 25% [37]! These large vibrational contributions arise from the dependence of S_{vib} on frequency [37], and the complexes have many low frequency modes. Although rotational and translational entropies are little affected by the oxidation state of a given complex, both vibrational and electronic entropies can change significantly. Thus, gas-phase redox processes in metal complexes can be accompanied by large and variable entropy changes that depend on the metal and ligand types. These intrinsic ΔS values are expected to be more or less maintained when the couple is studied in solution, except that entropy changes associated with solvation now become significant. By careful statistical mechanical analysis, intramolecular sources of seemingly anomalous entropy changes can be deduced [37]. For example, the positive entropy change that evidently accompanies the oxidation of ferrocene in the gas phase can be accommodated by the contributions of various degrees of freedom to $\Delta S_i°$ (changes in vibrational frequencies account for approximately one half of the total) [38].

4.5. ROLE OF THERMODYNAMICS IN THE KINETICS OF ELECTRON TRANSFER

The importance of accurate values for $\Delta G_i°$ is seen when gas-phase electron-transfer rates are determined as a function of driving force in cross reactions [8]. As predicted by theory [39], rate constants for cross reactions depend markedly on the exoergicity of the reactions. In some cases, a clear increase in rate constants has been observed for increasing driving force for gas-phase cross reactions. The theoretical analysis of this kinetic behavior requires accurate values of $\Delta G_{et}°$ for construction of the relevant potential surfaces [21].

5. Conclusions

The emphasis of this article has been on information that can be obtained about the bonding and solvation of transition metal compounds when accurate measures of electron attachment and ionization energies are available. The gas-phase $\Delta G_a°$ and $\Delta G_i°$ values provide direct thermochemical data on the intrinsic redox properties of these complex polyatomic species. The data can be used to determine the dependence of BDE values on central metal oxidation state for a variety of complexes. The thermochemical cycles also lead to solvation energetics that can be compared to predictions of models that could previously be tested in less direct ways only. As more related data on other gas-phase organometallic compounds become available, our understanding of organometallic redox thermochemistry will certainly increase even further.

6. Acknowledgments

The contributions of students and postdoctoral associates are gratefully acknowledged (C. Christ, P. Sharpe, M. Ryan, M. N. Khan, and K. Maxwell). M. Burk (DuPont) and A. Siedle (3M) provided some of the ruthenocene derivatives in Figure 2. Professor S. Nelsen provided the hydrazine alkyl reference compound in Figure 1. This work has benefitted from a close collaboration with Prof. J. Eyler (U. of Florida). Studies in the area of inorganic gas-phase ion chemistry are currently funded by the National Science Foundation (CHE 9008663).

References

(1) Kotz, J. C. In *Advances in Organic Electrochemistry*; Plenum: New York, 1986.
(2) Sharpe, P.; Richardson, D. E. *Coord. Chem. Rev.* **1989**, 93, 59.
(3) Sharpe, P.; Richardson, D. E. *Inorg. Chem.* **1990**, 29, 2779.
(4) Sharpe, P.; Richardson, D. E. *J. Am. Chem. Soc.* **1991**, in press.
(5) Richardson, D. E.; Christ, C. S.; Sharpe, P.; Ryan, M. F.; Eyler, J. R. In *Bond Energetics in Organometallic Compounds*, T. Marks, Ed.; ACS Symposium Series 428; American Chemical Society, 1990.
(6) For recent reviews of the technique see (a) Buchanan, M. V.; Comisarow, M. B. In *Fourier Transform Mass Spectrometry*; Buchanan, M. V., Ed.; ACS Symposium Series 359; **1987** pp 1-21. (b) Marshall, A. G. *Acc. Chem. Res.* **1985**, 18, 316.
(7) Harrison, A. *Chemical Ionization Mass Spectrometry*; CRC: Boca Raton, 1983.
(8) Richardson, D. E.; Christ, C. S.; Sharpe, P.; Eyler, J. R. *J. Am. Chem. Soc.* **1986**, 109, 3894.
(9) Eyler, J. H.; Baykut, G. *Trends in Analytical Chemistry* **1986**, 5, 44.
(10) Evans, S.; Green, M. L. H.; Jewitt, B.; King, G. H.; Orchard, A. F. *J. Chem. Soc. Faraday Trans. 2* **1974**, 70, 356. (b) Evans, S.; Hamnett, A.; Orchard, A. F.; Lloyd, D. R. *Discuss. Faraday Soc.*, **1972**, 54, 227.
(11) Lichtenberger, D. L.; Kellog, G. E. *Acc. Chem. Res.* **1987**, 20, 379.
(12) Green, J. C. *Struct. Bonding* **1981**, 43, 37.
(13) Lias, S. G.; Jackson, J. A.; Argentar, H.; Liebman, J. F. *J. Org. Chem.* **1985**, 50, 334.

(14) Lias, S. G.; Ausloos, P. *J. Am. Chem. Soc.* **1978**, *100*, 6027.

(15) Kebarle, P.; Chowdhury, S. *Chem. Rev,* **1987**, *87*, 513.

(16) Nelsen, S. F.; Rumack, D.; Meot-Ner (Mautner), M. *J. Am. Chem. Soc.* **1988**, *110*, 6303.

(17) Meot-Ner (Mautner), M. *J. Am. Chem. Soc.* **1989**, *111*, 2830.

(18) Bruice, J.; Eyler, J. R., unpublished results.

(19) Ryan, M.; Bruice, J., unpublished results.

(20) Eyler, J. R.; Richardson, D. E. *J. Am. Chem. Soc.* **1985**, *107*, 6130-31.

(21) Richardson, D. E. *J. Phys. Chem.* **1986**, *90*, 3697-3700.

(22) Sharpe, P.; Christ, C. S.; Eyler, J. R.; Richardson, D. E. *Int. J. Quantum Chem., Quantum Chem. Symp.* **1988**, *22*, 601-610.

(23) Hotop, H.; Lineberger, W. C. *J. Phys. Chem. Ref. Data* **1985**, *14*, 731.

(24) Lichtenberger, D. L., private communication.

(25) Gassman, P.; Winter, C. *J. Am. Chem. Soc.*, **1988**, *110*, 6130.

(26) Ryan, M.; Maxwell, K.; Khan, M.; Richardson, D. E., unpublished results.

(27) Richardson, D. E.; Christ, C.S.; Sharpe, P.; Eyler, J.R. *Organometallics* **1987**, 1819-1821.

(28) Richardson, D. E. *Inorg. Chem.* **1990**, *29*, 3213.

(29) (a) McClure, D. S. In *Some Aspects of Crystal Field Theory*, Dunn, T. M.; McClure, T. M.; Pearson, R. G., Eds.; Harper and Row: New York **1965** (b) Figgis, B. N. *Introduction to Ligand Fields*; Interscience: New York, 1966; Ch. 5. (c) Burdett, J. K. *Molecular Shapes*; Wiley-Interscience: New York, 1980; Ch. 10.

(30) Stevens, A. E.; Beauchamp, J. L. *J. Am. Chem. Soc.* **1981**, *103*, 190.

(31) Trasatti, S. *Pure Appl. Chem.* **1986**, *58*, 955.

(32) Born, M. Z. *Phys.* **1920**, 1, 45.

(33) For a concise review of Born models applied to solvation, see Gomez-Jeria, J. S.; Morales-Lagos, D. *J. Phys. Chem.* **1990**, 94, 3790.

(34) Sharpe, P.; Richardson, D. E., unpublished results.

(35) Janousek, B. K.; Brauman, J. I. In *Gas Phase Ion Chemistry*, Volume 2; Academic Press: New York, 1979, pp 53-83.

(36) Nakamoto, K. *Infrared and Raman Spectra of Inorganic and Coordination Compounds*, 4th Edition; Wiley-Interscience, 1986; p 260.

(37) Richardson, D. E.; Sharpe, P. *Inorg. Chem.* **1991**, *30*, 1412.

(38) Ryan, M.; Eyler, J. R.; Richardson, D. E., in preparation.

(39) Marcus, R. A.; Sutin, N. *Biochem. Biophys. Acta* **1985**, *811*, 265.

(40) Lias, S. G.; Bartmess, J. E.; Liebman, J. F. Holmes, J. L.; Levin, R. D.; Mallard, W. G., Eds. *"Gas-Phase Ion and Neutral Thermochemistry"*; American Institute of Physics: New York, **1988**.

GAS-PHASE ACIDITIES OF TRANSITION-METAL CARBONYL AND TRIFLUOROPHOSPHINE HYDRIDES

AMY E. STEVENS MILLER, THOMAS M. MILLER
Phillips Laboratory, Ionospheric Effects Division (PL/GPID) *
Hanscom Air Force Base, Massachusetts 01731-5000
United States

ABSTRACT. The gas-phase acidities of thirteen transition-metal carbonyl and trifluorophosphine complexes have been determined. The complexes are all strong gas-phase acids, with most of the trifluorophosphine complexes comparable in acidity to fluorosulfonic and triflic acids. Acidities increase across a row and decrease down a column in the periodic table. The periodic trends and substituent effects are discussed in view of the dependence of the acidity on the metal–hydrogen bond enthalpies and electron affinities of the resulting metal radicals. Solvation energies are also discussed.

1. Introduction

1.1 GAS-PHASE ACIDITIES OF INORGANIC AND ORGANIC ACIDS

The acidities of main-group inorganic and organic molecules in the gas-phase have been known for some time, although the list of compounds for which this important property has been determined continues to grow. The acidities provide information on periodic trends and substituent effects in the acidities, and can be used to determine electron affinities, bond energies, and energetics of ion solvation [1-8].

An introduction to the energetics of gas-phase acidities is provided by the thermo-dynamic cycle of Figure 1, which links the enthalpy change for the acidity, ΔHacid[AH], to the homolytic bond enthalpy, D[A–H], and the electron affinity of A, EA[A].

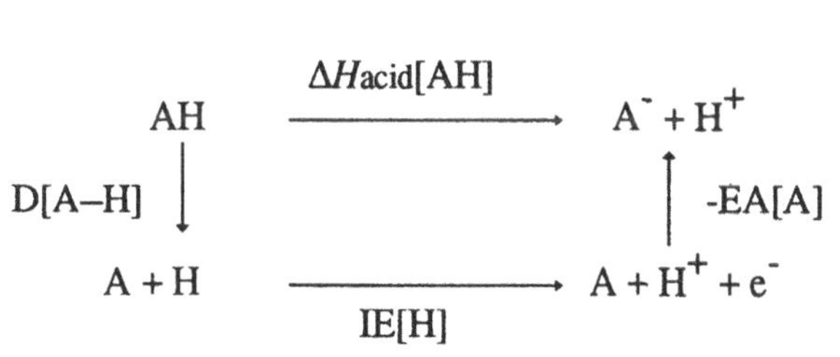

$$
\begin{array}{ccc}
\text{AH} & \xrightarrow{\ \Delta H\text{acid[AH]}\ } & \text{A}^- + \text{H}^+ \\[2pt]
D\text{[A–H]} \downarrow & & \uparrow\ \text{-EA[A]} \\[2pt]
\text{A} + \text{H} & \xrightarrow[\text{IE[H]}]{} & \text{A} + \text{H}^+ + \text{e}^-
\end{array}
$$

Figure 1. Thermodynamic cycle linking ΔHacid[AH] to D[A–H], EA[A], and IE[H].

J. A. Martinho Simões (ed.), Energetics of Organometallic Species, 253–267.
© 1992 *Kluwer Academic Publishers.*

254

The thermodynamic cycle of Figure 1 can be used to derive equation 1, in which 313.6 kcal/mol has been substituted for the ionization energy of H, IE[H].

$$\Delta H\text{acid}[A-H] = D[A-H] - EA[A] + 313.6 \text{ kcal/mol} \qquad (1)$$

A careful examination Figure 1 reveals that the use of IE[H] and EA[A] are not strictly correct: those quantities imply an energy, rather than an enthalpy, for the steps in question. An alternative viewpoint is that these quantities are correct for a 0 K electron, rather than a 298 K electron whose thermal enthalpy is 1.48 kcal/mol. However, in equation 1 the electron's thermal enthalpy would appear as an additive term from the ionization of H, and as a subtractive term from the electron attachment of A. More generally stated, as long as free electrons are not created or destroyed in a given reaction, the electron thermal enthalpy can be ignored in the overall energetics, as given here by equation 1.

Once a number of "benchmark" acidities are established by the combination of bond energies and electron affinities, it is straight-forward to determine additional gas-phase acidities either by equilibrium or bracketing methods. A selection of acidities for main-group inorganic and small organic acids is given in Figure 2.

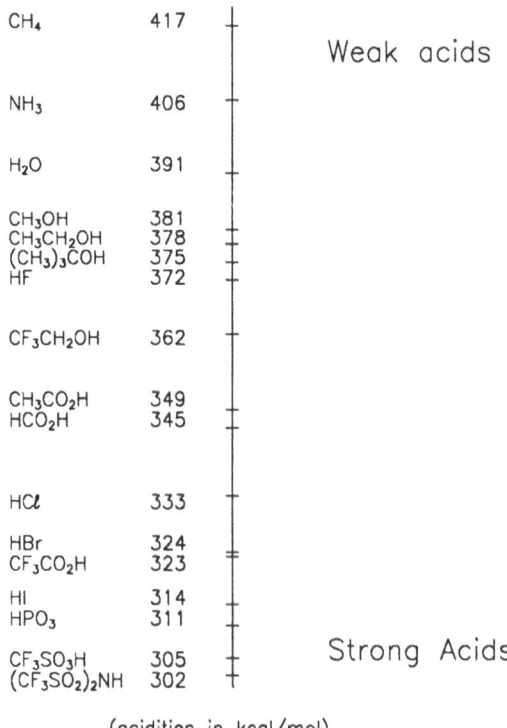

(acidities in kcal/mol)

Figure 2. Gas-phase acidities of representative inorganic and organic acids [4-8].

Several examples [6,7] serve to illustrate how the gas-phase acidities can be determined by equation 1, and the quantitative scale of Figure 2 derived. The first example is the gas-phase acidity of methane, CH_4, a molecule we would all recognize as a poor acid. Accurate determination of the bond enthalpy gives a value for $D[CH_3-H]$ of 104.8 ($\pm$0.1) kcal/mol, and laser photoelectron spectroscopy provides the electron affinity, $EA[CH_3]$, to be 1.8 ($\pm$0.7) kcal/mol. These quantities are substituted into equation 1 to yield a 416.6 ($\pm$ 0.8) kcal/mol gas-phase acidity for CH_4, making it one of the weakest gas-phase acids known. The gas-phase acidity, ΔHacid, is a large number, since it takes a large amount of energy to separate CH_4 into CH_3^- and a free proton, H^+. A weak acid is the result of a strong homolytic bond to hydrogen and/or a small electron affinity of the resulting A radical.

A second example comes from the acidity of HI; $D[H-I]$ of 71.3 kcal/mol and $EA[I]$ of 70.5 kcal/mol provide ΔHacid[HI] = 314.4 ($\pm$ 0.1) kcal/mol. HI is a strong gas-phase acid, and the gas-phase acidity a smaller quantity, reflecting the relatively less energetically costly process of separating HI into I^- and H^+. A strong acid has a weak homolytic bond, and/or a large electron affinity of the resulting A radical.

Our research [9-11] has been to determine the acidities of transition-metal hydride complexes. A major motivation in the work was the possibility of determining metal–hydrogen bond enthalpies, as indicated by equation 1. Recently we found that the trifluorophosphine complexes are very strong gas-phase acids, and the work has centered around them as possible electron-scavenger reagents.

2. Experimental

2.1 SYNTHESIS AND CHARACTER OF THE TRANSITION-METAL HYDRIDES

All of the metal carbonyl and trifluorophosphine complexes we studied were prepared by literature procedures [12-20]. The percarbonyl complexes tend to be air, heat, and light sensitive, but the trifluorophosphine complexes, with even partial substitution, are generally quite stable [21]. Although strong acids, the complexes do not have the same type of redox chemistry which makes most main-group acids, such as fluorosulfonic or trifluoroacetic, quite caustic. If wet, the PF_3 complexes do seem to etch glass slowly--presumably by slow hydrolysis of the PF_3 ligands to make HF.

2.2 DETERMINATION OF GAS-PHASE ACIDITIES

The acidities of the metal-hydrides were determined by bracketing experiments. In these reactions, the conjugate base of an acid with known acidity is produced (often from an anhydride, ester, or other reagent less caustic than the acid) and allowed to react with the metal hydride. Observation of fast proton transfer shows the metal hydride to be the stronger acid than the reference neutral. Generally the metal anions were not reacted with the neutral acid, for a number of reasons. First, the reference acids are often extremely toxic and caustic, and second, the metal anions are difficult to generate cleanly. We have been able to examine the reactions of $Mn(CO)_5^-$ with reagent acids, determining the acidity of $HMn(CO)_5$ by both forward and reverse proton-transfer rates, the ratio of which determines the equilibrium constants for proton transfer [11].

We also were able to observe the proton abstraction from $(CF_3SO_2)_2NH$ by reaction with $Co(PF_3)_4^-$ [11]. These reactions were completely consistent with energetics based on proton abstraction from the hydrides, and give confidence that the bracketing experiments for the other species are accurate.

2.3 INSTRUMENTATION FOR DETERMINING ION-MOLECULE REACTIONS

The ion-molecule reactions were done at Caltech on an ion-cyclotron resonance spectrometer (ICR), or at the Phillips Laboratory using a selected-ion flow tube (SIFT). Both the techniques and these particular instruments are well-described in the literature [22, 23], and an extensive description is not warranted. It is useful to describe the advantages and disadvantages of each method for these types of experiments.

The ICR spectrometer operates at very low pressures--10^{-7} Torr of metal hydride would be high for these experiments. The ICR technique is therefore most useful for examining metal complexes with low vapor pressures. The SIFT operates with a helium buffer with a pressure about 0.6 Torr, and reagent vapors only about 10^{-4} Torr, but at a very high (600 ℓ/s) flow rate. We found that $HRe(CO)_5$ and $HMn(PF_3)_5$ were at the limit of the reagent flow requirements, even though their vapor pressures are a few Torr. Clearly many of the complexes we would like to examine are even less volatile.

An advantage of the SIFT technique is that the primary ion (usually the conjugate base of a reference acid) is injected into the helium flow, and so the ion-molecule reactions occur in a situation in which only the primary ion and metal hydride reagent are present, and kinetic determinations are quite simple to perform based solely on decrease of the primary ion signal. This is not possible in the ICR--the reagent for producing the primary ion will always be present, and often the metal anion produced by proton transfer will then undergo reaction with it. These secondary reactions are often multiple ligand displacements, halogen atom transfer, or nucleophilic displacements by the metal anion [24]. The double-resonance technique [22], a clear advantage of ICR for conclusively identifying the primary ion precursor of any particular secondary ion, was used to identify reaction pathways, but in essentially no instances were we able to determine accurate kinetics for the reactions of the transition-metal hydrides using the ICR spectrometer.

A final advantage of using the ICR spectrometer is that it can be rapidly changed between trapping and drift modes, and positive and negative ion modes. This makes it very easy to determine mass spectra of any reagent, and was particularly useful in identifying the purity of samples of the thermally unstable carbonyls $H_2Fe(CO)_4$ and $HCo(CO)_4$ [9].

3. Results and Discussion

3.1 GAS-PHASE ACIDITIES OF TRANSITION-METAL HYDRIDES

The gas-phase acidities of thirteen carbonyl and trifluorophosphine complexes, determined by bracketing reactions, are given in Table 1. Also in Table 1 are the thermodynamic data for the reference acids, and the literature citations for both reference acids and the metal-hydride acidities.

Table 1. Acidity scale used for the determination of the gas-phase acidities of the transition-metal compounds, with the relative positions of the transition-metal hydrides also given.

acid	ΔHacid	ΔGacid	reference
HF	371.5 ± 0.2	365.7 ± 0.5	[6]
CHCl$_3$	357.1 ± 6.3	349.3 ± 6.0	[6]
HCO$_2$H	345.2 ± 2.3	338.2 ± 2.0	[6]
HNO$_2$	338.2 ± 4.3	330.5 ± 4.6	[6]
HCl	333.4 ± 0.2	328.0 ± 0.5	[6]
HRe(CO)$_5$	bracketed position		[10]
CHF$_2$CO$_2$H	330.0 ± 2.3	323.5 ± 2.0	[6]
CF$_3$COCH$_2$COCH$_3$	328.4 ± 4.1	322.0 ± 2.0	[6]
CHCl$_2$CO$_2$H	327.3 ± 2.6	320.8 ± 2.0	[6]
HBr	323.5 ± 0.1	318.2 ± 0.4	[6]
CF$_3$CO$_2$H	322.9 ± 4.1	316.3 ± 2.0	[8]
CCl$_3$CO$_2$H	319.9 ± 2.9	312.8 ± 2.0	[8]
HMn(CO)$_5$	321 ± 3	314 ± 2	[11]
CF$_3$C(O)SH	320	313.4 ± 2	[8]
CF$_3$C(O)CH$_2$C(O)CF$_3$	318	311.5 ± 2	[8]
H$_2$Fe(CO)$_4$	317 (estimated position, see text)		
	bracketed between HBr and HI		[9]
HMn(CO)$_4$(PF$_3$)	bracketed position		[11]
HRe(PF$_3$)$_5$	≥ 315 (by electron attachment)		[11]
HI	314.4 ± 0.1	309.3 ± 0.4	[6]
HCo(CO)$_4$	312 (estimated position, see text)		
	≤314 (bracketed position)		[9]
HPO$_3$	311 ± 4	304 ± 4	[4]
FSO$_3$H	307	300 ± 2	[5]
HMn(CO)$_3$(PF$_3$)$_2$,	bracketed positions		[11]
HMn(CO)$_2$(PF$_3$)$_3$,			
HMn(CO)(PF$_3$)$_4$,			
HMn(PF$_3$)$_5$,			
HIr(PF$_3$)$_4$			
CF$_3$SO$_3$H	305	298.3 ± 2	[5,8]
HCo(PF$_3$)$_4$,	bracketed positions		[11]
HRh(PF$_3$)$_4$			
(CF$_3$SO$_2$)$_2$NH	302	294.5 ± 2	[8]

Several items in the table bear some additional comment. At the time we did the work with $H_2Fe(CO)_4$ and $HCo(CO)_4$, HI was the only acid below HBr and CF_3CO_2H for which a quantitative acidity had been determined. The bracketing of these two carbonyl hydride acidities was therefore not very accurate. It becomes useful for the discussions in this paper to make estimates of the acidities of $H_2Fe(CO)_4$ and $HCo(CO)_4$. We begin by examining the pKa in aqueous and acetonitrile solutions. The pKa of $HMn(CO)_5$ is 7.1 in water [25] and 14.1 in acetonitrile [26]; the pKa of $H_2Fe(CO)_4$ is ∿4.2 in water [28,29] and 11.4 in acetonitrile [30]; and the pKa of $HCo(CO)_4$ is <0.4 [28] in water and 8.3 in acetonitrile [30]. The differences in the pKa between any of these two complexes are relatively insensitive to the solvent. We can take an average <u>difference</u> between the $HMn(CO)_5$ and $H_2Fe(CO)_4$ pKa of ∿2.8, and an average <u>difference</u> between the $HMn(CO)_5$ and $HCo(CO)_4$ pKa of ∿6.3. The pKa are related to the free energies by equation 2,

$$\Delta G_{acid} = 2.303\ RT\ pKa$$
$$= 1.364\ pKa\ (kcal/mol), \tag{2}$$

so that these pKa differences correspond to approximate differences in ΔG_{acid} of 4 kcal/mol between $HMn(CO)_5$ and $H_2Fe(CO)_4$, and 9 kcal/mol between $HMn(CO)_5$ and $HCo(CO)_4$. We will assume that the entropy and solvation differences make negligible contributions to these differences. Therefore, the ΔG differences are reflected in the differences in the gas-phase ΔH values. Using $\Delta H_{acid}[HMn(CO)_5]$ of 321 kcal/mol (Table 1) therefore provides a estimates for ΔH_{acid} of $H_2Fe(CO)_4$ (317 kcal/mol) and $HCo(CO)_4$ (312 kcal/mol). These estimates are consistent with the values based on bracketing reactions; both estimates and bracketed positions are presented in Table 1.

Although lengthy discussion is not within the scope of this article, we have also examined electron-attachment reactions of thermal (298 K) electrons to each of the metal hydrides in Table 1 [9,11]. These reactions are useful in that when the electron affinity of the metal radical and the thermal electron enthalpy (1.48 kcal/mol) together exceed the metal–hydrogen bond enthalpy, electron attachment to the hydride can produce the 18-electron metal anion and a free hydrogen atom. This energetic requirement can be seen from equation 1 as equivalent to the requirement that $\Delta H_{acid} \leq$ 315 kcal/mol. All acidities for the hydrides given in Table 1 were confirmed with respect to this limit by electron attachment data. The acidity for one other hydride, $HRe(PF_3)_5$, was not determined by proton-transfer reactions, but the electron attachment showed only $HRe(PF_3)_4{}^-$. Since no 18-electron $Re(PF_3)_5{}^-$ is produced, the acidity limit $\Delta H_{acid} \geq$ 315 kcal/mol is determined and given in Table 1.

For places where more than one metal complex is placed within a single bracketed position, as is particularly noticable between FSO_3H and CF_3SO_3H, no relative order could be determined for the metal hydrides. No energetic ordering is therefore implied by the order in which the several metal complexes have been listed in the bracketed position.

An expansion of the gas-phase acidity scale over the strong acid region is given in Figure 3.

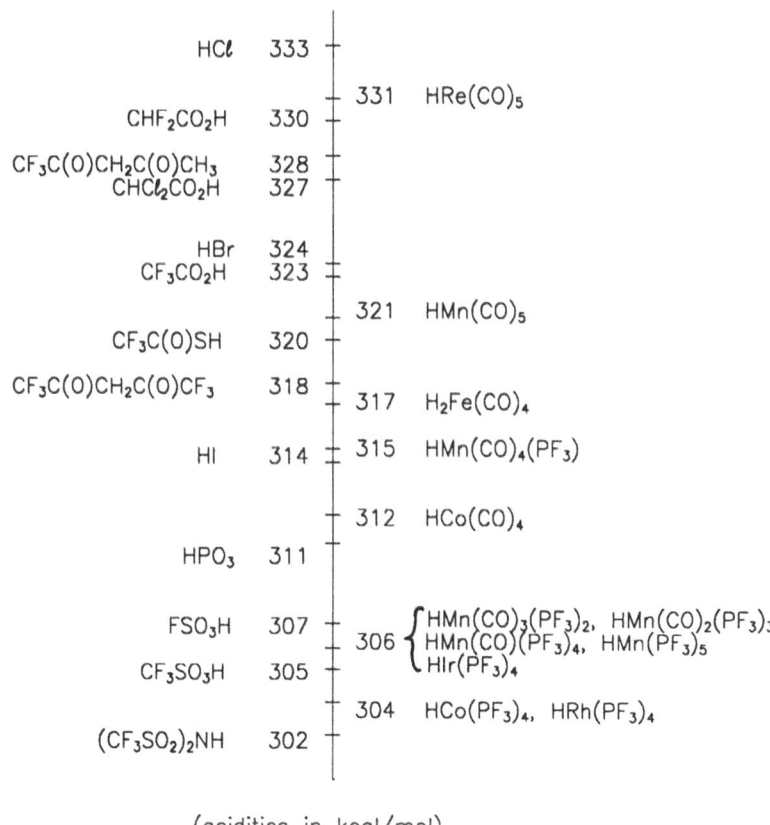

(acidities in kcal/mol)

Figure 3. Expansion of the gas-phase acidity scale over the strong acid region [4-11].

3.1.1 Periodic Trends and Substituent Effects. Table 1 and Figure 3 illustrate several periodic trends in the gas-phase acidities of the transition-metal hydrides. First, there are examples of an increase in acidity moving to the right in the periodic table. These include the increase of about 4 kcal/mol from HMn(CO)₅ to H₂Fe(CO)₄, and then about 5 kcal/mol to HCo(CO)₄; there is also an increase of about 2-4 kcal/mol from HMn(PF₃)₅ to HCo(PF₃)₄. Only one example of this trend is available for third-row metals, indicating an increase of at least 8 kcal/mol from HRe(PF₃)₅ to HIr(PF₃)₄.

Third-row metal substitution decreases the acidity of the complex, with examples given by HRe(CO)₅ [10 kcal/mol less acidic than HMn(CO)₅], HRe(PF₃)₅ [at least 8 kcal/mol less acidic than HMn(PF₃)₅], and HIr(PF₃)₄ [2-4 kcal/mol less acidic than

HCo(PF$_3$)$_4$]. Within current ability to bracket the acidities, the only example of second row substitution shows HRh(PF$_3$)$_4$ quite close, within 3 kcal/mol, in acidity to HCo(PF$_3$)$_4$.

PF$_3$ substitution for CO increases the acidity, as shown by the HMn(CO)m(PF$_3$)n series, and HCo(CO)$_4$ compared to HCo(PF$_3$)$_4$. The HMn(CO)m(PF$_3$)n series provides an interesting finding--the acidity increases by ~6 kcal/mol for the first PF$_3$ substitution, ~8 kcal/mol for the second substitution, but then the acidity shows no increase (within the present 2 kcal/mol bracketing ability) on successive substitutions! Whether this is true for the cobalt and rhenium series remains to be seen. Certainly on complete PF$_3$ substitution, the total increase of <16 kcal/mol in the rhenium series and ~8 kcal/mol in the cobalt series indicate that the ~7 kcal/mol increase per PF$_3$ cannot be sustained in these cases either.

In considering these periodic and substituent effects, it's useful to consider the acidity in terms of the metal–hydrogen bond enthalpy and electron affinity, as given by equation 1. These effects are considered in the next section.

3.2 THERMODYNAMICS

3.2.1 *Bond Enthalpies and Electron Affinities.* This research was initiated to determine metal-hydrogen bond energies according to the cycle of Figure 1. Our belief was that the electron affinities of the metal radicals would be forthcoming, principally by laser photoelectron detachment. Such has not in fact been the case, due in part to difficulties in making thermal or cold metal anion sources, the large electron affinities requiring photodetachment with ultra-violet photons, the relatively weak photodetachment signals (from competition of photodetachment and photodissociation, and small detachment cross-sections) and undoubtedly the lack of interest in determining the electron affinities for lack of the acidity data.

Some data do exist which can serve to illustrate the energetics of equation 1. First, although there is some variation in the determinations, a value for $D[(CO)_5 Mn–H]$ of 59 kcal/mol is the value selected by Martinho Simões and Beauchamp [31]. An EA[Mn(CO)$_5$] of 2.2 ± 0.3 eV is derived by substitution of this value and the acidity into equation 1.

Assuming ΔHacid[(CO)$_4$CoH] to be 312 kcal/mol, and using the value for $D[(CO)_4 Co–H] \simeq 54$ kcal/mol [31], provides an estimate for EA[Co(CO)$_4$] $\simeq 2.4$ eV. Considering the acidities in terms of the bond enthalpy and electron affinity contributions suggests that the increase in acidity moving to the right in the periodic table, here illustrated by HMn(CO)$_5$ to HCo(CO)$_4$, is due both to a weakening of the homolytic bond and an increase in the electron affinity.

There are no data for the gas-phase bond energies for any of the other metal hydrides. Tilset and Parker [27] used acidities, oxidation potentials of the anions, and the H$_2$/H$^+$ electrochemical couple to determine free energies for metal–hydrogen bond breaking for a number of hydrides, including HRe(CO)$_5$ and H$_2$Fe(CO)$_4$. These data were determined in acetonitrile solution, and give free energies somewhat larger than most estimates for the bond enthalpies, but probably give reasonable estimates of the differences in bond enthalpies which could be used to derive additional electron affinities by equation 1.

We can provide some insight into the changes which occur from one species to another. First, Meckstroth and Ridge [32] examined the ion chemistry of $(CO)_5 Mn–Re(CO)_5$. Of interest to the present work is their studies on the reaction of the species with thermal electrons, equation 3:

$$(CO)_5 Mn–Re(CO)_5 + e^- \text{ (thermal, 298 K)} \longrightarrow$$
$$Mn(CO)_5^- + Re(CO)_5 \quad (11\%)$$
$$Re(CO)_5^- + Mn(CO)_5 \quad (5\%)$$
$$(CO)_9 ReMn^- + CO \quad (84\%). \quad (3)$$

The unusual feature is that the attachment produces both $Mn(CO)_5^-$ and $Re(CO)_5^-$ in quite similar percentages. Fragmentation where two pathways differ only by which of the two fragments captured the electron is possible only if the electron affinities of the two fragments are "very nearly" equal--certainly within 0.1 eV and undoubtedly closer. This means $EA[Re(CO)_5] \simeq EA[Mn(CO)_5]$. From equation 1 we can then see that the acidity difference of 10 kcal/mol between $HMn(CO)_5$ and $HRe(CO)_5$ is due entirely to the difference in the homolytic bond energies, as given by equation 4:

$$D[(CO)_5 Re–H] = D[(CO)_5 Mn–H] + 10 \text{ kcal/mol}$$
$$= 69 \text{ kcal/mol.} \quad (4)$$

No error bars are given for the bond enthalpy; the error in the difference between the two acidities is ~ 1 kcal/mol, and error in the difference in electron affinities ~ 2 kcal/mol at most, but the uncertainty in the absolute bond energy for the Mn–H easily ~ 5 kcal/mol. This difference of 10 kcal/mol between the $HRe(CO)_5$ and $HMn(CO)_5$ bond energies can be compared to the work of Tilset and Parker [27], who find a free energy difference of 6.7 kcal/mol, and work of Ziegler et al. [33], who calculated a bond energy difference of 13.6 kcal/mol. The very limited data here indicate the decrease in acidity with third-row metal substitution is due to an increase in the metal–hydrogen homolytic bond strengths.

Another game can be played by assuming that for the same metal and oxidation state that the metal–hydrogen homolytic bond energies are the same. Work by Tilset and Parker [27] on a number of hydrides in which per-methyl cyclopentadienyl is substituted for cyclopentadienyl, or in which trimethylphosphine, triphenylphosphine, or trimethylphosphite are substituted for CO show this assumption is valid within 2 kcal/mol, and more often 1 kcal/mol. Studies on gas-phase protonated species by us [34] also show this to be a reasonable assumption. In this work, then, we can assume the metal–hydrogen bond energies do not change with PF_3 for CO substitution. That is to say, the increase in acidity on PF_3 substitution for CO in the $HMn(CO)_m(PF_3)_n$ series, and between $HCo(CO)_4$ and $HCo(PF_3)_4$ is presumed entirely due to an increase in electron affinity of the metal radical. Using the values for $EA[Mn(CO)_5]$ and $EA[Co(CO)_4]$ derived above yields estimates for the electron affinities summarized in Table 2.

Table 2. Estimated electron affinities (in eV) for the $Mn(CO)_m(PF_3)_n$, $Co(CO)_4$ and $Co(PF_3)_4$ radicals.

species	EA	species	EA
$Mn(CO)_5$	2.2	$Mn(CO)(PF_3)_4$	2.85
$Mn(CO)_4(PF_3)$	2.46	$Mn(PF_3)_5$	2.85
$Mn(CO)_3(PF_3)_2$	2.85	$Co(CO)_4$	2.4
$Mn(CO)_2(PF_3)_3$	2.85	$Co(PF_3)_4$	2.66

Although these are sizable electron affinities, they in no way rival the electron affinities of the radicals from main-group superacids such as FSO_3H and CF_3SO_3H, which are more on the order of 5 eV [35]. The metal electron affinities increase somewhat with initial PF_3 substitution, but soon reach a maximum value. Data for the mixed $HCo(CO)_m(PF_3)_n$ and $HRe(CO)_m(PF_3)_n$ complexes should provide some information on the generality of this conclusion.

Two questions need to be answered: First, why is the effect of a PF_3 substitution for CO so small? At first glance it seems reasonable to compare it to the effect on acidities of CF_3 substitution for CH_3. In CH_3CH_2OH and CF_3CH_2OH, for example, CF_3 substitution for CH_3 causes a 16 kcal/mol increase in the acidity, as seen from Figure 2. A similarly large increase on substitution is evident from comparison of CH_3CO_2H to CF_3CO_2H--with a 26 kcal/mol increase in the acidity of the fluorinated acid! The CF_3 group acts as a highly efficient σ-withdrawing group, delocalizing charge onto the highly electronegative fluorines. PF_3 substitution for CO must act in a very different fashion. Most of the anionic charge in the metal complexes can be thought of as "in" the d-orbitals which π-backbond into the π* CO or π-symmetry PF_3 acceptor orbital. The π-acceptor ability of PF_3 puts electron density into the π-system--anti-bonding with respect to the fluorine lone-pair electrons, thus limiting its π-acceptor ability. This effect is supported by a crystal structure of the related cobalt complex $[K(222)]^+[Co[P(OCH_3)_3]_4]^-$, which shows the Co–P bond is shortened, but the P–O bond increased from the distances in neutral cobalt complexes [36]. Of course, there may in fact be no special electronic effect at all--a simple increase in polarizability of PF_3 as compared to CO may be responsible for the increase in electron affinity, in the way CH_3 for H substitution increases the acidities of H_2O and the alcohols.

Second, why does the electron affinity (and acidity) stop increasing with successive substitution? At present we can speculate that it has to do with the high symmetry of the metal d-orbitals which carry the "extra" charge. As long as a single substitution effectively delocalizes the d-orbital onto the ligand, subsequent substitution will only act to compete for the charge. Such a diminishing effect can be seen in the successive CH_3 substitutions in the H_2O, CH_3OH, CH_3CH_2OH and $(CH_3)_3COH$ series shown in Figure 2.

3.2.2 *Solvation Effects.* In all of the above discussions, we have ignored the most overwhelming feature of Figure 3, namely, the transition-metal hydrides are much stronger acids in the gas phase than they are in solution. $HRe(CO)_5$ and $HMn(CO)_5$ provide among the best examples of this, since both solution acidities [25, 26, 30] and gas-phase acidities [10, 11] are accurately determined. The aqueous acidity of $HMn(CO)_5$ is pKa = 7.1, making it comparable in acidity to a weak acid such as acetic. In the gas-phase however, it is comparable in acidity to the strong acids HBr and CF_3CO_2H. Another example is from $HCo(CO)_4$, which in aqueous solution is comparable to HNO_2 and $HC\ell$ in acidity [28], but a stronger acid than HI in the gas phase. Although the transition-metal hydrides retain their relative acidity ordering with respect to one another, in moving to the gas phase the metal hydrides on the right in Figure 3 are all shifted downward by some 20 to 30 kcal/mol relative to the reference organic and inorganic acids on the left of Figure 3. To help understand this change, we can use the thermodynamic cycle of Figure 4.

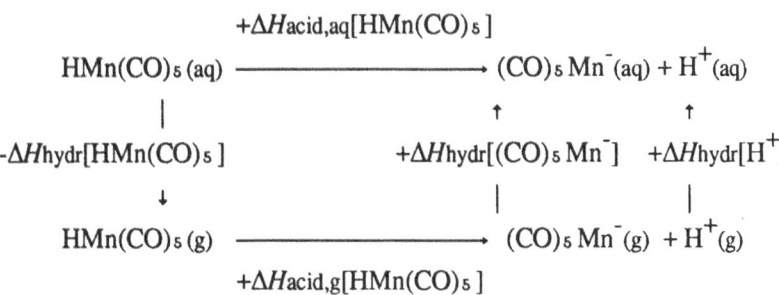

Figure 4. Relationship between the gas-phase and aqueous acidities of $HMn(CO)_5$, as linked by the hydration enthalpies of $HMn(CO)_5$, $Mn(CO)_5^-$, and H^+.

The $HMn(CO)_5$ example shown in the figure provides a unique opportunity to quantify the effects of the change in gas-phase versus solution acidities, since much of its thermodynamics are known; thus the hydration enthalpy of $Mn(CO)_5^-$ is the only unknown in the cycle of Figure 4, and can be found by equation 5.

$$\Delta H_{hydr}[Mn(CO)_5^-] = \Delta H_{acid,aq}[HMn(CO)_5] - \Delta H_{acid,g}[HMn(CO)_5] + \Delta H_{hydr}[HMn(CO)_5] - \Delta H_{hydr}[H^+] \qquad (5)$$

A ΔG_{acid} of +9.7 kcal/mol can be found from equation 2 and the known pKa of 7.1 [25]. $\Delta S_{acid,aq}$ is estimated as +20 kcal/mol by comparison to other acids, and in combination with $\Delta G_{acid, aq}$ provides $\Delta H_{acid,aq}[HMn(CO)_5] = +16$ kcal/mol, making the dissociation of aqueous $HMn(CO)_5$ a moderately <u>endothermic</u> reaction. The value for $\Delta H_{acid,g}[HMn(CO)_5]$ of 321 kcal/mol is taken from Table 1. $\Delta H_{hydr}[HMn(CO)_5]$ is not known directly. However, the enthalpy of vaporization of $HMn(CO)_5$ has been determined to be 9.324 kcal/mol, and it is reported that the solubility of $HMn(CO)_5$ is only 1.25×10^{-4} mol/ℓ [25]. Thus the hydration of gaseous $HMn(CO)_5$ can be taken as the two steps: an exothermic condensation to the liquid and an endothermic solvation of the liquid $HMn(CO)_5$, or a net estimate for ΔHhydr[$HMn(CO)_5$] of ~ 0 ($\pm$ 5) kcal/mol. Various estimates for ΔHhydr[H^+] have been made; we chose -262 kcal/mol (the variations are about $\pm$ 10 kcal/mol) [9]. Substitution of these quantities into equation 5 yields a hydration enthalpy for $Mn(CO)_5^-$ of a mere -43 kcal/mol! Notice that even with some variation in the hydration of $HMn(CO)_5$ or H^+, this number will still correspond to a hydration enthalpy very much --a factor of 2--smaller than for common monatomic ions as Cl^-. Such an effect is a dramatic illustration of the large size and poor solvation of $Mn(CO)_5^-$.

Energetics of the other acids are not as available--many of the PF_3 complexes are fully dissociated in aqueous solution [21], as is $HCo(CO)_4$ [28], making it impossible to estimate the ΔHacid, aq. Small solvation enthalpies for the transition-metal anions of the PF_3 complexes undoubtedly make the transition-metal hydrides poorer acids in solution than the comparable gas-phase acids FSO_3H and CF_3SO_3H.

The effect of large size of the ion decreasing the ion solvation energy can be quantified by considering the Born theory of solvation [37], in which the change in energy for N_i ions of radius r_i and uniform charge q_i on moving from a vacuum (dielectric 1) to a medium of dielectric D is given by equation 6.

$$\Delta H\text{solvation} = -\frac{N_i\, q_i{}^2}{2r_i}\left\{1 - \frac{(1 - LT)}{D}\right\} \tag{6}$$

Additional quantities in equation 6 are L, a small, empirical temperature dependence for the dielectric ($L = 4.63 \times 10^{-3}$ for water) and the temperature T (298K). Substitution of the dielectric of water, 78.53, and one mole of ions of unit charge gives a simplified expression (with appropriate unit conversion) for the relation of the hydration enthalpy to ion radius:

$$\Delta H\text{hydr} = -\frac{165.7}{r_i\,(\text{Å})}\ (\text{kcal/mol}) \tag{7}$$

The ion radius obtained from equation 7 is a "solvation radius", and for anions is typically about 0.85 Å larger than the crystal radius of the ion. Substitution of ΔHhydr[$Mn(CO)_5^-$] derived by equation 5 yields r_i[$Mn(CO)_5^-$] = 3.85 Å. Although the errors in ΔHhydr[$Mn(CO)_5^-$] are inconsequential, they do, because of the inverse relationship, make a commensurately large change in r_i[$Mn(CO)_5^-$]: with hydration energies of -35 to -55 kcal/mol, the hydrated ion radius varies from 4.7 to 3.1 Å. Even so, these are still large radii compared to a monatomic anion as Cl^-. Special types of interaction with the solvent, particularly hydrogen-bonding of the anion, could act to increase the magnitude of the hydration enthalpy. Such interactions are not possible for the organometallic anions, and the model is in very good agreement with expectations for the ion radius based on the crystal geometry of $Mn(CO)_5^-$, as shown in Figure 5.

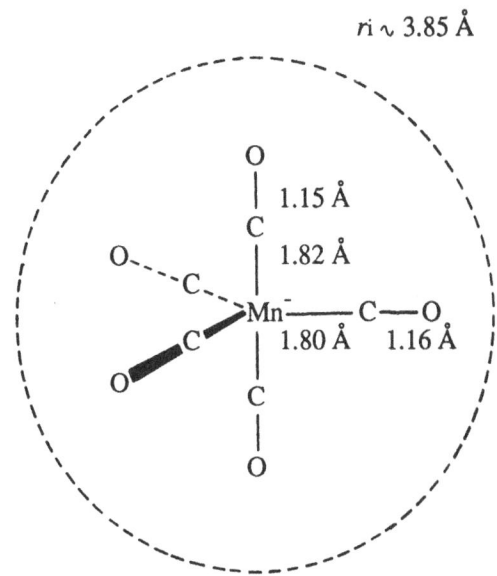

$r_i \sim 3.85$ Å

Figure 5. Idealized trigonal bipyramidal geometry and average bond lengths (from the crystal structure [38]) and estimated solvation radius of Mn(CO)$_5{}^-$.

4. Conclusions

The complexes HCo(CO)$_4$, HCo(PF$_3$)$_4$, HRh(PF$_3$)$_4$, HIr(PF$_3$)$_4$, HMn(CO)$_3$(PF$_3$)$_2$, HMn(CO)$_2$(PF$_3$)$_3$, HMn(CO)(PF$_3$)$_4$, and HMn(PF$_3$)$_5$ are all "gas-phase superacids", comparable in acidity to triflic acid, and among the strongest acids known. The transition-metal hydrides are generally strong gas-phase acids, as a result of relatively small metal-hydrogen homolytic bond energies and relatively large electron affinities. Relatively weak solution acidities for the complexes are largely the result of poor solvation of the bulky metal anions, although the intrinsic acidity of several of the complexes is so great as to make them strong aqueous acids as well.

Present data show the acidities increase from the middle to right of the transition series, and decrease with third-row metal (but not necessarily second-row metal) substitution. Data for the manganese series show that acidities increase ~ 7 kcal/mol with initial PF$_3$ for CO substitution, but the effect is evident only for the first two substitutions. The limits of the present data suggest that the acidity increase to the right in the periodic table is due to increasing electron affinities of the metal fragments, which overrides the decreasing bond strengths. The decrease in acidity going down a column in the periodic table is due to an increase in the homolytic bond energies (with the electron affinities relatively constant). The acidity increases with PF$_3$ substitutions are due primarily to increases in the electron affinities.

Periodic trends, substituent and solvation effects in the acidities will be an interesting topic to explore as we are able to expand the diversity of metal complexes for which we have data. Photoelectron spectroscopy on the metal anions, and determination of anion heats of formation by collision-induced dissociation [39] should eventually provide accurate metal–hydrogen bond enthalpies and heats of formation of the parent hydrides.

5. Acknowledgements

This work was begun at Caltech with Jack Beauchamp, with support from the United States Department of Energy. This work has been continued with support from the National Science Foundation (NSF/EPSCoR-OU-88 and NSF/ROW/CAW CHE-9008860), the donors of the Petroleum Research Fund administered by the American Chemical Society, and the United States Air Force through the Geophysics Scholar program and the University Resident Research Program. Debra Ewing and Amy Kawamura Lambrecht were responsible for aspects of the metal-hydride syntheses. We thank Darryl DesMarteau for his donation of $(CF_3 SO_2)_2 NH$, and Bob Taft for providing us with the acidities of reference compounds. We also thank the Phillips Laboratory group, John Paulson, Michael Henchman, Al Viggiano, Bob Morris, and Jane Van Doren, for their interest and hospitality.

6. References

*   Permanent address: Department of Chemistry and Biochemistry and Department of Physics and Astronomy, University of Oklahoma, Norman, OK 73019-0370.

[1]   Bartmess, J. E.; McIver, R. T., Jr. The Gas-Phase Acidity Scale. In *Gas Phase Ion Chemistry*; Bowers, M. T., Ed.; Academic Press: New York, 1979; Vol. 2.

[2]   Bartmess, J. E.; Scott, J. A.; McIver, R. T., Jr. *J. Am. Chem. Soc.* **1979**, *101*, 6056.

[3]   Moylan, C. R.; Brauman, J. I. *Annu. Rev. Phys. Chem.* **1983**, *34*, 187.

[4]   Viggiano, A. A.; Morris, R. A.; Dale, F.; Paulson, J. F.; Henchman, M. J.; Miller, T. M.; Miller, A. E. S. *J. Phys. Chem.* **1991**, *95*, 1275.

[5]   Viggiano, A. A.; Henchman, M. J.; Dale, F.; Deakyne, C. A.; Paulson, J. F. *J. Am. Chem. Soc.*, submitted for publication.

[6]   Lias, S. G.; Bartmess, J. E.; Liebman, J. F.; Holmes, J. L.; Levin, R. D.; Mallard, W. G. Gas-Phase Ion and Neutral Thermochemistry. *J. Phys. Chem. Ref. Data* **1988**, *17*, Suppl. No. 1.

[7]   Bartmess, J. E., personal communication, 1987.

[8]   Kahfel, I.; Taft, R. W., personal communication 1990.

[9]   Stevens Miller, A. E.; Beauchamp, J. L., *J. Am. Chem. Soc.* **1991**, *113*, 0000.

[10]   Stevens Miller, A. E.; Kawamura, A. R.; Miller, T. M. *J. Am. Chem. Soc.* **1990**, *112*, 457.

[11]   Stevens Miller, A. E.; Miller, T. M., *J. Am. Chem. Soc.*, to be submitted.

[12]   Davison, A.; Faller, J. W. *Inorg. Chem.* **1967**, *6*, 845.

[13]   Farmery, K.; Kilner, M. *J. Chem. Soc. A* **1970**, 634.

[14] Sternberg, H. W.; Wender, I.; Friedel, R. A.; Orchin, M. *J. Am. Chem. Soc.* **1953**, *75*, 2717.

[15] Warner, K. E.; Norton, J. R. *Organometallics* **1985**, *4*, 2150.

[16] Miles, W. J.; Clark, R. J. *Inorg. Chem.* **1968**, *7*, 1801.

[17] Udovich, C. A.; Clark, R. J. *Inorg. Chem.* **1969**, *8*, 938.

[18] van der Ent, A.; Onderdelinden, A. L. *Inorg. Syn.* **1973**, *14*, 92.

[19] Bennett, M. A.; Patmore, D. J. *Inorg. Chem.* **1971**, *10*, 2387.

[20] Head, R. A.; Nixon, J. F.; Sharp, G. J.; Clark, R. J. *J. Chem. Soc., Dalton* **1975**, 2054.

[21] Kruck, T. *Angew. Chem., Int. Ed. Engl.,* **1967**, *6*, 53.

[22] Beauchamp, J. L. *Annu. Rev. Phys. Chem.* **1971**, *22*, 527.

[23] Viggiano, A. A.; Paulson, J. F. *J. Chem. Phys.* **1983**, *79*, 2241.

[24] Stevens, A. E., Ph. D. Thesis, California Institute of Technology, 1981.

[25] Hieber, W.; Wagner, G. Z. *Naturforsch.* **1958**, *13B*, 339.

[26] Kristjánsdóttir, S. S.; Norton, J. R., unpublished work, cited in [27].

[27] Tilset, M.; Parker, V. D. *J. Am. Chem. Soc.* **1989**, *111*, 6711. Tilset, M.; Parker, V. D. *J. Am. Chem. Soc.* **1990**, *112*, 2843 (erratum).

[28] Hieber, W.; Hübel, W. Z. *Elektrochem.* **1953**, *57*, 235.

[29] Galembeck, F.; Krumholz, P. *J. Am. Chem. Soc.* **1971**, *93*, 1909.

[30] Moore, E. J.; Sullivan, J. M.; Norton, J. R. *J. Am. Chem. Soc.* **1986,** *108*, 2257.

[31] Martinho Simões, J. A.; Beauchamp, J. L. *Chem. Rev.* **1990**, *90*, 629, and references therein.

[32] Meckstroth, W. K.; Ridge, D. P. *J. Am. Chem. Soc.* **1985,** *107*, 2281.

[33] Zeigler, T.; Tschinke, V.; Becke, A. *J. Am. Chem. Soc.* **1987**, *109, 1351*.

[34] Stevens, A. E.; Beauchamp, J. L. *J. Am. Chem. Soc.* **1981**, *103*, 190.

[35] Viggiano, A. A.; Paulson, J. F.; Dale, F.; Henchman, M. *J. Phys. Chem.* **1987**, *91*, 3031.

[36] Protasiewicz, J. D.; Theopold, K. H.; Schulte, G. *Inorg. Chem.* **1988**, *27*, 1133.

[37] Moelwyn-Hughes, E. A. *Physical Chemistry*, 2nd ed."; Pergamon Press: New York, 1961.

[38] Frenz, B. A.; Ibers, J. A. *Inorg. Chem.* **1972**, *11*, 1109.

[39] See Sunderlin, L. S.; Wang, D.; Squires, R. R. *J. Am. Chem. Soc.*, submitted.

BOND STRENGTHS IN TRANSITION METAL CARBONYL ANIONS

LEE S. SUNDERLIN and ROBERT R. SQUIRES
The Department of Chemistry
Purdue University
West Lafayette, IN 47907 USA

ABSTRACT. Translational energy thresholds for collision-induced dissociative loss of carbonyl ligands from metal carbonyl anions $M(CO)_n^-$ (M = V, Cr, Mn, Fe, Co, and Ni) have been measured in a flowing afterglow-triple quadrupole apparatus. These thresholds can be used to derive metal-carbonyl bond strengths for these species. The sequential M-CO bond strengths vary widely, emphasizing the importance of measuring sequential rather than average bond strengths. The bond energies can be combined with other thermochemistry to give neutral metal-carbonyl bond strengths, ionization potentials, metal-alkene bond energies, and limits on other metal-ligand bond strengths. The 14- and 16-electron metal carbonyl anions display systematically weaker bonds than the 13-, 15-, and 17-electron species.

1. Introduction

As prototype organometallic compounds, transition metal carbonyls have been traditional subjects of thermochemical investigation for several decades. [1,2] Calorimetric measurements of average M-CO bond enthalpies in binary metal carbonyls and many of their derivatives provide an essential framework for our current understanding of periodic trends in metal-ligand bond strengths. [2,3,4] However, growing interests in the properties and reactivity of coordinatively-unsaturated metal carbonyl fragments [5,6,7] have led to an increased need for information about *sequential* rather than average M-CO bond energies. Measurements of this type are almost exclusively the domain of thermokinetic experiments, wherein activation energies for loss of one or more CO ligands from a metal center are determined and then equated with the thermochemical bond dissociation energies. [8,9,10] For highly unsaturated (<16 valence electron) $M(CO)_n$ systems, gas-phase techniques are usually required because of the extreme reactivity of these species towards even "inert" solvents. Mass spectrometry has made many important contributions in this regard, especially where transition metal carbonyl ions are

269

J. A. Martinho Simões (ed.), Energetics of Organometallic Species, 269–286.
© 1992 *Kluwer Academic Publishers.*

concerned. For example, the differences in the measured appearance energies of metal carbonyl ion fragments obtained from electron- or photoionization have been used to estimate the sequential M-CO bonds in chromium, [11] molybdenum, [11] tungsten, [11] iron, [11, 12] and nickel [12] carbonyl anions. Further, the metal carbonyl bond energies in $M(CO)_n^-$ negative ions can be used to derive M-CO bond strengths in the corresponding neutral complexes by combining them with measured electron affinities (EAs) via simple thermochemical cycles. [2]

A relatively new tool for investigating ion thermochemistry, at least in its applications to metal-ligand bond strengths, involves the measurement of translational energy thresholds for collision-induced dissociation (CID) of kinetically excited ions in a mass-selected beam. [13] We recently described an application of this approach to the determination of the sequential M-CO bond energies in $Fe(CO)_n^-$, n=1-4 and $Ni(CO)_n^-$, n=1-3. [14] A key aspect of this study was the availability of electron affinities for each of the $Fe(CO)_n$ and $Ni(CO)_n$ fragments that not only allowed us to derive estimates of the corresponding neutral metal carbonyl bond strengths, but also provided a means to compute the total binding energies for $Fe(CO)_4^-$ and $Ni(CO)_3^-$ from which average bond strengths could be computed for comparison to the individual values. We review here the general procedures and protocols, scope and limitations of these measurements, and present the preliminary results of an extension of the CID threshold method to other metal carbonyl negative ions. Homologous and periodic trends in the sequential bond strengths are described, along with selected applications of these data in deriving additional thermochemical quantities.

2. Instrumentation and Experimental Procedures

2.1 THE FLOWING AFTERGLOW-TRIPLE QUADRUPOLE INSTRUMENT [15]

Threshold energies for collision-induced dissociation reactions are meaningful only if the internal energy content of the reactant is known exactly (ie. as in state-selected experiments), or if the reactant ions are thermalized at a known temperature. For this reason, the flowing afterglow ion source is a key element of these experiments. The system consists of a 100 cm x 7.6 cm i.d steel flow tube interposed between an ion source region and a differentially-pumped detector chamber that houses the triple quadrupole mass analyzer (Figure 1). Purified helium enters the instrument at the upstream end, and flows through the tube at relatively high velocity (*ca.* 100 m/s) under the influence of a Roots blower. Typical operating pressures are in the range 0.2-1.0 Torr. Under these conditions the ions formed in the source region undergo rapid thermal equilibration with the flow tube walls and are cooled to a room temperature (298 K) internal energy distribution within 20-30 cm of the source. Transition metal ions are produced by dissociative electron impact with volatile metal carbonyls using either an electron emission filament or a DC discharge. With the DC discharge source, a *ca.* 5:1 He:Ar mixture is used

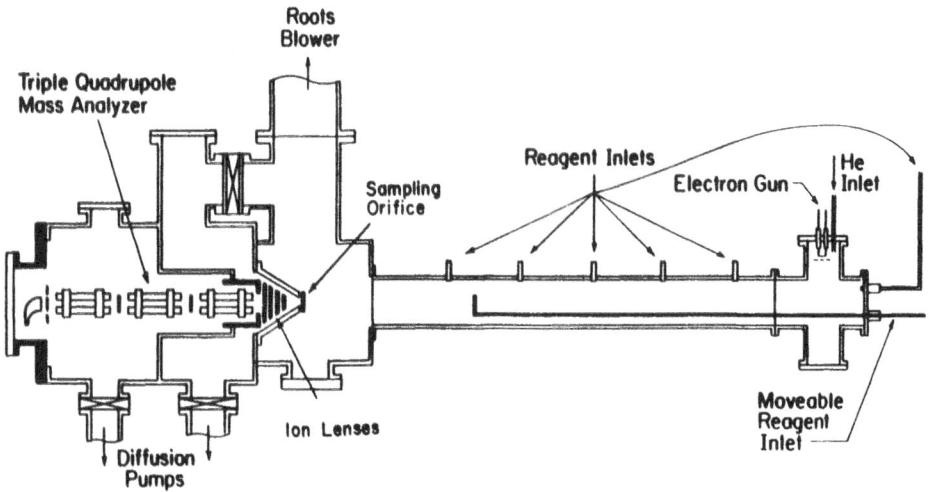

Triple Quadrupole Mass Analyzer

Roots Blower

Sampling Orifice

Reagent Inlets

Electron Gun

He Inlet

Ion Lenses

Diffusion Pumps

Moveable Reagent Inlet

The Flowing Afterglow - Triple Quadrupole

Figure 1. The flowing afterglow-triple quadrupole apparatus.

instead of pure He. In order to generate some of the more highly dissociated $M(CO)_n^-$ fragments such as $Fe(CO)^-$, a 90-150 V drift field is imposed between the ion source region and the flow reactor. This has the effect of promoting decomposition of the larger fragments by energetic collisions with the bath gas as the ions traverse the drift field. Yields of metal carbonyl fragments drop dramatically with the extent of decomposition, such that insufficient intensity was available for some of the smaller fragments to measure CID thresholds. Ions in the flow tube are gently extracted and then focussed into an Extrel triple quadrupole mass analyzer. The desired reactant ion is selected with the first quadrupole (Q1) and injected into the rf-only, gas-tight central quadrupole (Q2) with an axial kinetic energy determined by the Q2-rod offset voltage. Inert target gases such as argon or xenon are admitted to Q2 at pressures in the range of 0.02-0.10 mTorr. Fragment ions resulting from CID are efficiently contained in Q2 and then extracted into the third quadrupole (Q3) for mass analysis. Ion detection is carried out with a conversion dynode and an electron multiplier operating in pulse-counting mode.

2.2 COLLISION-INDUCED DISSOCIATION THRESHOLDS. MEASUREMENT AND ANALYSIS

In order to measure a CID threshold, the signal intensity of a dissociation product ion is monitored as a function of the Q2-rod offset voltage. To aid in collection of fragment ions, the offset voltage of the third quadrupole is also ramped at a value 3-15 volts higher than that of Q2, and the intervening lens is held at a constant attractive potential. The collision target pressure in Q2, which was either argon or xenon in the present experiments, is maintained low enough (< 0.05 mTorr) to ensure that the majority of reactant ions undergo only one activating collision. This allows the nominal axial kinetic energy to be equated with the maximum collisional energy deposition to the dissociating ions. A plot of the dissociation cross section or normalized fragment ion yield versus the center-of-mass collision energy $E_{CM} = E_{LAB}[M_{target}/(M_{target}+M_{ion})]$ gives an ion appearance curve from which the CID threshold may be determined.

The activation energy for the dissociation may be deconvoluted from the ion appearance curve by means of a fitting procedure based on the assumed model function given by eq 1, [13,16,17]

$$I(E) = I_o [(E - E_T)^n / E] \tag{1}$$

where $I(E)$ is the intensity of the product ion at center-of-mass collision energy E, E_T is the desired threshold energy, I_o is a scaling factor, and n is an adjustable parameter. Optimization is carried out by an iterative procedure [18] in which n, I_o, and E_T are varied so as to minimize the deviations between the experimental and calculated appearance curves in the steeply rising portion of the threshold region. Convoluted into the fit are the reactant ion kinetic energy distribution approximated by a Gaussian function with a 2 eV width in the laboratory frame, and a Doppler broadening function developed by Chantry to account for the random thermal motion of the neutral target. [19] The CID threshold, E_T, derived in this way is considered to correspond to a thermal activation energy for production of room temperature (298 K) products from thermalized, room temperature reactants. [14,20,21]

The energy range over which the data is fit can have a large effect on the derived threshold. At the lowest energies, the ion appearance curve shows significant tailing, which is attributed to translational excitation of the ions in the first quadrupole or to internal excitation due to collisions outside the interaction region. Fitting this part of the data results in abnormally high values for n and low values for E_T. The data at energies higher than the steeply rising portion of the threshold should also not be included in the fit because equation 1 is no longer appropriate. Fitting to too high an energy gives abnormally low values for n and high values for E_T. Choosing the appropriate energy range is a critical part of the threshold fitting procedure, and is the only part not automated by computer. Thresholds that are more sensitive to the precise energy range have higher error limits because of a higher standard deviation in the

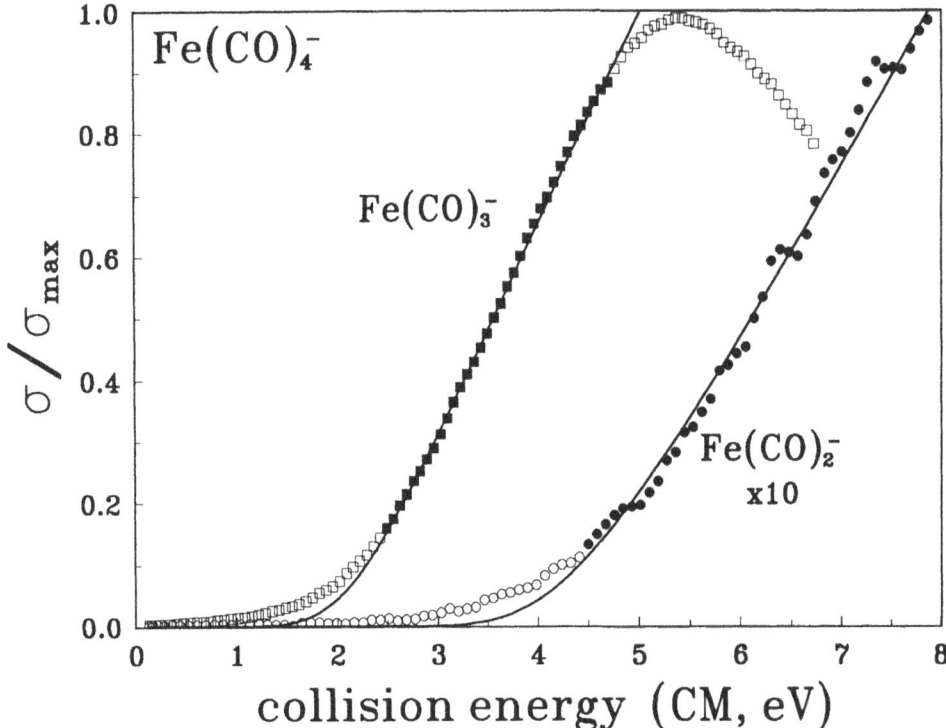

Figure 2. Appearance curves for products from CID of Fe(CO)$_4^-$ as a function of kinetic energy. The solid lines are model appearance curves calculated using eq 1 and convoluted as discussed in the text. The eq 1 parameters are (n = 1.71, E$_T$ = 1.84) for Fe(CO)$_3^-$ and (n = 1.75, E$_T$ = 3.71) for Fe(CO)$_2^-$. The solid symbols indicate the energy range over which the data was fit.

thresholds taken from individual data sets. In Figures 2-6, the typical energy ranges over which the data is fit are denoted by filled symbols.

3. Results

A typical set of results, those for the iron carbonyl anions, are described in detail in this section. The data analysis for the other metal ions is similar.

The products observed from CID of Fe(CO)$_4^-$ with argon and xenon (rare gases, Rg) correspond to loss of one to three carbonyl ligands, reactions 2-4.

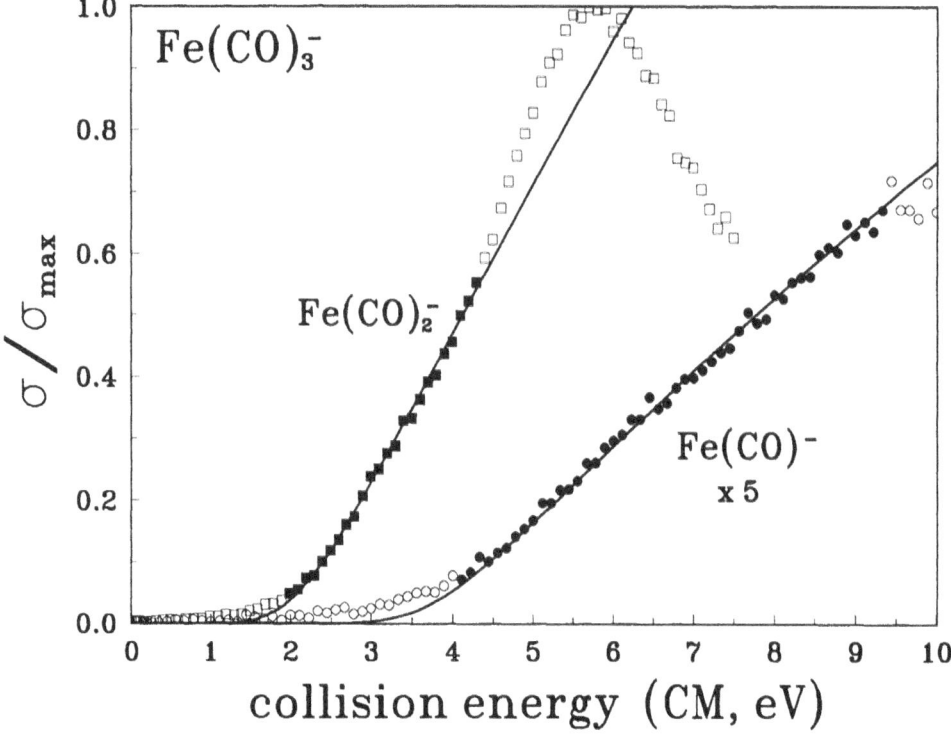

Figure 3. Appearance curves for products from CID of $Fe(CO)_3^-$ as a function of kinetic energy. The solid lines are model appearance curves calculated using eq 1 and convoluted as discussed in the text. The eq 1 parameters are ($n = 1.77$, $E_T = 1.77$) for $Fe(CO)_2^-$ and ($n = 1.60$, $E_T = 3.37$) for $Fe(CO)^-$. The solid symbols indicate the energy range over which the data was fit.

$$Fe(CO)_4^- \xrightarrow{\text{Rg}} \begin{cases} \rightarrow Fe(CO)_3^- + CO & (2) \\ \rightarrow Fe(CO)_2^- + 2\ CO & (3) \\ \rightarrow Fe(CO)^- + 3\ CO & (4) \end{cases}$$

Relative cross sections for reactions 2 and 3 are plotted as a function of center-of-mass collision energy in Figure 2 (Rg = Ar). The relative cross sections for reactions 5 and 6, reactions 7 and 8, and reaction 9 are plotted as a function of kinetic energy in Figures 3, 4, and 5, respectively (Rg = Ar).

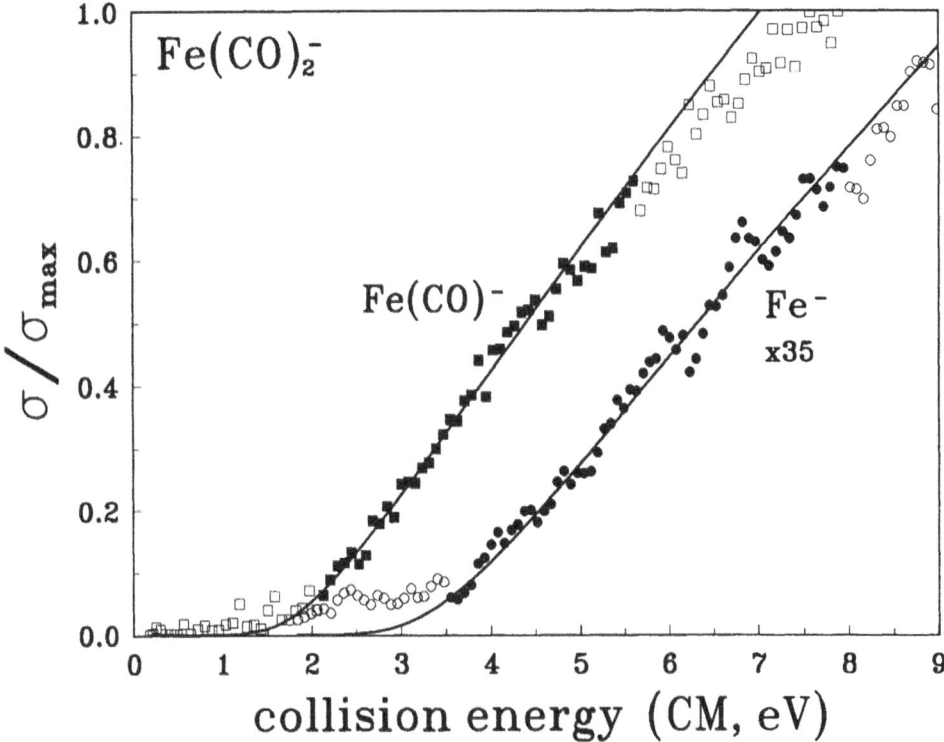

Figure 4. Appearance curves for products from CID of $Fe(CO)_2^-$ as a function of kinetic energy. The solid lines are model appearance curves calculated using eq 1 and convoluted as discussed in the text. The eq 1 parameters are ($n = 1.72$, $E_T = 1.56$) for $Fe(CO)^-$ and ($n = 1.7$, $E_T = 2.95$) for Fe^-. The solid symbols indicate the energy range over which the data was fit.

$$Fe(CO)_3^- \xrightarrow{\text{Rg}} \begin{cases} \rightarrow Fe(CO)_2^- + CO & (5) \\ \rightarrow Fe(CO)^- + 2\,CO & (6) \end{cases}$$

$$Fe(CO)_2^- \xrightarrow{\text{Rg}} \begin{cases} \rightarrow Fe(CO)^- + CO & (7) \\ \rightarrow Fe^- + 2\,CO & (8) \end{cases}$$

$$Fe(CO)^- \xrightarrow{\text{Rg}} Fe^- + CO \qquad (9)$$

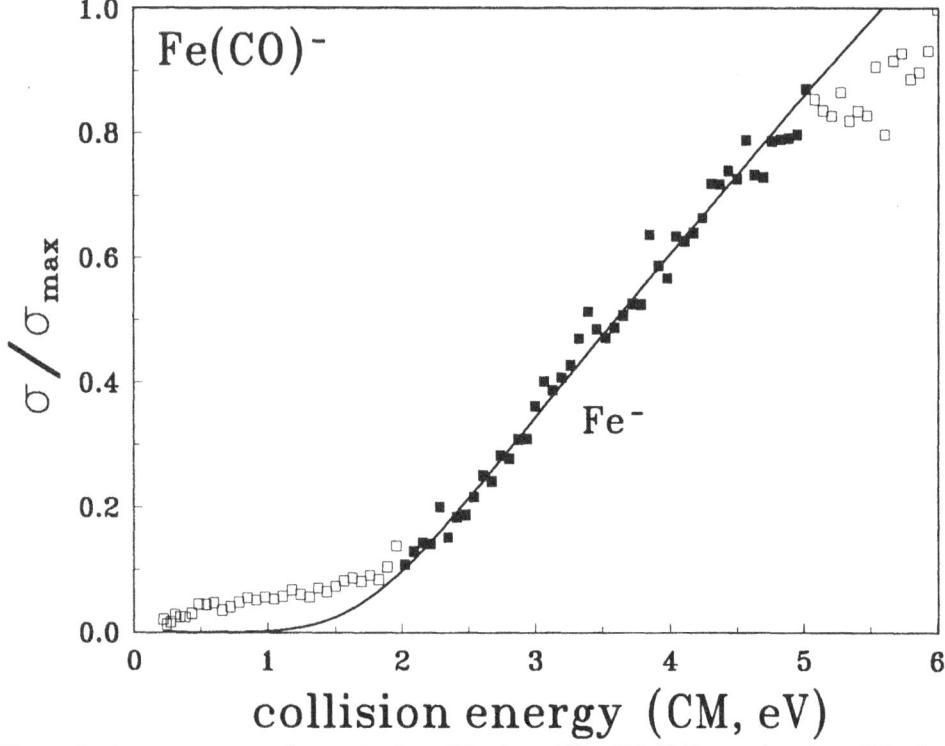

Figure 5. Appearance curve for production of Fe⁻ from CID of Fe(CO)⁻ as a function of kinetic energy. The solid line is a model appearance curve calculated using eq 1 and convoluted as discussed in the text. The eq 1 parameters are $n = 1.72$ and $E_T = 1.48$. The solid symbols indicate the energy range over which the data was fit.

As an example of the threshold behavior of an early transition metal system, we show in Figure 6 data for CID of $Cr(CO)_5^-$, reactions 10 and 11. The general behavior is very similar

$$Cr(CO)_5^- \xrightarrow{\text{Rg}} \begin{cases} Cr(CO)_4^- + CO & (10) \\ Cr(CO)_3^- + 2\ CO & (11) \end{cases}$$

to that for the iron systems, although the thresholds in this case are somewhat higher.

The maximum cross sections estimated for these reactions are in the range 0.2-24 Å², where single CO loss is always the dominant process, and multiple dissociations display successively lower yields. Cross sections of this order of magnitude are typical for low-energy CID processes involving negative ions. [13]

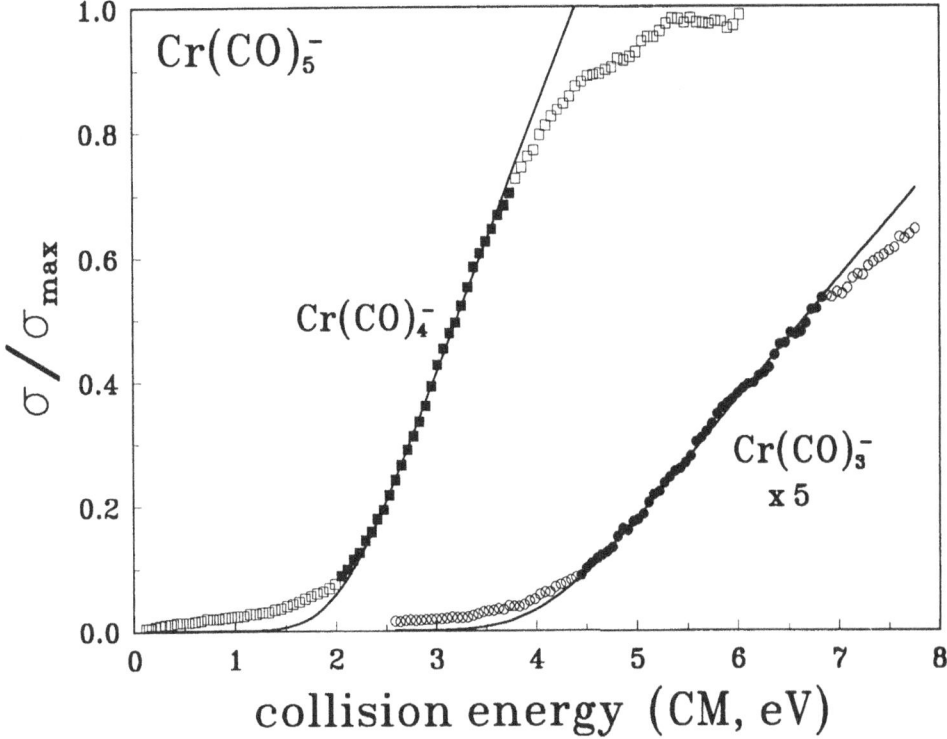

Figure 6. Appearance curve for products from CID of $Cr(CO)_5^-$ as a function of kinetic energy. The solid lines are model appearance curves calculated using eq 1 and convoluted as discussed in the text. The eq 1 parameters are (n = 1.60, E_T = 1.89) for $Cr(CO)_4^-$ and (n = 1.53, E_T = 3.85) for $Cr(CO)_3^-$. The solid symbols indicate the energy range over which the data was fit.

3.1. THRESHOLD DETERMINATIONS

The optimized fitting parameters for the iron carbonyl anions used with equation 1 are listed in Table 1, and some of the corresponding fits are shown in Figures 2-5. The error limits listed are standard deviations for the parameters optimized for the individual data sets. The standard deviation of E_T is a good estimate for the accuracy of the derived bond dissociation energies.

In order to test the effect of the mass of the neutral target, both Ar and Xe were used as target gases for some reactions. For the systems studied here, there are no systematic differences between thresholds derived using the two different targets. Almost every reaction optimized with values of n from 1.6 to 1.8. This suggests that $n \approx 1.7$ is applicable to most or all of the reactions studied. Data for a few reactions were insufficiently robust to allow n to optimize. We therefore average thresholds determined when n is held equal to 1.5, 1.7, and 1.9 for these reactions.

Table I. Fitting parameters[a]

Rxn	Rg	E_T (eV)	n
2	Ar	1.84 ± 0.10	1.71 ± 0.16
2	Xe	1.78 ± 0.12	1.76 ± 0.19
3	Ar	3.71 ± 0.10	1.75 ± 0.16
3	Xe	3.81 ± 0.48	1.7 ± 0.2
5	Ar	1.77 ± 0.15	1.77 ± 0.28
6	Ar	3.37 ± 0.30	1.60 ± 0.13
7	Ar	1.56 ± 0.06	1.72 ± 0.18
8	Ar	2.95 ± 0.26	1.7 ± 0.2
9	Ar	1.48 ± 0.17	1.72 ± 0.15

[a]Optimal fitting parameters for eq 1 with m = 1.

E_T for reaction 2 is equivalent to the measured bond energy $D(CO)_3Fe^--CO]$, provided there are no barriers to dissociation in excess of the endothermicity (see discussion below). The values for Rg = Ar (1.84 ± 0.10 eV) and Rg = Xe (1.78 ± 0.12 eV) can be averaged to give $D[(CO)_3Fe^--CO] = 1.81 ± 0.11$ eV (41.7 ± 2.5 kcal/mol). E_T for reaction 3 (Rg = Ar), 3.71 ± 0.10 eV corresponds to the energy needed to remove two carbonyl ligands, leading to $D[(CO)_2Fe^--CO] = 3.71 - 1.81 = 1.90 ± 0.15$ eV. The threshold for reaction 3 with Rg = Xe, 3.81 ± 0.49 eV, is consistent with this result but not of sufficient precision to be included in the final determination of thermochemistry. The threshold for reaction 5, 1.77 ± 0.15 eV, is also a measurement of $D[(CO)_2Fe^--CO]$. The two determinations combine to give $D[(CO)_2Fe^--CO] = 1.84 ± 0.15$ eV (42.4 ± 3.5 kcal/mol), where the error limit is estimated. The threshold for reaction 7 gives $D[(CO)Fe^--CO] = 1.57 ± 0.14$ eV (36.2 ± 3.2 kcal/mol), where the uncertainty is taken to be twice the 0.07 eV precision of the threshold measurement because no corroborating data could be obtained. The threshold for reaction 6, 3.37 ± 0.30 eV, is consistent with the sum of the latter two bond strengths but of insufficient precision to be included.

The results for a similar analysis of the other transition metal anions can be used to derive the further thermochemical values listed in Table II. The results for vanadium, manganese, and cobalt should be considered preliminary. Comparisons of the present results to the limited number of previous metal-carbonyl bond strength determinations and to theoretical calculations of metal-carbonyl bond strengths [22,23] have already been discussed, [14,24] and will not be repeated here.

4. Discussion

4.1. CAVEATS

Certain features of these experiments could compromise the correspondence between the measured CID thresholds and the true bond energies. For example, efficient electron detachment could cause a competitive shift [25] in the measured thresholds by suppressing the cross sections for ligand loss. Electron detachment cannot be observed directly in the present experiment, since the quadrupole collision chamber does not trap electrons. Indirect evidence suggests that while electron detachment may occur in some of the reactions studied, it apparently does not interfere with the threshold measurements for loss of one or two CO ligands. [14]

The thermochemistry derived above assumes that products are formed in their ground electronic state. With metal carbonyls, this may not always be the case if the reactant and product ions have different spin. Evidence discussed previously [14] suggests that reactions of metal carbonyls that do not conserve spin are not even necessarily inefficient, nor do they necessarily have a barrier when they are inefficient. In the absence of further information, it is assumed that the products of the CID reactions examined in this work are in their electronic ground state, *i.e.* that adiabatic dissociation prevails.

Broadening of the threshold appearance curve due to the reactant ion internal energy distribution can have a significant effect (≈ 2 kcal/mol) on the analysis of CID reactions. [20] The effect of this broadening is smaller in the present case, since the region very near the threshold, which is most affected by energy broadening, is not included in the fit. Fits that take explicit account of the reactant vibrational energy distribution differ by less than 0.01 eV from those that do not. [14]

4.2. BOND STRENGTHS SUMS

The bond strength sum $D[Fe^--4CO]$ can be calculated using equation 12

$$D[Fe^--4CO] = - \Delta H_f[Fe(CO)_5] + \Delta H_f(Fe^-) + 5\Delta H_f(CO)$$

$$- D[(CO)_4Fe-CO] + EA[Fe(CO)_4] -5kT \qquad (12)$$

and literature thermochemistry. [14] The result is $D[Fe^--4CO] = 147.8 \pm 7.6$ kcal/mol. The sum of the metal-carbonyl bond strengths in $Fe(CO)_4^-$ measured here is 153.6 ± 6.5 kcal/mol, giving a difference of 6 ± 10 kcal/mol. The discrepancy suggests either that $EA[Fe(CO)_4]$ is at the upper end of the 2.4 ± 0.3 eV range, or that the bond strengths measured here are high by an average of 1-2 kcal/mol. However, the two results are within error limits of each other. This indicates that the experimental method gives results that are not systematically in error by a significant amount.

The sum of all three nickel carbonyl bond strengths in $Ni(CO)_3^-$ can be calculated using an equation analogous to eq 12. The result is $D[Ni^--3CO] = 114.3 \pm 1.3$ kcal/mol. Subtraction of $D[(CO)Ni^--2CO] = 81.9 \pm 5.8$ kcal/mol from the present study gives $D[Ni^--CO] = 32.4 \pm 5.8$ kcal/mol. In the absence of data for some of the smaller metal carbonyl fragments, only certain bond strength sums can be derived (Table II). Unfortunately, only the lower limits $EA[V(CO)_6] \geq 0.53$ eV, [24] $EA[Cr(CO)_5] \geq 2.3$ eV, [24] and $EA[Co(CO)_4] \geq D[(CO)_4Co-H] = 2.3$ eV [26, 27] are available, leading to lower limits on the corresponding metal carbonyl bond strength sums. The value for cobalt, $D[Co^--2CO] \geq 96 \pm 7$ kcal/mol, clearly stands out. If correct, it would indicate that the average of the two cobalt-carbonyl bond strengths in $Co(CO)_2^-$ is roughly the same as the strongest of all the bonds in Table II. Nothing in the present data suggests that $Co(CO)_2^-$ should have two unusually strong metal-carbon bonds, although it is not impossible. Confirmation of the two experiments used to derive $EA[Co(CO)_4]$, particularly the problematical metal-hydrogen bond strength, [2] would be highly desirable. The analogous result for manganese, $D[Mn-2CO] = 51 \pm 11$ kcal/mol, indicates that the metal-carbon bond strengths in $Mn(CO)_2$ are relatively weak, while for $V(CO)_3$ and $Cr(CO)_2$, the average bond strengths can only be shown to be ≥ 24 kcal/mol.

4.3. THERMOCHEMISTRY FOR NEUTRAL METAL CARBONYL FRAGMENTS

The anion thermochemistry can be combined with the known electron affinities (EAs) of neutral iron carbonyls [28] to give bond strengths for the neutral species using equation 13. [29]

$$D[(CO)_nM-CO] = D[(CO)_nM^--CO] + EA[M(CO)_n] - EA[M(CO)_{n+1}] \qquad (13)$$

In some cases this results in increased uncertainties because the error limits for the anion bond strength and two EAs are combined in deriving the neutral bond strength. The results are given in Table II.

More complete sets of electron affinities for the vanadium, chromium, manganese, and cobalt carbonyls would be of great value, since they would allow calculation of additional neutral metal carbonyl bond strengths. Measurements of this type are forthcoming. [30]

An equation analogous to eq. 13 can be used to derive ionization potentials for the neutral metal carbonyls from the neutral bond strengths and metal carbonyl cation bond strengths in the literature. [20, 31] Details of these calculations for iron and nickel have been given previously. [14]

Table III. Average Metal-Carbonyl Anion Bond Strengths.

Electron Count	Average Bond Strength (kcal/mol)
18	40
17	41
16	33
15	43
14	31
13	38

4.4. BOND STRENGTH TRENDS

The neutral metal carbonyl bond strengths are typically weaker than those in the anions. The bond strengths in the vanadium carbonyl anions are on average the weakest observed. These observations are consistent with less π-backbonding in both the neutral complexes and the vanadium carbonyl anions, which have fewer electrons available to back bond to multiple CO ligands. [22, 32] Insufficient data are available for the neutral metal carbonyls to discuss any general trends, although the wide variation in bond strengths clearly indicates that the individual bond strengths deviate significantly from the average.

Previous treatments of the trends in metal-ligand bond strengths have made use of a model wherein a metal-ligand bond energy is approximated as the difference between an "intrinsic" bond strength (which should be relatively independent of the metal or number of ligands) and a promotion energy that is necessary to reconfigure the metal in a state suitable for bonding to the ligand. Applying this simple model to the neutral carbonyls of the first transition series metals suggests that the intrinsic metal-carbonyl bond strength in this series is probably equal to the maximum bond strengths observed, $\approx$40-47 kcal/mol. Weaker bonds are then attributed to the necessity for electronic or geometric rearrangement upon adding a carbonyl. For example, D(Fe-CO) is particularly weak, 8.1 ± 3.5 kcal/mol. The ground state of Fe(CO) is calculated to be a triplet derived from the $3d^7 4s$ (^3F) state of Fe, [23] which is 34 kcal/mol above the ground state of Fe. The low bond strength thus correlates with the energy needed to promote the iron atom into a state suitable for bonding.

Applying the intrinsic bond strength-promotion energy model to the metal carbonyl anions suggests that the intrinsic M-CO bond energy in metal carbonyl anions is also in the range of 41-45 kcal/mol. While more conclusive analysis of the periodic trends in the metal carbonyl anion bond strengths must await more complete data, one trend is clear from the data in Table III, which gives the average metal-carbonyl bond strength as a function of electron count. There is a definite odd-even alternation in the bond strengths, with the 14- and 16- electron complexes displaying consistently weaker bonds than the 13-, 15-, and 17-electron complexes. This nominally correlates with high promotion energies for the 12- and 14-electron complexes. In the absence of more detailed theoretical calculations on metal carbonyl anions, [33] any discussion of the nature of the promotions involved will be speculative. However, it is known that the ground states of the bare metal anions are $V^-(^5D)$, $Cr^-(^6S)$, $Fe^-(^4F)$, $Co^-(^3F)$, and $Ni^-(^2D)$. [34] The 17- and 18-electron complexes should be doublets and singlets, respectively. Therefore, the number of changes of spin needed to go from the ground state of M^- to the largest stable carbonyl (17- or 18-electron species) is two for V^-, Cr^-, and Mn^-, one for Fe^- and Co^-, and zero for Ni^-. Promotion energy associated with spin changes that accompany the addition of a carbonyl to 12- and 14-electron metal carbonyl anions may thus account for the weak bond strengths in the 14- and 16-electron vanadium, manganese, and cobalt carbonyl anion systems. Obtaining a value for $D[(CO)Co^- - CO]$ would be a particularly useful test of this hypothesis. Changes in spin alone do not account for every bond that is weaker than 40 kcal/mol, however, since $D(Ni^- - CO)$ is relatively low even though both Ni^- and $NiCO^-$ are doublets. [33] In this case, promotion energy associated with orbital rehybridization must also be taken into account.

4.5. RELATED METAL-LIGAND BOND STRENGTHS

Another method for obtaining new thermochemistry is to combine the measured enthalpy of disruption [3] of an organometallic species that contains a metal carbonyl fragment

$$\Delta H_{disr}[M(CO)_nL_m] = \Delta_fH(M) + n\Delta_fH(CO) + m\Delta_fH(L) - \Delta_fH[M(CO)_nL_m] \quad (14)$$

with the neutral M-CO bond energies listed in this work to derive bond strengths for the other ligands. Illustrative results derived using this procedure include $D[(CO)_4Fe-C_2H_4] = 36.5 \pm 3.6$ kcal/mol, [35] $D[(CO)_3Fe-C_4H_6] = 62 \pm 9$ kcal/mol, [35] $D[(CO)Fe-2(C_4H_6)] = 111.1 \pm 9$ kcal/mol, [35] $D[(CO)Fe-2(C_6H_8)] = 116 \pm 9$ kcal/mol, [35] and $D[(CO)_3Fe-C_8H_8] = 56.7 \pm 9.8$ kcal/mol, [36] (C_4H_6 = 1,3-butadiene, C_6H_8 = cyclohexa-1,3-diene, and C_8H_8 = 1,3,5,7-cyclooctatetraene). These results indicate that bonds from iron to one, two, or four olefinic groups are somewhat weaker than bonds to the same number of carbonyl ligands. The individual metal-carbonyl bond strengths can also be used to derive heats of formation for the metal carbonyl fragments. [14] These data will be particularly useful for deriving additional metal-ligand bond energies when further calorimetrically determined heats of formation for organometallic metal carbonyl complexes become available.

The metal-carbonyl bond strengths listed in Table II can also be used to derive bond strength estimates for other ligands on the basis of previous investigations of gas-phase ligand exchange reactions involving $M(CO)_n^-$ ions. [37] Examples of the thermochemistry to be derived from these types of reactions have been noted. [14]

5. Conclusions

Energy-resolved collision-induced dissociation has been used to determine the metal-carbonyl bond energies given in Table II. These can be combined with literature thermochemistry where available to derive a wide variety of other thermochemical quantities of interest, including metal-carbonyl bond strengths, heats of formation and ionization potentials for the neutral metal carbonyls, as well as thermodynamic data for other organometallic species which contain metal carbonyl fragments, in particular species where carbonyl ligands are replaced with alkene ligands. The sequential M-CO bond strengths vary widely, emphasizing the importance of measuring sequential rather than average bond strengths. An emerging pattern in the bond strengths as a function of electron count has been described wherein the 14- and 16-electron $M(CO)_n^-$ ions display systematically weaker bonds than the 13-, 15-, and 17-electron ions. Future extensions and refinements in the measurement of sequential metal-carbonyl bond strengths by threshold measurements of collision-induced dissociation processes, when combined with reliable electronic structure calculations, should make the origins of this pattern more clear.

Acknowledgment

This work was supported by the Department of Energy, Office of Basic Energy Science.

References

1. Cotton, F. A.; Wilkinson, G. *Advanced Inorganic Chemistry 5th Ed.*; Wiley: New York, 1988. Collman, J. P.; Hegedus, L. S.; Norton, J. R.; Finke, R. G. *Principles and Applications of Organotransition Metal Chemistry*; University Science Books: Mill Valley, CA, 1987. Kochi, J. K. *Organometallic Mechanisms and Catalysis*; Academic Press: New York, 1978.
2. Simões, J. A. M.; Beauchamp, J. L. Chem. Rev. **1990**, 90, 629.
3. Connor, J. A. Topics Curr. Chem. **1977**, 71, 71.
4. Pilcher, G.; Skinner, H. A. in *The Chemistry of the Metal-Carbon Bond*; F. R. Hartley and S. Patai, Eds.; Wiley: New York, 1982; Chapter 2.
5. Elian, M.; Hoffmann, R.; Inorg. Chem. **1975**, 14, 1058. Hoffmann, R.; Angew. Chem., Int. Ed. Engl. **1982**, 21, 711.

6. Tolman, C. A. Chem. Soc. Rev. **1972**, 1, 337. Casey, C. P.; Cyr, C. R. J. Am. Chem. Soc. **1973**, 95, 2248. Mitchener, J. C.; Wrighton, M. S. J. Am. Chem. Soc. **1981**, 103, 975. Whetten, R. L.; Fu, K.; Grant, E. R. J. Chem. Phys. **1982**, 77, 3769. Whetten, R. L.; Fu, K.; Grant, E. R. J. Am. Chem. Soc. **1982**, 104, 4270.

7. Weitz, E. J. Phys. Chem. **1987**, 91, 3945.

8. Siefert, E. E.; Angelici, R. J. Organomet. Chem. **1967**, 8, 374-376.

9. Lewis, K. E.; Golden, D. M.; Smith, G. P. J. Am. Chem. Soc. **1984**, 106, 3905-3912.

10. Day, J. P.; Basolo, F.; Pearson, R. G. J. Am. Chem. Soc. **1968**, 90, 6927-6933.

11. Pignataro, S.; Foffani, A.; Grasso, F.; Cantone, B. Z. Phys. Chem. (Frankfurt) **1965**, 47, 106.

12. Compton, R. N.; Stockdale, J. A. D. Int. J. Mass Spectrom. Ion Phys. **1976**, 22, 47.

13. Graul, S. T.; Squires, R. R. J. Am. Chem. Soc. **1990**, 112, 2517. Paulino, J. A.; Squires, R. R. J. Am. Chem. Soc. **1991**, 113, 5573. Lifshitz, C.; Wu, R.; Tiernan, T. O. J. Chem. Phys. **1978**, 68, 247. Armentrout, P. B.; Beauchamp, J. L. Chem. Phys. **1980**, 50, 21. Hales, D. A.; Armentrout, P. B. J. Cluster Sci. **1990**, 1, 27.

14. Sunderlin, L. S.; Wang, D.; Squires, R. R. J. Am. Chem. Soc., accepted for publication.

15. Graul, S. T.; Squires, R. R. Mass Spectrom. Rev. **1988**, 7, 263.

16. Chesnavich, W. J.; Bowers, M. T. J. Phys. Chem. **1989**, 83, 900.

17. Sunderlin, L. S.; Armentrout, P. B. Int. J. Mass Spec. Ion Processes **1989**, 94, 149, and references therein.

18. Ervin, K. M.; Armentrout, P. B. J. Am. Chem. Soc. **1985**, 83, 166.

19. Chantry, P. J. J. Chem. Phys. **1971**, 55, 2746.

20. Schultz, R. H.; Crellin, K. C.; Armentrout, P. B. J. Am. Chem. Soc. **1991**, 113, 8590.

21. Holmes, J. L.; Lossing, F. P.; Mayer, P. M. J. Am. Chem. Soc. **1991**, 113, 9723.

22. Ziegler, T.; Tschinke, V.; Ursenbach, C. J. Am. Chem. Soc. **1987**, 109, 4825. Blomberg, M. R. A.; Siegbahn, P. E. M.; Lee, T. J.; Rendell, A. P.; Rice, J. A. J. Chem. Phys., submitted. Veillard, A. Chem. Rev. **1991**, 91, 743-766.

23. Barnes, L. A.; Rosi, M.; Bauschlicher, C. W. J. Chem. Phys. **1991**, 94. 2031.

24. Wang, D. PhD Thesis, Purdue University, 1990.

25. Lifshitz, C.; Long, F. A. J. Chem. Phys. **1964**, 41, 2468.

26. Stevens Miller, A. E.; Beauchamp, J. L. J. Am. Chem. Soc. **1991**, 113, 8765.

27. Bronshtein, Y. E.; Gankin, V. Y.; Krinkin, D. P.; Rudkovskii, D. M. Russ. J. Phys. Chem. (English Trans.) **1966**, 40, 802.

28. Engelking, P. C.; Lineberger, W. C. J. Am. Chem. Soc. **1979**, 101, 5569.

29. This equation strictly applies to bond strengths at 0 K. It applies at 298 K if the difference between the heat capacities of $[M(CO)_n^- + M(CO)_{n+1}]$ and $[M(CO)_{n+1}^- + M(CO)_n]$ is negligible, which should hold true for these systems.

30. Bengali, A. A.; Casey, S. M.; Cheng, C.; Dick, J. P.; Fenn, P. T.; Villalta, P. W.; Leopold, D. G. J. Am. Chem. Soc., submitted.

31. Distefano, G. J. Res. Natl. Bur. Stand. **1970**, 74A, 233.

32. Caulton, K. G.; Fenske, R. F. Inorg. Chem. **1968**, 7, 1273. Pierloot, K.; Verhulst, J.; Vanquickenborne, L. G. Inorg. Chem. **1989**, 28, 3059.

33. Bauschlicher, C. W.; Barnes, L. A.; Sanghoff, S. R. Chem. Phys. Lett. **1988**, 151, 391.

34. Hotop, H.; Lineberger, W. C. J. Phys. Chem. Ref. Data **1985**, 14, 731. Mn⁻ is unbound, but for the present argument we shall interpolate an $s^2 d^6\ ^5D$ state as the most reasonable ground state.

35. Brown, D. L. S.; Connor, J. A.; Leung, M. L.; Paz-Andrade, M. I.; Skinner, H. A. J. Organomet. Chem. **1976**, 110, 79.

36. Connor, J. A.; Demain, C. P.; Skinner, H. A.; Zafarani-Moattar, M. T. J. Organomet. Chem. **1979**, 170, 117.

37. McDonald, R. N.; Chowdhury, A. K.; Schell. P. L. J. Am. Chem. Soc. **1984**, 106, 6095. McDonald, R. N.; Schell, P. L. Organometallics **1988**, 7, 1820. McDonald, R. N.; Chowdhury, A. K.; Jones, M. T. J. Am. Chem. Soc. **1986**, 108, 3105. Pan, Y. H.; Ridge, D. P., J. Am. Chem. Soc. **1989**, 111, 1150. Pan, Y. H.; Ridge, D. P., J. Am. Chem. Soc., submitted, and references therein. Gregor, I. K.; Inorg. Chim. Acta **1990**, 176, 19. VanOrden, S. L.; Pope, R. M.; Buckner, S. W. Organometallics **1991**, 10, 1089.

FUNDAMENTAL GAS PHASE STUDIES OF THE MECHANISM AND THERMOCHEMISTRY OF ORGANOMETALLIC REACTIONS

J.L. Beauchamp[1] and Petra A. M. van Koppen[2]
[1]California Institute of Technology 127-72
Pasadena, California 91125
[2]Department of Chemistry
University of California
Santa Barbara, California 93106

ABSTRACT. A variety of experimental techniques, including FT-ICR, ion beam, and measurement of kinetic energy release distributions in reverse sector instruments, have been applied to examine the mechanism and energetics of organometallic reactions in the gas phase. These studies have provided a wealth of thermochemical data for organometallic species and have revealed important features of the potential energy surfaces associated with the reactions of hydrocarbons with atomic transition metal ions. The experimental methods and theoretical models used in these investigations are reviewed and presented along with representative results from recent studies.

1. Introduction

The past twenty years have led to the development of an impressive array of experimental methods to conduct fundamental gas phase studies of the mechanisms and energetics of organometallic reactions. While detailed studies of the reactions of neutral species are now being undertaken, the majority of these investigations have involved reactions of ionic species, which in comparison to neutrals are more readily manipulated and detected using mass spectrometric techniques. These developments have stemmed in part from the importance of transition metal sites as active centers which provide low energy pathways for the selective catalytic transformation of small molecules into useful products.

Considerable interest in the subject of C-H bond activation at transition-metal centers has materialized in the past several years, stimulated by the observation that even saturated hydrocarbons can react with little or no activation energy under appropriate conditions. Interestingly, gas phase studies of the reactions of saturated hydrocarbons at transition-metal centers were reported as early as 1973 [1]. More recently, ion cyclotron resonance and ion beam experiments have provided many examples of the cleavage of *both* C-H and C-C bonds of alkanes by transition-metal ions in the gas phase [2]. These gas phase studies have provided a surfeit of highly speculative reaction mechanisms. Conventional mechanistic probes, such as isotopic labeling, have served mainly to indicate the complexity of "simple" processes such as the dehydrogenation of alkanes [3,4]. More detailed studies using ion beam methods [5], multiphoton infrared laser activation [6] and the determination of kinetic energy release distributions [7,8],

J. A. Martinho Simões (ed.), Energetics of Organometallic Species, 287–320.
© 1992 Kluwer Academic Publishers.

have revealed important features of the potential energy surfaces associated with these reactions.

A wide range of reactant species have been investigated in gas phase studies of organometallic reactions, including atomic metal ions, organometallic fragments and complexes, and metal clusters. The majority of work by far has involved studies of atomic metal ions, since these reactants can, perhaps naively, be considered the "simplest" reaction centers with which to conduct fundamental studies. As a result this review will focus attention on these systems and only briefly consider reactions of ligated metals. The thermochemistry of organometallic species, especially individual bond dissociation energies, play a central role in these studies. Such data are necessary to understand periodic trends in reactivity, the energetic viability of proposed reaction intermediates, and product stabilities.

A starting point for considering the energetics of oxidative addition processes in general and hydrocarbon activation in particular is the potential energy surface or reaction coordinate diagram shown in Figure 1. In neutral systems a significant barrier to reaction is usually associated with the activation of C-H or C-C bonds at coordinatively unsaturated transition metal centers. Ions, on the other hand, interact with neutrals via charge-induced dipole forces creating a chemically activated adduct, often with sufficient excitation energy to overcome intrinsic barriers for insertion into C-H or C-C bonds, as shown in Figure 1.

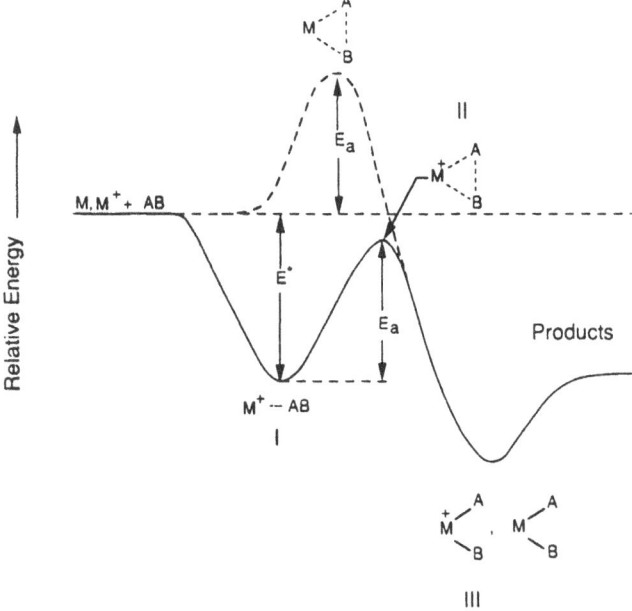

Figure 1. Generalized reaction coordinate diagram for the oxidative insertion of a metal atom or ion into the bond of a species AB, where A and B can represent molecular fragments.

The initial oxidative addition product III in Figure 1 may undergo further rearrangement, leading to the elimination of small molecules and the *exothermic* formation of stable organometallic products. These reactions are regarded as being *facile* if they occur efficiently at thermal energies. This implies that all interconnecting transition states are lower in energy than the reactants. Typically, elimination of molecular hydrogen, alkanes, and alkenes are facile processes. Reagent electronic or translational excitation can promote *endothermic* reactions, the simplest processes resulting from cleavage of the newly formed bonds in III. These processes typically yield radical products.

The experimental methods [2] which have been used in these studies include ion cyclotron resonance (ICR) mass spectrometry and its Fourier transform derivative (FT-ICR), flowing afterglow (FA), tandem mass spectrometry (or beam techniques), and kinetic energy release measurements. The first three techniques permit the measurement of reaction rates, k(T), and branching ratios at thermal energies. The beam method allows the measurement of the absolute probability for reaction as a function of kinetic energy [5,9]. This probability is given in terms of an energy dependent cross section $\sigma(E)$, the effective area which the reactants present to one another. Kinetic energy release measurements are, as the name implies, direct studies of the kinetic energy of the products of reaction [5,10,11]. These studies provide insight into the potential energy surface in the exit channel of the reaction.

The main strength of the beam method is the ability to vary precisely the relative kinetic energy of the reactants. This permits determination of thresholds for endothermic reactions, which in the absence of a barrier for the reverse process, directly yield reaction thermochemistry and heats of formation of organometallic fragments. In the case of exothermic reactions, which comprise the majority of identified reactions, ion beam experiments are no longer a useful technique for determining reaction thermochemistry. It is for just these cases, however, that the detailed analysis of kinetic energy release distributions can be a valuable source of mechanistic information as well as data relating to reaction exothermicities and metal ligand bond dissociation energies.

2. Experimental Methods

The majority of quantitative studies of the reactions of atomic transition metal ions with small molecules have been conducted by Armentrout and coworkers [12,13] using an ion beam apparatus of the type generalized in Figure 2, equipped with an octopole ion guide to constrain ions in the collision region. This instrument is a highly improved version of an earlier beam machine used in our laboratory. One of the keys to the success of these experiments has been the use of a surface ionization source shown in the inset of Figure 2. Metal halide salts are evaporated from a heated tube furnace onto a resistively heated rhenium ribbon (1800 - 2500 ^{O}K) where they thermally decompose and the resulting metal atoms are surface ionized. Atomic metal ions are extracted, mass analyzed in a magnetic sector, and allowed to interact with neutral molecules in a collision cell with relative kinetic energies in the range 0.01-10.0 eV. The collision region comprises either a simple gas cell (as used in many earlier experiments) or an octopole ion guide which passes through a chamber containing the neutral reactant of interest. Ion-molecule reactions occur in a well-defined region and at a pressure low enough that products are the result of single ion-neutral encounters. These experiments yield an absolute reaction cross section, $\sigma(E_T)$, a direct measure of the probability of the reaction at E_T. Cross sections are directly related to rate constants by k(T) = <σv>, where the brackets indicate integration over a Maxwell-Boltzmann distribution of velocities v. The ability of the ion beam experiment to measure $\sigma(E_T)$ rather than k(T) is its primary

290

distinguishing feature when compared with FA, ICR and FT-ICR methods. One advantageous feature of the beam technique is that the ion source is physically separated from the interaction region. This allows tremendous versatility in the means used to produce ions and has made possible the study of reactions of atomic metal ions in specific electronic states. The marriage of an instrument of this type with a drift cell to separate ions in different electronic states by their differences in mobility ("excited state chromatography") as developed recently by Bowers and Kemper [14]) should provide a powerful combination for studies of state specific reactivity in future investigations.

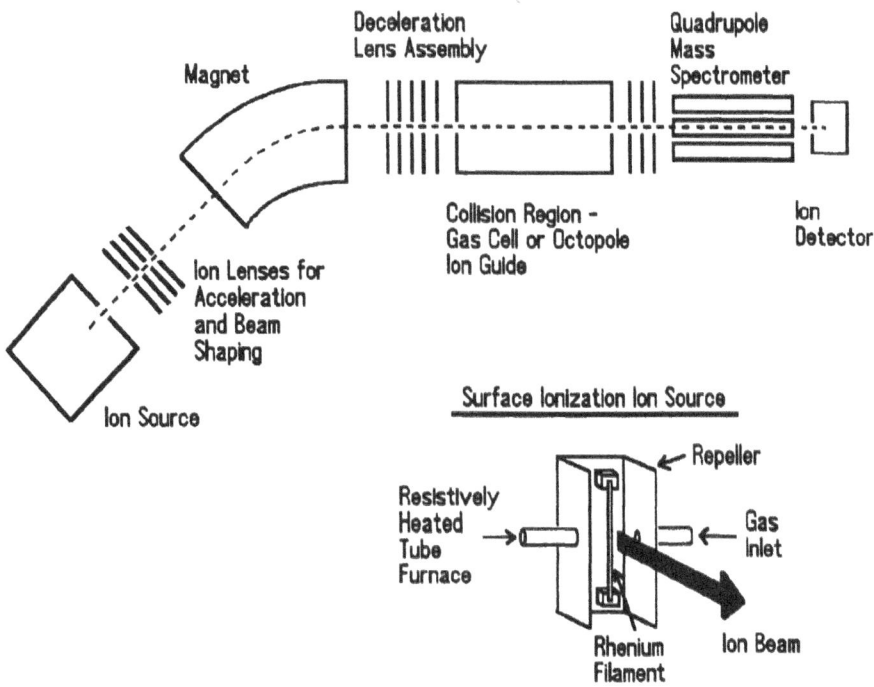

Figure 2. Schematic illustration of an ion beam apparatus for studies of ion molecule reactions. Ions are accelerated from the source, mass analyzed in the magnetic sector, and decelerated to the desired laboratory energy before entering the collision cell. Products are detected with a quadrupole mass spectrometer. The inset shows the surface ionization source used to generate atomic transition metal ions.

With the surface ionization source it is generally assumed that the reactant ion internal state distribution is characterized by the source temperature and that the majority of the reactant ions are in their ground electronic state. This contrasts with the uncertainty in reactant state distributions when transition metal ions are generated by electron impact fragmentation of volatile organometallic precursors [15] or by laser evaporation and ionization of solid metal targets [16]. Many examples have now been recorded of modified reactivities due to electronic excitation of atomic transition metal ions [17].

The apparatus and techniques of ion cyclotron resonance spectroscopy have been described in detail elsewhere [18]. Ions are formed, either by electron impact from a volatile precursor, or by laser evaporation and ionization of a solid metal target, and allowed to interact with neutral reactants. Freiser and co-workers have refined this experimental methodology with the use of elegant collision induced dissociation experiments for reactant preparation and the selective introduction of neutral reactants using pulsed gas valves [19]. Irradiation of the ions with either lasers or conventional light sources during selected portions of the trapped ion cycle makes it possible to study ion photochemical processes [20]. In our laboratory we have utilized multiphoton infrared laser activation of metal ion-hydrocarbon adducts to probe the lowest energy pathways of complex reaction systems [6]. Freiser and co-workers have utilized dispersed visible and UV radiation from conventional light sources to examine photochemical processes involving organometallic fragments [21]. The recent development of external ion source instruments makes it possible to combine high pressure or "dirty" sources with FT-ICR spectrometers employing superconducting magnets. Ions can be transferred through several stages of differential pumping and injected into a trapped ion ICR cell which is maintained under high vacuum. Schwarz and coworkers have used energetic particle bombardment techniques to generate atomic metal ions in experiments of this type [22]. Smalley's group has pioneered the combined use of FT-ICR spectroscopy and pulsed supersonic nozzle techniques to prepare and study the reactions of size selected transition metal cluster ions [23].

A very powerful technique for obtaining information relating to potential energy surfaces for organometallic reactions involves the determination of kinetic energy release distributions for product ions [7]. This experimental methodology, along with the development of theoretical models employing phase space theory to analyze measured distributions, has been highly refined by Bowers and coworkers at the University of California at Santa Barbara. The instrument used for these studies, a VG ZAB-2F reversed geometry double focusing mass spectrometer, is shown schematically in Figure 3.

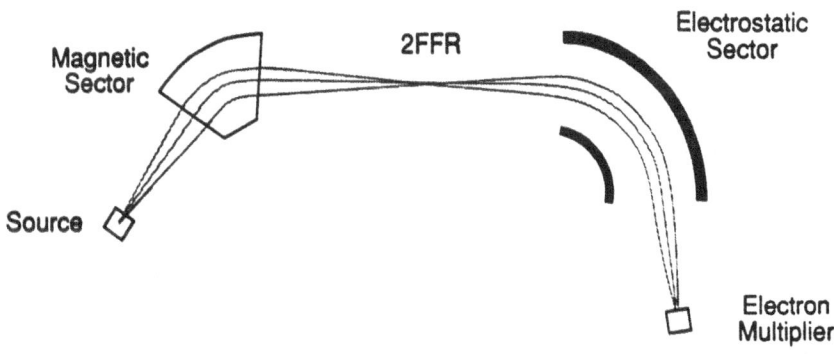

Figure 3. Reverse geometry double focusing mass spectrometer used for kinetic energy release measurements. Adducts are formed in the ion source and mass analyzed using the magnetic sector. The electric sector is used to determine product kinetic energies resulting from decomposition in the second field free region.

Adducts of atomic metal ions with neutral reactants are extracted from a high pressure ion source and mass analyzed using the magnetic sector. Metastable decomposition processes occurring in the second field free region between the magnetic and electric sectors are observed by scanning the electric sector. The kinetic energy release distributions for these processes are obtained by differentiating the measured peaks [24]. Since volatile organometallic complexes are used as atomic metal ion sources in these experiments, the possible role of excited state reactions are a major concern. Determining the electronic state of the metal ion has not been a problem for the systems studied thus far. Excited electronic states of cobalt ions can be relaxed if necessary by adding a collision gas to the ion source. $Co(CO)_3NO$ is known to give ~ 66% excited state Co^+ under electron impact conditions ($\geq$ 40 eV) [14]. Cobalt cyclopentadienyl dicarbonyl, however, gives 90% ground state Co^+ with electron impact ($\leq$ 40 eV). Greater signal intensities are obtained with the nitrosyl complex due to its higher vapor pressure and as a result it is normally employed. However, if excited states are suspected to be a problem, a comparison of results with both compounds can help to resolve ground state contributions to the products of interest. In the formation of a chemically activated intermediate starting with an excited state, the excess energy will most likely become available as vibrational excitation on the ground electronic surface of the system. This in turn yields higher dissociation rates. Even a minor excited state component can make a major contribution to the observed signal when the dissociation rate matches the temporal requirements for detection with high sensitivity. This is met when dissociation occurs in the second field free region with high probability.

In addition to the experimental methods pertinent to the work described herein, Farrar and co-workers have described a crossed beam investigation of the decarbonylation of acetaldehyde by atomic iron ions [25]. Studies of this type, which provide product energy and angular distributions, will provide detailed insights into the mechanisms of organometallic reactions. Several investigators, including Depuy [26], McDonald [27], Squires [28], and Weisshaar [29], have utilized flowing afterglow techniques to study organometallic reactions.

These experimental methods all complement each other in terms of the type of information they provide and the accessible range of experimental parameters such as pressure and temperature. Many of these experimental techniques yield thermochemical data for organometallic ions and neutrals.

3. Experimental Studies of Gas Phase Organometallic Reactions

3.1. ICR and Ion Beam Studies of the Reactions of Atomic Metal Ions

3.1.1. *Endothermic Reactions.* The reaction of atomic nickel ion with molecular hydrogen to yield NiH^+ is substantially endothermic. Reaction cross sections for this process, measured using the ion beam apparatus shown in Figure 2 (using a simple collision cell in this case), are displayed in Figure 4 for reactions 1 and 2 with HD as the neutral.

$$\begin{array}{lll} & \rightarrow \quad NiH^+ + D & (1) \\ Ni^+ + HD & & \\ & \rightarrow \quad NiD^+ + H & (2) \end{array}$$

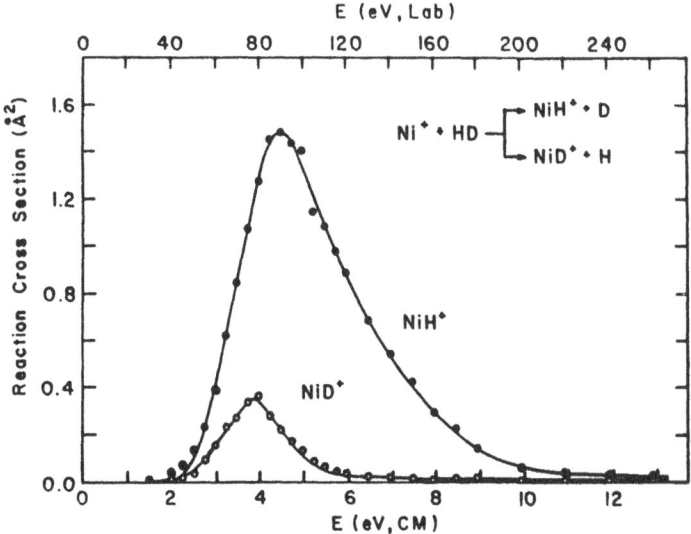

Figure 4. Variation of cross sections with kinetic energy for the reactions of Ni⁺ with HD.

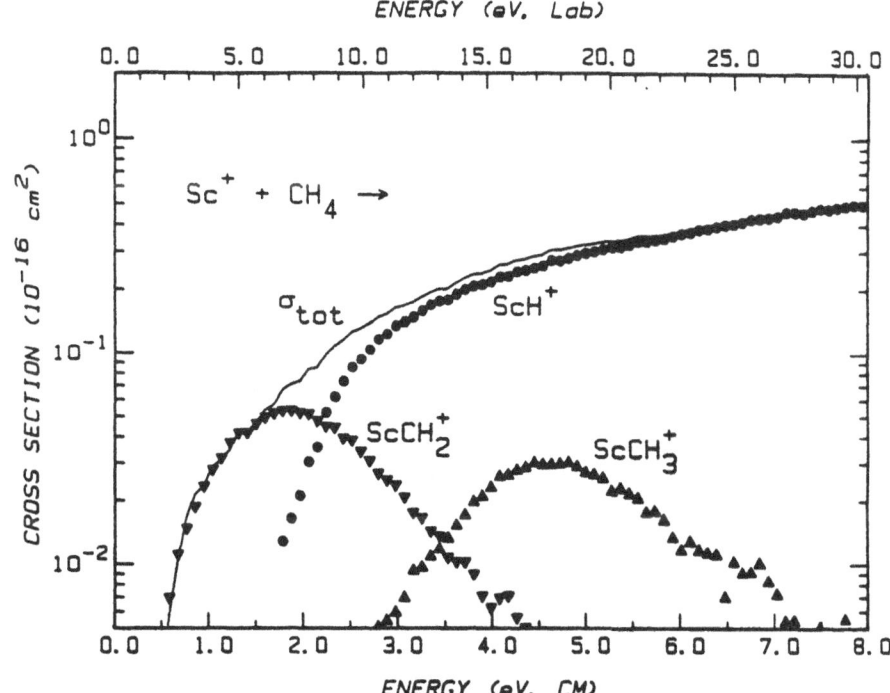

Figure 5. Variation of cross sections with kinetic energy for reactions of Sc⁺ with methane. Data from reference 5.

TABLE 1. Bond Dissociation Energies in Organometallic Fragments[a]

M	M^+-H	M^+-CH_3	M^+-CH_2	M^+-CH	M-CH_3	M-H
Sc	57(2)	59(3)	98(5)		32(7)	48(4)
Ti	54(3)	54(2)	93(4)	122(4)	46(7)	
V	48(2)	50(2)	80(3)	115(2)	37(9)	41(4)
Cr	32(2)	30(2)	54(2)	75(8)	41(7)	41(3)
Mn	48(3)	51(2)	71(3)		30(4)	30(4)
Fe	50(2)	58(2)	83(4)	101(7)	37(7)	46(3)
Co	47(2)	49(4)	78(2)	100(7)	46(3)	46(3)
Ni	40(2)	45(2)	75(2)		55(3)	58(3)
Cu	22(3)	30(2)	64(2)		58(2)	61(4)
Zn	55(3)	71(3)			19(3)	20(1)
Y	62(1)	59(1)	95(3)			
Zr	55(3)					
Nb	54(3)		109(7)	145(8)		
Mo	42(3)					53(5)
Ru	41(3)	54(5)				56(5)
Rh	36(3)	47(5)	91(5)	102(7)		59(5)
Pd	47(3)	59(5)				56(6)
Ag	16(3)					54(3)

[a]All values kcal mol[-1] at 300 K with uncertainties in parentheses. Values are derived mainly from ion beam studies, and taken from the summary in reference 30, which contains references to original citations.

An analysis of the threshold region for reaction 1 yields a Ni^+-H bond dissociation energy (Table 1). Determination of a threshold involves assuming a functional form for the variation of the cross section with energy in excesss of the threshold value and folding in the velocity distribution of the neutral target molecules (Doppler broadening) to fit the experimental data.

Examination of endothermic reactions in more complex systems has yielded a wide range of bond dissociation energies for neutral and ionic organometallic fragments [30]. For example, the reaction of atomic scandium ions with methane leads to three products as shown in Figure 5 and generalized in reactions 3-5.

$$M^+ + CH_4 \longrightarrow \begin{cases} MH^+ + CH_3 & (3) \\ MCH_2^+ + H_2 & (4) \\ MCH_3^+ + H & (5) \end{cases}$$

The data in Figure 5 are recorded with the guided ion beam apparatus described in the previous section. All three reactions are observed for the early first row atomic metal ions, Sc+, Ti+, V+ and Cr+. Carbene formation is not observed for the late metals Fe+, Co+ and Ni+. This difference in reactivity can be rationalized by electronic requirements for activation of hydrogen and methane, where the early metals react primarily by an insertion mechanism and the late metals react by a direct process [5].

3.1.2. *Exothermic Reactions.* Endothermic reactions, such as illustrated by the data in Figures 4 and 5, have total cross sections which at most are a few square angstroms. As noted in the introduction, facile exothermic reactions can occur with cross sections and reaction rates which approach theoretical collision limits, even at thermal ion energies.

Atomic cobalt ions react with methane and ethane only in processes which have substantial barriers (even though the dehydrogenation of ethane by Co+ is exothermic by 11 kcal mol^{-1}). Reactions with propane and isobutane give quite different results, however, with reaction cross sections varying with translational energy as shown in Figure 6 for isobutane. Loss of H_2 and CH_4 is observed at low energies in both systems (reactions 6-9).

$$Co^+ + C_3H_8 \begin{cases} \longrightarrow & Co(C_3H_6)^+ + H_2 \quad (6) \\ \longrightarrow & Co(C_2H_4)^+ + CH_4 \quad (7) \end{cases}$$

$$Co^+ + i\text{-}C_4H_{10} \begin{cases} \longrightarrow & Co(C_4H_8)^+ + H_2 \quad (8) \\ \longrightarrow & Co(C_3H_6)^+ + CH_4 \quad (9) \end{cases}$$

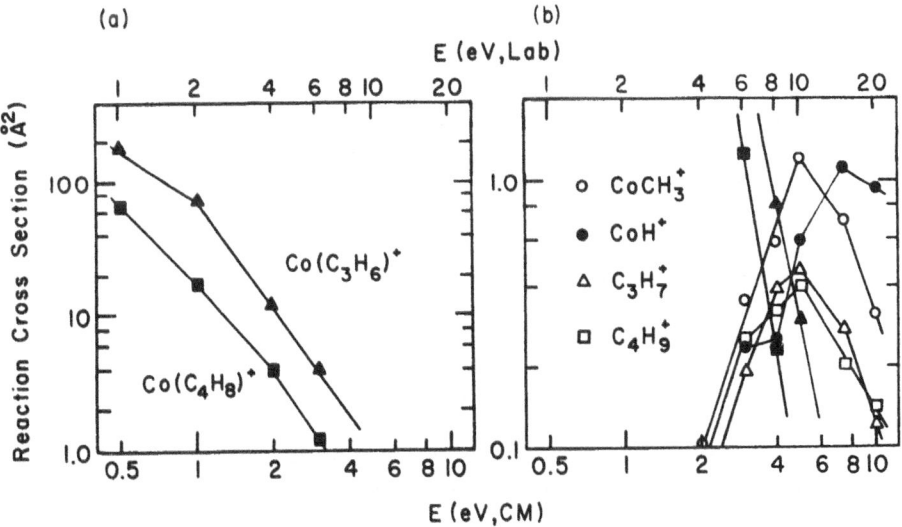

Figure 6. Variation of cross sections with kinetic energy for reaction of Co$^+$ with isobutane. (a) Products formed with large cross section at low energy are the result of exothermic reactions. (6) Endothermic products are observed with small cross sections at higher energy.

The total cross sections in Figure 6 are comparable to the predicted collision limits, and isobutane reacts efficiently on nearly every encounter. Reaction with propane, on the other hand, occurs on roughly 1 out of 10 collisions with Co⁺. The behavior of isobutane is typical of the reactions of larger hydrocarbons with Co⁺.

The mechanisms of hydrocarbon activation processes such as those indicated by reactions 6-9 in which not only C-H but also C-C bonds are cleaved have received a considerable amount of attention. Only recently has a clear picture of these processes begun to emerge, with details of the potential energy surfaces being revealed mainly by a careful theoretical analysis of kinetic energy release distributions with concomitant explanation of the observed reaction efficiencies. These results are described in greater detail below.

Scheme 1 for the reaction of Co⁺ with propane summarizes the simplest mechanisms for processes which can result in the elimination of H_2 and CH_4. Initial insertion of the metal ion into the primary or secondary C-H bond can be followed by β-hydrogen transfer and elimination of H_2. Either insertion into the primary C-H bond and β-methyl transfer or insertion into the C-C bond followed by β-hydrogen transfer and elimination of methane can lead to the other product in this system.

For reactions 6-9 to be facile, the overall process must be exothermic and no intermediate can be higher in energy than the energy of the reactants. The reaction coordinate diagrams shown in Figure 7 for these processes are consistent with this requirement. Evidence has been presented which supports the depicted mechanism for methane loss, namely that C-H activation is the first step in these processes [31]. The effect of the initial C-H insertion barrier on the kinetic energy release distribution is discussed in detail below (section 3.5.6).

As noted in the experimental section, ICR studies are typically not carried out at higher ion kinetic energies. An attempt is usually made to insure that observed reactions involve mainly ground state reactants at thermal energies. As a result the majority of the reactions studied using this technique have been exothermic processes. An interesting example of such reactions is the recent observation that several third row atomic metal ions can activate methane [32,33]. A trapped ion FT-ICR study of the sequential reactions of W⁺ with methane is shown in Figure 8. Reactions 10-13 occur in rapid succession, with rate constants of 1.2, 3.0, 1.8, and 1.0 x 10⁻¹⁰ cm³ molecule⁻¹ sec⁻¹, respectively.

$$W^+ + CH_4 \quad \rightarrow \quad W(CH_2)^+ + H_2 \qquad (10)$$

$$W(CH_2)^+ + CH_4 \quad \rightarrow \quad W(CH_2)_2^+ + H_2 \qquad (11)$$

$$W(CH_2)_2^+ + CH_4 \quad \rightarrow \quad W(CH_2)_3^+ + H_2 \qquad (12)$$

$$W(CH_2)_3^+ + CH_4 \quad \rightarrow \quad W(CH_2)_4^+ + H_2 \qquad (13)$$

These reactions are of interest for several reasons, not the least of which is the fact that methane activation is of considerable economic importance. None of the atomic metal ions from the first two transition series undergo related reactions at thermal energies in their ground electronic states. As noted above for Sc⁺, formation of the carbene can be promoted at higher translational energies for several metal ions. In addition, electronically excited chromium ions react efficiently with methane to yield a metal carbene [34].

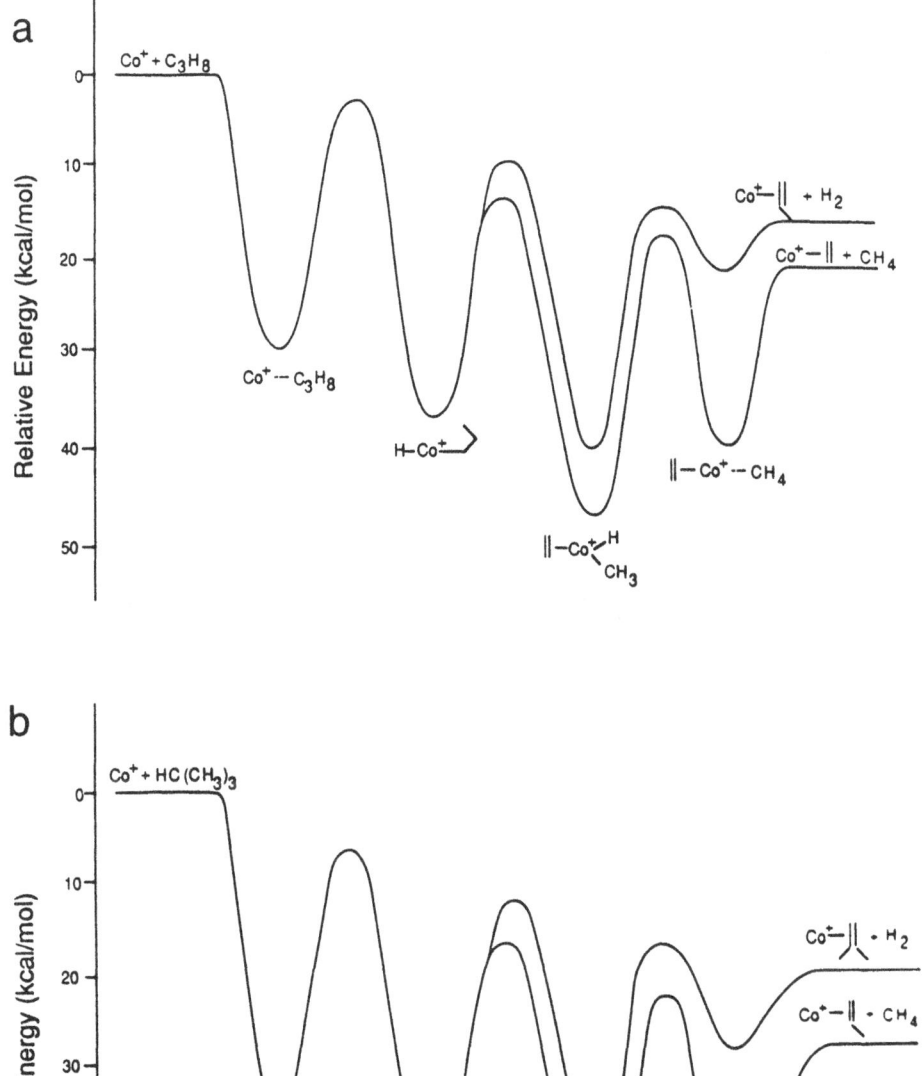

Figure 7. Reaction coordinate diagrams for the reactions of atomic cobalt ions with (a) propane and (b) isobutane.

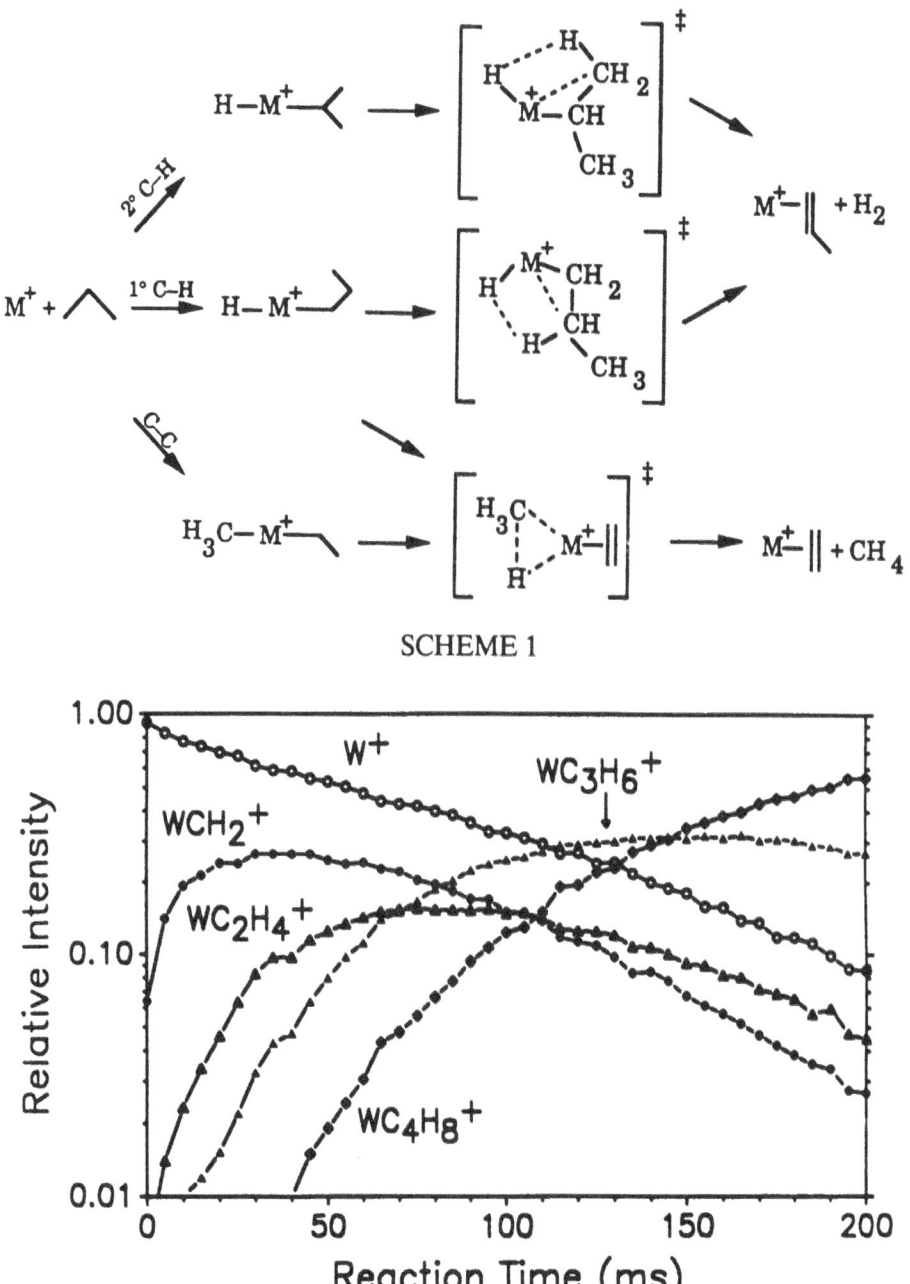

SCHEME 1

Figure 8. FT-ICR trapped ion study of the sequential reactions of laser desorbed atomic tungsten ions with methane. Methane pressure is 3.5×10^{-6} torr.

3.2.   Summary of Ion Thermochemical Data

Representative bond dissociation energy data for organometallic fragments are summarized in Table I. Several trends in these data warrant comment. Not surprisingly, Cr^+ forms unusually weak bonds to ligands such as H and CH_3. This results from the disruption of the unusually stable half filled 3d shell in the atomic ion to form a σ-bond. Periodic trends of the metal hydride ion bond dissociation energies are shown in Figure 9 for first, second, and third row metals, respectively, where experimental results are compared to the results of *ab initio* generalized valence bond dissociation consistent configuration interaction (GVB-DC-CI) calculations [35]. The agreement is generally very good, with the experimental results tending to be slightly higher than the theoretical values. A careful analysis of the wave functions indicates that first row metals use a mixture of the 4s and 3d orbitals to form strong σ-bonds. The second row metals form strong σ-bonds with only minor participation of the 5s orbital. It has been noted that the bond strengths can be correlated with the promotion energy from the ground state to the lowest state derived from a $4s^1 3d^n$ configuration, which is thus assumed to form a strong bond to hydrogen [36]. The correlation is somewhat better with a more appropriately defined promotion energy in which exchange terms are accounted for and suggests intrinsic metal hydrogen bond dissociation energies of around 60 and 64 kcal mol^{-1} for first and second row metals, respectively. The comparison of promotion energies with bond dissociation enthalpies has also been used to correlate metal-carbon dissociation enthalpies and to determine intrinsic bond enthalpies for first row transition metal-methyl, metal-carbene, and metal carbyne ions, 57, 101, and 130 kcal mol^{-1}, respectively. The successive increases of approximately 44 and 29 kcal mol^{-1} represent the additional bond strength associated with π-bonding in proceeding from single to double and from double to triple bonds, respectively. The value of 101 kcal mol^{-1} for the metal carbenes indicates that reaction 4 remains slightly endothermic even in the most favorable case for first row metals, the enthalpy change for the process $CH_4 \rightarrow CH_2 + H_2$ being 110 kcal mol^{-1}.

Rather limited experimental bond dissociation energy data are available for third row metals. At this point more is known about the carbenes than the metal hydrides. The available results for carbenes are compared to recent MR-CISD theoretical calculations with relativistic effective core potentials in Figure 10 (Irikura, K.K.; Goddard, W.A.; Beauchamp, J.L.; to be published). The calculated values appear low by around 10%, but the trend is reproduced reasonably well.

It is of particular interest to note that the metal ion methyl bond dissociation energies in Table I are often *greater* than the metal hydrogen bond energies. In part this can be attributed to the greater polarizability of the methyl group compared to hydrogen and the effect which this has on stabilizing the positive charge. Since $D^o[H\text{-}H]$ and $D^o[CH_3\text{-}H]$ are approximately equal, the relative nickel ion hydrogen and methyl bond strengths are confirmed by the observation that reaction 14, which is exothermic by 9 kcal mol^{-1}, of nickel hydride ion with methane is a fast process [37]. Reaction 14 is one of the few processes involving a first or second row metal ion in which methane is known to undergo a facile reaction at a transition metal center. The analogous reaction with FeH^+ is not observed, even though it is exothermic by 10 kcal mol^{-1} [38].

$$NiH^+ + CH_4 \rightarrow NiCH_3^+ + H_2 \qquad (14)$$

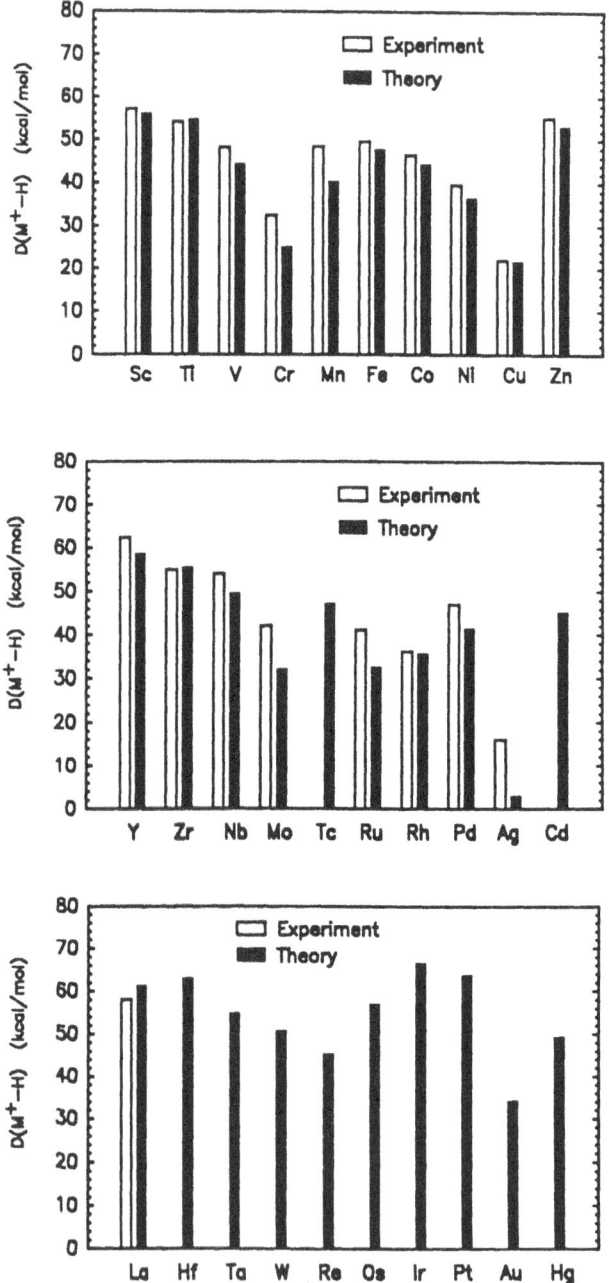

Figure 9. Periodic trends in experimental (where available) and theoretical transition metal hydride ion bond dissociation energies for first, second and third row metals. Theoretical and experimental results are in generally good agreement.

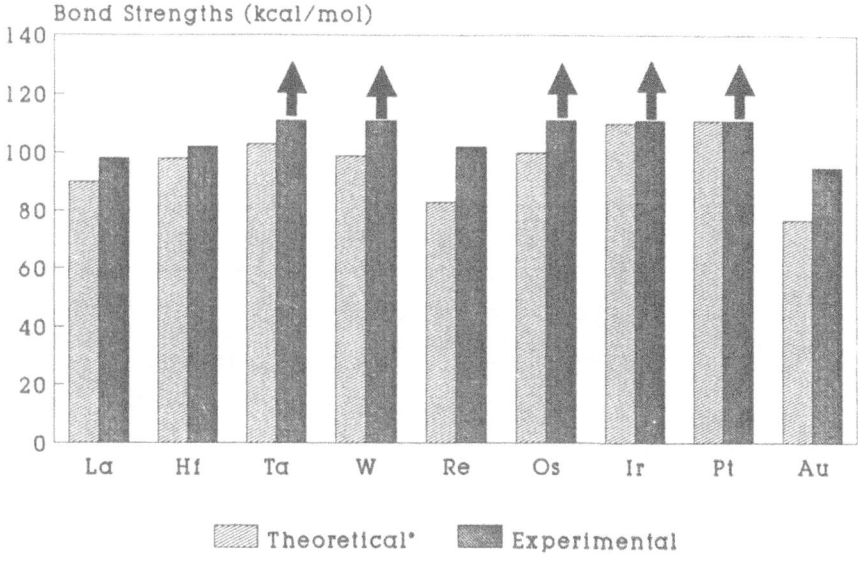

Figure 10. Comparison of theoretical and experimental (reference 33) third row metal ion carbene bond dissociation energies. The arrows indicate that only a lower limit is known to the experimental value.

The addition of several ligands to the metal system can reverse relative metal methyl and metal hydrogen bond dissociation energies. For example, reaction 15 is observed to be a fast exothermic process, indicating that the rhodium hydride bond energy in the product exceeds the rhodium methyl bond energy in the reactant ion [39].

$$CpRh(CH_3)(CO)^+ + H_2 \rightarrow CpRhH(CO)^+ + CH_4 \quad (15)$$

3.3. Periodic Trends in Atomic Metal Ion Reactivity

Figure 11 displays a periodic table in which the metal ions studied to date are classified as to their reactivity *in facile exothermic processes* with saturated hydrocarbons. All reactive metal ions will dehydrogenate hydrocarbons. The first row metal ions mainly remove a single H_2 molecule from the neutral species. Multiple dehydrogenation processes are more prevalent with second and third row metals. While the dehydrogenation of cyclohexane by rhodium ions to lose three hydrogen molecules in a single bimolecular collision is perhaps not surprising in view of the stability of the likely metal ion-benzene product, the mechanism of the reaction and the structure of the product in which three molecules of hydrogen are removed from *propane* by atomic niobium ions at thermal energies is not obvious. The latter process is reminiscent of the reactions of hydrocarbons on metal surfaces.

302

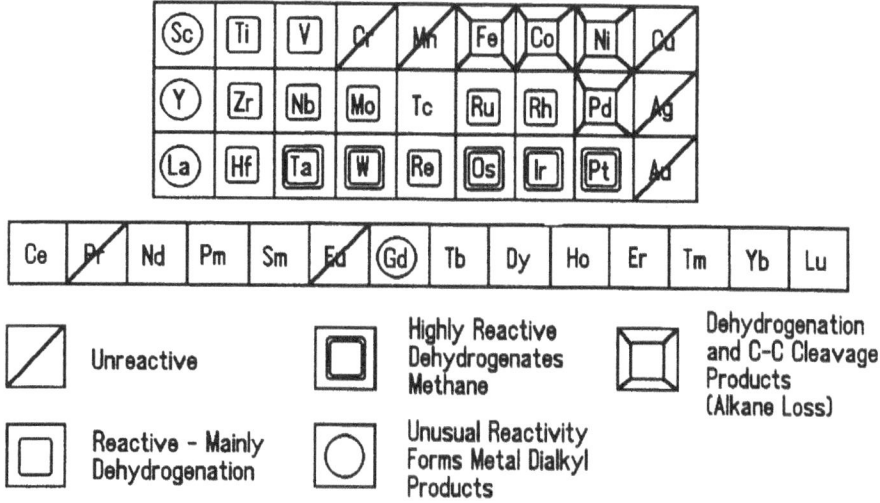

Figure 11. Periodic trends in reactivity of transition metal ions. Several lanthanides are also included for comparison.

The late first row metals iron, cobalt and nickel, along with palladium in the second row, are different in that cleavage of carbon-carbon bonds of saturated hydrocarbons to yield metal ion-olefin product ions with loss of smaller alkanes are prevalent and often dominant processes. It is still not known why this mode of reaction can sometimes be the preferred pathway. With reference to Scheme 1, a preference for insertion into C-C rather than C-H bonds, or a preference for β-alkyl over σ-hydrogen transfer, can lead to these alkane loss products.

Scandium, yttrium, lanthanum and gadolinium ions all exhibit unusual behavior in that saturated hydrocarbons are cleaved to yield metal dialkyl products with olefin loss [40,41]. A mechanism for this process is suggested in Scheme 2 for the reaction of Sc$^+$ with n-butane. The suggested reaction pathways are supported by isotopic labelling. Why do these four metal ions exhibit similar reactivity? The answer rests in an appraisal of the ground state electronic configurations of these ions. Sc$^+$, Y$^+$, and La$^+$ all have ground states derived from s^1d^1 configurations, with two valence electrons available to form σ-bonds. Gd$^+$ has a ground state derived from a $f^7s^1d^1$ configuration. In addition to the stability associated with a half filled shell in this case, the diffuse nature of f orbitals render them ineffective in forming strong σ-bonds. This group of atomic ions are thus different in that they are capable of forming a maximum of *two* strong σ-bonds. While oxidative addition processes are reasonable, the absence of additional σ-bonding electrons makes it inappropriate to suggest any subsequent reaction steps which require the formation of additional σ-bonds between hydrogen or carbon and the metal center.

The lack of reactivity of several of the transition metal ions can also be rationalized on the basis of their electronic structures. In the first row, Cr$^+$ and Cu$^+$ have ground states derived from stable d^5 and d^{10} configurations. Mn$^+$, with a ^7S ground state derived from an s^1d^5 configuration, can form one moderately strong

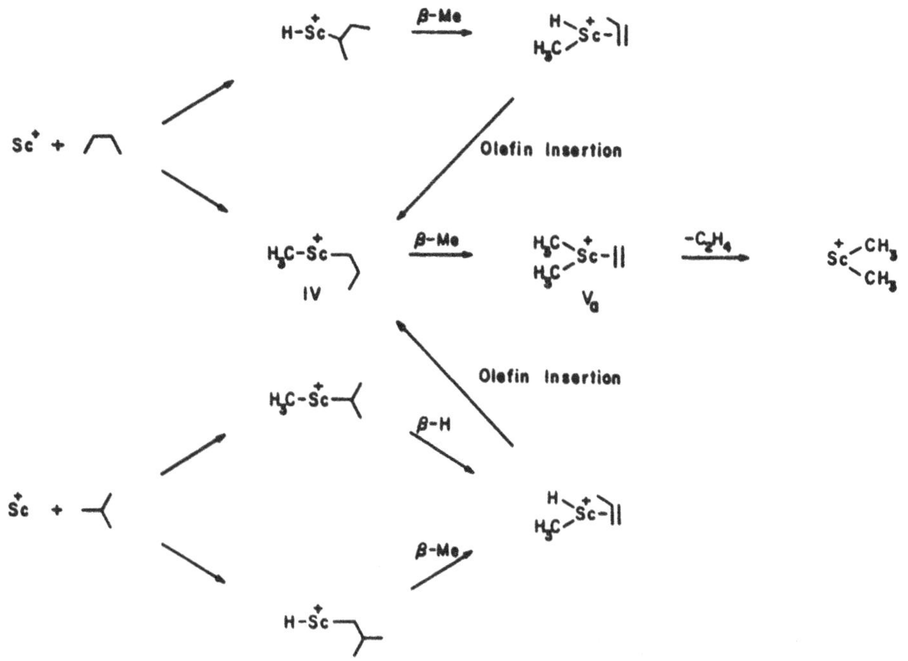

H-Sc⁺$\diagdown$⟨

β-Me →

H$\diagdown$Sc⁺-||
H₃C

Sc⁺ + ⟨⟩

Olefin Insertion

H₃C-Sc⁺⟩

IV

β-Me →

H₃C$\diagdown$Sc⁺-||
H₃C

Vₐ

-C₂H₄ →

Sc⁺$\diagdown$CH₃
$\diagup$CH₃

Olefin Insertion

H₃C-Sc⁺⟨

β-H

H$\diagdown$Sc⁺-||
H₃C

Ṡc⁺ + ⟨

β-Me

H-Sc⁺⟩

SCHEME 2

σ-bond, but cannot form a second without disruption of the d^5 subshell. Similar behavior might be expected for Mo⁺, since it also has a ^6S ground state derived from a d^5 configuration, and similar excitation energies to the first excited electronic state (^6D, d^4s^1 with state splittings of 1.52 and 1.59 eV for Cr⁺ and Mo⁺, respectively). Even though reaction cross sections are small, Mo⁺ reacts to dehydrogenate alkanes [42]. The difference in reactivity can be attributed to the difference in size of the Mo⁺ orbitals with respect to Cr⁺. The larger size of the d orbitals reduces the d-d exchange energy and increases the s-bond energies between the metal and carbon or hydrogen, allowing exothermic insertion into C-H and possibly C-C bonds.

The trend of increasing reactivity continues into the third row, with tungsten ions reacting as shown in Figure 8 even with methane. Due to the increased stability of the s orbitals in the third row, the ^6D ground state of W⁺ is derived from a d^4s^1 configuration, with the ^6S (d^5) state coming at higher energy. As noted in Figure 11 the additional third row metals Ta, Os, Ir and Pt all react with methane to eliminate H_2, presumably forming metal carbenes. As is the case for tungsten (Figure 8), several of the third row metal ions react sequentially with methane to yield a product ion with the formula $M(C_4H_8)^+$ which is stable with respect to further reaction. A variety of studies support a metallocyclopentane structure for this species. The reaction energetics associated with the formation of such a species are demanding, and require that the metal carbon bond dissociation energies in the product must *each* exceed 70 kcal mol⁻¹.

The only process in common to the reactive metal ions indicated in Figure 11 is the ability to dehydrogenate alkanes. Any hope of discerning a general mechanism for this process fades rapidly when deuterium labelled substrates are employed. For example, scandium, nickel and palladium ions dehydrogenate n-butane in highly specific and distinct processes (Scheme 3). It might naively be expected that the favorable dehydrogenation process would involve removal of H_2 from across the central C-C bond in a 1,2-process. For the three examples considered, this is observed cleanly only in the case of Pd^+ [17]. This particular metal ion has an unusually high Lewis acidity, which likely results in hydride abstraction as a first step in the dehydrogenation process and makes initial attack at the secondary position highly favorable. Nickel ions very cleanly remove hydrogen from both ends of the molecule (1,4-mechanism) [3]. and Sc^+ undergoes a 1,3-dehydrogenation process [40]. As noted above the unique reactivity of Sc^+ is attributed to the availability of only two valence electrons on the metal center and the restrictions which this imposes on the stability of reaction intermediates and products [40]. Although the multiple dehydrogenation processes make an examination of reaction specificity difficult, the ions Ti^+, V^+, Ru^+, and Rh^+ all appear predominantly by a 1,2 process [17, 43].

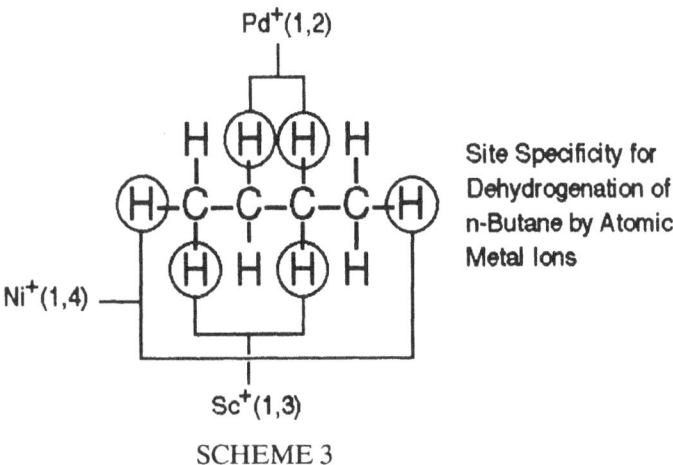

SCHEME 3

A particular dehydrogenation pathway may not be available for all substrates. Thus it is found that Co^+ ions, which preferentially remove hydrogen from longer chain hydrocarbons by a 1,4 process, dehydrogenate propane and isobutane cleanly by a 1,2-mechanism to yield propylene and isobutylene, respectively, as shown in Figure 7 [3].

3.4. Reaction Intermediates and Their Lifetimes

The lifetimes of reaction intermediates formed by interaction of metal ions with hydrocarbons are of particular interest. In several instances, quantitative studies of the lifetimes of reaction intermediates have been carried out by examining the collisional stabilization of adducts using a variety of buffer gases [44]. Intermediates formed by attachment of Ti^+ and V^+ to n-butane could be collisionally stabilized, with measured upper limits to the unimolecular decomposition rates of 1.2×10^7 sec^{-1} and 1.5×10^5 sec^{-1}, respectively. Studies of isotope effects have led to the inference that *the rate limiting step leading to product formation is the initial insertion of the metal ion into a C-H bond* in these systems.

The development of techniques to measure real time product formation (and reactant disappearance) rates would be an important advance since distinct intermediates would have different dissociation rates. It might thus be possible to discern different processes such as insertion into C-H and C-C bonds when they are competitive. Limited information can be gleamed from a comparison of metastable yields measured in the second field free region of the reverse geometry instrument shown in Figure 3 to product distributions measured with the ion beam apparatus. In the case of reactions 8 and 9, comparable yields are observed, suggesting a common intermediate. In the case of n-butane, however, methane loss is not as prominent relative to loss of hydrogen and ethane in the metastable yields. The methane loss process (probably involving initial insertion into the secondary C-H position as the rate limiting step) is faster than loss of hydrogen and ethane. The latter processes can both be accommodated by initial insertion into the primary C-H bond as a rate limiting step.

3.5. Kinetic Energy Release Distributions as a Probe of the Potential Energy Surfaces for Organometallic Reactions.

3.5.1. *General Considerations.* Statistical product kinetic energy release distributions can be used to determine gas phase bond dissociation energies for organometallic species. The amount of energy appearing as product translation can be used to infer details of the potential energy surfaces primarily in the region of the exit channel and has implications for the ease with which the reverse association may occur.

To illustrate how the amount of energy released to product translation for a given reaction pathway may reflect specific details of the potential energy surface, consider the two hypothetical surfaces in Figure 12. The interaction of a metal ion M^+ with a neutral molecule A can result in the formation of a chemically activated adduct, MA^+. In the absence of collisions, the internal excitation may be utilized for molecular rearrangement and subsequent fragmentation. In Figure 12, the adduct MA^+ is shown to dissociate along two different potential energy surfaces, designated Type I and Type II, yielding products MB^+ and C. For a reaction occurring on a Type I surface, a smooth transition in the exit channel, without a barrier for the reverse association, allows for complete energy randomization prior to dissociation. This results in a statistical product kinetic energy release distribution. A typical example is a simple bond cleavage. Statistical phase-space theory [44,45] has been successful in modeling translational energy release distributions for reactions occurring on this type of potential energy surface. In a statistical kinetic energy release distribution, the relative probability for a given product kinetic energy maximizes near zero and drops off rapidly with increasing energy as shown on the upper right hand portion of Figure 12. The average kinetic energy release for a large molecule will generally be much less than the total reaction exothermicity, ΔH, since the energy of the system in excess of that necessary for dissociation will be statistically divided between all the modes.

A reaction occurring on a Type II surface involves a barrier with an activation energy (E_{ar}) for the reverse association. In this case, the rate of product formation is too fast to allow for complete energy randomization, giving rise to a non-statistical kinetic energy release distribution. In the absence of coupling between the reaction coordinate and other degrees of freedom after the molecule has passed through the transition state, all of the reverse activation energy would appear as translational energy of the separating fragments. Accordingly, the translational energy release distribution would be shifted from zero by the amount E_{ar}. More typically however, some coupling does occur, yielding a broader

TYPE I (No Barrier for Reverse Association Reaction)

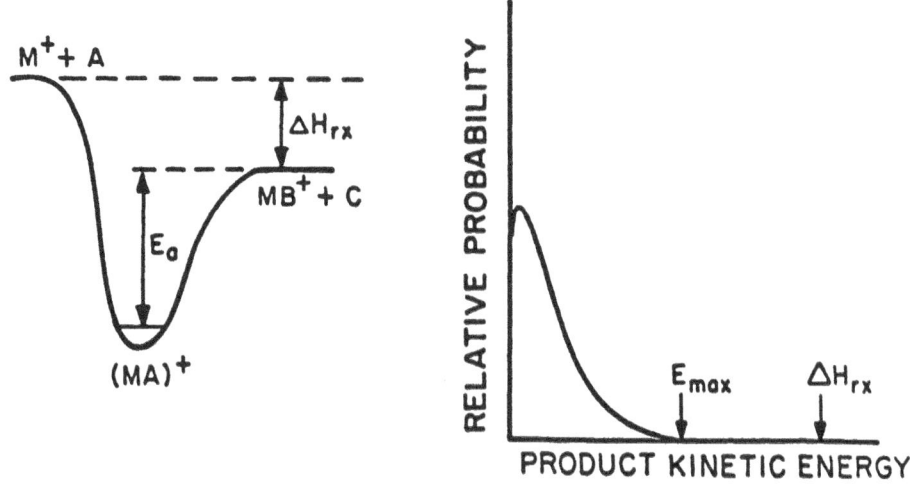

TYPE II (Large Barrier for Reverse Association Reaction)

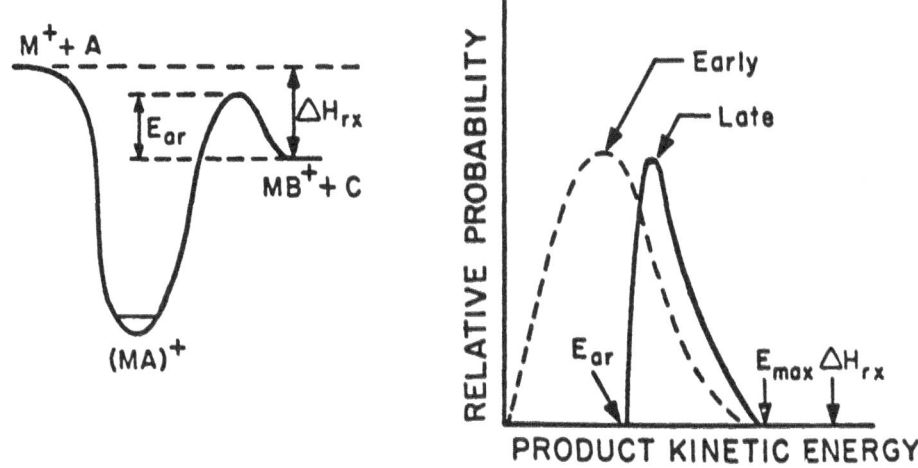

Figure 12. Relationship of kinetic energy release distributions to general features of the potential energy surface. For a type I surface (no barrier for reverse reaction), kinetic energy release distributions can often be modeled using statistical theory. Type II surfaces (large barrier for reverse association reaction) are more complex and often give large non-statistical kinetic energy releases.

distribution shifted to lower energy. This type of surface is often associated with complex reactions which involve the simultaneous rupture and formation of several bonds in the transition state. Statistical phase space theory is not applicable to these systems. With either potential energy surface, the maximum kinetic energy release, E_{max}, places a lower limit on the reaction exothermicity.

Complex reactions may involve the formation of several reaction intermediates which are local minima on the potential energy surfaces, separated by barriers. In such systems, the shape of the kinetic energy release distribution can be determined by the exit channel even though the rate of reaction will generally be determined by an earlier transition state. However, barriers near the asymptotic energy of the reactants can affect the kinetic energy release distribution by imposing dynamic constraints on the system, even when they are remote from the exit channel [46,47]. This is discussed further below.

Studies of kinetic energy release distributions have implications for the reverse reactions. Notice that on a Type II surface, the association of ground state MB^+ and C to form MA^+ cannot occur at thermal energy. In contrast, on a Type I potential energy surface the reverse association can occur to give the adduct MA^+, which does not have sufficient energy to yield the reactants M^+ and A. Although the reaction is nonproductive, it is possible in certain cases to determine that adduct formation did occur by use of isotopic labeling and observing isotopically mixed products or by collisional stabilization at high pressures.

3.5.2. *Detailed Analysis of Kinetic Energy Release Distributions for Type I Surfaces using Phase Space Theory.* The model for the statistical phase space theory calculations begins with Equation 16,

$$M^+ + A \quad \xrightarrow{F^{orb}(E,J)} \quad [MA^+(E,J)]^* \quad \xrightarrow{k_f(E,J)} \quad \text{products} \qquad (16)$$

where $F^{orb}(E,J)$ is the flux through the orbiting transition state of the formation reaction and yields the initial E,J distribution of $(MA^+)^*$. The fraction of ions decomposing in the second field-free region (2FFR) of the instrument, between the magnet and the electrostatic analyzer (ESA), is given for channel i, by

$$P_i(E,J,t_r) = \exp[-k_i(E,J)(t_1 + t_r)] - \exp[-k_i(E,J)(t_2 + t_r)] \qquad (17)$$

where t_r is the time spent in the ion source after formation of the adduct, t_1 the flight time from the ion source to the exit of the magnet (entry of the 2FFR), and t_2 the flight time from the ion source to the entrance of the ESA (exit of the 2FFR). The unimolecular rate constants, $k_i(E,J)$ are given by Equation 18

$$k_i(E,J) = F_i^{orb}(E,J) / \rho(E,J) \qquad (18)$$

where $F_i^{orb}(E,J)$ is the total microcanonical flux through the orbiting transition state leading to products in channel i, and $\rho(E,J)$ is the microcanonical density of states of the $[MA^+(E,J)]^*$ complex. The fraction of molecules at energy E and angular momentum J decaying through the orbiting transition state to yield products i with translational energy E_t is given by Equation 19,

$$P_i(E,J;E_t) = F_i^{orb}(E,J;E_t)/F_i^{orb}(E,J) \qquad (19)$$

where $F_i^{orb}(E,J;E_t)$ is the microcanonical flux which leads to products i with energy E_t. Finally, the fraction of molecules that decay via channel i rather than some other channel is given by the expression 20. Combining these equations,

$$\gamma_i(E,J) = k_i(E,J)/\Sigma_i k_i(E,J) \tag{20}$$

averaging over the initial Boltzman energy distributions of the reactants and the angular momentum distribution of the $[MA^+]^*$ collision complex, and normalizing yield the probability for forming products in channel i with translational energy E_t (Equation 21).

$$P_i(E_t) =$$

$$\frac{\int_0^\infty dE \exp(-E/k_B T) \int_0^{J_{max}} dJ 2J F^{orb}(E,J) P_i(E,J,t_r) P_i(E,J;E_t)\gamma_i(E,J)}{\int_0^\infty dE \exp(-E/k_B T) \int_0^{J_{max}} dJ 2J F^{orb}(E,J) P_i(E,J,t_r)\gamma_i(E,J)} \tag{21}$$

For simplicity, the term $\gamma_i(E,J)$ was set equal to unity. In special cases where the rate constant for one reaction channel may have a very strong J dependence relative to another, this term can have an effect. However, for all systems considered here the effect will be small. The expression for $P_i(E_t)$ should also be averaged over the distribution of source residence times, $P_i(t_r)$. This term has been shown to have an effect of only a few percent in similar systems and hence is usually ignored. Further, little is generally known about the rate determining transition states along the reaction coordinate for the organometallic systems studied. Hence for simplicity it is assumed that $P_i(E,J,t_r)$ is a constant. This is a reasonable assumption since the kinetic energy distribution will not depend strongly on the detection time window for dissociating MA^+ complexes which have a narrow range of internal energies relative to the reaction exothermicity. A number of kinetic energy release distributions have been measured as a function of the accelerating voltage, over the range which maintains reasonable sensitivity and resolution, from 3 kV to 8 kV. This changes the time window by a factor of 1.6. The distributions were identical, confirming this contention. For ion complexes formed with a broad range of internal energy the time window is taken into account and will be discussed further below.

In order to calculate kinetic energy release distributions, structures and vibrational frequencies for the various species are required. These are taken from the literature where possible, or estimated from literature values of similar species. The details of the kinetic energy release distributions are found to vary only weakly with structure or vibrational frequencies over the entire physically reasonable range for these quantities. The distributions are strongly dependent on the total energy available to the dissociating complex, and hence in our model to the ΔH^o of reaction. Often all heats of formation of product and reactants are well known except one, the organometallic product ion. This quantity can then be used as a parameter and varied until the best fit with experiment is obtained.

3.5.3. *Chemically Activated Species with Well Characterized Internal Energies.*
Atomic cobalt ions react with isobutane to yield two products as indicated in
reactions 8 and 9, which involve the elimination of hydrogen and methane to yield
cobalt ion complexes with isobutylene and propylene, respectively [8]. Kinetic
energy release distributions for these processes are shown in Figures 13 and 14. An
attempt to fit the experimental distribution for dehydrogenation of isobutane by
Co^+ with phase-space theory is included in Figure 13. The observed disagreement
supports a Type II surface for this process. Most studies to date of the
dehydrogenation of alkanes by group 8, 9 and 10 first row metal ions exhibit kinetic
energy release distributions which are characterized by large barriers for the reverse
association reactions. This is consistent with the failure to observe the reverse
reaction as isotopic exchange processes when D_2 interacts with metal olefin
complexes.

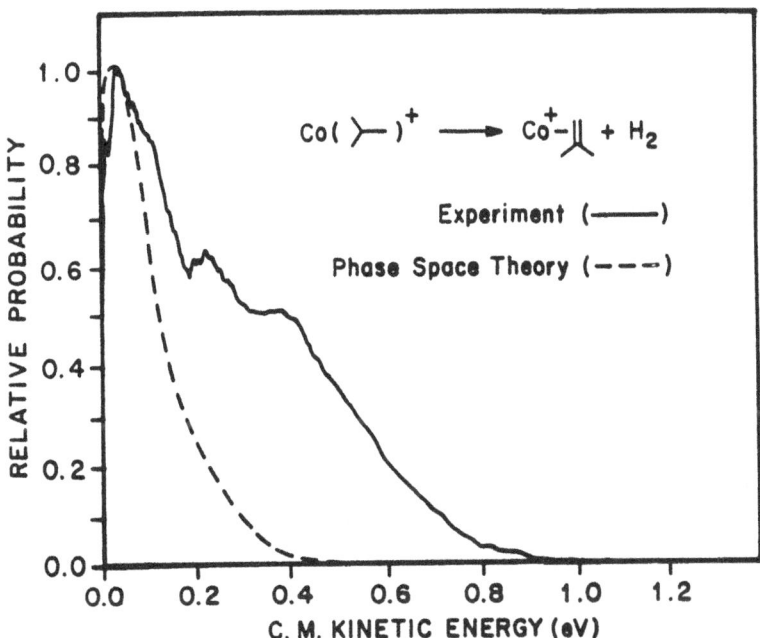

Figure 13. Kinetic energy release distribution for loss of hydrogen in the reaction of
atomic cobalt ions with isobutane. The observed energy release is higher than
predicted by statistical theory.

In contrast to the results obtained for dehydrogenation reactions, kinetic
energy release distributions for alkane elimination processes can usually be fit with
phase space theory. Results for the loss of methane from reaction 8 of Co^+ with
isobutane are shown in Figure 14. In fitting the distribution calculated using phase
space theory to the experimental distribution the single important parameter in
achieving a good fit is the reaction exothermicity, which in the case of reaction 9
depends on the binding energy of propylene to the cobalt ion in the reaction
products. As shown in Figure 14, a best fit is achieved with a bond dissociation
energy of 1.91 eV at 0 ^{O}K (2.08 eV or 48 kcal mol^{-1} at 298 ^{O}K). An analysis of the
kinetic energy release distribution for reaction 22 of Co^+ with cyclopentane [8]

yields an identical value for the binding energy of propylene to an atomic cobalt ion.

$$Co^+ + cyclo\text{-}C_5H_{10} \rightarrow Co(C_3H_6)^+ + C_2H_4 \qquad (22)$$

Bond energies derived from these and other studies [48] of this type are summarized in Table 2.

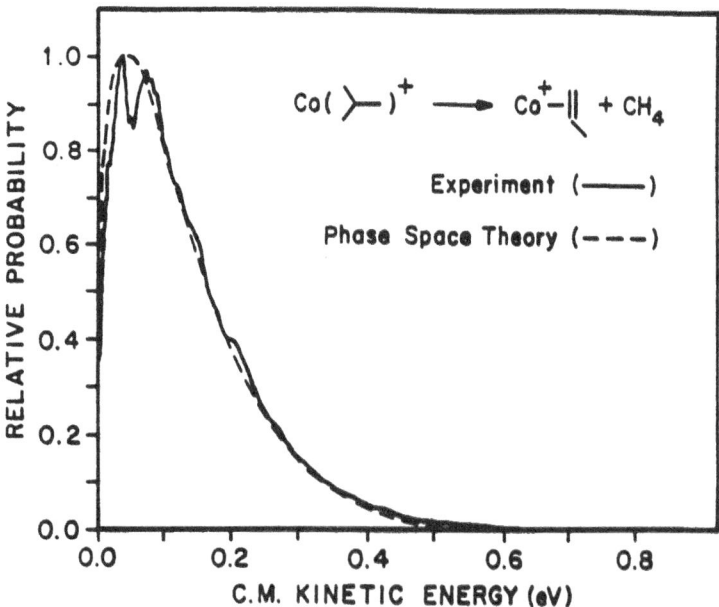

Figure 14. Kinetic energy release distribution for loss of methane in the reaction of atomic cobalt ions with isobutane. Statistical phase space theory calculations match the experimental distribution.

As is the case for reaction 22, elimination of a π-donor or n-donor base from a coordinately unsaturated metal center would generally be expected to proceed with a Type I potential surface (no barrier for the reverse association reaction). The validity of the phase space analysis for such processes is not surprising. The statistical theory analysis for alkane elimination (e.g. reaction 9) indicates a loose transition state is operative in the exit channel. This requires that the alkane being eliminated is strongly interacting with the transition metal center, almost certainly via a significant energy well on the potential energy surface prior to product formation. This suggests an important feature of the reaction coordinate diagram indicated in Figure 7 for the reaction of Co$^+$ with isobutane. The kinetic energy release distribution for methane elimination is determined entirely by the dissociation of the methane adduct, and *is not useful in identifying the presence or determining the height of a reverse activation barrier which might be associated with oxidative addition.* For this reason, the features in the exit channel for methane elimination are represented by a dashed line. The substantial kinetic energy release observed for hydrogen elimination and the failure of a statistical analysis to reproduce the observed distribution is a clear signature of the barrier shown in the

Table 2. Summary of Thermochemical Data Derived from Measurement and Analysis of Kinetic Energy Release Distributions.

Species	$D_0^0(D_0^{298})^a$	ΔH_{f0}^0(kcal mol^{-1})	Reference
Co$^+$-CO	31(34)	224	8
Co$^+$-(ethene)	42(46)	255	c
Co$^+$-(propene)	44(48)	247	8
Co$^+$-(CH$_3$)$_2$	105^b(110)	247	8
Co$^+$ (cyclobutane ring)	81^b(86)	274	48
Fe$^+$-CO	26(30)	227	48
Fe$^+$-(ethene)	35(39)	260	c
Fe$^+$-(propene)	37(41)	252	c
Fe$^+$ (cyclopentadiene ring)	50(56)	265	50
Fe$^+$ (cyclohexadiene ring)	66(72)	246	50
Fe$^+$ (cyclobutane ring)	85(90)	268	48

^{a}The error on values of D_0^0 is on the order of $\pm$ 5 kcal mol^{-1}, reflecting the sensitivity of the fit between theory and experiment. The value in parenthesis is the estimated value for 298 K.
^{b}The bond energy is the sum of the two bonds forming Co$^+$ and 2CH$_3$ from Co(CH$_3$)$_2^+$ and Co$^+$ and trimethylene from cobaltacyclobutane ion.
^{c}Values quoted in reference 48 (van Koppen, P.A.M.; Bowers, M.T.; Beauchamp, J.L., unpublished results).

exit channel for this process in Figure 7b. Detailed considerations of isotopically labeled species suggest that the low energy component of the distribution results from initial insertion into the primary C-H bond, and the high energy component results from attack at the secondary position.

The initial well associated with the formation of an adduct of the Co$^+$ with isobutane is included in Figure 7b. The chemical activation associated with the formation of such an adduct is essential in overcoming the intrinsic barrier associated with insertion into a C-H bond. In comparison to larger hydrocarbons, the weaker interaction of ethane with first row group 8-10 metal ions is apparently insufficient to overcome intrinsic barriers for insertion. This would explain the failure to observe dehydrogenation of ethane by these metal ions, even though the process is known to be exothermic. The reactions of the isomeric butanes and larger hydrocarbons with cobalt ions and with other reactive metal ions as well are generally facile. Reaction rate constants and cross sections approach the Langevin or collision limit. Propane is intermediate in behavior between isobutane and ethane, and this gives rise to interesting behavior which is further discussed below.

It should be apparent that phase space fitting of kinetic energy release distributions yields important thermochemical information for exothermic reactions with no reverse activation barrier. As another example, Co^+ ions decarbonylate acetone (reaction 23) yielding a dimethyl cobalt ion as

$$Co^+ + (CH_3)_2CO \rightarrow Co(CH_3)_2^+ + CO \qquad (23)$$

the product. A phase space theory fit of the kinetic energy release distribution for this process indicates the total energy of process 24 is 4.55 eV at 0 K (4.77

$$Co(CH_3)_2^+ \rightarrow Co^+ + 2CH_3 \qquad (24)$$

eV or 110 kcal mol^{-1} at 298 K) [8]. Ion beam threshold measurements indicate the bond strength of Co^+-CH_3 is 49.1 kcal mol^{-1} [49]. From our measurement we obtain, by difference, a value of 61 kcal mol^{-1} for the second bond energy. One interesting conclusion that can be drawn from this result is that insertion of Co^+ into a C-C bond in saturated alkanes should be exothermic by about 20 kcal mol^{-1}. However, the fact that the reaction is exothermic does not guarantee that the process will be observed.

3.5.4. *Examples of Chemically Activated Systems where the Reverse Reaction is Known to be Facile.* The dehydrogenation of cyclopentene by atomic iron ions (Reaction 25), is known to be reversible [50]. In the presence of D_2, the product of reaction 25,

$$Fe^+ + cyclopentene \rightarrow Fe(C_5H_6)^+ + H_2 \qquad (25)$$

presumed to be an iron cyclopentadiene complex, rapidly undergoes isotopic hydrogen exchange. This suggests that there is no barrier for the addition of H_2 to $Fe(C_5H_6)^+$, which corresponds to the first step in the reverse of reaction 25. In accordance with the considerations outlined in the previous section, a statistical kinetic energy release distribution should be observed for this system. As shown in Figure 15, the experimental distribution for this process can be fit very closely using statistical phase space theory, which yields a bond dissociation energy $D_0^0(Fe^+$-$C_5H_6) = 50 \pm 5$ kcal mol^{-1}. A reaction coordinate diagram for this system is shown in Figure 16. The initial interaction of the metal ion with the olefin leads to a chemically activated intermediate with around 35 kcal mol^{-1} internal energy. Insertion into the allylic C-H bond is followed by β-hydrogen transfer and elimination of H_2. The reverse of the final steps accomodates isotopic exchange with D_2. The overall reaction is exothermic by 30 kcal mol^{-1} due to the strong binding of cylopentadiene to Fe^+. This example is particularly interesting in that dehydrogenation of alkanes by Fe^+ and other group 8 metal ions generally yields non-statistical kinetic energy release distributions. Apparently a full range of behavior can be observed, depending on the ligand environment. When bound to cyclopentadiene, oxidative addition of hydrogen to the metal center is facile. This is not generally the case when Fe^+ is ligated by a single olefin.

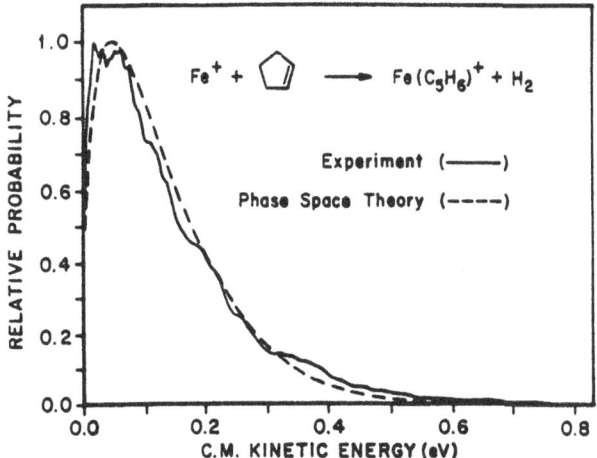

Figure 15. Kinetic energy release distribution for the dehydrogenation of cyclopentene by atomic iron ions. Statistical phase space calculations provide a reasonable fit to the experimental distribution.

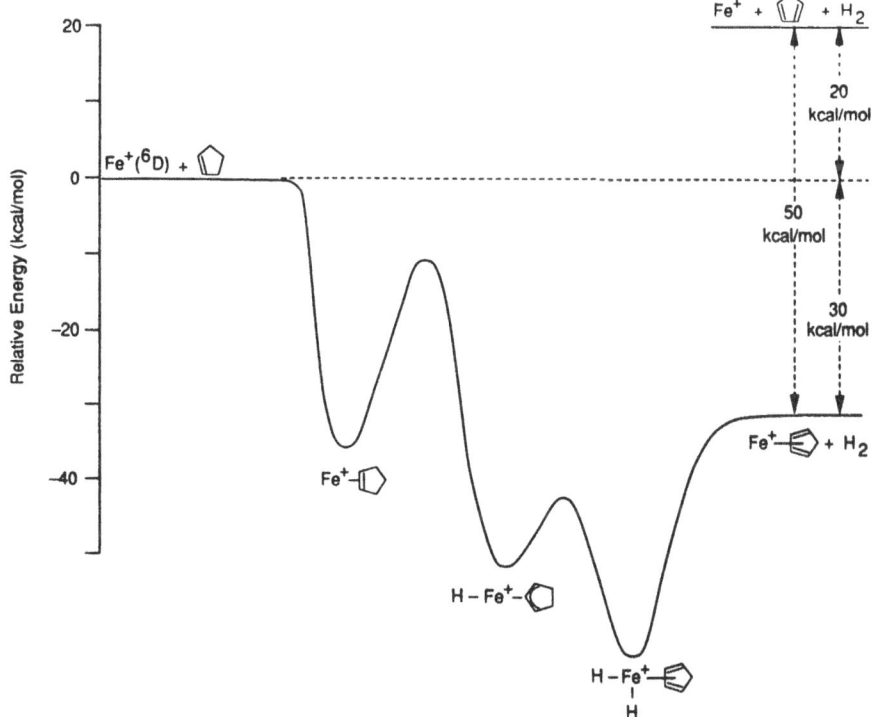

Figure 16. Reaction coordinate diagram for the dehydrogenation of cyclopentene by atomic iron ions.

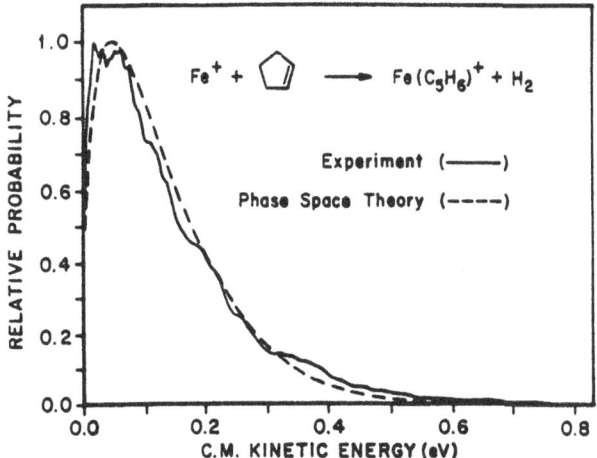

Another example of a hydrogen elimination process which is both reversible and exhibits a statistical kinetic energy release distribution is the dissociation of the cyclopentadienyl rhodium isopropyl ion, reaction 26 [51]. In this case the reaction of the product ion, presumed to be a π-allyl species, with D_2 leads to the incorporation of four deuterium atoms into the complex.

$$CpRhCH(CH_3)_2^+ \rightarrow CpRh(C_3H_5)^+ + H_2 \qquad (26)$$

3.5.5. *Quantitative Studies of Ions Formed with a Broad Range of Internal Energies.* The above studies have all involved chemically activated systems in which the internal energy of the decomposing ion is defined within a narrow range. In all of our earlier investigations we had worked with the assumption that the energy content of the decomposing ion had to be precisely known in order to extract quantitative results from the application of phase space theory to reproduce kinetic energy release distributions. We now realize that if the method of ion formation produces a broad distribution of internal energies, then the experiment will select a particular internal energy (since the time window for observation is dictated by the ion flight time through the second field free region of the VG ZAB-2F reversed geometry mass spectrometer). Statistical theory is used to accurately calculate the internal energies of ions being observed, and this energy is then used to analyze the kinetic energy release distribution to extract the enthalpy change for the observed process. This technique has been used, for example, to determine the sequential bond energies of the metal carbonyl ions $Mn(CO)_x^+$ (Table 3), and appears to be generally applicable for determination of the thermochemical properties of both ionic and neutral species.

Table 3. Bond Dissociation Energies for Manganese Carbonyl Ions

Species $Mn(CO)_x^+$	$D[Mn(CO)_{x-1}^+\text{-}CO]^a$
$Mn(CO)_6^+$	32 ± 5
$Mn(CO)_5^+$	16 ± 3
$Mn(CO)_4^+$	20 ± 3
$Mn(CO)_3^+$	31 ± 6
$Mn(CO)_2^+$	$<25^b$
$Mn(CO)^+$	$>7^b$

[a] kcal mol^{-1} at 0 K
[b] The sum of the last two bond energies is 32 kcal mol^{-1}. This is determined from the known heat of formation of $Mn(CO)_5^+$ by subtracting the first three bond energies for this species.

Manganese carbonyl ions are of particular interest since the spin state of the system changes from septet for Mn$^+$ to singlet for the fully coordinated Mn(CO)$_6^+$. With the exception of the fully coordinated species, all of the Mn(CO)$_x^+$ ions undergo rapid exchange of CO with labelled ^{13}CO. Not surprisingly, then, the Mn(CO)$_x^+$ species with x = 3-6 all exhibit kinetic energy release distributions which appear statistical. These ions, however, were formed by electron impact fragmentation of various precursor ions and as a result had a broad distribution of internal energies. According to the phase space description, the statistical kinetic energy release distribution is a strong function of the amount of energy above threshold in the energized molecule, E‡, which in the dissociation of the ions is just the difference between the bond dissociation energy and E*, the internal energy of the adduct. The lifetime requirements for the observation of metastables are stringent. They must decompose during the time they are in the second field-free region, a window of approximately 5-20 μsec after extraction from the source. The lifetime requirement in turn will serve to *select* ions with a narrow range of E* which are detected with high sensitivity. If E* is too high, ions will decompose while still in the source region. If E* is too low, ions will reach the detector without decomposing. These considerations can be quantified using statistical kinetic theory to determine decomposition rates as a function of ion internal energy. These calculations yield values of E* which can be employed in fitting statistical phase space theory to experimental kinetic energy release distributions. Typical results are shown in Figure 17 for x = 4.

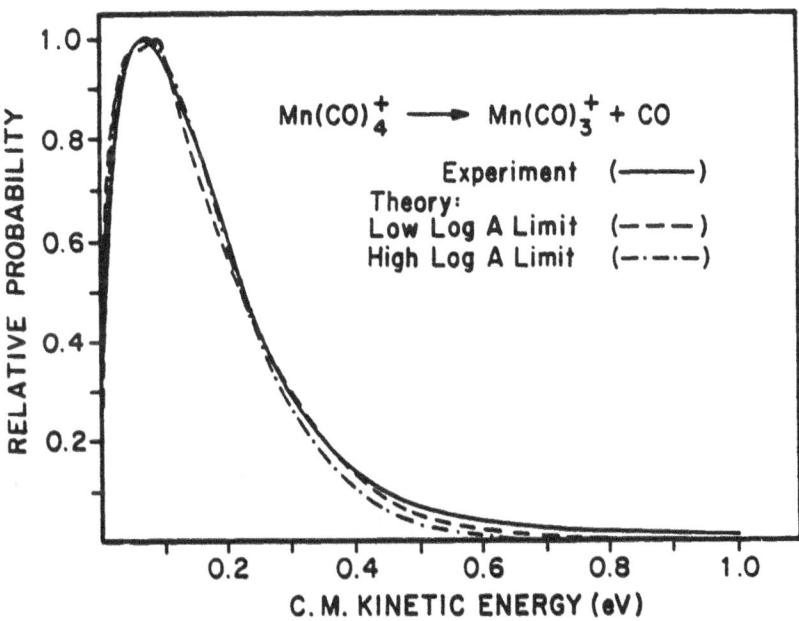

Figure 17. Kinetic energy release distributions for loss of CO from Mn(CO)$_4^+$.

The two calculated curves in Figure 17 correspond to logA values of 15.3 and 13.0, which in turn yield bond dissociation energies of 23 and 17 kcal mol^{-1}, respectively. These results are arrived at by iteration, with the major uncertainty in the analysis being the A factors for the reactions. The values chosen represent a range which likely includes the actual value. As a result the bond energy for the example given is taken to be 20 ± 3 kcal mol^{-1}.

The two smallest ions, $Mn(CO)_2{}^+$, and $MnCO^+$ decompose too rapidly to be detected as metastables. It is of interest that this analysis works best with large ions, which require large excess energies to decompose within the temporal constraints imposed by the experiment. This in turn gives large and easily measured kinetic energy releases.

3.5.6 *Coupling of Multiple Transition States by Dynamical Constraints: Effect on Kinetic Energy Release Distributions.* In cases where exoergic reactions are inefficient, i.e. the reaction cross section is significantly less than the collision cross section, the rate determining transition state must be located and included in the theoretical model. Qualitatively, an initial electrostatic well followed by a tight transition state, near in energy to the reactant energy (as shown in Figure 1), can restrict the flow of reactants to products. Quantitative statistical phase space theory modeling using this simple idea has been done for a number of systems, including reactions of Co$^+$ with propane as noted above. Both dehydrogenation and demethanation of Co(propane)$^+$ are known to be exoergic but inefficient reactions, the sum of them occurring at 8% of the collision rate.

The reaction can be schematically written as in Equation 27.

$$M^+ + C_3H_8 \quad \underset{k_{orb}}{\overset{k_{collision}P(E,J)}{\rightarrow}} \quad [M(C_3H_8)^+]^* \overset{k^{\ddagger}(E,J)}{\rightarrow} \text{products} \qquad (27)$$

The rate constant for product formation can be written as indicated in Equation (29)

$$k_{products} = k_{collision} \int_E \int_J P(E,J)[k^{\ddagger}(E,J)/[k^{\ddagger}(E,J) + k_{orb}(E,J)]dEdJ \qquad (28)$$

where $k_{collision}$ is the Langevin collision rate constant for formation of $[Co(C_3H_8)^+]^*$, $P(E,J)$ is the distribution of energy and angular momentum states of $[Co(C_3H_8)^+]^*$ created by passage of the reactants through the orbiting transition state in the entrance channel and k_{orb} and $k^{\ddagger}(E,J)$ are, respectively, the microscopic rate constants for passage through the orbiting transition state leading back to reactants, and the tight transition state that eventually leads to products.

The probability of product formation is simply $k_{products}/k_{collision}$. This ratio is calculated using statistical phase space theory as a function of the barrier height of the rate limiting transition state. A barrier 0.11 eV below the asymptotic energy of the reactants reproduces the experimental probability of product formation for the inefficient Reaction 7. The kinetic energy release distribution for methane elimination is *narrower* than statistical if an orbiting transition state in the entrance channel is assumed (shown as the *unrestricted* phase space kinetic energy release distribution in Figure 18). The kinetic energy release distribution was recalculated by implementing a tight transition state for insertion into a C-H bond located 0.11 eV below the asymptotic energy of the reactants. The resulting distribution fits well with the experimental distribution and is shown as the *restricted* phase space calculations in Figure 18.

If the microcanonical (E,J) values of the $k_{products}/k_{collision}$ calculation are examined it becomes clear that the principal effect of the C-H insertion transition state is to bias against formation of products from collisions with high J values. This is a reasonable result since angular momentum increases the energy of the tight transition state more than it does the loose orbiting transition state. Loss of the higher J portion of the (E,J) distribution results in products with less kinetic energy than would have been expected if the full (E,J) distribution were operative. This

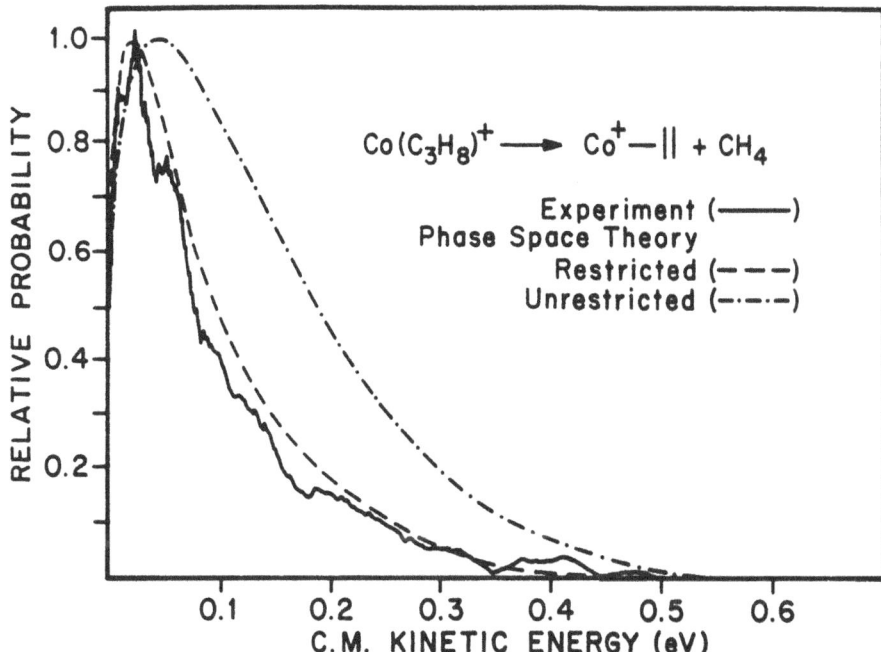

Figure 18. Kinetic energy release distribution for loss of methane in the reaction of atomic cobalt ion with propane. The experimental distribution is "cold" relative to the predictions of phase space theory with no consideration of a barrier in the entrance channel. Dynamical constraints imposed by the barrier result in better agreement of calculated and experimental distributions ("restricted" model).

example is presented to express a cautionary note in analyzing kinetic energy release distributions. If the dynamical constraints had not been considered in fitting phase space theory to the data in Figure 18, an erroneously low bond dissociation energy would have resulted. Inefficient reactions which proceed at only a fraction of the collision rate can signal the possible importance of these factors.

4. Summary and Conclusion

The range of experimental techniques available for studying the energetics and mechanisms of organometallic reactions in the gas phase are able to provide information relating to details of the potential energy surfaces for these processes. However, even the simplest reactions have very complicated potential energy surfaces and some significant mysteries remain concerning the course of these chemical transformations. Whereas theoretical calculations have provided a detailed understanding of the energetics and electronic structures of various organometallic fragment ions, they have still not provided useful insights into the potential surfaces for processes such as insertion into C-H and C-C bonds. Recent calculations have considered these processes for neutral atoms [52], but similar results for ions are not yet available. Theory predicts that cobalt, nickel, rhodium and palladium atoms have a barrier for C-C insertion which is 14-20 kcal mol^{-1} higher than the barrier for C-H insertion. This is explained in part by the difference in directionality between bonds to methyl groups and to hydrogen atoms. The overall process is not favorable, mainly as a result of the predominance of $d^n s^2$ configurations contributing to the ground electronic states of the metal atoms considered. Unlike the situation for atomic transition metal ions, the chemical activation resulting from interaction of these metal atoms with hydrocarbons is insufficient to overcome the intrinsic barriers associated with insertion processes. These predictions are corroborated by recently reported results for interactions of transition metal atoms with small hydrocarbons in the gas phase. Weisshaar and coworkers [53] observe that neutral Fe, Co, Ni, and Cu atoms are unreactive with alkanes (specifically propane and butane) in the gas phase. Adduct formation appears to occur with alkenes and Ni. The absence of reactivity is consistent with modest barriers for both C-H and C-C insertion. Isolated metal atoms in matrices are generally unreactive with added hydrocarbons. Photoexcitation often results in reaction, indicating that excited states derived from different configurations can insert into C-H bonds of hydrocarbons [54]. Continued theoretical and experimental studies of this type, both for neutrals and ions, will eventually provide a thorough understanding of hydrocarbon transformations at transition metal centers.

Reverse geometry double focusing mass spectrometers are available in many laboratories around the world. These instruments facilitate measurement of kinetic energy release distributions as described in this review, and other groups have already started reporting such results for organometallic reactions [55]. These experiments provide both mechanistic and, in favorable circumstances, thermochemical information. The basic computer programs for carrying out statistical phase space calculations are available from the Quantum Chemistry Program Exchange. As a result, this experimental methodology can be widely applied in many laboratories to determine heats of formation and bond dissociation energies for organometallic intermediates. In view of the results presented in this review, it is desirable to have additional information available, such as reaction efficiencies or knowledge relating to the reverse reaction. When carefully analyzed, these experiments provide bond dissociation energies with accuracy comparable to other thermochemical kinetic methods. The experiments described above for manganese carbonyl bond energies, in which reactants have broad internal energy distributions, is perhaps of greatest interest for future studies. We are presently attempting to measure the cobalt-carbon bond energy in co-enzyme B_{12}, to demonstrate the generality of the method, even for relatively large molecules. The deprotonated molecular ion of this species can be generated by fast atom bombardment. The most intense metastable observed from this ion involves homolytic cleavage of the cobalt-carbon bond with a kinetic energy release distribution that is not very different in magnitude from the ones presented above. Basically this results from the requirement of a nearly constant excess energy *per degree of freedom* to produce dissociation at a particular rate. Hence the methodology can be applied to macromolecules without difficulty. These preliminary results indicate that the cobalt-carbon bond is somewhat stronger than solution measurements suggest [56].

5. Acknowledgments

This work has been funded by the National Science Foundation through Grants No. CHE87-11567 and CHE91-08318 (Caltech) and CHE88-17201 (M.T. Bowers, Santa Barbara). In addition we thank the donors of the Petroleum Research Fund, administered by the American Chemical Society, for additional support. This manuscript is contribution number 8360 from the Division of Chemistry and Chemical Engineering at the California Institute of Technology.

References

1. Muller, J.; Goll, W. Chem. Ber. 1973, 106, 1129.
2. This field is well represented in the book Gas Phase Inorganic Chemistry; Russell, D. H., Ed.; Plenum: New York, 1989.
3. Houriet, R.; Halle, L. F.; Beauchamp, J.L.; Organometallics, 1983, 2, 1818.
4. Beauchamp, J. L. ACS Symp. Ser. 1987, 333, 11.
5. Armentrout, P. B.; Beauchamp, J. L. Acc. Chem. Res. 1989, 22, 315.
6. Hanratty, M. A.; Paulson, C. M.; Beauchamp, J. L. J. Am. Chem. Soc. 1985, 107, 5074.
7. Hanratty, M.; Beauchamp, J. L.; Illies, A. J.; Bowers, M. T. J. Am. Chem. Soc. 1985, 107, 1788.
8. Hanratty, M. A.; Beauchamp, J. L.; Illies, A. J.; van Kopppen, P. A. M.; Bowers, M. T. J. Am. Chem. Soc. 1988, 110, 1.
9. Ervin, K. M.; Armentrout, P. B. J. Chem. Phys. 1985, 83, 166.
10. Jarrold, M. F.; Illies, A. J.; Bowers, M. T. Chem. Phys. 1982, 91, 2573.
11. Kirchner, J. J.; Bowers, M. T. J. Phys. Chem. 1987, 91, 2573.
12. Armentrout, P.B.; Beauchamp, J.L. J. Amer. Chem. Soc. 1981, 103, 784.
13. Ervin, K.M.; Armentrout, P.B. J. Chem. Phys. 1985, 83, 166.
14. Kemper, P.R.; Bowers, M.T.; J. Am. Chem. Soc., 1990, 112, 3231.
15. Halle, L.F.; Armentrout, P.B.; Beauchamp, J.L. J. Amer. Chem. Soc. 1981, 103, 962.
16. Kang, H.; Beauchamp, J.L. J. Phys. Chem. 1985, 89, 3364.
17. For a recent example, see Mandich, M.L.; Halle, L.F.; Beauchamp, J.L. J. Am. Chem. Soc. 1984, 106, 4403.
18. Beauchamp, J.L. Ann. Rev. Phys. Chem. 1971, 22, 527.
19. Cody, R.B.; Burnier, R.C.; Freiser, B.S. Anal. Chem. 1982, 54, 96.
20. Thorne, L.R. Beauchamp, J.L. Ions and Light (Gas Phase Ion Chemistry, Vol. 3), Bowers, M.T., ed., Academic Press, New York, 1983.
21. Hettich, R.L.; Freiser, B.S. J. Amer. Chem. Soc. 1986, 108, 2537.
22. Schwarz, H.; Acct. Chem. Res., 1989, 22, 282.
23. Alford, J.M.; Weiss, F.D.; Laaksonen, R.T.; Smalley, R.E.; J. Phys. Chem. 1986, 90, 4480
24. Jarrold, M.F.; Illies, A.J.; Kirchner, N.J.; Wagner-Redeker, W.; Bowers, M.T.; Mandich, M.L.; Beauchamp, J.L. J. Phys. Chem. 1983, 87, 2313, and references therein.

320

25. Sonnenfroh, D.M.; Farrar, J.M. <u>J. Amer. Chem. Soc.</u> 1986, <u>108</u>, 3521.
26. Walba, D.M.; Depuy, C.H.; Grabowski, J.J.; Bierbaum, V.M. <u>Organometallics</u> 1984, <u>3</u>, 498.
27. McDonald, R. N.; Schell, P.L.; McGhee, W.D. <u>Organometallics</u> 1984, <u>3</u>, 182.
28. Lane, K.R.; Sallans, L.; Squires, R.R. <u>J. Am. Chem. Soc.</u> 1986, <u>108</u>, 4368.
29. Tonkyn, R.; Weisshaar, J.C. <u>J. Phys. Chem.</u>, 1986, <u>90</u>, 2305.
30. Simoes, J.A. M.; Beauchamp, J.L.; <u>Chem. Rev.</u>, 1990, <u>90</u>, 629.
31. van Kopen, P.A.M.; Brodbelt-Lustig, J.; Bowers, M.T.; Dearden, D.V.; Beauchamp, J.L.; Fisher, E.; Armentrout, P.B.; <u>J. Am. Chem. Soc.</u>, 1990, <u>112</u>, 5663.
32. Irikura, K.K.; Beauchamp, J.L.; <u>J. Am. Chem. Soc.</u> 1991, <u>113</u>, 2769.
33. Irikura, K.K.; Beauchamp, J.L.; <u>J. Phys. Chem.</u>, 1991, <u>95</u>, 8344.
34. Halle, L.F.; Armentrout, P.B.; Beauchamp, J.L.; <u>J. Am. Chem. Soc.</u>, 1981, <u>103</u>, 962
35. Schilling, J.B.; Goddard, W.A., III; Beauchamp, J.L. <u>J. Am. Chem. Soc.</u> 1986, <u>108</u>, 582.
36. Armentrout, P.B.; Halle, L.F.; Beauchamp, J.L. <u>J. Am. Chem. Soc.</u> 1981, <u>103</u>, 6501.
37. Carlin, T.J.; Sallans, L.; Cassady, C.J.; Jacobson, D.B.; Freiser, B.S., <u>J. Am. Chem. Soc.</u> 1983, <u>105</u>, 6320.
38. Halle, L.F.; Klein, F.S.; Beauchamp, J.L. <u>J. Am. Chem. Soc.</u> 1984, <u>106</u>, 2543.
39. Beauchamp, J.L.; Stevens, A.E.; Corderman, R.R. <u>Pure and Appl. Chem.</u> 1979, <u>51</u>, 967.
40. Tolbert, M.A.; Beauchamp, J.L.; <u>J. Am. Chem. Soc.</u>, 1984, <u>108</u>, 2162.
41. Schilling, J.B.; Beauchamp, J.L.; <u>J. Am. Chem. Soc.</u>, 1988, <u>110</u>, 15.
42. Schilling, J.B.; Beauchamp, J.L.; <u>Organometallics</u>, 1988, <u>7</u>, 194.
43. Tolbert, M.A.; Beauchamp, J.L.; <u>J. Am. Chem. Soc.</u>, (1986), <u>108</u>, 7509.
44. Chesnavich, W. J.; Bowers, M. T. <u>J. Am. Chem. Soc.</u> 1977, <u>99</u>, 1705.
45. Chesnavich, W. J.; Bass, L.; Su, T.; Bowers, M. T. <u>J. Chem. Phys.</u> 1981, <u>74</u>, 2228.
46. van Kopen, P.A.M.; Brodbelt-lustig, J.; Bowers, M.T.; Dearden, D.V.; Beauchamp, J.L.; Fisher, E.; Armentrout, P.B.; <u>J. Am. Chem. Soc.</u>, 1991, <u>112</u>, 5663.
47. van Kopen, P.A.M.; Brodbelt-lustig, J.; Bowers, M.T.; Dearden, D.V.; Beauchamp, J.L.; Fisher, E.; Armentrout, P.B.; <u>J. Am. Chem. Soc.</u>, 1991, <u>113</u>, 2359.
48. van Koppen, P. A. M.; Jacobson, D. B.; Illies, A.; Bowers, M. T.; Hanratty, M.; Beauchamp J. L. <u>J. Am. Chem. Soc.</u> 1989, <u>111</u>, 1991.
49. Georgiadis, R.; Fisher, E.R.; Armentrout, P.B. <u>J. Am. Chem. Soc.</u> 1989, <u>111</u>, 4251.
50. Dearden, D.V.; Beauchamp, J.L.; van Koppen, P.A.M.; Bowers, M.T.; J. Am. Chem Soc., 1990, 112, 9372.
51. Beauchamp, J .L.; Stevens, A. E.; Corderman, R. R. <u>Pure and Appl. Chem.</u> 1979, <u>51</u>, 967.
52. Blomberg, M.R.A.; Siegbahn, P.E.M., Nagashima, U.; Wennerberg, J.; <u>J. Am. Chem. Soc.</u> 1991, <u>113</u>, 424.
53. Ritter, D.; Weisshaar, J.C.; <u>J. Am. Chem. Soc.</u> 1990, <u>112</u>, 6425.
54. Ozin, G.A.; <u>Acc. Chem. Res.</u> 1977, <u>10</u>, 21.
55. Schulze, C.; Schwarz, H.; <u>Int. J. Mass Spectrom. Ion Proc.</u> 1989, <u>88</u>, 291.
56. Halpern, J.; Kim, S.-H.; Leung, T.W.; <u>J. Am. Chem.Soc.</u> 1984, <u>106</u>, 8317.

GUIDED ION BEAM STUDIES OF THE ENERGETICS OF ORGANOMETALLIC SPECIES

P. B. ARMENTROUT and D. E. CLEMMER
Department of Chemistry
University of Utah
Salt Lake City, UT 84112
USA

ABSTRACT. Dissociation energies for a variety of bonds between transition metals and simple ligands comprised of hydrogen, carbon, nitrogen, and oxygen have been measured by using the technique of guided ion beam mass spectrometry. For species that form covalent bonds with the metals, the bond energies are found to correlate with the energy necessary to promote the bare metal atom or ion into an electron configuration suitable for bonding. The trends in the bond energies of M^+ to isoelectronic series of ligands, CH_3, NH_2 and OH or CH_2, NH and O, are used to quantify the contributions of covalent and dative bonding for these ligands. The strength of the dative interactions is found to depend on the number of electron lone-pairs on the ligands and the number of empty or half-filled d orbitals on the metal. The concept of an intrinsic metal-ligand bond energy is defined and used to consider the strengths of bonds in coordinatively saturated metal-ligand compounds. Initial studies that address the metal-ligand bond energies in such species more directly include measurements of the sequential bond energies of $M(CO)_x^+$ (x = 1-6), $M(H_2O)_x^+$ (x = 1-4) and $M(CH_4)_x^+$ (x = 1-4). Variations in these bond energies as the number of ligands increases are interpreted by considering how the valence electrons on the metal reorganize to accommodate the ligand shell. Finally, studies of the thermochemistry of proposed reactive intermediates for the reactions of Fe^+ and Co^+ with small alkanes are discussed with an emphasis on understanding the potential energy surfaces for C-H and C-C bond activation processes.

1. Introduction

Since its first application to the organometallic chemistry of transition metals [1], ion beam techniques have provided information about the thermochemistry and bonding of coordinatively unsaturated organometallic molecules. Such highly reactive species can be studied by these gas-phase methods even though they do not obey the 18-electron rule and are not isolable in macroscopic quantities. In this report, we examine our contributions to organometallic thermochemistry and focus on

321

J. A. Martinho Simões (ed.), Energetics of Organometallic Species, 321–356.
© 1992 *Kluwer Academic Publishers.*

three types of studies. The first is studies of the periodic trends in metal ligand bond energies. Since they are not restricted to studying saturated organometallic complexes, ion beam techniques can be used to elucidate details of metal-ligand bonding by varying the identify of the metal while maintaining the same ligand shell or by systematically varying the ligand for the same metal. Second, effects due to the degree of unsaturation can be studied simply by changing the number of ligands on the metal. The third type of experiment concerns our efforts to better understand the mechanisms by which atomic gas-phase metal ions activate C-C and C-H bonds in alkanes by characterizing the energetics of the postulated intermediates in such reactions.

2. Experimental Section

2.1. INSTRUMENTATION

The experimental methods used in our laboratory to measure gas phase bond dissociation energies have been discussed in detail before [2-6], and involve the use of a "guided" ion beam tandem mass spectrometer [2,7]. Ion beam instruments are two back-to-back mass spectrometers with a reaction zone in between, and an ion source and an ion detector at either end. In ion beam experiments, reactant ions are created, mass selected in the first mass spectrometer, accelerated to a particular kinetic energy, and reacted with a neutral reagent. In the second mass spectrometer, reactant and product ions are mass separated and their absolute intensities detected. The reaction zone is designed so that reactions occur over a well-defined path length and at a pressure low enough that all products are the result of *single* ion-neutral encounters. (Deviations from single collision conditions are easily verified and corrected by performing studies at several different neutral pressures [2,8,9].) In our apparatus, the interaction region is surrounded by an rf octopole ion beam "guide" [10] which ensures efficient collection of all ions. The sensitivity of the detector (a secondary electron scintillation ion detector [11]) is sufficiently high that all ions are detected with near 100% efficiency.

To present the results of such experiments in the most meaningful way, the raw data of an ion beam experiment, intensities of the reactant and product ions as a function of the ion kinetic energy in the laboratory frame, are converted to an absolute reaction cross section as a function of the kinetic energy in the center-of-mass frame, $\sigma(E)$. A cross section is the effective area that the reactants present to one another and is a direct measure of the probability of the reaction at a given kinetic energy. It is easily related to a rate constant, $k = \sigma v$ where v is the relative velocity of the reactants. Conversion of intensities to cross sections is readily performed by using a Beer's law type of formula [2]. A center-of-mass energy scale is used because this accounts for the fraction of the laboratory ion energy that is tied up in motion of the entire reaction system through the laboratory. This fraction must be conserved and is unavailable to induce chemical reactions. Conversion of the laboratory ion energy to the center-of-mass frame energy involves a simple mass factor (except at very low energies where truncation of the ion beam must be accounted for) [2]. For accurate thermochemistry, particular attention must also be paid to a determination of the absolute zero of energy. In our laboratories, this is routinely measured by a retarding potential analysis which has been verified by time-of-flight measurements [2] and comparisons with theoretical cross sections [12].

2.2. REACTIONS

To measure metal-ligand bond dissociation energies (BDEs) of a species like ML^+, we measure the cross section for reaction (1) or (2) while varying the kinetic energy available to the reactants, as outlined above.

$$M^+ + RL \longrightarrow ML^+ + R \tag{1}$$

$$ML^+ + Xe \longrightarrow M^+ + L + Xe \tag{2}$$

Metal-ligand BDEs for neutral species can be obtained from processes such as reaction (3).

$$M^+ + RL \longrightarrow ML + R^+ \tag{3}$$

The collision-induced dissociation (CID) reaction (2) is intrinsically endothermic, and the neutral reagent in reactions (1) and (3) is chosen such that these processes are endothermic. (A table of suitable reagents is provided in reference 13.) The thermochemical information of interest can then be found from the thresholds for these reactions, E_0, as shown in the following equations:

$$D^\circ(M^+\text{-}L) = D^\circ(R\text{-}L) - E_0(1) \tag{4}$$

$$D^\circ(M^+\text{-}L) = E_0(2) \tag{5}$$

$$D^\circ(M\text{-}L) = D^\circ(R\text{-}L) + IE(R) - IE(M) - E_0(3) \tag{6}$$

These equations assume that there is no activation energy in excess of the reaction endothermicity (i.e. there is no reverse activation barrier). This is usually a reasonable assumption due to the strong long-range ion-induced dipole potential [14]. Thus, exothermic ion-molecule reactions are generally observed to proceed without an activation energy, and endothermic ion-molecule reactions generally proceed once the available energy exceeds the thermodynamic threshold [6]. Exceptional behavior can be observed when there are restrictions due to spin or orbital angular momentum conservation [6,15]. One means of avoiding errors due to such exceptions is to verify that the same BDE is measured in more than one reaction system.

We have explicitly tested this assumption in several reactions where the thermochemistry is well established [4,5,16-19]. No activation barriers are found in any of these cases, although the observation of

the true thermodynamic threshold can require extremely good sensitivity
[4]. Direct checks for organometallic species are less common, but BDEs
obtained from our experiments compare well with values available from
other experiments [20-25] and from ab initio theory [26-31].

2.3. DATA ANALYSIS

In analyzing the cross sections for processes (1) - (3), it is important
to characterize all sources of energy that can be used to drive the
reaction. Characterization of the translational energy of the reagents
is intrinsic to the ion beam method, while an assessment of the *internal*
energy of the reactants can be more difficult, at least for the ionic
reactant since ion formation is an energetic process. (The neutral
reagents have well characterized temperatures of 298 K.) For atomic
metal ions, electronic energy is the only kind of internal energy
possible. The effects of such electronic energy on ion beam studies of
organometallic thermochemistry have been thoroughly discussed elsewhere
[32-34]. For polyatomic organometallic ions, internal energy can be
electronic, vibrational, and rotational. Our recent studies of such
species have been enabled by the construction of a flowing afterglow ion
source that is capable of producing thermalized ions [7]. In this
source, ions undergo ~10^5 collisions with the He flow gas. Several
studies [7,9,35,36] have verified that thermalized (298 K) ions are
routinely produced under these conditions. Exceptional behavior
includes the observation of excited electronic states of SF^+ and CF_2^+
(which can be quenched by adding O_2 to the flow) [37] and vibrational
modes of NO_2^+ (which can be quenched by adding CO_2 to the flow) [38].
The final step in obtaining the bond energies by equations (4) - (6)
is the determination of the reaction threshold. Theory [39,40] and
experiment [3-6,16-19,41,42] indicate that cross sections for
endothermic reactions can be modelled by equation (7),

$$\sigma(E) = \sigma_o \sum_i g_i \, (E + E_i - E_0)^n/E \qquad (7)$$

which involves an explicit sum of the contributions of individual
reactant states (vibrational, rotational, or electronic), denoted by i,
with energies E_i and populations g_i. (Early versions of this equation
did not include the sum of states.) Here, σ_o is a scaling factor, E is
the relative kinetic energy, n is an adjustable parameter, E_0 is the 0 K
threshold for reaction of the lowest electronic and vibrational levels
of the reactants. Before comparison with the experimental data, this
model is convoluted with the neutral and ion kinetic energy distributions as described previously [2]. The σ_o, n, and E_0 parameters are
then optimized by using a non-linear least squares analysis to give the
best reproduction of the data. Error limits for E_0 are calculated from
the range of threshold values obtained for different data sets with
different values of n and the error in the absolute energy scale.
The accuracy of the thermochemistry obtained by this modeling
procedure is dependent on a variety of experimental parameters. This
has recently undergone an extensive discussion elsewhere [6] and the

interested reader is referred to this work for further information. An example of the pitfalls, however, is illustrated by results for the reaction of $Fe^+(^6D)$ with O_2, shown in Figure 1. Part a of this figure shows that the state-specific data can be reproduced nicely by equation (7). The threshold obtained leads to a bond energy for FeO^+ of 341 ± 12 kJ/mol, in good agreement with measurements from other chemical reactions in our laboratories, 341 ± 6 kJ/mol [43], and within experimental error of the best literature value, 326 ± 14 kJ/mol (see discussion in reference [43]). Part b shows an alternate interpretation of the energy dependence of the cross section involving a linear rise from threshold. While such a model reproduces the data in the sharply rising portion of the cross section, it does not model the data at the lowest energies even when the distribution of kinetic energies is included. The threshold obtained is 46 kJ/mol higher in energy than that obtained from the analysis shown in part a. This leads to a FeO^+ BDE of 294 kJ/mol, inconsistent with the values above. Note that unless the data in the threshold region is sufficiently precise, the difference in these two analyses could not be assessed and the thermochemistry would be in error.

Early experiments in our laboratory have demonstrated that accounting for the vibrational energy of the metal-ligand complex is a very important consideration in the accurate measurement of bond energies by collision-induced dissociation. This is a particularly significant energy contribution in organometallic complexes due to the many low-lying vibrational modes (corresponding to hindered rotations of the ligands and ligand-metal-ligand bends) whose number increases as the number of ligands increases. In our recent study on the CID of $Fe(CO)_x^+$ (x = 1-5) [9], we found that reasonable bond energies could only be obtained when the distribution of vibrational energy was included explicitly, as in equation (7). Significant systematic errors in the analysis are introduced if the threshold is simply corrected by the average vibrational energy, and very large errors are obtained if the vibrational energy is ignored.

3. Periodic Trends in Covalent M-L Bond Energies

Over the past several years, we have endeavored to systematically measure bond energies for covalent metal-ligand interactions in the simplest possible complexes, those containing only a single ligand. In particular, we have been interested in the periodic trends in these BDEs and how the strength of the bonding varies with the ligand. In this section, we discuss these results, concentrating on those ligands for which the data is more recent. Species such as M^+-H, M^+-CH$_3$, M^+-CH$_2$, M^+-CH and M^+-O have been discussed previously [13,25,32,44-48] and will be reviewed briefly here. We restrict ourselves to a discussion of first row transition metals since not as many BDEs for second and third row metal-ligand complexes have been studied. Those that have been measured have been discussed previously [13,32,44,45].

326

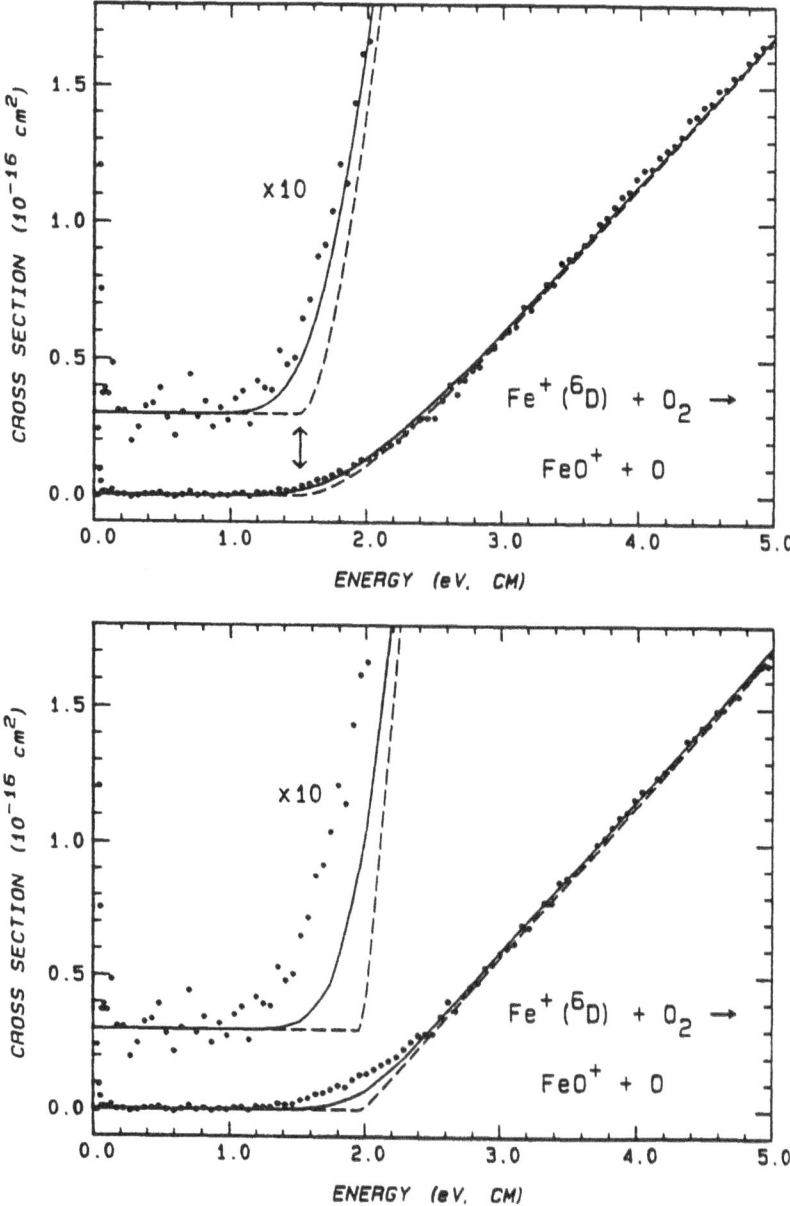

Figure 1. Cross section for reaction of $Fe^+(^6D)$ with O_2 as a function
of translational energy in the center-of-mass frame. In part a, the
dashed line is the empirical model given by equation (7) ($n = 1.85$, $E_0 =$
1.51 eV, $E_i = 0$, and $i = 1$). In part b, the dashed line is a linear
model with $E_0 = 1.99$ eV. In both parts, the full line is the model
convoluted over the experimental kinetic energy distributions.

3.1. COVALENT METAL-LIGAND SINGLE BONDS: M^+-H and M^+-CH_3

Table 1 reports our values for MH^+ [49-55] and MCH_3^+ BDEs
[3,33,41,56-61], and these are plotted across the periodic table in
Figure 2. It can be seen that these values are similar for each metal,
a result that is easily rationalized by formation of a single covalent
bond in both species. Further, the strongest bonds are for the early
metals, Sc^+ and Ti^+, while the weakest BDEs are for Cr^+ and Cu^+. These
latter metals form weak bonds to H and CH_3 because they have electron
configurations corresponding to a half-filled and filled 3d subshell,
respectively. To form a covalent bond, these ions must disrupt these
stable electron configurations and rearrange the electrons in order to
allow formation of a strong covalent bond.

3.1.1. *Promotion Energies*. The energetics associated with such electron
reorganization is called the promotion energy, E_p. A correlation
between E_p and $D°(MH^+)$ was first suggested by Armentrout, Halle, and
Beauchamp [62] and later refined by Mandich *et al.* [63] and Elkind
and Armentrout [45]. For the H and CH_3 ligands, we define E_p as the
energy required to take the ground state of the metal ion and excite it
into an effective electronic state having a singly occupied orbital
suitable for bonding to the ligand. The energy associated with spin
decoupling this bonding electron from the remaining nonbonding electrons
is also included in E_p. The idea is that when the M^+-L bond is cleaved,
the bonding electron which remains with M^+ has a 50:50 chance of being
high-spin or low-spin coupled to the nonbonding electrons on M^+.

TABLE 1. Bond Dissociation Energies of Transition Metal Ions with
Ligands that can Form a Single Covalent Bond (kJ/mol)[a]

M	E_p	M^+-H	M^+-CH_3	M^+-NH_2	M^+-OH[b]
Sc	15	239(9)[c]	247(13)[d]	356(7)[e]	509(13)
Ti	27	227(11)[f]	225(8)[g]	356(13)[e]	476(12)
V	68	202(6)[h]	209(10)[i]	307(10)[j]	446(14)
Cr	192	136(9)[k]	127(7)[l]	209(17)[m]	309(13)
Mn	56	203(14)[n]	215(10)[o]	264(20)[m]	343(24)
Fe	47	208(6)[p]	242(10)[q]	271(14)[m]	323(11)
Co	79	195(6)[r]	205(13)[s]	257(9)[t]	307(13)
Ni	133	166(8)[r]	188(10)[s]	233(8)[t]	246(19)
Cu	292	92(13)[r]	124(7)[s]	202(13)[t]	

[a]Values are at 298 K with uncertainties in parentheses.
[b]References 97 and 98. [c]Reference 49. [d]Reference 41.
[e]Reference 95. [f]Reference 50. [g]References 56 and 57.
[h]Reference 51. [i]Reference 3. [j]Reference 94. [k]Reference 52.
[l]References 58 and 59. [m]Reference 98. [n]Reference 53.
[o]Reference 60. [p]Reference 54. [q]Reference 33. [r]Reference 55.
[s]Reference 61. [t]Reference 96.

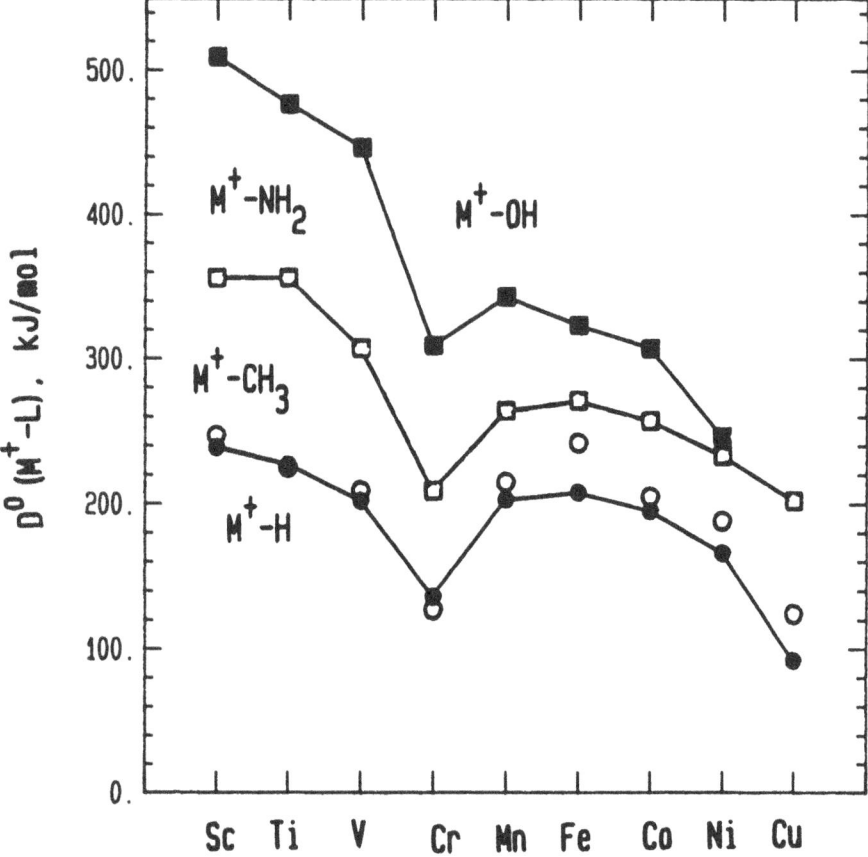

Figure 2. Transition metal ion bond energies to H (closed circles), CH₃ (open circles), NH₂ (open squares), and OH (closed squares) across the first row of the periodic table.

Empirically, the best correlation between the ML^+ BDEs and E_p is found for a $4s^1 3d^{n-1}$ electron configuration where the 4s electron is spin-decoupled from the 3d electrons. (Calculation of the promotion energy for this configuration is easily performed as the average electronic energy of the high-spin and excited low-spin $4s^1 3d^{n-1}$ states. These energies can be looked up in the appropriate tables [64,65], or the compilation of E_p values given by Elkind and Armentrout [45] or Carter and Goddard [66] can be used. For convenience, promotion energies taken from these references are listed in the tables where required.) The bond energies for the first row transition metal- hydride and -methyl ions versus this definition of E_p are shown in Figure 3. It can be seen that this correlation allows the periodic trends observed in Figure 2 to be accounted for systematically.

Alternate definitions of E_p can also be considered. For instance, the spin exchange energy can be ignored, but this leads to a much worse correlation with bond energy. This is most easily illustrated by $Mn^+(4s^1 3d^5)$, which has a promotion energy of zero if one ignores the exchange energy but a promotion energy of 56 kJ/mol if it is included. If the former definition is used, the BDEs for MnH^+ and $MnCH_3^+$ do not correlate well with those for the other metal ions. Elkind and Armentrout [45] also considered bonding to the $3d\sigma$ orbital in a $3d^n$ or $4s^1 3d^{n-1}$ electron configuration. No correlation between BDEs and $E_p(3d^n)$ was found, but a reasonable correlation existed for $E_p(4s^1 3d^{n-1})$. This suggests that the $4s^1 3d^{n-1}$ configuration is important in describing the bonding of first row transition metal hydride ions, but that the bonding orbital on the metal has both 4s and $3d\sigma$ character. This conclusion agrees with results of ab initio calculations [26-28,67,68]. These theoretical results also find that 3d character is most significant on the left side of the periodic table.

3.1.2. *Intrinsic Bond Energies.* One significant feature in Figure 3 is that the maximum MH^+ and MCH_3^+ BDEs are about 240 kJ/mol (the intercepts of the linear regression lines are 236 $\pm$ 7 and 244 $\pm$ 8 kJ/mol, respectively) and occur when $E_p = 0$. This is comparable to the most commonly cited value for metal-hydride BDEs of ~250 kJ/mol [69] and to several specific metal-hydride BDEs in condensed phase complexes: for example, $HCo(CO)_4$ and $[HCo(CN)_5^{3-}]_{aq}$ [70,71], and $[H_2IrCl(CO)(PPh_3)_2]$ [72]. As discussed elsewhere [32,45], this maximum BDE can be thought of as the "intrinsic" metal-hydrogen or metal-methyl BDE, i.e. the dissociation energy for *any* metal-hydride or metal-methyl bond in the absence of electronic and steric effects. For atomic metal ions, the BDE can be weakened by electronic effects, as represented by the correlations to promotion energy. For ligated metals and on metal surfaces, a directional and sterically unhindered metal orbital having a single electron that is electronically decoupled from other electrons must be prepared for efficient bonding. In such cases, the equivalent of the atomic electronic promotion can be achieved by the electronic environment (the ligands and nearest neighbors) surrounding the binding site.

The observation that the MCH_3^+ BDEs are slightly stronger than the MH^+ BDEs, Figures 2 and 3, is noteworthy since metal-carbon bonds are generally much weaker than metal-hydrogen bonds in condensed phase

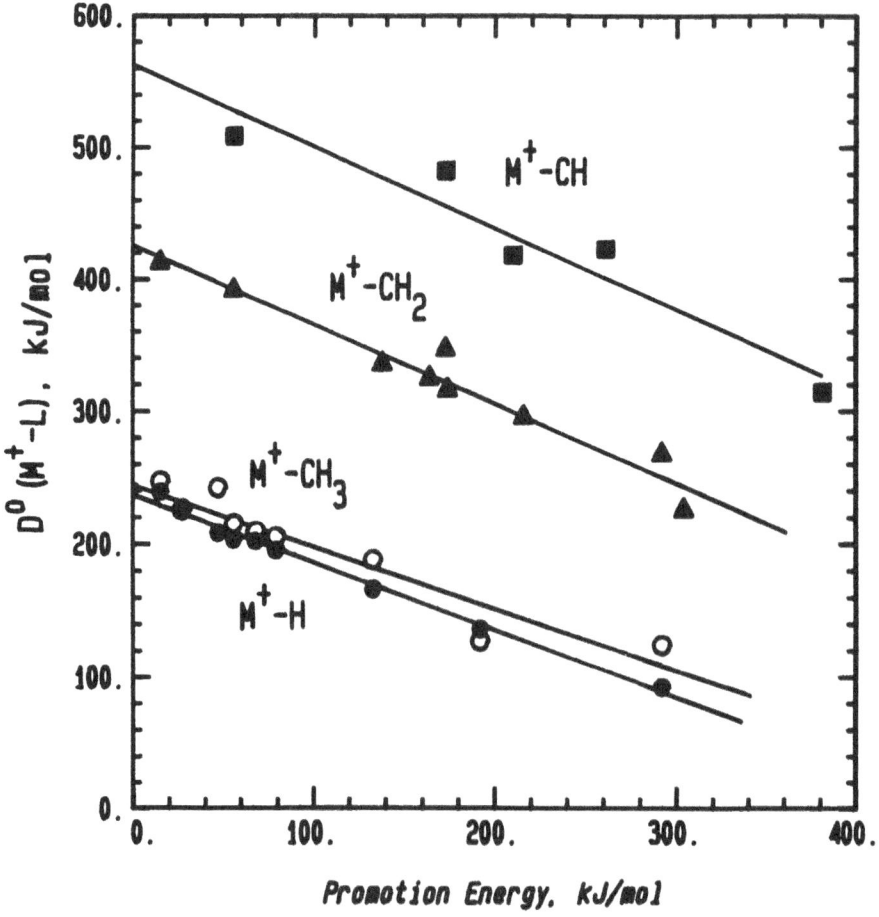

Figure 3. Transition metal ion ligand bond energies vs. the atomic metal ion promotion energy to a $4s^1 3d^{n-1}$ spin-decoupled state (see text). Results for H (closed circles), CH_3 (open circles), CH_2 (closed triangles) and CH (closed squares) ligands are shown and taken from Tables 1, 3, and 4. The lines are linear regression fits to the data.

organometallic complexes. In the gas phase, isolobal arguments [73] and detailed ab initio calculations [29,30,74] show that both H and CH_3 can form strong covalent bonds with transition metals. For gas-phase ions, it has been suggested [75] that the MCH_3^+ BDEs may be stronger than MH^+ BDEs due to the increased polarizability of the methyl group compared to a hydrogen atom, an effect that can contribute about 20-25 kJ/mol to the bond energy [30]. For the condensed phase species, Halpern has suggested that steric effects are largely responsible for the weak M-C bond strengths [69]. Calculations show that the reduction of steric strain in cationic species can be attributed in part to electronic effects [74]. Possible contributions due to agostic [76] M··H-C interactions have also been suggested [77], but there is no evidence of such interactions in M^+-CH_3 in either experimental data [32] or calculations [29].

Finally, we note that the slopes of the lines in Figure 3 are not unity, or in other words, the sum of the BDEs and E_p does not yield a constant value. Rather the slopes of these lines are close to -0.5, showing that the sum of $D^\circ(M^+$-$L)$ and $E_p/2$ is nearly a constant equal to the intrinsic bond energy. The physical meaning of the 1/2 slope is not completely clear, although it may be a further indication that the 4s and $3d\sigma$ orbitals mix to form the best covalent metal bond.

3.1.3. *Metal Ion-Ethyl and -Propyl Bond Energies.* Limited data also exist for the BDEs of larger alkyl groups with atomic metals ions. There is some ambiguity about the structures of these species since a species with the $MC_nH_{2n+1}^+$ formula could exist in two energetically similar structures, the metal-alkyl ion, M^+-C_nH_{2n+1}, or the hydrido-alkene complex, H-M^+-C_nH_{2n}. Unfortunately, the thermochemistry obtained for these species cannot unequivocally differentiate between these possibilities. In the following discussion, we simply cite the thermochemistry that we have measured for the $MC_nH_{2n+1}^+$ species as the bond energy for dissociation to M^+ + the alkyl group.

In general, we find that the bond energies for metal-ethyl and -propyl cations are similar to the metal-methyl ion BDEs listed in Table 1. Specifically, $D^\circ(Ti^+$-$C_2H_5)$ = 220 ± 7 kJ/mol [57] is nearly identical to $D^\circ(Ti^+$-$CH_3)$, while the bonds to ethyl for V^+, 238 ± 13 kJ/mol [78], and Cr^+, 146 ± 9 kJ/mol [59], are slightly stronger than those to methyl; and that for Fe^+ is slightly weaker, 219 ± 11 kJ/mol [33]. While specific BDEs have not been reported in the cases of Co^+ and Ni^+, there are indications that the ethyl and methyl BDEs are comparable in the former case, while the Ni^+-ethyl bond is stronger than the Ni^+-methyl bond [61]. The only reliable measurements of metal ion-propyl BDEs that exist are for Cr^+ [59] where the bond energy to 1-C_3H_7 is found to be 134 ± 6 kJ/mol and that for 2-C_3H_7 is 119 ± 6 kJ/mol, again similar to $D^\circ(Cr^+$-$CH_3)$.

3.2. COVALENT METAL-LIGAND SINGLE BONDS: MH and MCH_3 NEUTRALS

In an effort to explore how much influence the positive charge has on this thermochemistry, we have recently made concerted efforts to measure the bond energies of neutral transition metal ligand complexes,

TABLE 2. Bond Dissociation Energies of Neutral Transition Metals with Ligands that Form a Single Covalent Bond (kJ/mol)[a]

M	E_p	M-H[b]	M-CH$_3$
Sc	158	212(13)[c]	134(29)[d]
Ti	108	205(9)[e]	192(29)[d]
V	63	221(13)[c]	155(38)[d]
Cr	45	206(10)[c]	159(8)[f]
Mn	241	126(18)[g]	40-125[g]
Fe	113	157(8)[h]	155(29)[d]
Co	65	194(13)[i]	191(13)[i]
Ni	22	252(8)[i]	231(14)[i]
Cu	0	258(11)[i]	243(8)[i]
Zn	387	82(4)[j]	81(14)[k]

[a]Values are at 298 K with uncertainties in parentheses.
[b]These values have been critically reviewed in reference 25.
[c]Chen, Y.; Clemmer, D. E.; Armentrout, P. B., work in progress.
[d]Reference 13. [e]Reference 48. [f]Reference 59. [g]Reference 79.
[h]Reference 80. [i]References 61 and 81. [j]K. P. Huber and G. Herzberg, *Molecular Spectra and Molecular Structure IV, Constants of Diatomic Molecules.* (Van Nostrand-Reinhold: New York, 1979). [k]Reference 4.

in particular the diatomic metal hydrides. Table 2 lists values for the bond energies of MH and MCH$_3$ [48,59,61,79-81]. We might imagine that ionic and neutral species with the same number of valence electrons should have similar BDEs [82]. While in some cases, for instance ScH and TiH$^+$ or TiH and VH$^+$, this is true, in others, such as MnH and FeH$^+$ or NiH and CuH$^+$, the BDEs disagree substantially. The reason that there is no good correspondence between the ionic and neutral species is that these comparisons involve *isovalent* species (i.e. those having the same number of valence electrons), but not necessarily *isoelectronic* species (i.e. those where the electrons are in the same orbitals). Consider the case of Mn and Fe$^+$, which have ground state electron configurations of $4s^23d^5$ and $4s^13d^6$, respectively. Based on the promotion energy arguments outlined above, this should clearly make a difference in the BDEs.

3.2.1. *Promotion Energies.* The periodic trends in these neutral metal-ligand BDEs can be understood by noting that the strongest bonds are for Cu and Ni, while Mn and Zn have the weakest. The first two metals have a ground state (Cu) and a very low-lying excited state (Ni) with $4s^13d^{n-1}$ electron configurations, while the latter two metals have $4s^23d^{n-2}$ configurations where the 3d subshell is half-filled or filled. Thus, promotion energy can again be used to help quantify the periodic trends in these neutral BDEs. Table 2 shows that the metals having the highest bond energies have the lowest values for E_p (defined in the same manner as for the ions). Plots of $D°(M-H)$ and $D°(M-CH_3)$ vs E_p show the same general trends as Figure 3 [25,32,48], but the correlations are not as good. This may be because the 4p electrons on the metal are involved in the bonding to a greater extent in the neutral species [83,84] or because the accuracies of the neutral BDEs values are not yet good enough to provide a definitive comparison.

3.2.2. *Intrinsic Bond Energies*. The intrinsic BDEs obtained from such promotion energy plots are 243 kJ/mol for MH and 213 kJ/mol for MCH_3. This intrinsic BDE for the neutal metal-hydrides is virtually the same as that for the cations, illustrating that charge has little effect on these covalent bond strengths. The observation that the intrinsic BDE for the neutral metal-methyl species is somewhat less than for MH (exactly the opposite effect observed for the cations) is consistent with contributions to the bonding due to the polarizability of CH_3 vs H. While the MCH_3 BDEs are somewhat less than the MH BDEs, the decrease is not nearly as great as is often observed in the condensed phase. Our results demonstrate that this discrepancy between gas-phase and condensed-phase results is not simply an effect of the charge on the metal.

3.3. COVALENT METAL-LIGAND MULTIPLE BONDS

3.3.1. M^+-CH_2 and M^+-CH. Values measured for M^+-CH_2 and M^+-CH BDEs are listed in Tables 3 and 4 [56,57,79,81,85-89]. For all metals, these bond energies are stronger than the M^+-CH_3 BDEs, consistent with the formation of double bonds with CH_2 and triple bonds with CH. If these are truly covalent interactions, then we expect them to again correlate with the promotion energy, although now two or three metal electrons need to be spin-decoupled from the nonbonding electrons. These promotion energies are also listed in Tables 3 and 4 and are taken from the calculations of Carter and Goddard [66]. In all cases, these are the values for promotion and electron decoupling for a $4s^13d^{n-1}$ electron configuration. Correlations with alternate definitions of E_p are not as satisfactory, as discussed previously [46]. Figure 3 shows that the correlation between the bond energies for M^+-CH_2 and E_p is again very good. Data for $D^o(M^+-CH)$ is too sparse to be definitive, but the results still suggest a reasonable correlation.

One interesting aspect of the comparison of the correlations for the MH^+, MCH_3^+, MCH_2^+ and MCH^+ molecules of Figure 3 is the nearly parallel slopes of the linear regression lines to the data. This suggests that the bonding characteristics of these species are similar, consistent with the suggested covalent nature of the bonding in all cases. Additionally, we find that the intrinsic bond energies for formation of double and triple bonds in the M^+-CH_2 and M^+-CH systems are 425 ± 14 and 562 ± 22 kJ/mol, respectively. This intrinsic metal-carbon double BDE is close to that for the saturated species, $D^o[(CO)_5Mn^+=CH_2] = 435 \pm 13$ kJ/mol [90]. Also the intrinsic metal-ligand BDEs of M^+-H, M^+-CH_3, $M^+=CH_2$ and $M^+\equiv CH$ increase proportionally to the organic analogs, CH_3-H, CH_3-CH_3, $CH_2=CH_2$, and $CH\equiv CH$, respectively [13,91]. The ratios of these bond energies are approximately 1.0:1.0:1.7:2.3 in the metal-ligand case and 1.0:0.9:1.6:2.2 for the organic species.

TABLE 3. Bond Dissociation Energies of Transition Metal Ions with Ligands that can Form Double Covalent Bonds (kJ/mol)[a]

M	E_p	M^+-CH_2	M^+-NH	M^+-O	M^+-C
Sc	15	412(22)[b]	498(10)[c]	693(11)[d]	322(6)[d]
Ti	56	391(15)[e]	466(12)[c]	668(7)[d]	391(23)[d]
V	138	335(14)[f]	415(15)[g]	582(10)[d]	381(4)[d]
Cr	304	225(8)[h]		363(12)[i]	
Mn	216	295(13)[j]		288(13)[i]	
Fe	173	346(6)[k]	255(21)[l*]	341(6)[m]	
Co	164	324(10)[n]		324(6)[i]	
Ni	174	315(8)[n]		268(7)[i]	
Cu	292	267(7)[n]		160(14)[i]	

[a]Values are at 298 K with uncertainties in parentheses. Values derived from work other than ion beam data are marked by an asterisk. [b]Reference 85. [c]Reference 95. [d]Reference 111. [e]References 56 and 57. [f]Reference 86. [g]Reference 94. [h]References 58 and 87. [i]Reference 47. [j]Reference 79. [k]Reference 88. [l]Reference 112. [m]Reference 43. [n]Reference 81.

TABLE 4. Bond Dissociation Energies of Transition Metal Ions with Ligands that can Form Triple Covalent Bonds (kJ/mol)[a]

M	E_p	M^+-CH	M^+-N	M^+-O
Ti	56	508(14)[b]	501(13)[c]	668(7)[d]
V	173	482(6)[e]	449(6)[f]	582(10)[d]
Cr	381	314(33)[g]		363(12)[h]
Fe	261	423(29)[i*]		344(6)[j]
Co	210	418(29)[i*]		324(6)[h]

[a]Values are at 298 K with uncertainties in parentheses. Values derived from work other than ion beam data are marked by an asterisk. [b]References 56 and 57. [c]Reference 95. [d]Reference 111. [e]Reference 86. [f]Reference 94. [g]Reference 58. [h]Reference 47. [i]Reference 89. [j]Reference 43.

3.3.2. *Metal Ion-Ethylidene and -Propylidene Bond Energies.* Metal-ethylidene ions, M^+=$CHCH_3$, have been observed in five systems, Sc [41], V [78], Cr [59], Ni and Cu [61], but only in the latter four could thermochemical data be obtained. The thermochemistry cited below uses a heat of formation for $CHCH_3$ of 330 ± 8 kJ/mol, a value based on ab initio calculations of Pople et al. [92] and Trinquier [93] and roughly confirmed by experiment [88]. In the cases of V and Cr, the metal-ethylidene ion BDEs are 311 ± 21 and 218 ± 13 kJ/mol, respectively, close to the values for the respective metal-methylidene ion BDEs, Table 3. In contrast, the BDEs between Cr^+ and 1-propylidene, 151 ± 13 kJ/mol, and 2-propylidene, 163 ± 13 kJ/mol, are much weaker.

This difference could be due to barriers in the reactions that form these latter species [59], and is an observation that deserves additional study. Similar differences are also observed for $D^o(Ni^+=CHCH_3)$ = 236-286 kJ/mol and $D^o(Cu^+=CHCH_3) \approx 111$ kJ/mol compared with the $Ni^+=CH_2$ and $Cu^+=CH_2$ BDEs.

3.4. COVALENT AND DATIVE METAL-LIGAND BONDS

3.4.1. *M^+-NH_2 and M^+-OH*. Metal-amide and -hydroxide ion BDEs obtained in our work [94-98] are listed in Table 1. Our metal-amide BDEs are within experimental error of the two alternate literature measurements, $D^o(Fe^+-NH_2)$ = 280 ± 50 kJ/mol and $D^o(Co^+-NH_2)$ = 272 ± 33 kJ/mol [99]. Our hydroxide BDEs agree well with those of Michl and coworkers [22] for Ti, V, Cr, Mn, and Co, but our values differ from theirs for $ScOH^+$ (367 ± 13 kJ/mol) and $NiOH^+$ (177 ± 13 kJ/mol). Our value for $D^o(Sc^+-OH)$ is in better agreement with a theoretical calculation [100]. The Fe^+-OH BDE is a somewhat unusual case that is undergoing further study in our laboratories. Our preliminary value of 323 ± 11 kJ/mol compares nicely to that of Murad, 322 ± 19 kJ/mol [101], and is bracketed by those of Michl and coworkers, 357 ± 13 [22], and Cassady and Freiser, 305 ± 13 kJ/mol [102].

Our M^+-NH_2 and M^+-OH BDEs are compared with values for M^+-CH_3 in Figure 2. While CH_3, NH_2, and OH are similar groups that form only single covalent bonds in organic molecules, it can be seen that the metal-amide BDEs are consistently higher than the metal-methyl BDEs, and the metal-hydroxide BDEs are larger still. The most obvious difference in thinking about the bonding in these three types of metal-ligand complexes is that the NH_2 and OH ligands have lone-pair electrons which can be used to form dative bonds by donating into metal d orbitals. The hydroxide group has the potential for a stronger dative bonding interaction since it has has two lone-pairs of electrons, while NH_2 has only one. It can also be seen that the enhancements in the NH_2 and OH BDEs decrease as one moves to the right in the periodic table, Figure 2. This is easily explained by considering the occupation of the 3d orbitals on the metal ion. For the early metals, the dative bonds are formed by donating the nitrogen or oxygen lone-pair electrons into *empty* 3d orbitals, while for the late metal ions, the 3d orbitals are already partially occupied. The conclusion that there is both covalent and dative bonding in these unsaturated gas-phase molecules has been confirmed by recent calculations on M^+-NH_2 (M = Sc - Mn) and Sc^+-OH [100,103-105]. These theoretical results go on to more fully characterize the electronic nature of these metal-ligand bonds.

3.4.2. *Correlation with Promotion Energies*. A more quantitative understanding of the strength of the dative bonding interaction can be gained by correlation with promotion energy, Figure 4. We use the same definition of E_p as for the M^+-H and M^+-CH_3 species since this clearly accounts for the periodic trends in the covalent bonding, Figure 3. The most obvious characteristic of this plot is that the M^+-NH_2 and M^+-OH BDEs do not correlate with E_p nearly as well as the M^+-H and M^+-CH_3 BDEs do; however, the deviations are instructive and easily rationalized.

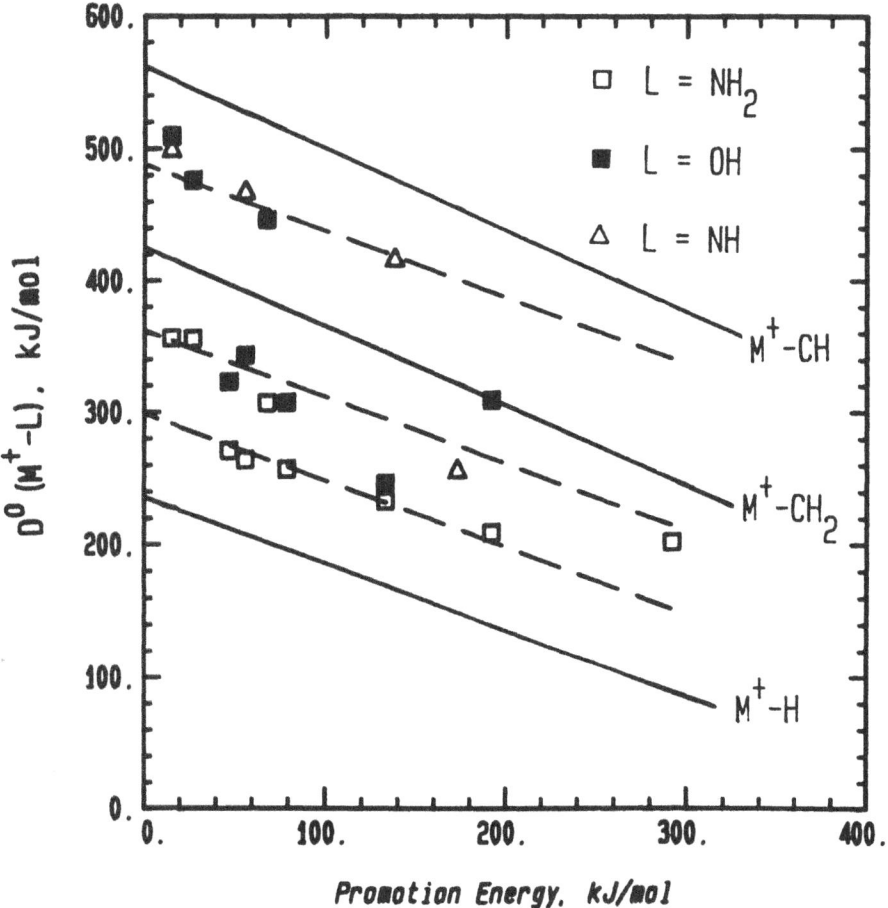

Figure 4. Transition metal ion ligand bond energies vs. the atomic metal ion promotion energy to a $4s^1 3d^{n-1}$ spin-decoupled state (see text). Results for NH_2 (open squares) and OH (closed squares) ligands are taken from Table 1. Results for NH (open triangles) ligands are taken from Table 3. The full lines are the linear regression fits to the MH^+, MCH_2^+ and MCH^+ data shown in Figure 3. The dashed lines are the MH^+ correlation line plus 63, 126, and 252 kJ/mol, where 126 kJ/mol is the average energy for a full dative bond, see text.

For the early metals, Sc, Ti, and V, the M^+-NH_2 BDEs are an average of 117 ± 12 kJ/mol stronger than $D°(M^+$-$H)$, and approach the strength of the M^+-CH_2 correlation. (MH^+ species are used for comparison here to eliminate the contribution of the polarizability to the M^+-L single BDE. This leads to a more consistent characterization of the dative bond energy than comparison to the MCH_3^+ species.) The hydroxide BDEs for these three metals are another 137 ± 17 kJ/mol stronger and approach the M^+-CH bonds in strength. Thus, while the strength of the dative bonds does not match covalent π bonds, the average dative bond contributes about 127 ± 17 kJ/mol for these early transition metal species. Note that these three metal ions have $4s^1 3d^{n-1}$ electron configurations where there are two or more empty 3d orbitals that can form a full dative bond by accepting electron density from the lone-pair electrons on the N or O atoms. Thus, the bond orders of the M^+-NH_2 and M^+-OH species approach 2 and 3 for the early metal ions.

For the later transition metals (Mn, Fe, Co, and Ni), the amide bond energies show a correlation that lies about halfway between that for M^+-H and that for the early metal amide ions. A similar observation is true for the late metal hydroxide ions, such that the correlation for the late metal-hydroxides is similar to the early metal-amides. The clear implication is that the dative bonding for the late transition metals, where the metal orbitals accepting electron density are all half-filled, is only half as strong as for the early metals, where these orbitals are empty. Thus, the bond orders of the M^+-NH_2 and M^+-OH species approach 1.5 and 2 for the late metal ions. On average, a dative bond contributes 62 ± 10 kJ/mol to the bond strength of these late transition metal species, or about half that for the early transition metal species. Overall, this analysis suggests that a full dative bond contributes an average of 126 ± 13 kJ/mol to the bond energies in these metal-amide and -hydroxide ions. The dashed lines in Figure 4 show the M^+-H correlation enhanced by $1/2$, 1 and 2 times this dative bond contribution.

The correlation between the metal-hydroxide ion BDEs and E_p shown in Figure 4 can be contrasted with one published by Michl and coworkers [22]. These workers chose to use a different definition of E_p for the early metals ($3d^n$) and the late metals ($4s^1 3d^{n-1}$). While a rough correlation is achieved, this is not retained for the BDEs of $ScOH^+$ and $NiOH^+$ given in Table 1. Further, the intrinsic bond energy that is obtained from their correlation is about 675 kJ/mol, a value that is much too high to be physically reasonable.

3.4.3. *Exceptions*. A couple of species that deserve further comment are $NiOH^+$, $CrNH_2^+$, $CrOH^+$, and $CuNH_2^+$. Figure 2 and Table 1 show that $D°(Ni^+$-OH) is no stronger than $D°(Ni^+$-$NH_2)$, indicating that there is only one dative interaction contributing to the bonding. Since a promoted Ni^+ has a $4s^1 3d^8$ electron configuration with two half-filled 3d orbitals, this observation can be rationalized by suggesting that the $3d\sigma$ orbital prefers to be singly occupied in order to form a strong covalent bond (consistent with $4s$-$3d\sigma$ hybridization), leaving only one singly occupied $3d\pi$ orbital for a dative interaction.

This rationale is also consistent with the BDE of $CrNH_2^+$ if we presume that the promoted $Cr^+(4s^13d^4)$ prefers an empty $3d\sigma$ orbital (or equivalently a singly occupied $4s$-$3d\sigma$ hybrid bonding orbital). Thus, the two $3d\pi$ orbitals on Cr^+ are singly occupied, such that NH_2 can form only one-half dative bond and thus falls in line with the late transition metal amides. Alternatively, $1\ ^1/_2$ dative bonds could be formed by keeping one $3d\pi$ orbital empty, but the occupied $3d\sigma$ orbital would then compromise the strength of the sigma bond. Calculations [105] find that the latter configuration corresponds to the ground state of $CrNH_2^+$ while the former configuration is 45 kJ/mol higher in energy. Such results show some of the limitations of the empirical correlations and the need for accurate theoretical calculations in understanding the bonding of these organometallic species in more detail. Similar considerations hold for the BDE of $CrOH^+$ which is 100 kJ/mol stronger than $D^o(Cr^+$-$NH_2)$, and thus falls in between the correlations for early and late transition metals, Figure 4. (Alternatively, one can note that the amide and hydroxide BDEs for Sc^+, Ti^+, V^+, and Cr^+ do correlate with one another although the slopes of these correlations are very different from those for the M^+-H, M^+-CH_3, M^+-CH_2, and M^+-CH BDEs.)

Finally, $D^o(Cu^+$-$NH_2)$ is quite a bit larger than what is expected from this correlation with promotion energy, suggesting that this molecule may bond by an alternate scheme. A likely possibility is that NH_2 donates its lone-pair electrons into the empty $4s$ orbital of ground state $Cu^+(3d^{10})$. In fact, the Cu^+-NH_2 BDE is quite close to that calculated for $D^o(Cu^+$-$NH_3)$ = 217 [106] or 206 [107] kJ/mol, a species where there is no covalent interaction. (By using an RRKM analysis, we have measured that $D^o(Cu^+$-$NH_3)$ is stronger than that for Co^+-NH_3 by about 5 kJ/mol [96], consistent with the theoretical difference of 6 kJ/mol [106]. If this is tied to the measurement of $D^o(Co^+$-$NH_3)$ by Marinelli and Squires [23], then we obtain an experimental estimate for $D^o(Cu^+$-$NH_3)$ of ~251 kJ/mol.)

3.4.4. *Comparison to Condensed-Phase BDEs*. In the condensed-phase, metal-amides are also found to exhibit such dative π bonding as evidenced by the geometry of these molecules [108]. Comparison of our intrinsic M^+-NH_2 and M^+-OH bond energies with M-N and M-O bond energies for coordinatively saturated species is limited since the data for the latter species is scant. For $M[N(CH_3)_2]_4$ complexes where M = Ti and Zr, average M-N bond energies of 339 and 381 kJ/mol, respectively, have been reported [109]. These values fall close to the intrinsic bond energy for a covalent and full dative bond (similar to the early transition metal-amide ions). When M = Mo, however, the mean M-N BDE falls to 255 $\pm$ 5 kJ/mol [110], a value similar to the intrinsic BDE for a single covalent bond with no dative contribution. This decrease in bond energy is qualitatively consistent with the relative BDEs of the early vs late metal-amide ions.

3.4.5. M^+-NH, M^+-O, M^+-N, and M^+-C. Mixtures of covalent and dative bonding are also apparent in the BDEs between metal ions and ligands capable of forming multiple covalent bonds. For the early transition metals, Sc, Ti, V and Cr, comparison of the BDEs for the isoelectronic series of ligands, CH_2, NH and O, in Table 3 and Figure 5 shows that dative bonds must be contributing to the bond strengths of these species. For the M^+-NH complexes of these metals, the BDEs are plotted vs E_p in Figure 4 and can be seen to have a correlation nearly the same as that for M^+-OH (even though the definition of E_p is different since NH can form two covalent and one dative bond while OH forms one covalent and two dative bonds). Harrison and coworkers [100] have previously commented on the rough equivalence of these effectively triple-bonded species. For these early metal oxide ions, Table 4 shows that MO^+ BDEs are even stronger than the covalent triple bond in M^+-CH or M^+-N. This has been discussed in detail elsewhere [47]. The ability of the lone-pair of electrons on the oxygen atom to increase the bond strength in these species is further demonstrated by comparison with the MC^+ BDEs [111], Table 3. While both C and O have two-unpaired valence electrons (and thus ^3P ground states), the carbon atom does not have the lone-pair of electrons and therefore forms bonds more comparable in strength to CH_2.

For the later transition metals, Mn, Fe, Co, and Ni, the metal-methylidene and metal-oxide ion BDEs are nearly identical, Figure 5, suggesting that the dative interaction is absent. The weak BDE found for CuO^+ and NiO^+ relative to the MCH_2^+ BDEs has been suggested to be a result of more ionic character in these metal-oxide ions [47]. The lone metal-nitrene ion BDE available for these metals, $D°(Fe^+$-NH$)$, has been measured by Buckner et al. [112]. This value seems anomalously low compared with $D°(FeCH_2^+)$ and $D°(FeO^+)$, Table 3 and Figure 5. Clearly, more information about these late transition metal nitrenes is required before any analysis of the periodic trends can be made.

3.4.6. *Metal-Vinyl Ions*. Another ligand that can show both covalent and dative bonding interactions is vinyl, C_2H_3. Metal ion-vinyl BDEs have been measured for Ti^+, 334 ± 24 kJ/mol [57]; for V^+, 368 ± 21 kJ/mol [3]; for Cr^+, 247 ± 10 kJ/mol [59]; and for Fe^+, 252 ± 10 kJ/mol [88]. On average, the three early metal ions have BDEs that are 128 ± 33 kJ/mol stronger than the MH^+ BDEs, nearly the same dative bond enhancement found above. For the late transition metal ion, Fe^+, the enhancement is substantially smaller (only 44 kJ/mol), in qualitative agreement with the trends found above.

3.5. METAL-LIGAND COMPLEXES WITH TWO COVALENTLY BOUND LIGANDS

While the determination of individual M-H and M-R bond strengths is clearly useful, a quantitative understanding of oxidative addition and reductive elimination processes requires a knowledge of the second metal ligand BDE as well, i.e. the bond strengths of RM-R, HM-R and HM-H. Unfortunately, the experimental determination of such second BDEs is more difficult than the first BDE for radical species like hydrogen atoms and alkyls. Nevertheless, this type of information has now been obtained for several systems [41,57,78,113-116], Table 5.

In the cases of Sc^+, Ti^+, V^+ and Fe^+, there are good reasons to believe that the information listed pertains to a species containing two covalent metal ligand bonds. For Co^+ and Ni^+, these ions are observed

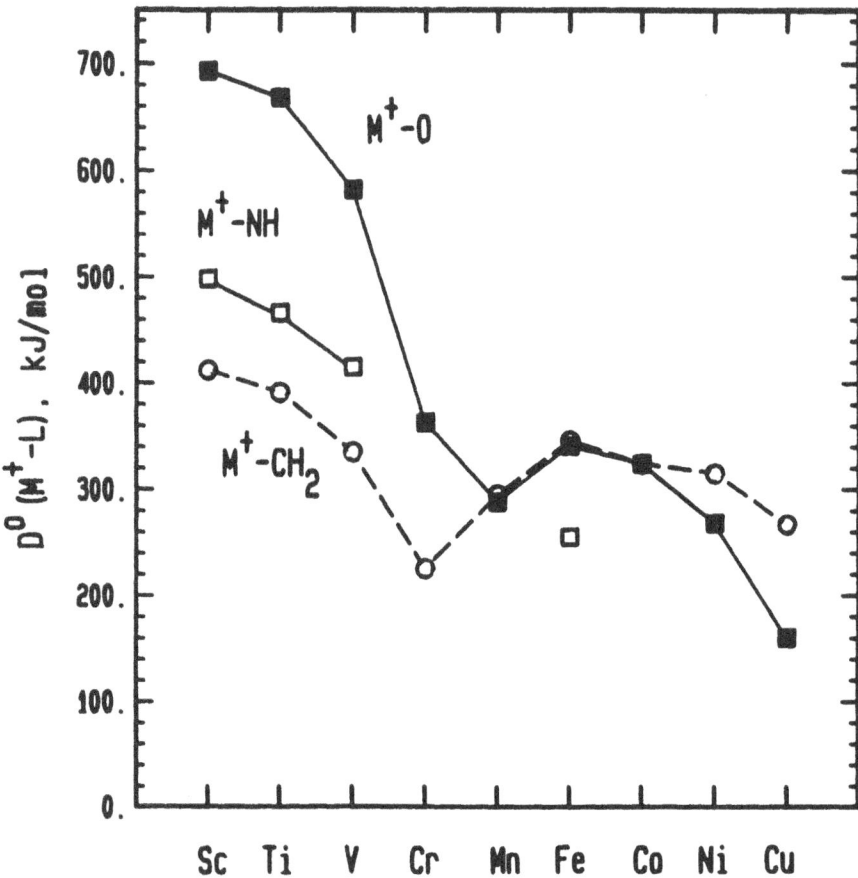

Figure 5. Transition metal ion bond energies to CH_2 (open circles), NH (open squares), and O (closed squares) across the first row of the periodic table.

TABLE 5. First and Second Bond Dissociation Energies (kJ/mol)[a]

M-R	M^+-R[b]	RM^+-R	Sum
Sc-H	239(9)	244(15)	483(13)[c]
Sc-H(CH$_3$)[d]	239(9)	264(13)	503(9)[e]
V-H(CH$_3$)[d]	202(6)	187(26)	389(25)[f]
Sc-CH$_3$	247(13)	242(15)	489(7)[e]
Ti-CH$_3$	225(8)	270(26)	495(25)[g]
V-CH$_3$	209(10)	201(23)	410(21)[f]
Fe-CH$_3$	242(10)	180(11)	422(15)[h]
Co-CH$_3$	205(13)	255(25)	460(21)[i,j]
Ni-CH$_3$	188(10)	>210	>398[j,k]

[a]Values are at 298 K with uncertainties in parentheses.
[b]Values are from Table 1. [c]Reference 41. [d]For these values,
$D°(M^+$-$R) = D°(M^+$-$H)$ and $D°(RM^+$-$R) = D°(HM^+$-$CH_3)$. [e]Reference 113.
[f]Reference 78. [g]Reference 57. [h]Reference 114. [i]Reference 115.
[j]These values may correspond to an $M^+ \cdot C_2H_6$ structure; see text.
[k]Reference 116.

to undergo an exothermic decarbonylation reaction with acetone to form
$MC_2H_6^+$ [116], a product that has reasonably been assigned as having a
dimethyl structure. This observation implies that the sum of the two
metal-methyl BDEs exceeds 398 kJ/mol. Hanratty et al. [115] have gone
on to study this product ion in the Co^+ case by kinetic energy release
measurements and have determined the sum of the two metal-methyl BDEs to
be 460 ± 21 kJ/mol assuming the dimethyl structure. A reanalysis
assuming that the structure is a $Co^+ \cdot C_2H_6$ adduct provides a Co^+-ethane
BDE of ~96 kJ/mol [117]. This agrees with a direct measurement in
our laboratory of this latter quantity, 98 ± 5 kJ/mol [118]. At this
point, no further studies have been performed to clarify the structure
of $NiC_2H_6^+$ formed by decarbonylation of acetone.

It has been noted before [13] that a reasonable correlation is
obtained between the sum of the bonds in $M(CH_3)_2^+$ and E_p for formation of
two covalent bonds as listed in Table 3. While the data is still
sparse, this correlation shows a line with a slope similar to those
observed for MH^+, MCH_3^+, and MCH_2^+ in Figure 3. Further, the intrinsic
dimethyl BDE, 508 kJ/mol, is close to twice that of the single methyl
BDE, 244 kJ/mol. This correlation also finds that the value listed in
Table 5 for the sum of the bond energies in $Co(CH_3)_2^+$ seems high. This
further suggests that this species probably has a $Co^+ \cdot$ethane adduct
structure instead.

4. Sequential Bond Energies

Recent technological developments in our laboratory and others
[7,119,120] have enabled the measurement of the sequential bond
energies of metal ions with several ligands attached. Most of these
studies involve ligands that bond primarily by dative (essentially,
electrostatic) interactions, in part because of the relative ease with
which they are prepared and characterized. We are primarily interested
in two interrelated facets of these sequential BDEs: 1) how do the bond
energies vary as the metal approaches saturation (either electronically,
in terms of reaching a stable 18-electron valence configuration, or
sterically, in terms of reaching a complete solvent shell), and 2) can
these BDEs help elucidate how the valence electrons on the metal
reorganize to accommodate the ligand shell. The latter consideration
may allow an understanding of the types of ligand spheres that put the
metal in configurations that are suitable for making additional covalent
or dative metal-ligand bonds or for activating particular kinds of
chemistry.

4.1. M^+-CO SEQUENTIAL BOND ENERGIES

Table 6 lists experimental sequential bond energies for the $(CO)_x M^+$-CO
molecules (where M = Cr, Mn, Fe, and Ni). Results for the chromium
[121] and iron [9] systems were obtained in our laboratories by
measuring the thresholds for the CID reactions (2). Results for the
manganese system were obtained by Dearden *et al.* [122] from analysis
of kinetic energy release distribution (KERD) measurements of metastable
decomposition. Results for the nickel system were obtained by
photoionization measurements of Distefano [123]. Except for Ni, it
is clear that the trends observed in these bond energies are not due
solely to simple electrostatic interactions since if this were the case,
the bond energies would be expected to gradually decrease as the ligand
sphere is increased due to ligand-ligand repulsions. To understand the
bonding in these molecules, we must account for the electronic
reorganization that occurs at the metal center as CO ligands are added.

TABLE 6. Sequential Bond Energies of $M^+(CO)_x$ (kJ/mol)[a]

x	M = Cr^b	Mn^c	Fe^c	Ni^e
1	86(8)	> 29	131(8) $[153(8)]^f$	203(14)
2	85(8)	<105	151(14)	150(14)
3	52(12)	130(25)	66(5)	128(11)
4	62(10)	84(13)	103(7)	43(3)
5	63(12)	67(13)	112(4)	
6	130(20)	134(21)		

[a] Values are at 0 K with uncertainties in parentheses.
[b]Reference 121. [c]Reference 122. [d]Reference 9. [e]Reference 123.
[f]The value chosen depends on whether the observed dissociation
  is to the ^4F or ^6D (bracketed value) state of Fe^+; see text.

4.1.1. *Fe⁺-CO Dissociation*. Of all these metal carbonyl ions, Fe^+-CO has been studied the most. We measure that this molecule dissociates when 153 ± 8 kJ/mol of energy is added at 0 K, in reasonable agreement with other work (see discussion in reference 9). Figure 6 shows the potential energy surfaces for dissociation of this species based on the molecular parameters calculated by Barnes, Rosi, and Bauschlicher [124]. (Note that there are *no* repulsive curves in this figure, in contrast to other qualitative potential energy surfaces that have been shown for such metal-carbonyl complexes [122]. The behavior shown here is correct since the bonds dissociate heterolytically [125].) This figure shows that a spin change is required for the low-spin $FeCO^+(^4\Sigma^-)$ ground state to dissociate to the high-spin $Fe^+(^6D)$ + CO ground state. This means that there is a possible ambiguity in our measurement of the CID threshold for $FeCO^+$; namely, does our measured threshold correspond to formation of $Fe^+(^6D)$ + CO, a spin-forbidden process, or does it occur along the spin-allowed pathway to produce excited state $Fe^+(^4F)$ + CO? While there is a great deal of information in the literature concerning this bond energy, none of it allows this question to be answered definitively; however, we believe that this energy corresponds to the spin-allowed dissociation pathway, as discussed elsewhere [9]. Thus, the measured dissociation threshold needs to be corrected for the splitting between the ^4F excited state and ^6D ground state of Fe^+, resulting in a Fe^+-CO bond energy of 131 ± 8 kJ/mol, Table 6.

4.1.2. *Iron Carbonyl Ions*. The ground state of $Fe(CO)_2^+$ has been calculated to be $^4\Sigma_g^-$ [124], such that no spin change occurs when another carbonyl ligand is added to $FeCO^+$. Consequently, the bond energy of $(CO)Fe^+$-CO, 151 ± 14 kJ/mol, is nearly the same as Fe^+-CO dissociating along the spin-allowed pathway, 153 ± 8 kJ/mol. We anticipate another spin change must occur as more CO ligands are added, however, since $Fe(CO)_5^+$ is expected to have doublet ground state based on removing a single electron from the singlet $Fe(CO)_5$ species. We have recently suggested [9] that this spin change occurs between $Fe(CO)_2^+$ and $Fe(CO)_3^+$, i.e. that the latter molecule as well as $Fe(CO)_4^+$ and $Fe(CO)_5^+$ have doublet ground states. This is based on the observation that $D^0[(CO)_2Fe^+$-CO] is substantially lower than the BDEs for the other iron carbonyls, Table 6. This can be explained by again referring to Figure 6 and imagining that the spin-allowed dissociation process along a doublet surface requires an energy comparable to $D^0[(CO)_3Fe^+$-CO] $\approx D^0[(CO)_4Fe^+$-CO] ≈ 108 kJ/mol, dissociations that both conserve spin in this proposal. A weak bond energy would then indicate that $Fe(CO)_3^+$ in a low-spin doublet ground state dissociates via the formally spin-forbidden process to form $Fe(CO)_2^+$ in its high-spin quartet ground state. Since no further electronic reorganization is required as two more CO ligands are added to $Fe(CO)_3^+$, these bonds are relatively strong compared to that for $Fe(CO)_3^+$. In essence, the first three CO ligands pay all the energy costs associated with reorganizing the electronic character at the Fe^+ center (the promotion energy), and the fourth and fifth CO ligands can bond with minimal interference. It will be of interest to verify these ideas with further experimental and theoretical work.

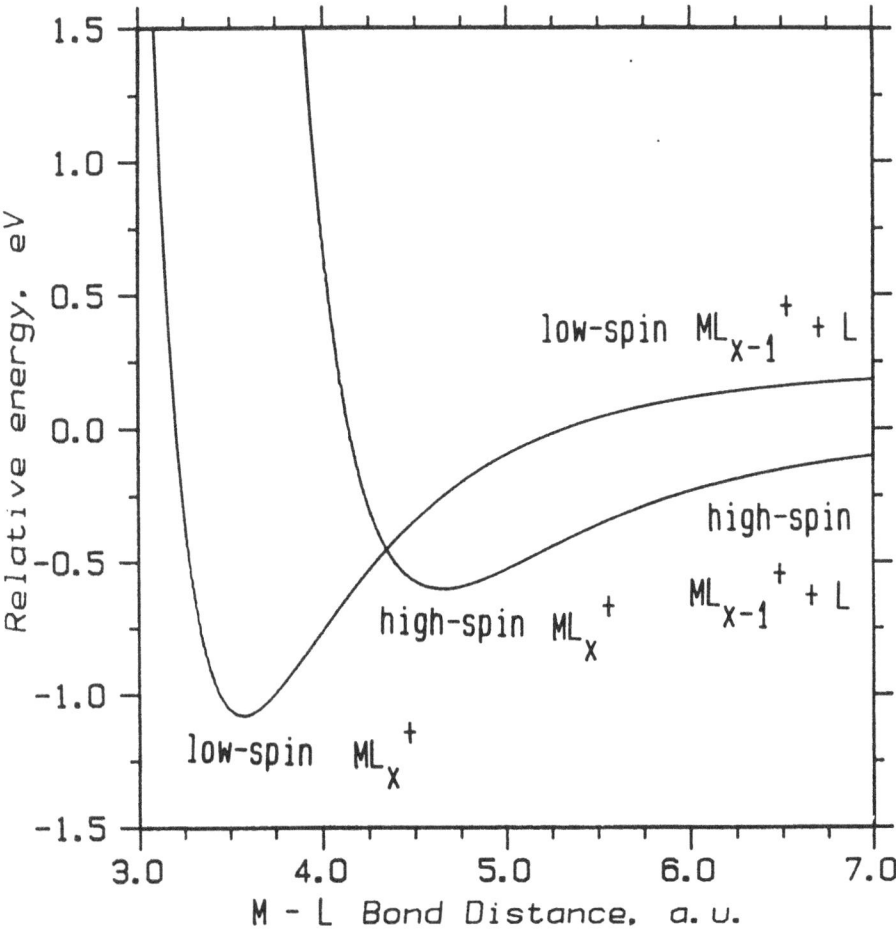

Figure 6. Potential energy curves for dissociation of a metal-ligand complex with both a high-spin and low-spin electronic state. The curves are Morse potentials with molecular parameters appropriate for dissociation of $FeCO^+$ as calculated in reference 124.

4.1.3. *Chromium Carbonyl Ions*. For the chromium carbonyl ions, we start by noting that the ground state of Cr^+ is $^6S(3d^5)$. Theoretical results [124] indicate that $CrCO^+$ and $Cr(CO)_2^+$ have $^6\Sigma^+$ and $^6\Sigma_g^+$ ground states, respectively. This is consistent with the similar bond energies for $CrCO^+$ and $Cr(CO)_2^+$, Table 6. Addition of one or two CO molecules to Cr^+ fails to induce a spin change, in contrast to the result for Fe^+, because the lowest lying quartet state of Cr^+ is 233 kJ/mol above the ground state [64], while the $Fe^+(^4F)$ excited state is only 29 kJ/mol above the $Fe^+(^6D)$ ground state.

As additional carbon monoxide molecules are added to Cr^+, the spin should eventually change since we anticipate that the saturated complex, $Cr(CO)_6^+$, should have a doublet ground state based on removing a single electron from the singlet neutral species. By using similar considerations as those discussed above for the iron carbonyl ions, the bond energies in Table 6 suggest that the tri- and tetracarbonyl chromium ions may have quartet ground states while the penta- and hexacarbonyl chromium ions may have doublet ground states. Thus, likely places for the spin changes are for dissociation of $Cr(CO)_3^+$ and $Cr(CO)_5^+$. Again verification of this suggestion by further experimental and theoretical work is clearly needed.

4.1.4. *Manganese Carbonyl Ions*. Mn^+ has a $^7S(4s^13d^5)$ ground state and the ground state of $Mn(CO)_6^+$ is known to be $^1A_{1g}$ [126,127]. Thus, there are three spin changes as the six carbonyl ligands are added to this metal ion. Theoretical results [124] indicate that $MnCO^+$ has a $^7\Sigma^+$ ground state, while the dicarbonyl has a $^5\Pi_g$ ground state, although a $^7A_1$ state is only 15 kJ/mol higher in energy. This result is consistent with the energy of the lowest lying quintet state of Mn^+, 113 kJ/mol above the ^7S state [64], being intermediate compared with the excitation energies for Fe^+ and Cr^+. The available experimental BDEs, Table 6, do not allow an assessment of whether there is a spin change here or not. The bond energy patterns are suggestive that there may be spin changes for dissociation of $Mn(CO)_4^+$ and $Mn(CO)_5^+$. Dearden *et al.* [122] attempted to study where the spin-changes occur in this system by examining rates of CO exchange. Except for the saturated $Mn(CO)_6^+$ complex, the exchange rates were fast and thus provided no information.

4.1.5. *Nickel Carbonyl Ions*. The nickel system is the only one where the BDEs decrease monotonically as additional CO ligands are added. This is the trend that might be expected for purely electrostatic interactions. The reason is simple. Ni^+ has a $^2D(3d^9)$ ground state and $Ni(CO)_4^+$ is also expected to have a doublet ground state, such that there should be no spin changes as CO ligands are added to the atomic metal ion.

4.2. M^+-H_2O SEQUENTIAL BOND ENERGIES

Recently, two independent studies [22,23] reported thermochemistry for $M(H_2O)_x^+$ ($x = 1,2$). The most surprising result of these studies was that the second water binding energy was found to be greater than the first for most of the first row transition metal ions. This behavior differs from observations regarding alkali [128] and coinage metal ions [20] where the bond strengths decrease gradually as additional water molecules are added. Calculations by Rosi and Bauschlicher [31] verified these experimentally observed trends although the theoretical second bond energies were systematically closer to the first bond energies (an average difference of 5 ± 10 kJ/mol for V^+, Cr^+, Fe^+, Co^+, Ni^+ and Cu^+) than those found experimentally (an average difference of 25 ± 11 kJ/mol for these same metals).

We have now also measured the binding energies of water molecules to ions of the first transition series by CID and extended this work to the third and fourth water ligands [36]. Our preliminary results for $M(H_2O)_x^+$ ($x = 1,2$) are given in Table 7 and with few exceptions agree with values from the Michl [22] or Squires [23] measurements. We find that the second metal ion-H_2O BDEs for V^+, Cr^+, Fe^+, Co^+, Ni^+ and Cu^+ are an average of 12 ± 17 kJ/mol stronger than the first BDE, in good agreement with the theoretical results. Further, we find that $D°(H_2ONi^+\text{-}OH_2)$ is slightly weaker than $D°(Ni^+\text{-}OH_2)$, also in agreement with theory, but varying from the previous experimental observations [22,23]. Our BDEs for $M(H_2O)_x^+$ ($x = 3,4$), Table 7, are still undergoing more extensive evaluation (hence the relatively large uncertainties), but those for $Cu(H_2O)_x^+$ ($x = 3,4$) are in good agreement with the equilibrium measurements of Castleman and coworkers [20] and the CID results of Michl and coworkers [119]. We note that this agreement is obtained only when the vibrational energy of the ions is accounted for explicitly, as discussed in section 2.3.

The calculations of Rosi and Bauschlicher [31] find that the bonding between M^+ and H_2O molecules is primarily electrostatic, and that the second metal ion-water BDE can be greater than the first due to 4s-3d hybridization and promotion effects, as originally postulated by Marinelli and Squires [23]. In essence, the metal ion must be promoted to an electron configuration suitable for forming a strong dative metal-

TABLE 7. Sequential Bond Energies of $M^+(H_2O)_x$ (kJ/mol)[a]

M	x = 1	2	3	4
Ti	172(6)	137(10)	153(20)	125(20)
V	140(5)	153(4)	81(20)	80(20)
Cr	127(7)	148(6)	50(20)	65(20)
Mn	115(10)	74(19)[b]		
Fe	128(5)	164(4)	76(20)	82(20)
Co	159(5)	170(4)	68(20)	88(32)
Ni	180(4)	164(4)	71(20)	67(20)
Cu	157(4)	167(4)	67(4)	65(7)
			64(1)[c]	65(1)[c]

[a]All values, from reference 36 except as indicated, are at 0 K with uncertainties in parentheses.
[b]Reference 23. Values correspond to an unspecified temperature.
[c]Reference 20. Values are adjusted to 0 K.

ligand bond (by removing metal electron density from the bond axis). This promotion energy can be "paid" by the first ligand such that no more electronic reorganization is necessary for binding the second ligand, and thus its bond energy is stronger. This is the same concept used to understand the BDEs for $FeCO^+$ and $Fe(CO)_2^+$, discussed above.

A notable exception to this order of bond energies is Mn^+, where the second water is bound by ~40 kJ/mol less than the first. This is because the $^7S(4s^13d^5)$ ground state of Mn^+ is sufficiently stable that 4s-3d hybridization and promotion cost too much energy, and thus the $Mn(H_2O)_2^+$ complex is forced into a bent configuration (a calculated O-Mn-O bond angle of 94° [31]) where there is extensive ligand-ligand repulsion that weakens the second bond.

Due to the preliminary nature of the energetics associated with addition of a third and fourth water molecule to the metal ion, these values will not be discussed at length. If these values are confirmed, however, it is clear that there are interesting periodic trends in these sequential BDEs. While these bonds are generally weaker than the first two metal-water BDEs (a result that is probably due to increased ligand-ligand repulsion as the size of the ligand sphere increases [129]), there are metals that present exceptions. These variations are likely to be due to electronic reorganization, as discussed for the CO sequential bond energies. Indeed, Figure 7 shows that the pattern in the sequential bond energies of H_2O and CO to Fe^+ are fairly similar. Such comparisons of the sequential BDEs for different ligands should help identify and quantify the factors that influence such sequential bond energies.

4.3. M^+-HYDROCARBON BOND ENERGIES

In addition to exploring the bond energies of organometallic complexes that involve nontraditional oxidation states of the metal and nontraditional numbers of ligands, the gas-phase techniques discussed here also allow studies of nontraditional ligands. Such studies can help elucidate the nature of metal-ligand interactions by allowing a systematic variation of the molecular properties of the ligand.

4.3.1. *$M(CH_4)_x^+$ Sequential Bond Energies.* A good example of an unconventional ligand is methane. In contrast to carbon monoxide and water, CH_4 does not have a lone-pair of electrons that is readily accessible for forming dative bonds to M^+ nor does it have a dipole moment (like H_2O) or a strong dipole moment derivative (like CO). Thus, methane is expected to form weaker bonds than these ligands. Table 8 lists preliminary values for the sequential bond energies of $Fe(CH_4)_x^+$ and $Co(CH_4)_x^+$ as determined in our laboratories [130,131]. Our experiments provide no indications that these ions involve anything but intact methane ligands. To our knowledge, no other measurements of the thermochemistry for $M(CH_4)_x^+$ molecules have been reported.

Comparison of these values to those for other sequential metal-ligand BDEs shows that the pattern of bond energies for cobalt is very similar to that for $Co(H_2O)_x^+$, Table 7. On average, the Co^+-CH_4 BDEs are 57 ± 14% of the Co^+-H_2O BDEs, suggesting that similar electronic

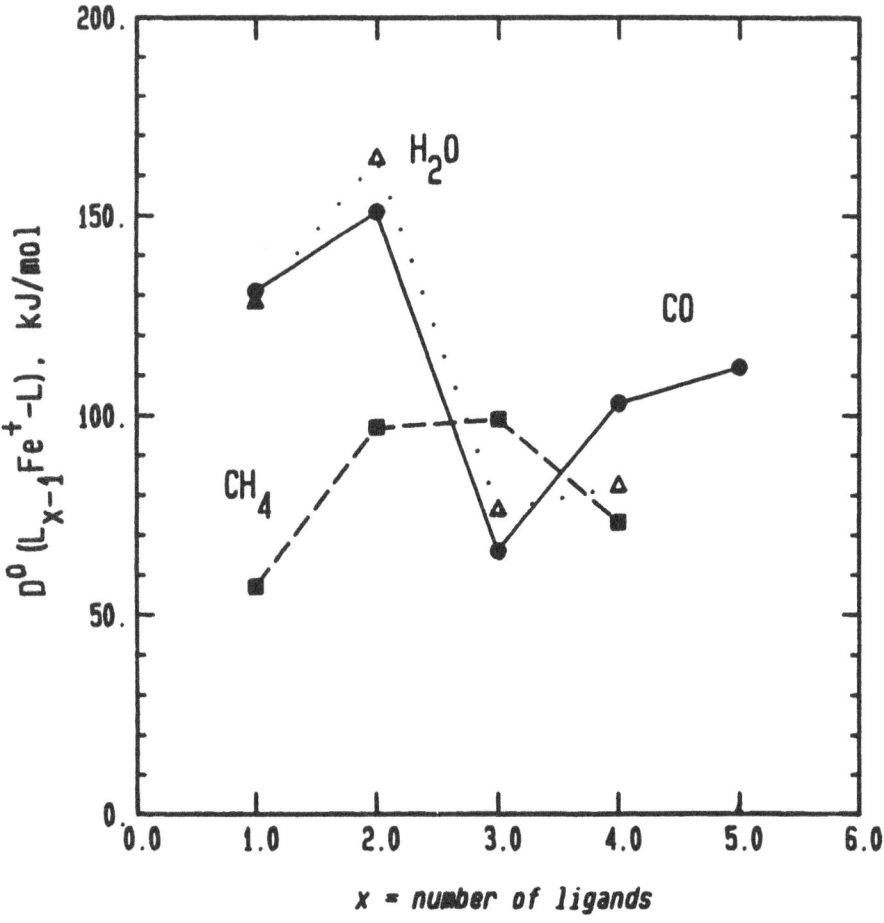

Figure 7. Sequential bond energies for FeL_x^+ complexes where L = CO (closed circles), H_2O (open triangles), and CH_4 (closed squares).

TABLE 8. Sequential Bond Energies of $M^+(CH_4)_x$ (kJ/mol)[a]

x	M = Fe	Co
1	57(3)	89(10)
2	97(4)	95(10)
3	99(6)	27(10)
4	73(6)	66(10)

[a]Values are at 0 K with uncertainties in parentheses.

considerations hold for both types of complexes. In contrast, the pattern of BDEs for iron is very different, as illustrated in Figure 7. This suggests that the promotion energy effects are very different for methane, a very weak-field ligand, as compared with CO and H_2O. As noted above, calculations find that $Fe(CO)^+$, $Fe(CO)_2^+$ [124] and $Fe(H_2O)_2^+$ [31] have quartet ground states while $Fe(H_2O)^+$ has a sextet ground state with a low-lying quartet state only 3-18 kJ/mol higher in energy [31]. Thus, $Fe(CH_4)^+$ almost certainly has a sextet ground state in which the BDE is weak due to the repulsion between the ligand and the 4s electron in the $^6D(4s^1 3d^6)$ ground state of Fe^+. This explanation is consistent with the relative bond strengths of Fe^+-CH_4 vs Co^+-CH_4, since the $^3F(3d^8)$ ground state of Co^+ has an empty 4s orbital. A possible explanation for why both the second and third Fe^+-CH_4 BDEs are stronger than the first is that $Fe(CH_4)_2^+$ and $Fe(CH_4)_3^+$ have quartet ground states (associated with the $^4F(3d^7)$ electronic state of Fe^+) that dissociate to products in quartet states, corresponding to an excited asymptote in the case of $Fe(CH_4)_2^+$. Note also that these BDEs are comparable to $D^o(Co^+-CH_4)$ and $D^o(CH_4Co^+-CH_4)$, which probably involve binding to $Co^+(^3F, 3d^8)$. The contrast in this BDE pattern with those for CO and H_2O, Figure 7, may also suggest that the spin change between quartet and doublet complexes postulated for these ligands does not occur for the CH_4 complexes until the fourth methane ligand is added.

4.3.2. *Metal-Ethane and -Propane Ions*. We have also used CID studies to measure the bond energies of Co^+ and Fe^+ ligated by the larger alkanes, ethane and propane. Again there are no indications that these BDEs refer to anything but simple adducts between the metal ion and an intact alkane molecule. The 0 K BDEs are found to increase systematically as the size of the alkane increases: $D^o(Co^+-C_2H_6) = 98 \pm 5$ kJ/mol, $D^o(Co^+-C_3H_8) = 126 \pm 6$ kJ/mol [118], $D^o(Fe^+-C_2H_6) = 64 \pm 6$ kJ/mol [114], and $D^o(Fe^+-C_3H_8) = 75 \pm 4$ kJ/mol [132,133]. This is consistent with an increase in the polarizability of the alkane, $\alpha = 2.6$, 4.4 and 6.2 Å^3 for CH_4, C_2H_6, and C_3H_8, respectively [134], although the BDEs are not directly proportional to the polarizability (in particular, the metal ion-methane BDEs are higher than might be expected based on the ethane and propane BDEs). This result indicates that the metal-ligand bond lengths change as the ligand changes (presuming a purely electrostatic interaction).

4.3.3. *Metal-Ethene and -Propene Ions*. Finally, CID methods have also been used to measure the BDEs for metal ion-alkene complexes. Our results at 0 K are $D°(Co^+-C_2H_4) = 180 \pm 7$ kJ/mol, $D°(Co^+-C_3H_6) = 183 \pm 6$ kJ/mol [118], $D°(Fe^+-C_2H_4) = 167 \pm 6$ kJ/mol [133], and $D°(Fe^+-C_3H_6) = 166 \pm 6$ kJ/mol [88]. These values compare favorably to those measured via KERD studies by van Koppen *et al.* [24]: 176, 184, 146, and 156 ± 21 kJ/mol, respectively. Compared with the metal ion-alkane BDEs, the alkene bond energies are substantially higher. While the polarizability of the alkenes is actually slightly less than the comparable alkane [134], the ability of the alkenes to donate its pair of π electrons to the metal center clearly allows a much stronger bond to be formed.

5. Reaction Intermediates and Isomers

In addition to simple metal-ligand bond cleavages (CID), collisional activation of ionic metal-ligand complexes can lead to more complicated decompositions depending on the complexity of the ligand. In several recent studies, we have measured the thresholds for such chemistry in order to provide thermochemical information regarding reaction intermediates that might be formed in the bimolecular chemistry between an atomic metal ion and the ligand. In particular, such "threshold collisional activation" (TCA) experiments have been used to study the potential energy surfaces associated with atomic metal ion reactions with hydrocarbons.

5.1. METAL DIMETHYL VS METAL ETHANE IONS

One example of such a study is our comparison of $Fe(CH_3)_2^+$ and $Fe^+ \cdot C_2H_6$ [114], both species that could be important intermediates in the interaction of Fe^+ with ethane. In this work, the iron dimethyl ion was generated by decarbonylating acetone with Fe^+ [116], while $Fe^+ \cdot C_2H_6$ was formed by collisional stabilization of Fe^+ and ethane in the high pressure environment of the flow tube. The ions made in this way showed distinct differences in their decomposition products upon collisional activation (CID). These differences were consistent with the proposed structures, although such a structural assignment cannot be made unequivocally by CID studies. Of particular interest here is the observation that the $Fe^+ \cdot C_2H_6$ isomer is more stable than $Fe(CH_3)_2^+$ by 15 ± 7 kJ/mol and that there must be a barrier between these two species that prevents conversion to the most stable isomer. This is not a necessary result since, as discussed in section 3.5, it seems likely that such a barrier is either absent or small for the case of $CoC_2H_6^+$.

5.2. METAL PROPENE VS METALLACYCLOBUTANE IONS

Another example of an ion beam study that can differentiate two different organometallic isomers is our work on $Fe^+ \cdot C_3H_6$ and $\overline{Fe-CH_2CH_2CH_2}^+$ [88]. The former ion could be made either by collisional stabilization of Fe^+ with propene or by dehydrogenation of propane. The latter species could be formed by collisional stabilization of Fe^+ with cyclopropane or by decarbonylation of cyclobutanone. Again, CID spectra showed large differences in the decomposition products of the two species, consistent with the proposed structures. As noted above, the Fe^+-propene BDE is 166 ± 6 kJ/mol, while decomposition of the metallacycle to Fe^+ + cyclopropane requires 133 ± 4 kJ/mol. Since propene is 36 ± 1 kJ/mol more stable than cyclopropane, the iron propene ion is 69 ± 7 kJ/mol more stable than the metallacycle, in good agreement with the value of 67 kJ/mol reported by van Koppen et al. [24].

5.3. THRESHOLD COLLISIONAL ACTIVATION OF METAL PROPANE IONS

One of the most interesting of these new TCA experiments is our study of $Fe^+ \cdot C_3H_8$ [132,133]. Several reactions were observed including the simple CID process to form Fe^+ + C_3H_8 beginning at 75 ± 4 kJ/mol. Also observed at low kinetic energies were the C-H and C-C bond activation products, $FeC_2H_4^+$ + CH_4 and $FeC_3H_6^+$ + H_2. In the bimolecular reaction of Fe^+ with C_3H_8 [33], these products are formed in exothermic and barrierless reactions, consistent with calculated heats of reaction of -85 and -41 ± 6 kJ/mol, respectively. Thus, formation of these products from $Fe^+ \cdot C_3H_8$ is calculated to have reaction thresholds of -10 and 34 ± 7 kJ/mol, respectively. What we oberve in the TCA experiment, however, is that these products have similar thresholds of 45 ± 12 kJ/mol, higher than the thermodynamic values for both processes. This energy was therefore assigned to the activation barrier for the rate limiting step, believed to be insertion of Fe^+ into a C-H bond of propane. This assignment was based on the conclusions of a previous study of the reaction of Co^+ with C_3H_8 [135], in which such a C-H bond activation transition state was identified. Using very different experimental methods, this study went on to measure that this transition state lay 11 ± 3 kJ/mol below the energy of Co^+ + C_3H_8.

A key feature of the TCA experiment is the realization that the failure to observe the thermodynamic threshold actually allows a more complete characterization of the potential energy surface for reaction. It is clear that similar experiments can be performed on a broad range of chemical systems.

6. Outlook

Ion beam techniques have proven to be one of the most reliable methods for obtaining thermochemical data of small gas-phase organometallic ions. Progress has been made in extending this methodology to the measurement of the energetics of simple neutral organometallics and of complexes with several ligands. The investigation of how metal-ligand bond energies vary as the degree of saturation (electronic and steric) is varied at the transition metal center should prove to be an active area of investigation in the coming years, as will studies of the details of the potential energy surfaces associated with simple organometallic reactions.

Acknowledgment

The authors thank their colleagues Dr. N. Aristov, Y.-M. Chen, N. F. Dalleska, Dr. J. L. Elkind, Dr. E. R. Fisher, Dr. R. Georgiadis, C. L. Haynes, Dr. K. Honma, Dr. F. Khan, Dr. S. K. Loh, Dr. R. H. Schultz, and Dr. L. S. Sunderlin for their contributions to these studies. The National Science Foundation has provided continuing support for this work.

References

1. P. B. Armentrout and J. L. Beauchamp, J. Am. Chem. Soc. 102, 1736 (1980).
2. K. M. Ervin and P. B. Armentrout, J. Chem. Phys. 83, 166 (1985).
3. N. Aristov and P. B. Armentrout, J. Am. Chem. Soc. 108, 1806 (1986).
4. R. Georgiadis and P. B. Armentrout, J. Am. Chem. Soc. 108, 2119 (1986).
5. K. M. Ervin and P. B. Armentrout, J. Chem. Phys. 84, 6738 (1986).
6. P. B. Armentrout, in *Advances in Gas Phase Ion Chemistry*, Vol. 1, edited by N. G. Adams and L. M. Babcock (JAI Press, Greenwich, Conn) in press.
7. R. H. Schultz and P. B. Armentrout, Int. J. Mass Spectrom. Ion Processes 107, 29 (1991).
8. D. A. Hales, L. Lian, and P. B. Armentrout, Int. J. Mass Spectrom. Ion Processes 102, 269 (1990).
9. R. H. Schultz, K. Crellin, and P. B. Armentrout, J. Am. Chem. Soc. 113, 8590 (1991).
10. E. Teloy and D. Gerlich, Chem. Phys. 4, 417 (1974).
11. N. R. Daly, Rev. Sci. Instrum. 31, 264 (1959).
12. J. D. Burley, K. M. Ervin, and P. B. Armentrout, Int. J. Mass Spectrom. Ion Processes 80, 153 (1987).
13. P. B. Armentrout, ACS Symp. Ser. 428, 18 (1990).
14. V. L. Talrose, P. S. Vinogradov, I. K. Larin, in *Gas Phase Ion Chemistry*; edited by M. T. Bowers, (Academic: New York, 1979) Vol. 1, p. 305.
15. P. B. Armentrout, L. F. Halle and J. L. Beauchamp, J. Chem. Phys. 76, 2449 (1982).
16. K. M. Ervin and P. B. Armentrout, J. Chem. Phys. 86, 2659 (1987).
17. M. E. Weber, J. L. Elkind, and P. B. Armentrout, J. Chem. Phys. 84, 1521 (1986).
18. J. L. Elkind and P. B. Armentrout, J. Phys. Chem. 88, 5454 (1984).
19. B. H. Boo and P. B. Armentrout, J. Am. Chem. Soc. 109, 3549 (1987).
20. P. M. Holland and A. W. Castleman, Jr., J. Am. Chem. Soc. 102, 6174 (1980); J. Chem. Phys. 76, 4195 (1982). K. I. Peterson, P. M. Holland, R. G. Keesee, N. Lee, T. D. Mark, and A. W. Castleman, Jr., Surf. Sci. 106, 136 (1981).
21. S. W. Buckner and B. S. Freiser, Polyhedron 7, 1583 (1988).
22. T. F. Magnera, D. E. David, and J. Michl, J. Am. Chem. Soc. 111, 4100 (1989).

23. P. J. Marinelli and R. R. Squires, J. Am. Chem. Soc. 111, 4101 (1989).

24. P. A. M. van Koppen, M. T. Bowers, J. L. Beauchamp, and D. V. Dearden, ACS Symp. Ser. 428, 34 (1990).

25. P. B. Armentrout and L. S. Sunderlin, in *Transition Metal Hydrides* edited by A. Dedieu, (VCH: New York), in press.

26. J. B. Schilling, W. A. Goddard, and J. L. Beauchamp, J. Am. Chem. Soc. 108, 582 (1986); 109, 5565 (1986); J. Phys. Chem. 91, 5616 (1987).

27. A. K. Rappe and T. H. Upton, J. Chem. Phys. 85, 4400 (1986).

28. L. G. M. Pettersson, C. W. Bauschlicher, S. R. Langhoff, and H. Partridge, J. Chem. Phys. 87, 481 (1987).

29. C. W. Bauschlicher, S. R. Langhoff, H. Partridge, and L. A. Barnes, J. Chem. Phys. 91, 2399 (1989).

30. J. B. Schilling, W. A. Goddard, and J. L. Beauchamp, J. Am. Chem. Soc. 109, 5573 (1987).

31. M. Rosi and C. W. Bauschlicher, J. Chem. Phys. 90, 7264 (1989); 92, 1876 (1990).

32. P. B. Armentrout and R. Georgiadis, Polyhedron 7, 1573 (1988).

33. R. H. Schultz, J. L. Elkind, and P. B. Armentrout, J. Am. Chem. Soc. 110, 411 (1988).

34. P. B. Armentrout, Ann. Rev. Phys. Chem. 41, 313 (1990).

35. E. R. Fisher and P. B. Armentrout, J. Chem. Phys. 94, 1150 (1991).

36. K. Honma, N. F. Dalleska, L. S. Sunderlin, R. H. Schultz, and P. B. Armentrout, work in progress.

37. E. R. Fisher, B. Kickel, and P. B. Armentrout, J. Chem. Phys. submitted for publication; work in progress.

38. D. E. Clemmer and P. B. Armentrout, J. Chem. Phys., submitted for publication.

39. See discussion in reference 3.

40. W. J. Chesnavich and M. T. Bowers, J. Phys. Chem. 83, 900 (1979).

41. L. Sunderlin, N. Aristov, and P. B. Armentrout, J. Am. Chem. Soc. 109, 78 (1987).

42. P. B. Armentrout and J. L. Beauchamp, J. Chem. Phys. 74, 2819 (1981); J. Am. Chem. Soc. 103, 784 (1981).

43. S. K. Loh, E. R. Fisher, L. Lian, R. H. Schultz, and P. B. Armentrout, J. Phys. Chem. 93, 3159 (1989).

44. P. B. Armentrout, in *Selective Hydrocarbon Activation: Principles and Progress*, edited by J. A. Davies, P. L. Watson, J. F. Liebman, and A. Greenberg, (VCH: New York, 1990), p. 467.

45. J. L. Elkind and P. B. Armentrout, Inorg. Chem. 25, 1078 (1986).

46. P. B. Armentrout, L. S. Sunderlin, and E. R. Fisher, Inorg. Chem. 28, 4436 (1989).

47. E. R. Fisher, J. L. Elkind, D. E. Clemmer, R. Georgiadis, S. K. Loh, N. Aristov, L. S. Sunderlin, and P. B. Armentrout, J. Chem. Phys. 93, 2676 (1990).

48. Y.-M. Chen, D. E. Clemmer, and P. B. Armentrout, J. Chem. Phys. 95, 1228 (1991).

49. J. L. Elkind, L. S. Sunderlin, and P. B. Armentrout, J. Phys. Chem. 93, 3151 (1989).

50. J. L. Elkind and P. B.Armentrout, Int. J. Mass Spectrom. Ion Processes **83**, 259 (1988).
51. J. L. Elkind and P. B. Armentrout, J. Phys. Chem. **89**, 5626 (1985).
52. J. L. Elkind and P. B. Armentrout, J. Chem. Phys. **86**, 1868 (1987).
53. J. L. Elkind and P. B. Armentrout, J. Chem. Phys. **84**, 4862 (1986).
54. J. L. Elkind and P. B. Armentrout, J. Am. Chem. Soc. **108**, 2765 (1986); J. Phys. Chem. **90**, 5736 (1986).
55. J. L. Elkind and P. B. Armentrout, J. Phys. Chem. **90**, 6576 (1986).
56. L. S. Sunderlin and P. B. Armentrout, J. Phys. Chem. **92**, 1209 (1988).
57. L. S. Sunderlin and P. B. Armentrout, Int. J. Mass Spectrom. Ion Processes **94**, 149 (1989).
58. R. Georgiadis and P. B. Armentrout, Int. J. Mass Spectrom. Ion Processes **89**, 227 (1989).
59. E. R. Fisher and P. B. Armentrout, J. Am. Chem. Soc. accepted for publication.
60. R. Georgiadis and P. B. Armentrout, Int. J. Mass Spectrom. Ion Processes **91**, 123 (1989).
61. R. Georgiadis, E. R. Fisher, and P. B. Armentrout, J. Am. Chem. Soc. **111**, 4251 (1989).
62. P. B. Armentrout, L. F. Halle, and J. L. Beauchamp, J. Am. Chem. Soc. **103**, 6501 (1981).
63. M. L. Mandich, L. F. Halle, and J. L. Beauchamp, J. Am. Chem. Soc. **106**, 4403 (1984).
64. C. Corliss and J. Sugar, J. Chem. Phys. Ref. Data **11** (1982).
65. C. E. Moore, Natl. Stand. Ref. Data Ser., Natl. Bur. Stand. **35**, Vol. I-III (1971).
66. E. A. Carter and W. A. Goddard, J. Phys. Chem. **92**, 5679 (1988). These authors use a somewhat different nomenclature than used here. The values of E_p equivalent to those in this paper are found in Table III of this reference are are called E(lost).
67. A. E. Alvarado-Swaisgood, J. Allison, and J. F. Harrison, J. Phys. Chem. **89**, 2517 (1985). A. E. Alvarado-Swaisgood and J. F. Harrison, J. Phys. Chem. **89**, 5198 (1985); *Ibid.* **92**, 2757 (1988).
68. M. A. Vincent, Y. Yoshioka, and H. F. Schaefer, J. Phys. Chem. **86**, 3905 (1982).
69. J. Halpern, Acc. Chem. Res. **15**, 238 (1982); Inorgan. Chim. Acta **100**, 41 (1985).
70. F. Ungvary, J. Organomet. Chem. **36**, 363 (1972).
71. B. de Vries, J. Catal. **1**, 489 (1962).
72. L. Vaska, Acc. Chem. Res. **1**, 335 (1968).
73. J.-Y. Saillard, and R. Hoffmann, J. Am. Chem. Soc. **106**, 2006 (1984).
74. T. Ziegler, V. Tschinke, and A. Becke, J. Am. Chem. Soc. **109**, 1351 (1987).
75. P. B. Armentrout and J. L. Beauchamp, J. Am. Chem. Soc. **103**, 784 (1981).
76. M. Brookhart and M. L. H. Green, J. Organomet. Chem. **250**, 395 (1983).
77. M. J. Calhorda and J. A. Simoes, Organometallics **6**, 1188 (1987).
78. N. Aristov, Thesis, University of California, Berkeley, 1986.

79. L. S. Sunderlin and P. B. Armentrout, J. Phys. Chem. **94**, 3589 (1990).

80. R. H. Schultz and P. B. Armentrout, J. Chem. Phys. **94**, 2262 (1991).

81. E. R. Fisher and P. B. Armentrout, J. Phys. Chem. **94**, 1674 (1990).

82. H. A. Skinner and J. A. Connor, in *Molecular Structure and Energetics: Physical Measurements*, edited by J. F. Liebman and A. Greenberg (VCH, New York, 1987), Vol. 2, p. 233.

83. J. Demuynck and H. F. Schaefer, J. Chem. Phys. **72**, 311 (1980).

84. S. P. Walch and C. W. Bauschlicher, J. Chem. Phys. **78**, 4597 (1983).

85. L. S. Sunderlin and P. B. Armentrout, J. Am. Chem. Soc. **111**, 3845 (1989).

86. N. Aristov and P. B. Armentrout, J. Phys. Chem. **91**, 6178 (1987).

87. R. Georgiadis and P. B. Armentrout, J. Phys. Chem. **92**, 7067 (1988).

88. R. H. Schultz and P. B. Armentrout, Organometallics, in press.

89. R. L. Hettich and B. S. Freiser, J. Am. Chem. Soc. **108**, 2537 (1986).

90. A. E. Stevens, Ph.D. Thesis, Caltech, Pasadena, California, 1981.

91. N. Aristov and P. B. Armentrout, J. Am. Chem. Soc. **106**, 4065 (1984).

92. J. A. Pople, K. Raghavachari, M. J. Frisch, J. S. Binkly, and P. v. R. Schleyer, J. Am. Chem. Soc. **105**, 6389 (1983).

93. G. Trinquier, J. Am. Chem. Soc. **112**, 2130 (1990).

94. D. E. Clemmer, L. S. Sunderlin, and P. B. Armentrout, J. Phys. Chem. **94**, 208 (1990).

95. D. E. Clemmer, L. S. Sunderlin, and P. B. Armentrout, J. Phys. Chem. **94**, 3008 (1990).

96. D. E. Clemmer and P. B. Armentrout, J. Phys. Chem. **95**, 3084 (1991).

97. D. E. Clemmer, N. Aristov, and P. B. Armentrout, J. Chem. Phys. manuscript in preparation.

98. D. E. Clemmer and P. B. Armentrout, work in progress.

99. S. W. Buckner and B. S. Freiser, J. Am. Chem. Soc. **109**, 4715 (1987).

100. J. L. Tilson and J. F. Harrison, J. Phys. Chem. **95**, 5097 (1991).

101. E. Murad, J. Chem. Phys. **73**, 1381 (1980).

102. C. J. Cassady and B. S. Freiser, J. Am. Chem. Soc. **106**, 6176 (1984).

103. A. Mavridis, K. Kunze, J. F. Harrison, and J. Allison, ACS Symp. Ser. **428**, 263 (1990).

104. A. Mavridis, F. L. Herrera, and J. F. Harrison, J. Phys. Chem. **95**, 6854 (1991).

105. S. T. Kapellos, A. Mavridis and J. F. Harrison, J. Phys. Chem. **95**, 6860 (1991).

106. S. R. Langhoff, C. W. Bauschlicher, H. Partridge, and M. Sodupe, J. Phys. Chem., in press.

107. H. J. Hoffman, P. Hobza, R. Cammi, J. Tomasi, and R. Zahradnik, J. Mol. Struct. (THEOCHEM) **201**, 339 (1989).

108. A thorough discussion of metal amide chemistry in solution can be found in: M. F. Lappert, P. P. Power, A. R. Sanger, and R. C. Srivastava, *Metal and Metalloid Amides* (Ellis Horwood Limited: West Sussex, England, 1980).

109. M. F. Lappert, D. S. Patil, and J. B. Pedley, J. C. S. Chem. Comm. 830 (1975).

110. J. A. Connor, G. Pilcher, H. A. Skinner, M. H. Chisholm, and F. A. Cotton, J. Am. Chem. Soc. 100, 7738 (1978).

111. D. E. Clemmer, J. L. Elkind, N. Aristov, and P. B. Armentrout, J. Chem. Phys. 95, 3387 (1991).

112. S. W. Buckner, J. R. Gord, and B. S. Freiser, J. Am. Chem. Soc. 110, 6606 (1988).

113. L. S. Sunderlin and P. B. Armentrout, Organometallics 9, 1248 (1990).

114. R. H. Schultz and P. B. Armentrout, J. Phys. Chem., in press.

115. M. A. Hanratty, J. L. Beauchamp, A. J. Illies, P. A. M. van Koppen, and M. T. Bowers, J. Am. Chem. Soc. 110, 1 (1988).

116. L. F. Halle, W. E. Crowe, P. B. Armentrout, and J. L. Beauchamp, Organometallics 3, 1694 (1984).

117. P. A. M. van Koppen, personal communication.

118. C. L. Haynes, E. R. Fisher and P. B. Armentrout, work in progress.

119. T. F. Magnera, D. E. David, D. Stulik, R. G. Orth, H. T. Jonkman, and J. Michl, J. Am. Chem. Soc. 111, 5036 (1989).

120. See the chapter by R. R. Squires, elsewhere in this volume.

121. F. Khan, D. E. Clemmer, R. H. Schultz and P. B. Armentrout, manuscript in preparation.

122. D. V. Dearden, K. Hayashibara, J. L. Beauchamp, N. J. Kirchner, P. A. M. van Koppen, and M. T. Bowers, J. Am. Chem. Soc. 111, 2401 (1989).

123. G. Distefano, J. Res. Natl. Bur. Stand. 74A, 233 (1970).

124. L. A. Barnes, M. Rosi, and C. W. Bauschlicher, J. Chem. Phys. 93, 609 (1990).

125. J. Simons and P. B. Armentrout, manuscript in preparation.

126. N. A. Beach and H. B. Gray, J. Am. Chem. Soc. 90, 5713 (1968).

127. J. K. Burdett, J. Chem. Soc. Faraday Trans. 2 70, 1599 (1974).

128. I. Dzidic and P. Kebarle, J. Phys. Chem. 74, 1466 (1970). S. K. Searles and P. Kebarle, Can. J. Chem. 47, 2619 (1969).

129. C. W. Bauschlicher, S. R. Langhoff, H. Partridge, J. E. Rice, and A. Komornicki, J. Chem. Phys. 95, 5142 (1991).

130. R. H. Schultz and P. B. Armentrout, manuscript in preparation.

131. C. L. Haynes and P. B. Armentrout, work in progress.

132. R. H. Schultz and P. B. Armentrout, J. Am. Chem. Soc. 113, 729 (1991).

133. R. H. Schultz and P. B. Armentrout, manuscript in preparation.

134. E. W. Rothe and R. B. Bernstein, J. Chem. Phys. 31, 1619 (1959).

135. P. A. M. van Koppen, J. Brodbelt-Lustig, M. T. Bowers, D. V. Dearden, J. L. Beauchamp, E. R. Fisher, and P. B. Armentrout, J. Am. Chem. Soc. 113, 2359 (1991).

A GUIDE TO DENSITY FUNCTIONAL THEORY AND ITS PRACTICAL APPLICATIONS TO THE ENERGETICS OF ORGANOMETALLIC SPECIES.

TOM   ZIEGLER
Department of Chemistry, University of Calgary
Calgary, Alberta, CANADA T2N 1N4

ABSTRACT. An evaluation is given of approximate Density Functional Theory as a practical tool in studies on organometallic energetics. The evaluation covers electronic excitations and ionizations, electron capture, conformational changes, molecular vibrations and bond energies as well as reaction profiles.

I. Introduction

Molecular energetics covers a variety of properties, **1**, encompassing electronic excitations and ionizations, electron capture, conformational changes, molecular vibrations and bond breakings as well as the flipping of an electronic spin. Most of these properties are important at one time or another in studies involving organometallic thermochemistry and kinetics. It is thus necessary in theoretical studies on organometalic energetics to make use of computational methods which are able to treat all the properties in **1** accurately, or at least on an "equal footing".

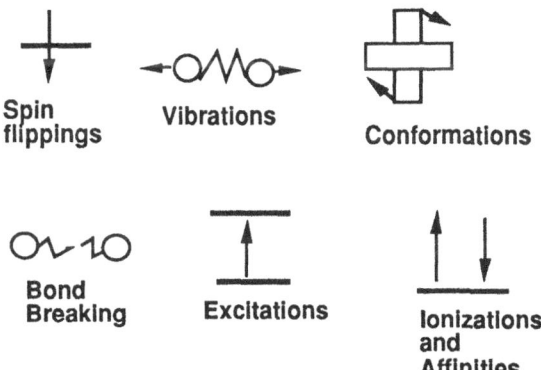

Spin flippings   Vibrations   Conformations

Bond Breaking   Excitations   Ionizations and Affinities

1

Traditionally most theoretical studies have been carried out by <u>ab initio</u> Hartree Fock methods or extensions including electron correlation. We shall discuss here an alternative approach based on Approximate Density Functional Theory[1] which over the past decade has emerged as a tangible and versatile computational method. It has been employed successfully to obtain thermochemical data[2,3] ; molecular structures[4,5] ; force fields and frequencies[6] ; assignments of NMR[7,8] -, photoelectron[9] -, E.S.R[10] - , and UV- spectra[9]; transition state

J. A. Martinho Simões (ed.), Energetics of Organometallic Species, 357–385.

structures as well as activation barriers[11] ; dipole moments[12] and other one-electron properties. Thus, approximate DFT is now applied to many problems previously covered exclusively by ab initio Hartree-Fock (HF) and post-HF methods. The recently acquired popularity of approximate DFT stems in large measure from its computational expedience which makes it amenable even to large size molecules at a fraction of the time required for HF or post-HF calculations. More importantly, perhaps, is the fact that expectation values derived from approximate DFT in most cases are better in line with experiment than results obtained from HF calculations. This is in particular the case for systems involving transition metals. An analysis of why approximate DFT affords more reliable results than HF has recently been published by Cook[13] and Karplus as well as Tschinke and Ziegler[14] .

2. General Theory

The total energy of an n-electron system can be written[15b] without approximations as

$$E = -\frac{1}{2}\sum_i \int \phi_i(\vec{r}_1)\nabla^2\phi_i(\vec{r}_1)d\vec{r}_1 + \sum_A \int \frac{Z_A}{|\vec{R}_A - \vec{r}_1|}\rho(\vec{r}_1)d\vec{r}_1$$

$$+\frac{1}{2}\int \frac{\rho(\vec{r}_1)\rho(\vec{r}_2)}{|\vec{r}_1 - \vec{r}_2|}d\vec{r}_1 d\vec{r}_2 + \sum_A \sum_{\neq B} \frac{Z_A Z_B}{|\vec{R}_A - \vec{R}_B|} + E_{xc} \qquad (1).$$

The first term in Eq. (1) represents the kinetic energy of n non-interacting[15] electrons with the same density $\rho(\vec{r}_1) = \sum \phi_i(\vec{r}_1)\phi_i(\vec{r}_1)$ as the actual system of interacting electrons. The second term accounts for the electron-nucleus attraction and the third term for the Coulomb interaction between the two charge distributions $\rho(\vec{r}_1)$ and $\rho(\vec{r}_2)$. The last term contains the exchange-correlation energy, E_{xc} . The exchange-correlation energy can be expressed in terms of the exchange-correlation energy densities[15], $\varepsilon_{xc}^{\gamma\gamma'}(\vec{r}_1)$, as

$$E_{XC} = \sum_\gamma \sum_{\gamma'} \int \rho_1^\gamma(\vec{r}_1)\varepsilon_{xc}^{\gamma\gamma'}(\vec{r}_1)d\vec{r}_1 \qquad (2),$$

where γ and γ' run over α as well as β spins . The functions $\varepsilon_{xc}^{\gamma\gamma'}(\vec{r}_1)$ contain all information about exchange and correlation between the interacting electrons as well as the influence[15b] of correlation on the kinetic energy. The functional form of the exact exchange-correlation energy densities, $\varepsilon_{xc}^{\gamma\gamma'}(\vec{r}_1)$, is not known. However, good approximations are available. The homogeneous electron gas has been particularly instrumental[16,17] in fostering useful approximate expressions for the exchange-correlation energy density . These expressions have $\varepsilon_{xc}^{\gamma\gamma'}(\vec{r}_1)$ as a simple function of the density and are referred to as local approximations . The simple HFS or $X\alpha$

method[16] retains only $\varepsilon_{xc}^{\gamma\gamma'}(\vec{r}_1)$ for $\gamma = \gamma'$, whereas the Local Density Approximations (LDA)[17] contains contributions from $\gamma = \gamma'$ as well as $\gamma \neq \gamma'$. Langreth[18a] and Mehl, Becke[18b] and Perdew[18c] have in a series of pioneering papers eliminated many of the shortcomings inherent in the local approximations by introducing correction terms based on electron density gradients. These theories are refered to as nonlocal. Nonlocal corrections are essential for a quantitative estimate of bond energies[18e-k] as well as metal-ligand bond distances[18l,m]. They are also of importance for other properties[18m].

The one electron orbitals, $\{\phi_i(\vec{r}_1); i=1,M\}$, of Eq. (1) are solutions to the set of one-electron Kohn-Sham equations[1]

$$[-\tfrac{1}{2}\nabla^2 + \sum_A \frac{Z_A}{|\vec{R}_A - \vec{r}_1|} + \int \frac{\rho(\vec{r}_2)}{|\vec{r}_1 - \vec{r}_2|} \, d\vec{r}_2 + V_{XC}] \, \phi_i(\vec{r}_1) = h_{KS} \, \phi_i(\vec{r}_1) = \varepsilon_i \, \phi_i(\vec{r}_1) \qquad (3)$$

where the exchange-correlation potential V_{XC} is given as the functional derivative of E_{XC} with respect to the density[1]

$$V_{XC}[\rho] = \frac{\delta E_{XC}[\rho]}{\delta \rho} \qquad (4).$$

It is relatively simple to derive an expression for the variational potential V_{XC} once a particular form for $\varepsilon_{xc}^{\gamma\gamma'}(\vec{r}_1)$ has been selected. With V_{XC} at hand the set of Kohn-Sham equations in Eq.(3) can be solved leading to a set of one electron Kohn-Sham orbitals $\{\phi_i(\vec{r}_1, i=1,M)\}$, from which the total energy E of Eq. (1) as well as other expectation values can be calculated. Scheme 1 presents in diagramatic form the relation between the various approximate DFT's.

3. Practical Implementations

The self-consistent (SCF) version of DFT, as formulated by Kohn and Sham[1], necessitates the solution of the Kohn-Sham equation given in (eq 3). This is accomplished in practice by deriving the potential V_{XC} from an approximate expression, $\tilde{E}_{XC}$, for the exact exchange-correlation energy, E_{XC}. The corresponding approximate Kohn and Sham equation reads

$$[-\tfrac{1}{2}\nabla^2 + V_N(\vec{r}_1) + V_C(\vec{r}_1) + \tilde{V}_{XC}(\vec{r}_1)] \, \phi_i(\vec{r}_1) = \tilde{h}_{KS} \, \phi_i(\vec{r}_1) = \varepsilon_i \, \phi_i(\vec{r}_1) \qquad (5a),$$

where

$$V_N(\vec{r}_1) = \sum_A \frac{Z_A}{|\vec{R}_A - \vec{r}_1|} \qquad (5b)$$

and

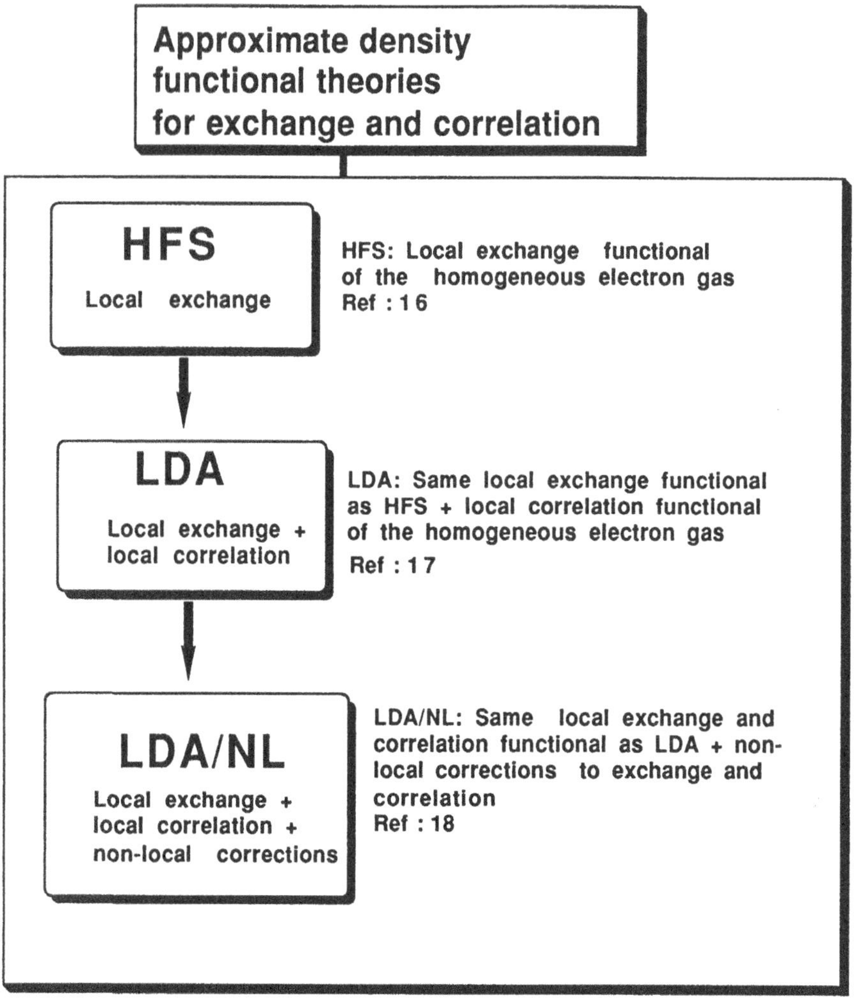

**Approximate density
functional theories
for exchange and correlation**

HFS

Local   exchange

HFS: Local exchange   functional
of the   homogeneous electron gas
Ref : 1 6

LDA

Local exchange +
local correlation

LDA: Same local exchange functional
as HFS + local correlation functional
of the homogeneous electron gas
Ref : 1 7

LDA/NL

Local exchange +
local correlation +
non-local   corrections

LDA/NL: Same   local exchange and
correlation functional as LDA + non-
local corrections to exchange and
correlation
Ref : 18

Scheme 1

$$V_C(\vec{r}_1) = \int \frac{\rho(\vec{r}_2)}{|\vec{r}_1 - \vec{r}_2|} \, d\vec{r}_2 \quad (5c)$$

whereas

$$\tilde{V}_{XC}[\rho(\vec{r}_1)] = \frac{\delta \tilde{E}_{XC}[\rho(\vec{r}_1)]}{\delta \rho(\vec{r}_1)} \quad (5d).$$

The set of solutions, $\{\phi_i(\vec{r}_1), i=1,n\}$, to (eq 5a) afford the electron density from which several expectation values can be evaluated, including the total energy.

It is customary in practical implementations to expand $\phi_i(\vec{r}_1)$ in terms of a known (basis) set of functions, $\{\chi_k(\vec{r}_1), k=1,M\}$, as

$$\phi_i(\vec{r}_1) = \sum_{k=1}^{k=M} C_{ik} \chi_k(\vec{r}_1) \quad (6).$$

The problem of solving the differential equation of (eq 5a) is now transformed into finding a set of eigenvectors $\{C_{ik}, k=1,n; k=1,M\}$ and corresponding eigenfunctions from the secular equation

$$\sum_{\mu=1}^{\mu=M} [F_{\nu\mu} - \varepsilon_i S_{\nu\mu}] = 0 \quad , \nu=1,M \quad (7a)$$

with

$$F_{\nu\mu} = \int \chi_\nu(\vec{r}_1) \tilde{h}_{KS} \chi_\mu(\vec{r}_1) d\vec{r}_1 \quad (7b)$$

and

$$S_{\nu\mu} = \int \chi_\nu(\vec{r}_1) \chi_\mu(\vec{r}_1) d\vec{r}_1 \quad (7c).$$

In the earliest implementation applied to molecular problems, Johnson[19] used scattered-plane waves[20] as a basis and the exchange-correlation energy was that of the HFS method (scheme 1). This SW-Xα method employed in addition an (muffin-tin) approximation[19] to the Coulomb potential of (eq 5c) according to which $V_C(\vec{r}_1)$ is replaced by a sum of spherical potentials around each atom. This approximation is well suited for solids for which the SW-Xα

method originally was developed. However, it is less appropriate in molecules where the potential around each atom might be far from spherical. The SW-Xα method is computationally expedient compared to standard ab initio techniques and has been used with considerably success[9] to elucidate the electronic structure in complexes and clusters of transition metals. However, the use of the muffin-tin approximation preclude accurate calculations of total energies. The method has for this reason not been successful[21] in studies involving molecular structures and bond energies.

The first implementations of self-consistent DFT, without recourse to the muffin-tin approximations, are due to Ellis and Painter[22], Baerends[23] et al., Sambe and Felton[24], Dunlap[25] et al. as well as Gunnarson[26] et al. Other implementations[27] and refinements have also appeared more recently. The accurate representation of $V_C(\vec{r}_1)$ is in general accomplished by fitting the molecular density to a set of one-center auxiliary functions[23,24] $f_\eta(\vec{r}_1)$, as

$$\tilde{\rho}(\vec{r}_1) \sim \sum_\eta a_\eta f_\eta(\vec{r}_1) \qquad (8)$$

from which $V_C(\vec{r}_1)$ now can be evaluated expediently by analytical[41] or numerical[46c] integration as

$$\tilde{V}_C(\vec{r}_1) \sim \sum_\eta a_\eta \int \frac{f_\eta(\vec{r}_2)}{|\vec{r}_1 - \vec{r}_2|} d\vec{r}_2 \qquad (9).$$

The matrix elements $F_{\nu\mu}$ and $S_{\nu\mu}$ can subsequently be obtained from numerical integration[28] as

$$F_{\nu\mu} = \sum_k \chi_\nu(\vec{r}_k) \tilde{h}_{KS} \chi_\mu(\vec{r}_k) W(\vec{r}_k) \qquad (10)$$

where $W(\vec{r}_k)$ is a weight factor[28]. The extensive use of numerical integration is amiable to modern vector-machines[29,30]. However, care must be exercised in order to calculate total energies accurately. This requirement has been met by the development of special algorithms[31] as well as new accurate integration schemes[28]. The adaptation of procedures based on numerical integration techniques makes it easy to deal even with complicated expressions for the potential, $\tilde{V}_{XC}(\vec{r}_1)$. The often intricate form of the potential $V_{XC}(\vec{r}_1)$ precludes on the other hand a direct analytical evaluation of $F_{\nu\mu}$. However, this problem can be side-stepped by fitting $V_{XC}(\vec{r}_1)$ to a set of auxiliary functions as

$$\tilde{V}_{XC}(\vec{r}_1) \sim \sum_\eta b_\eta g_\eta(\vec{r}_1) \qquad (11),$$

with the help of numerical integration techniques[24,25]. A substitution of (eq 11) and (eq 9) into (eq 7a) now allows for an analytical evaluation[24,25] of $F_{\nu\mu}$. The analytical procedure has the merit that advantage can be made of techniques already employed in ab initio methods. It also ensures accurate total energies in a relatively straightforward way. However, the price one must pay is the introduction of several sets of auxiliary functions.

A unique approach has lately been taken by Becke[32] in which (eq 5a) is solved directly without basis sets . This approach, which seems promising, was first applied to diatomic molecules and more recently to polyatomics. Alternative schemes have recently been proposed for diatomics[33]

The various SCF-schemes based on DFT are attractive alternatives to conventional ab initio methods in studies on large size molecules since the computational effort increases as n^3 with the number of electrons, n, as opposed to n^4 for the HF-method or n^5 for configuration-interaction techniques. The scope of Density Functional based methods has further been enhanced to include pseudo-potentials[34], relativistic-effects[35], as well as energy gradients of use in geometry optimization[4,36]. The existing program packages[29c,30c-f] are still not as user friendly, or readily available[29c,30c-f], as their ab initio counter-parts and much development work remains to be done. Information about the various implementations is given in Scheme 2.

4. Applications

4.1. BOND ENERGY CALCULATIONS

We shall begin our assessment of DFT-based methods by considering the calculation of bond dissociation energies. Results from this type of calculations should provide a clear indication of how well approximate DFT can account for molecular energetics. Of particular interest are the results obtained from the field of organometallic chemistry where the dearth of reliable experimental data is felt strongly. Accurate theoretical data in this area could afford a much needed supplement to the sparse available experimental data on metal-ligand bond energies, necessary for a rational approach to the synthesis of new transition metal complexes.

Table I compares bond energies calculated by various methods with experiment for a number of homonuclear diatomic molecules. The Hartree-Fock scheme (HF) is seen uniformly to under-estimate the bond energies. The discrepancy is especially large for the sample of multiple bonded molecules considered here. The disagreement is less pronounced for species with a single σ-bond[14]. It is by now very well known why the HF-scheme represents π-bonds so poorly[14]. Basically[14] , the molecular hole-correlation function $\bar{\rho}_{xHF}^{\gamma\gamma}(\vec{r}_1, s)$ differs considerably from its atomic counterpart for certain $\vec{r}_1$ where it is too diffuse as a function of the inter-electronic distance s.

The bond energies calculated by the HFS-method (HFS), in which E_{xc} is represented by the exchange energy of the homogeneous electron gas (Scheme 1) , are on the other hand in reasonable accord with experiment. The success of the HFS-scheme can be attributed[14] to the fact that $\bar{\rho}_{x\alpha}^{\gamma\gamma}(\vec{r}_1, s)$ changes little in the transition from atoms to molecule. We note that the HFS-scheme tend to overestimate bond energies slightly.

Methodology based on Approximate DFT

MS-Xα
1966

MS-Xα : Makes use of partial-waves as basis (20).
Relatively fast. Good for ionization potentials and excitation
energies (19). Total energies unreliable (21).
No geometry optimization. Full use of symmetry.
Has relativistic extension (35f). Make use of muffin-tin
approximation (20). Developed by K.H. Johnson (19) .

DV-Xα
1970

DV-Xα : Makes use of numerical atomic orbitals or STO's.
Avoids Muffin-tin approximation by fit of density (24).
Accurate total energies (27). Relativistic extension (35e).
Numerical integration of matrix elements by Diophantine
integration . Developed by Ellis and Painter (22).
Extensive improvements by Delley (D-MOL-program) including
new integration scheme (28c) and geometry optimization.

HFS-LCAO
1973

HFS-LCAO : Makes use of STO's . Accurate potentials (23).
Full use of symmetry. Relativistic extensions (35a,b). Highly
vectorized (29). Accurate total energies (31). Geometry
optimization (4). Accurate numerical integration (28a,d).
Many auxiliary property programs . Pseudo potentials (34a,b).
Embedding procedures (34d). Energy decomposition scheme
(31). Developed by Baerends,Snijders,Ravenek,Vernooijs and
te Velde (23,35,29)

LCGTO-LSD
1976

LCGTO-LSD : Make use of GTO's. Fit of exchange-correlation
and Coulomb potential $(_{24})$. Analytical calculation of matrix
elements (25). Accurate energies. Geometry optimization
(36b). Strongly vectorized (30b). First developed by
Dunlap (25) as well as Sambe and Felton (24). Extensive
improvements by Salahub (D-MON program)
and Andzelm (D-GAUSS-program) (36b).

NUMOL
1982

NUMOL : Unique basis-set free program (32). Accurate
numerical integration . Efficient generation of
Coulomb potential . Geometry optimization.
Developed by Becke (32).

Scheme 2

Table 1. Bond Dissociation Energies for Diatomic Molecules in eV.

	Expt.[a]	HF[b]	HFS[c]	LDA[d]	LDA-NL[e]
B_2	3.1	0.9	3.0	3.9	3.2
C_2	6.3	0.8	6.2	7.3	6.0
N_2	9.9	5.7	9.6	11.6	10.3
O_2	5.2	1.3	7.1	7.6	6.1
F_2	1.7	-1.4	3.3	3.4	2.2

[a] Ref. 32a. [b] Hartree-Fock calculations[32a].
[c] Hartree-Fock-Slater calculations[32a] with $\alpha_{ex} = .7$.
[d] LDA-calculations[18j] with $\alpha_{ex} = .666$.
[e] LDA calculations[18j] plus non-local corrections to exchange[18b] and correlation[18c].

       The HFS-scheme lacks correlation between electrons of different spins. This type of correlation is introduced in the LDA energy expression which contains exchange as well as correlation of the homogeneous electron gas (Scheme 1). It follows from Table I that the LDA method affords even larger bond energies than the HFS-scheme. The LDA method adds correlation between electrons of different spins to the HFS-energy expression. Correlation between electrons of different spins is roughly proportional to the number of spin-paired electrons. This type of correlation is as a consequence more important (stabilizing) in the molecule than in the constituting atoms since the molecule in most cases has more spin-paired electrons than the constituting atoms. Correlation will as a result increase the bond energy compared to the HFS-method. The tendency of either HFS or LDA to overbind is underlined further in Table II where we present theoretical values[18i] for the first CO-dissociation energy in a number of metal-carbonyls. Both HFS and LDA are seen to overestimate the M-CO bond strengths by nearly 100 %. A deviation of this magnitude is clearly unacceptable in reactivity studies. In fact, it seemed in the last part of the 70s and the early part of the 80s as if approximate Density Functional methods would be unable to deal with chemical energetics.

Table 2. First Metal-Carbonyl Dissociation Energy[a] in a Number of Metal Carbonyls

Molecule	HFS[b]	LDA[b]	LDA/NL[c]	Exp.[c]
$Cr(CO)_6$	278	276	147	162
$Mo(CO)_6$	226	226	119	126
$W(CO)_6$	247	249	142	166
$Ni(CO)_4$	194	192	106	104

[a]All energies in kJ/mol. [b]V.Tschinke,unpublished results. [c]Ref. 18i .

This situation was changed by the development of gradient based corrections due to Becke[18b],Perdew[18c] and others[18]. Thus, adding the non-local exchange[18b] and correlations[18c] corrections to the LDA-energy expressions (Scheme 1) improves considerably the agreement between theory and experiment, see LDA-NL results of Tables I-II. The largest effect comes from the non-local correction to the exchange whereas the influence of the non-local corrections to the correlation term is rather modest .

We conclude this section by a comparison between our calculated[18e,h] results for D(M-R) (R = H, CH3) and the few available experimental data ,Table 3. We find in general a good agreement with the experimental bond energies. Also, the stability order D(M-H) > D(M-CH3) in middle and late transition metal complexes supported by our theoretical study is consistent with data on organometallic reactions in which M-H and M-CH3 bonds are formed or broken. Thus, CO will readily insert into a M-CH3 bond whereas the corresponding insertions into M-H bonds are virtually unknown, and methyl has likewise a larger migratory aptitude toward most other ligands than hydride. The H2 molecule is known to add oxidatively and exothermically to several metal fragments where the corresponding oxidative additions of the H-Alkyl and Alkyl-Alkyl bonds are unknown and probably endothermic as a consequence of the weak M-R bond.

At this stage, it is important to comment on the role of the non-local correction to the exchange in the calculated bond energies. For middle and late transition metal complexes, the non-local correction reduces significantly the D(M-CH3) values (by 105 kJ mol^{-1} for CH3-Mn(CO)5) whereas the corresponding D(M-H) bond energies are decreased to a lesser extent (by 13 kJ mol^{-1} for H-Mn(CO)5).

Thus, it is apparent that Becke's non-local correction[18b] to the exchange is essential to assure the good agreement of the LDA/NL results with experiment, whereas the HFS and the LDA methods not only tend to give too large bond energies, but in some cases predict the wrong order for the M-H and M-CH3 bond strengths.

Table 3   Calculated and experimental values for the bond energies D(M-R) (R = H, CH$_3$) (kJ mol^{-1}).

M-H	LDA/NL	Exp.	M-CH$_3$	LDA/NL	Exp.
Cl$_3$Th-H	318.0	~335.[a]	Cl$_3$Th-CH$_3$	333.9	~335.[a]
Cl$_3$U-H	293.3	319.7[a]	Cl$_3$U-CH$_3$	302.1	302.9[a]
Cl$_3$Ti-H	250.7	—	Cl$_3$Ti-CH$_3$	267.5	—
Cl$_3$Zr-H	297.2	—	Cl$_3$Zr-CH$_3$	309.5	—
Cl$_3$Hf-H	313.5	—	Cl$_3$Hf-CH$_3$	326.6	—
H-Mn(CO)$_5$	225	213[b]	CH$_3$-Mn(CO)$_5$	153	153[b]
H-Tc(CO)$_5$	252	—	CH$_3$-Tc(CO)$_5$	178	—
H-Re(CO)$_5$	282	—	CH$_3$-Re(CO)$_5$	200	—
H-Co(CO)$_4$	230	238[c]	CH$_3$-Co(CO)$_4$	160	—
H-Rh(CO)$_4$	255	—	CH$_3$-Rh(CO)$_4$	190	—
H-Ir(CO)$_4$	286	—	CH$_3$-Ir(CO)$_4$	212	—

[a] Experimental bond energies from Ref. 37 correspond to Cp$_2$ MCl-R systems.   [b] Ref. 38.
[c] Ref. 39.

It seems at the present time that the LDA method augmented with non-local exchange and correlation corrections (LDA-NL) represents the most efficient and accurate method for the evaluation of bond energies within the Density Functional framework. Calculations on metal carbonyls[18i], binuclear metal complexes[40], alkyl and hydride complexes[18e,h] as well as complexes containing M-L bonds for a number of different ligands[18h],have shown that LDA-NL afford metal-ligand and metal-metal bond energies of nearly chemical accuracy ($\pm$ 5 Kcal mol^{-1}) . Becke[32d,e] has recently carried out accurate basis set free calculations on the bond energy in a number of smaller molecules. Calculations in which relativistic effects are taken into account have also appeared[41].

4.2 MOLECULAR STRUCTURES

The full optimization of molecular structures has been carried out routinely by ab-initio methods since the algorithm of analytical energy gradients was developed in the early 1970's. Ab-initio methods at the Hartree-Fock level usually yield satisfactory equilibrium structures for main group molecules[42]. More accurate electron-correlation calculations can also be carried out to improve the agreement between theory and experiment as long as the size of the molecule under investigation remains reasonably small. The size of systems that can be handled by such calculations has been constantly increasing for the past two decades due to the fast development of computer technology. The reasonable performance of the Hartree-Fock method on main group molecules seems not to be transferable to transition metal complexes[43]. Thus, a discrepancy of 0.1 Å between HF results and experimental findings is not uncommon for metal-ligand and metal-metal bonds . Faegri[44] et al. have pointed out that such discrepancies are most likely due to the inherent shortcomings of the HF method rather than basis set limitations. Thus post-HF methods are essentially necessary in the study of transition metal compounds. It is however unlikely that ab-initio methods beyond the Hartree-Fock level can be applied routinely to systems involving transition metal complexes in the foreseeable future.

Density functional theory (DFT) has been widely applied to the investigations of systems involving transition metal complexes, and it has proven to be a very successful alternative to the computationally more demanding ab initio methods[1]. Geometry optimizations by local DFT-methods such as the Hartree-Fock-Slater (HFS) scheme or the Local-Density-Approximation (LDA) are now feasible due to the successful implementation of analytical energy gradients[4-5,36]. The agreement between experimental geometries and structures optimized by DFT is in general good, as demonstrated by Versluis[4] and many others[4-5,36].

Table 4. Structures of transition metal complexes optimized by different methods[a]

Molecule	Bond(Å)	HFS	LDA-NL	HF	Expt.
C_5H_5NiNO	Ni-N	1.592	1.620	1.424	1.626
$Ni(CO)_4$	Ni-C	1.794	1.841	1.921	1.838
$Fe(CO)_5$	Fe-C_{ax}	1.774	1.817	2.047	1.807
	Fe-Ceq	1.798	1.814	1.874	1.827
$Fe(C_5H_5)_2$	Fe-Cp	1.60[i]	1.648	1.88[j]	1.65
$Cr(CO)_6$	Cr-C	1.868[l]	1.909	~2.00	1.914
MnO_4^-	Mn-O	1.603[a]	1.631		1.63[p]

[a]Reference to original data in Ref. 45

The superiority of DFT based methods over the HF scheme is most striking in the prediction of metal-ligand bond lengths , as shown in Table 4. The M-L bond distances optimized by the HFS method are 0.02~0.05Å shorter than the experimental values for the molecules listed in Table 4. On the other hand, ab initio calculations at the HF level yield bond distances which are too long by 0.05~0.2Å. The only exception is the HF calculation on C_5H_5NiNO, where the the Ni-N bond distances was found to be too short by 0.2Å, most likely due to the minimal STO-3G basis sets employed in the calculations. It is important to note that the HF method in the case of $Fe(CO)_5$ provides a distorted geometry for $Fe(CO)_5$ with the Fe-C_{ax} distance much longer than the Fe-C_{eq} distance ,whereas experiment and HFS finds the two Fe-CO distances to be about equal,Table 4.

The HFS method underestimates[46] systematically metal-ligand bond distances, and the same trend[46] is found in connection with calculations based on other local DFT methods such as LDA. Fortunately,the deficiencies of the local methods are seen largely to be eliminated by the non-local correction to exchange and correlation. Thus, the LDA-NL metal-ligand bond distances in Table 4 are within .01 Å of experiment. It is gratifying that the LDA-NL method not only affords better bond energies but also more accurate geometrical parameters.

4.3 POTENTIAL ENERGY SURFACES AND CONFORMATIONAL ANALYSIS

There have been a number of HFS or LDA studies on the relative stability of isomers and conformers. These include the relative energies of eclipsed and staggered conformations in $Fe(C_5H_5)_2$[46], C_2H_6[46] as well as a number of binuclear metal complexes[40,47,48] Conformational energy differences seem[46] to be well represented by HFS or LDA. The energetics for different coordination modes of ligands complexed to transition metals have been studied for phosphaalkene[49a],F_2 and H_2[49b], alkenes and alkynes[49c,d], BH_4^- [49e], O_2[49d], X_2, CX_2 (X=O,S,Se,Te)[49f,49g], carbocycles[49i] as well as H[49h]. A special energy decomposition scheme has been devised[31,41d,50,49c] which breaks down the coordination energy of a ligand into steric and electronic factors. This scheme makes it possible to explain interaction energies as well as conformational preference of a ligand in terms of well established chemical concepts. As examples, the decomposition analysis is able to provide estimates of the respective contributions from the σ,π, and δ components in binuclear complexes[41,47,48] with multiple bonds, **2a**; the relative importance of donation and backdonation in transition metal complexes of unsaturated ligands[49] such as olefins,**2b**; the contributions[49h] to the protonation energy,**2c**, from the pure electrostatic interaction of the proton as well as the charge rearrangement following the formation of the protonated complex

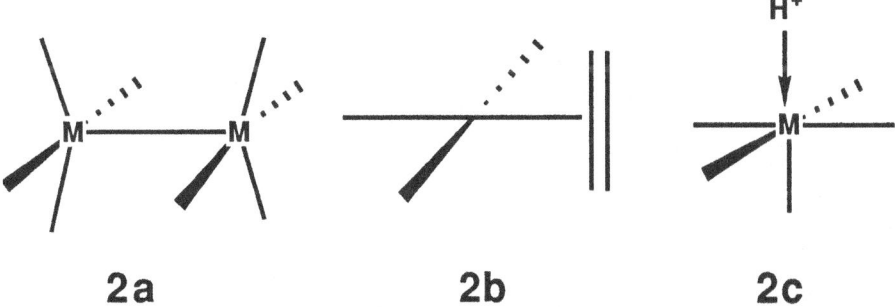

2a   2b   2c

The HF method represents conformational energies of saturated systems, such as rotation barriers in alkanes, quite well. It is not likely that HFS or LDA will afford substantially different results for this type of systems. However, examples are known among unsaturated systems in which HF and LDT based methods differ. Thus, HF[51a] finds $HCo(CO)_4$ of C_{2v} symmetry, with hydrogen in an equatorial position, to be more stable than the experimental observed C_{3v} structure where hydrogen is in the axial position. On the other hand, HFS[51b] finds the C_{3v} structure to be the more stable conformation.

4.4 EXCITATION ENERGIES

We shall in the following sections discuss calculations on energy differences associated with the processes of: a) excitation from one molecular state to another; b) ionization; c) electron attachment. The change in energy ΔE associated with these processes can be written as

$$\Delta E = \Delta E_k + \Delta E_{ne} + \Delta E_{ee} + \Delta E_x + \Delta E_c \tag{12}.$$

The first three terms in Eq. (12) represent the change in kinetic-, electron nuclear attraction-, and electron-electron repulsion energy, respectively, whereas the last two terms represent the change in exchange- and correlation energy, respectively.

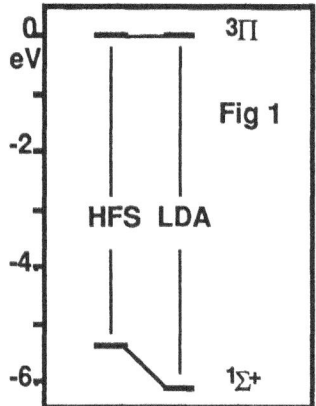

Figure 1. Comparative estimates of the first excitation energy of the CO molecule from HFS and LDA calculations. The energy of the triplet states is the reference level for the HFS and LDA energies

In our assessment of the HFS-, LDA-, and LDA/NL-methods, we shall consider as our first case the energy ΔE required to promote CO or CS from the $^1\Sigma$ ground state to the first excited triplet state $^3\Pi$. The HFS-method differs from LDA in that ΔE_c in Eq.(12) is neglected ($\Delta E_c = 0$). The contribution from ΔE_c to ΔE should however be positive, as $^1\Sigma$ compared to $^3\Pi$ has one additional pair of electrons with opposite spins, and thus[52] $E(^1\Sigma)$ has a larger negative contribution from E_c than $E(^3\Pi)$. Thus, we expect that ΔE from a HFS calculation should be smaller than ΔE calculated by the LDA-method, as indicated in Figure 1. This expectation is born out by our numerical calculations, Table 5. In fact, for CO and CS the excitation energies calculated by the HFS-method are too small compared to experiment whereas the larger ΔE values due to the LDA-method are more in line with experimental excitation energies, Table 5.

Table 5   Excitation Energies[a] (eV).

	HFS	LDA	LDA/NL	Exp
CO $(5\sigma \rightarrow 2\pi)$	5.37	6.12	6.18	6.3
CS $(5\sigma \rightarrow 2\pi)$	2.87	3.42	3.48	3.42
O_2 $(^3\Sigma_g \rightarrow {}^1\Delta_g)$	1.34	1.04	0.82	0.98

Reference to original data given in Ref. 53

Also considered in Table 5 is the energy ΔE required to promote O_2 from its $^3\Sigma_g$ ground state to the first excited state $^1\Delta_g$. For this process, the term ΔE_c in the expression of Eq.(12) should be negative, since $^1\Delta_g$ compared to $^3\Sigma_g$ has one additional pair of electrons with opposite spins. Neglecting the contribution to ΔE from ΔE_c, as it is done in the HFS-method, must now lead to too large an excitation energy, Table 5, whereas LDA in which correlation between electrons of different spins is taken into account provides a smaller value for ΔE (since ΔE_c is negative) more in line with the experimental excitation energy, Table 5. It is finally seen that non-local corrections give rise to a small improvement in the calculated excitation energies

The local HFS scheme has been used extensively to calculate excitation energies and many of the applications have been reviewed[9]. An indication of the high accuracy usually obtained by approximate DFT methods is given in Table 6. This table presents HFS calculation on the three first excitation in the tetrahedral d^0 oxo-complexes MnO_4^-, CrO_4^{2-} and VO_4^{3-}. In particular the electronic spectrum of MnO_4^- has served as an acid test for new theoretical methods. It should be mentioned that HF and post-HF methods so far have failed to calculate the excitation energies of MnO_4^- with near quantitative accuracy. Buijse[54] and Baerends have recently analyzed the error sources in HF-calculations on MnO_4^-.

Table 6. Singlet excitation[a] energies for tetrahedral d^0 oxo complexes in eV.

Transition	MnO_4^-		CrO_4^{2-}		VO_4^{3-}	
	E_s^{cal}[b]	E_s^{exp}	E_s^{cal}[b]	E_s^{exp}	E_s^{cal}[b]	E_s^{exp}
$t_1 \rightarrow 2e$	2.48	2.27	3.30	3.32	4.51	4.58
$4t_2 \rightarrow 2e$	3.96	3.47	4.58	4.53	5.71	5.58
$t_1 \rightarrow 5t_2$	4.15	3.99	4.90	4.86	6.15	6.15

[a] Original data from Ref[55]. [b] Calculated values based on the HFS method

It follows from our analysis, Table 5, that DFT calculations should include correlation and non local corrections (LDA/NL). However, most calculations[9] on excitation energies in transition metal complexes have to date been carried by the HFS method in which both correlation and non-local corrections are missing. This is likely to change in the future

4.5 IONIZATION POTENTIALS

Photoelectron spectroscopy (PES) emerged in the early 70s as a new and exciting technique with direct bearings on molecular orbital energies. It is thus not surprising that PES has served as a testing ground for new and increasingly sophisticated theoretical methods including DFT-based schemes. In fact, one of the first successful applications of the HFS-method in chemistry involved the assignment of photoelectron spectra. The early work based on the SW-Xα method has been reviewed by Case[9].

We shall now consider the energy ΔE required to remove an electron from a molecule. In the corresponding expression of Eq.(12), the contribution from ΔE_c to ΔE will be positive for the molecules considered here, since the neutral molecule has one additional pair of electrons with opposite spins compared to the ion. As indicated in Fig. 2, we must thus expect HFS ($\Delta E_c = 0$) to provide smaller ionization energies than LDA ($\Delta E_c > 0$).

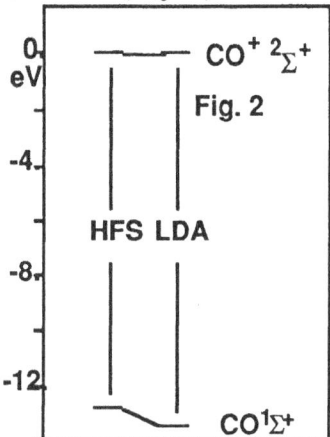

Figure 2. Comparative estimates of the first ionization energy of the CO molecule by HFS and LDA calculations. The energy of the ion is the reference level for the HFS and LDA energies

It follows from Table 7, where we compare the first two ionization energies for a number of molecules calculated by the two methods with experimental values, that HFS provides too small ionization energies, whereas the larger LDA values are more in line with experiment.

However, as shown in Table 7, the correlation correction to the HFS-method, while of the right sign, is not large enough to align LDA ionization energies with experiment, leaving some room — and need — for further improvement of the HFS-method. This improvement is accomplished by the LDA/NL scheme. The non-local corrections are seen for all the molecules to supply a small but significant improvement. A more detailed analysis of how non-local corrections improve on the calculated ionization potentials is given elsewhere[53].

Table 7   Ionization Energies (eV).

	HFS	LDA	LDA/NL	Exp[a]
N_2	14.60	15.05	15.24	15.60
	16.6	16.95	16.81	16.98
CO	12.81	13.50	13.85	14.01
	16.71	17.17	17.01	16.53
F_2	14.78	15.02	15.34	15.70
	18.08	18.35	18.58	18.98
H_2O	12.03	12.55	12.59	12.62
	13.95	14.46	14.60	14.75
NH_3	10.06	10.62	10.76	10.88
	15.59	15.98	16.37	16.0
CH_4	13.18	13.61	14.01	14.35
	21.20	21.50	22.00	23.00
C_2H_4	10.14	10.60	10.56	10.51
	11.84	12.22	12.75	12.85

The first and second ionization energies of each molecule are presented in the given order. All ionization energies are vertical ionization energies; all values presented are from ΔSCF calculations. [a] Original data in Ref. 53.

We present in Table 8 HFS-calculations on the first two ionization potentials for a number of organometallics. The agreement with experiment is in general quite good. It is to be expected that the more accurate LDA/NL scheme will replace the HFS method in calculation of ionization potentials.

4.6 ELECTRON AFFINITIES

The adiabatic electron affinity (EA_{ad}) of a molecule represents the decrease in energy of the molecular system when a single electron is added. It is in general a difficult property to obtain from experimental as well as computational techniques. The most accurate experimental estimates of electron affinities are obtained by laser photo-ionization of negatively charged systems. However, this technique is limited to anions that can be studied in the gas phase. It is still not routine to measure electron affinities accurately and values obtained by alternative techniques differ often considerably, even for relatively simple molecules .

Table 8. Ionization potentials (eV) for transition metal complexes

ML_n	HFS	Exp
$Ni(CO)_4$	9.3	8.9
	10.3	9.8
$Cr(CO)_6$	8.9	8.4
	10.3	13.4
$Fe(CO)_5$	9.0	8.6
	10.2	9.9
RuO_4	11.9	12.1
	12.9	12.9
Cp_2Fe	6.7	6.9
	6.7	7.2

Electron affinities offer also a considerable challenge from the computational point of view. In the first place, one must be able to describe accurately the difference in geometry between the neutral molecule M and the corresponding anion M^-. The computational method must in addition be able to account quantitatively for the difference in electron correlation energy between M and M^-. This difference is in most cases substantial due to the change in the number of electron pairs. Finally, the basis set employed must be flexible enough to describe M as well as M^- with a somewhat more diffuse charge distribution. It is thus not surprising that accurate electron affinities only are available for a quite limited number of molecules.

We compare in Table 9 calculated EA_{ad} values with experiment for a number of small well studied molecules. The comparison include[56] HFS, LDA as well as LDA with non-local exchange corrections (LSD/NL). Also included are results[57] based on HF, configuration interaction with all single and doubles (RCISD) as well as the equation of motion approach[58] (EOM). Several experimental EA_{ad} values have been published for each of the molecules in Table 9, covering quite a range. The experimental data selected represents what currently is considered to represent the best estimates. Flexible basis sets are crucial in theoretical evaluations of EA_{ad} and all calculations were carried out with extensive basis sets.

Table 9 . Adiabatic electron affinities (eV). Comparions between Theoretical and Experimental values

Molecule	HF	CISD[a]	EOM[a]	HFS[b]	LDA[b]	LDA/NL[b]	Exp.
CN	2.93	3.70	4.15	2.78	3.25	3.51	3.82
BO	1.41	2.35	2.97	1.25	1.96	2.33	3.1 2.2
OCN	2.28	3.21	-	2.83	3.19	3.35	3.6
N_3	0.89	2.15	-	2.12	2.50	2.65	2.7
NO_2	1.51	2.00		1.02	1.43	1.90	2.32

[a]Ref 57. [b]Ref.56

It follows from Table 9 that HF and HFS, in which correlation between electrons of different spins are absent, underestimate EA_{ad}. This is understandable as the lack of correlation is more crucial (destabilizing) for the negative ions A$^-$ than the neutral species. This is so since A$^-$ has one more pair of electrons than A. Thus the exclusion of correlation will underestimate the energy gap between A and A$^-$. The explicit inclusion of electron correlation greatly improves the theoretical results both in the case of the ab initio methods (CISD, EOM) as well as in the case of the DFT-based LDA-scheme. The addition of non-local exchange correction to LDA (LDA/NL) is seen further to bring the theoretical results in line with experiment. It seems clear that approximate-DFT calculations on EA_{ad} requires the inclusion of both electron correlation and non-local correction. However, at that level of approximate-DFT the calculated results are comparable in quality to ab initio values based on extensive configuration interaction (CISD and EOM). Approximate DFT has also been used to calculate atomic electron affinities.

4.7 MOLECULAR FORCE FIELDS

The construction of molecular harmonic force fields from experimental data (based on infrared and Raman spectroscopy) have benefited considerably over the past decade from new developments in computational chemistry. Thus, Pulay[59] has with his force method ,in which force fields and vibrational frequencies are evaluated from a numerical differentiation of analytical energy gradient calculated by *ab initio* methods, been able to study a number of small molecules. The studies by Pulay have more recently been augmented by investigations in which the second derivatives of the total energy with respect to nuclear displacements were calculated analytically[42].

Calculations of frequencies and force fields on the Hartree-Fock level of theory are now carried out almost routinely[42] although the basis sets employed in calculations on larger size molecules often are somewhat restricted as the computational effort in HF-calculations increases as M^4 with

Table 10. Frequecies obtained by HFS,HF and MP2 calculations (cm^{-1}).

Mol	Symmetry	HFS[a]	HF[b]	MP2[b]	EXP
	a_1	3814 (-18)	4070 (238)	3772 (-60)	3832
H_2O	a_1	1590 (-58)	1826 (178)	1737 (89)	1648
	b_1	3877 (-66)	4188 (245)	3916 (-27)	3943
	a_1	2631 (-91)	2918 (196)	2797 (75)	2722
H_2S	a_1	1178 (-37)	1368 (153)	1279 (64)	1215
	b_1	2644 (-89)	2930 (197)	2824 (91)	2733
	a_1	3314 (-192)	3690 (184)	3504 (-2)	3506
NH_3	a_1	953 (-69)	1207 (185)	1166 (64)	1022
	e	3468 (-109)	3823 (246)	3659 (82)	3577
	e	1564 (-127)	1849 (158)	1852 (161)	1691
PH_3	a_1	2313 (-139)	2666 (214)	2510 (58)	2452
	a_1	950 (-91)	1142 (102)	1079 (38)	1041
	e	2329 (-128)	2602 (145)	2526 (69)	2457
	a_1	2313 (-139)	3197 (60)	3115 (-22)	3137
	e	1467 (-100)	1703 (136)	1649 (38)	1567
CH_4	t_2	3102 (-56)	3302 (144)	3257 (99)	3158
	t_2	1236 (-121)	1488 (131)	1418 (61)	1357
	a_1	2187 (-190)	2233 (-144)	2323 (-14)	2377
	e	905 (-70)	1052 (77)	1005 (30)	975
SiH_4	t_2	2217 (-102)	2385 (66)	2337 (18)	2319
	t_2	777 (-168)	1016 (71)	956 (11)	945
	a_g	3063 (-90)	3344 (191)	3231 (78)	3153
	a_g	1647 (-8)	1856 (201)	1724 (69)	1655
	a_g	1348 (-22)	1499 (129)	1424 (54)	1370
	a_u	972 (-72)	1155 (111)	1083 (39)	1044
	b_{1g}	3138 (-94)	3394 (162)	3297 (65)	3232
C_2H_4	b_{1g}	1209 (-36)	1353 (108)	1265 (20)	1245
	b_{1u}	908 (-61)	1095 (126)	980 (11)	969
	b_{2g}	879 (-80)	1099 (140)	931 (-28)	959
	b_{2u}	3170 (-64)	3420 (186)	3323 (89)	3234
	b_{2u}	814 (-29)	897 (54)	873 (30)	843
	b_{3u}	3041 (-106)	3321 (174)	3222 (75)	3147
	b_{3u}	1399 (-74)	1610 (137	1523 (50)	1473

[a]Ref 60a. [b]Ref. 43

the number of basis functions M. Frequency and force field calculations in which correlated _ab initio_ methods have been employed are still restricted to small size molecules and the evaluation of first ,and in particular, second derivatives of the energy by such methods are rather demanding.

The recent implementations of energy derivatives[4,5,36] within the DFT framework makes it now tractable[60] to evaluate frequencies from a numerical differentiation of the energy gradients in the spirit of Pulay's force method. Harmonic frequencies calculated by the HF, MP2 and HFS methods for $H_2O, H_2S, NH_3, PH_3, CH_4, SiH_4$ and C_2H_4 are in Table 10 compared to experimental data, with deviation between calculated and experimental values given in parentheses. It is evident from Table 10 that the harmonic frequencies obtained by the HFS method[60a] are somewhat too small compared to experiment, whereas the HF method afford too high values. The average percentage deviation of the harmonic frequencies calculated by the HFS method varies from 1.9 % in the case of H_2O to 9.3 % for SiH_4. The corresponding deviation of the HF-results are for all molecules except SiH_4 higher. The mean absolute deviation of all frequencies listed in Table 10 is 5.0 % in the HFS case as opposed to 9% in the HF case. The frequencies calculated[43] by the MP2 method have a mean absolute deviation of 4 % and are thus slightly better than the DFT results. Salahub[60b] et al. have calculated frequencies for the same sample of molecules given in Table 10. Their results are surprisingly similar to the HFS-frequencies (± 30 cm^{-1}).

The data in Table 10 would indicate that DFT-based methods on the whole are able to furnish relatively accurate harmonic frequencies. They should thus over the next decade be able to supply valuable information about molecular force fields, in particular for transition metal complexes and metal clusters. In fact, Salahub[46] et al. have quite recently obtained frequencies for organic molecules interacting with metal clusters. This type of calculations can provide crucial information in connection with studies of chemisorption.

The use of DFT in studies on molecular force fields would be greatly enhanced by the employment of analytical second derivatives. Expressions[61] for the second and third derivatives can be derived in a manner[43] quite similar to that employed for HF-theory. However, a practical implementation of second derivatives within DFT has not yet been achieved.

4.8 REACTION PROFILES

We have in the previous sections documented that DFT-based methods are able to provide information on bond energies and molecular geometries of reasonable accuracy. We shall in the ensuing sections illustrate how the same methods can be used to trace energy profiles for elementary reaction steps involving organometallic compounds.

Insertion of Ethylene into the Cobalt-Hydrogen Bond

Experimental findings suggest that the migration of a hydride to a coordinated olefin group is very facile. In fact, the hydride-olefin insertion reaction has, with a few exceptions, rarely been directly observed. As a consequence, metal complexes containing both hydride and olefin are scarce and it has only in a few cases been possible to study the insertion process by experimental techniques . We have studied[62] the insertion process , **3a** $\rightarrow$ **3 b**, which is one of the key steps in the hydroformylation process .

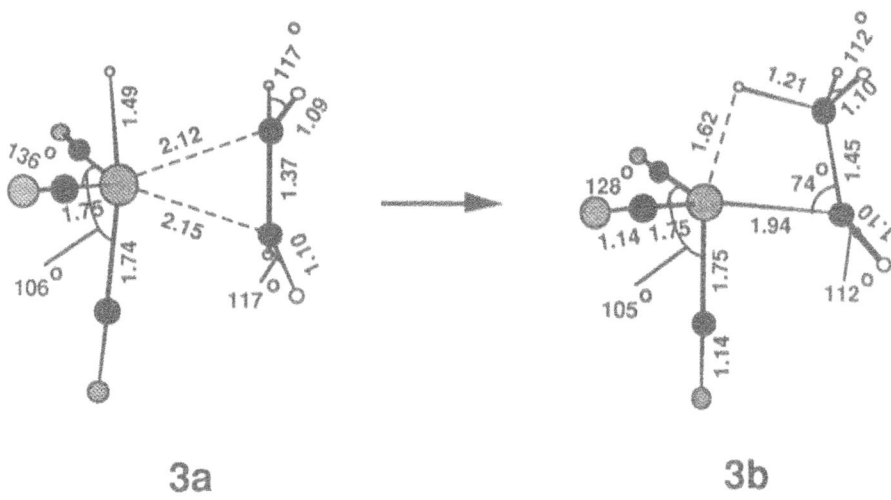

3a   3b

Figure 3 . Energy profile[a] of the hydride migration reaction **3a** → **3b** .

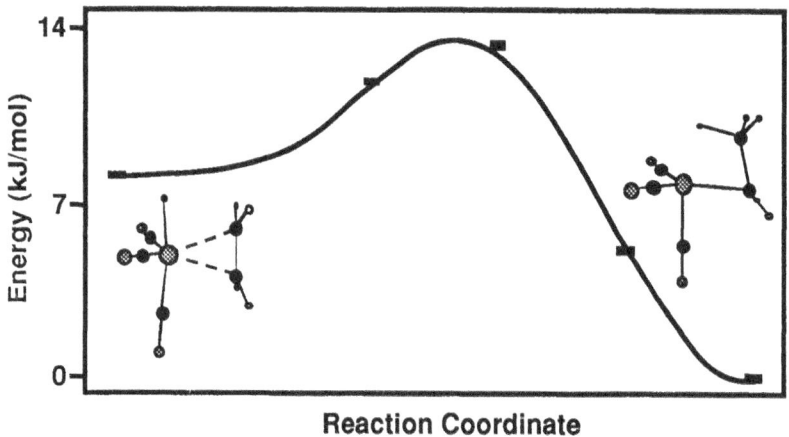

Reaction Coordinate

a) Reaction **3a** → **3b**. The energy zero point refers to structure **3b**

The energy profile for the reaction is given in Figure 3. We find the reaction to have a very small activation barrier ,$\Delta E^{\#}$,of only 6 kJ/mol and a modest reaction enthalpy of -8 kJ/mol. The product of the reaction is an ethyl-complex, **3b** , with a pronounced agostic interaction between the metal center and hydrogen. The ethyl-complex, **3b,** has a short metal-hydrogen distance of 1.62Å . This distance is only 0.12 Å longer than the Co-H bond distance in the parent complex **3a**. The bond length of the agostic hydrogen with the corresponding carbon atom of the ethyl ligand is, on the other hand, substantially elongated with R(H-C) = 1.21 Å in comparison to a normal ethyl C-H bond length of 1.11 Å. The C-C bond distance in the ethyl group is 1.45 Å, which is somewhat longer than the corresponding value in the parent structure **3a** (1.37 Å), but it is still substantially shorter than the C-C bond length of an undistorted ethyl group (1.53 Å). Also the bond angle Co-C$_{et}$-C$_{et}$ is with 74° considerably smaller than the expected value of about 109° for a normal ethyl ligand. The optimized molecular parameters of structure **3b** clearly indicate that the π-bond character of the initial ethylene ligand has been retained to a large extent. This is also underlined by the small calculated energy difference of 8 kJ/mol between the parent structure **3a** and complex **3b**. The modest activation barrier found in this study stems from the fact that the hydrogen atom is close to cobalt throughout the migration reaction.

Migratory Insertion of Alkyl Into the Co-CO Bond

Another important elementary reaction step is the migratory insertion of alkyls into the metal-carbonyl bond. We have modeled[51b] the migratory insertion process

$$RCo(CO)_4 \rightarrow RC(O)Co(CO)_3 \qquad (13),$$

which is an other important step in the hydroformylation cycle. The process in Eq. 13 could in principle proceed by an insertion of a CO into the Co-CH$_3$ bond of **4**, producing the coordinatively unsaturated complex, **5**, with the acyl group in an axial position. Alternatively, the methyl group might migrate to a cis-carbonyl thus affording the complex **6** with the acyl group in an equatorial position.

4   **5**   **6**

We find, perhaps not surprisingly, that the energy profile for the CO insertion, **4 → 5**, into the Co-CH$_3$ bond, Figure 4a , has a prohibitively high activation barrier of 200 kJ/mol. The CO insertion can as a consequence not be a viable mechanism for the process in Eq. 13.

The migration of CH$_3$ to the cis-CO ligand, **4 → 6**, was calculated, Figure 4b, to have an endothermicity, ΔH, of 71 kJ/mol and a very modest activation barrier, ΔE$^{\#}$, of only 9 kJ mol^{-1} for the reverse reaction **6 → 4**. Thus the CH$_3$ migration ,**4 → 6**, seems to be favored as the mechanism for the process in Eq.13. The calculated reaction enthalpy and activation barrier for 4 → 6 compare well with an earlier study[63] on the CH$_3$ to CO migration in CH$_3$Mn(CO)$_5$, where we found ΔH to be 75 kJ/mol and ΔE$^{\#}$ 11 kJ/mol. Our findings are also in agreement with a recent kinetic study by Roe[64], who found the rate constant of the methyl back migration of

Figure 4. (a) Energy profile for the insertion of CO into the Co-CH$_3$ bond, $4 \rightarrow 5$. (b) The energy profile for the migration of CH$_3$ to CO, $4 \rightarrow 6$. The zero point refer to **2** in both plots

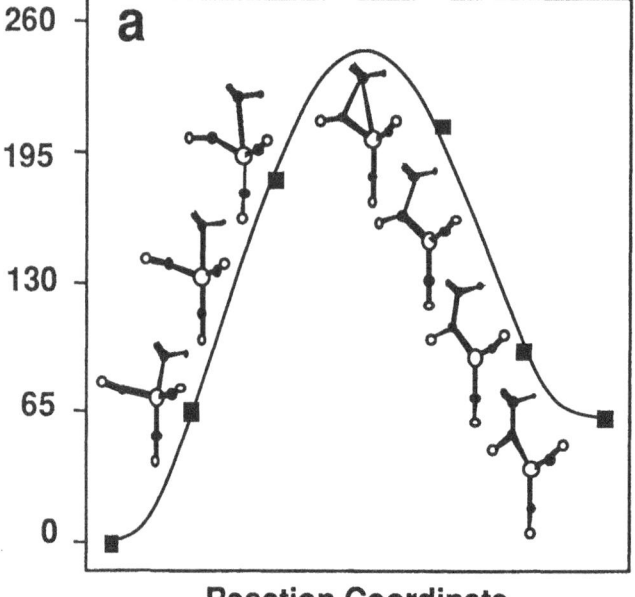

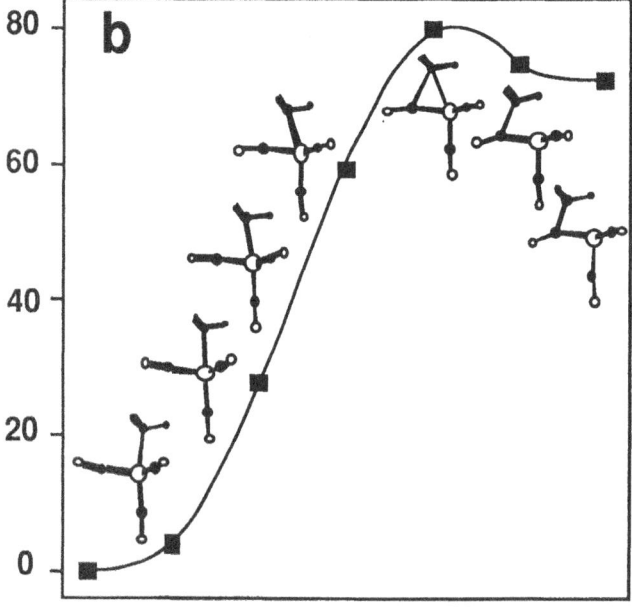

$CH_3C(O)Co(CO)_3$ to be considerably larger than the rate constant for the corresponding forward reaction. The structures in Figure 4b illustrate nicely how the methyl group can slide almost parallel along the cis C-Co bond onto the cis carbonyl carbon, while the remaining $Co(CO)_3$ framework stays almost unchanged. The 10 kJ/mol calculated for $\Delta E^{\#}$ in the present study is an upper bound to the actual value, and we can thus conclude that the methyl migration , $4 \rightarrow 6$, should proceed with a rather modest activation barrier.

Other Studies on Elementary Reaction Steps

Other studies on elementary reaction steps include oxidative addition[65] of H_2 to $Co(CO)_4$ as well as the oxidative addition[66] of H_2 and CH_4 to CpML (L=CO,PH_3;M=Rh,Ir). Polymerization of organosilanes[67] and halogen abstraction by metal carbonyl anions[68.] Investigations have in addition been carried out on the insertion of aldehydes into the Co-H bond[69] , the metathetical exchange reactions between Cp_2LuR (R=H,CH_3) and H_2 or CH_4[70] as well as the insertion[71] of olefins into the Lu-R bond (R=H,CH_3) . The optimization of transition states structures by DFT has been discussed in Ref. 11.

Acknowledgment : This investigation was supported by the Natural Sciences and Engineering Research Council of Canada (NSERC). We also acknowledge access to the Cyber-205 installations at the University of Calgary. I would like to thank Drs. L.Versluis;V.Tschinke;L.Fan and E.Folga for their contributions

---

REFERENCES

1. (a) R.G. Parr;W.Yang, Density-Functional Theory of Atoms and Molecules ; Oxford University Press,New York,1989.
    (b) E.S.Kryachko ;E.V. Ludena. Density Functional Theory of Many Electron Systems; Kluwer Press,Dordrecht,in press
    (c)T.Ziegler, Chem.Rev. in press (1991)
2. A.D.Becke, Int.J.Quantum Chem. , S23,599 (1989)
3. (a)T.Ziegler,V.Tschinke,L.Versluis,E.J.Baerends,W. Ravenek ,Polyhedron, 7, 1625 (1988)
4. L.Versluis,T.Ziegler, J.Chem.Phys. , 88,322 (1988)
5. R.Fournier,J.Andzelm,D.R.Salahub,J.Chem.Phys. , 90,6371(1989)
6. L.Fan,L.Versluis,T.Ziegler,E.J.Baerends,W.Ravenek, Int. J.Quantum Chem. S22,173(1988)
7. (a)W.Bieger,G.Seifert,H.Eschrig,G.Grossman,Chem.Phys. Lett. 115,275 (1985)
    (b) D.A.Freier,R.F.Fenske,Y.Xiao-Zeng, J.Chem.Phys. , 83,3526 (1985)
    (c) V.G.Malkin,Z.Zhidomirov, Zh.Strukt.Khim. ,29,32 (1988)
8. A.J. van der Est,P.B.Barker,E.E.Burnell,C.A.de Lange,J.G. Snijders, Mol.Phys , 56,1 (1985)

382

9.   D.A. Case,Annu.Rev.Phys.Chem. , 33,151(1982)
10.   (a) L.Noodleman ,J.G.Norman, J.Chem.Phys.   70,4903 (1979)
        (b) L. Noodleman, J.Chem.Phys.   , 74,5737 (1981)
        (c) L.Noodleman ,E.J.Baerends , J.Am.Chem.Soc.   ,106,2316(1984)
        (d) L. Noodleman, J.G. Norman,J.H. Osborne,A. Aizman, D.A. Case,
        J.Am. Chem. Soc. ,107,3418(1985)
11.   L.Fan,T.Ziegler, J.Chem.Phys. , 92,3645(1990)
12.   M.Trsic,T.Ziegler,W.G.Laidlaw, Chem.Phys. ,15,383(1976)
13.   M.Cook;M.Karplus J.Phys.Chem. ,91,31 (1987)
14.   V.Tschinke ;T. Ziegler. J.Chem.Phys., 93,8051 (1990)
15.   (a) A clear discussion of this point can be found in Ref. 15b
        (b) A.D.Becke . J.Chem.Phys. , 88,1053 (1988)
        (c) A.D.Becke. ACS Symposium Series , Number 394, Washington
        1989.pp. 165
16.   J.C.Slater Adv. Quantum Chem. 6,1 (1972)
17.   (a) O.Gunnarsson; L.Lundquist Phys.Rev. ,B10,1319 (1974)
        (b) O.Gunnarsson ; I. Lundquist. Phys.Rev. ,B13,4274 (1976)
        (c) O.Gunnarsson ;M.Johnson ; I.Lundquist, Phys.Rev. ,B20,3136(1979)
        (d) U.von Barth .; L. Hedin Phys.Rev. A 20,1693(1979)
18.   (a) D. C.Langreth and M. J. Mehl, Phys.Rev. B , 28, 1809, [1983]
        (b)   A.D. Becke, Phys. Rev., A33, 2786(1988).
        (c) J. P. Perdew, Phys. Rev., B33, 8822(1986). Also see the errotum, Phys. Rev.
        B34,7046(1986).
        (d) V. Tschinke and T. Ziegler, Can. J. Chem. 67, 460 (1988)
        (e) T. Ziegler, V. Tschinke and A. D. Becke, J. Am. Chem. Soc. 109,   1351(1987).
        (f) T. Ziegler, W. Cheng, E. J. Baerends and W. Ravenek,Inorg.Chem. 27,
        3458(1988).
        (g) T.Ziegler;V.Tschinke; L.Fan; A.D.Becke J.Am.Chem.Soc. 111,9177 (1989)
        (h) T. Ziegler, V. Tschenke, L. Versluis, E. J. Baerends and W. Ravenek,
        Polyhedron, 7, 1625(1988)
        (i) T. Ziegler, V. Tschinke and C. Ursenbach, J.Am.Chem.Soc. 109,   4825(1987).
        (j) A. D. Becke, ACS Symposium Series 1989, Number 394, Washington,1989.
        (k) A. D. Becke, Int.J.Quantum Chem. 1989,S23,599
        (l)   L.Fan; T.Ziegler ,J.Chem.Phys. 1991,94,6057
        (m) L.Fan; T.Ziegler ,J.Chem.Phys. ,submitted
19.   K.H.Johnson, J.Chem.Phys. 45,3085 (1966).
20.   J.Korringa,J. Physica. 13,392 (1947)
21.   D.R.Salahub;R.P.Messmer;K.H.Johnson Mol.Phys. 31,521 (1975)
22.   D.E.Ellis ;G.S.Painter Phys.Rev. B2,2887 (1970)
23.   E.J.Baerends;D.E.Ellis ;P.Ros Chem.Phys. 2,41 (1973)
24.   H.Sambe;R.H.Felton J.Chem.Phys. 62,1122 (1975)
25.   B.I.Dunlap ;J.W.D.Connolly ;J.F.Sabin J.Chem.Phys. 71,3396 (1977)

26. O.Gunnarson;J.Harris;R.O.Jones Phys.Rev. 15,3027 (1977)

27. (a) B.Delley ;D.E.Ellis J.Chem.Phys. 76,1949 (1982)

(b) G.L.Gutsev ;A.A.Levin Chem.Phys. 51,459 (1980)

(c) W.Bieger;G.Seifert ;G.Grossmann . Z.Chem. 24,156 (1984)

(d) F.W.Kutzler;G.S.Painter.Phys.Rev. B 37,285 (1988)

(e) M.A.Pederson Phys.Rev. B 38,3825 (1988)

28. (a) A.D.Becke J.Chem.Phys. 88,2547 (1988)

(b) P.M.Boerrigter ;G. t e Velde ;E.J. Baerends Int.J.Quantum Chem. 33,87 (1988)

(c) B.Delley J.Chem.Phys. 92,508 (1990)

(d) G. te Velde ;E.J.Baerends J.Comp.Physics, Submitted

(e) M.R.Pederson ;K.A.Jackson Phys.Rev.B 41,7453 (1990)

29. (a) W.Ravenek in Algorithms and Applications on Vector and Parallel Computers; Riele,H.J.J.; Dekker,Th.J.;van de Vorst,H.A. ,Eds.; Elsevier, Amsterdam, 1987

(b) W. Ravenek in : Scientific Computing on Supercomputers, eds. de Vreese,J. ; van Camp,P.E. New York,Plenum,198

(c) The LCAO-HFS program will be available from Polygen

30. (a) Wimmer,E; Freeman,A.;Fu,C.-L.;Cao,S. -H.;Delley,B. in Supercomputer Research in Chemistry and Chemical Engineering ;Jensen,K.F.; Truhlar,D.G., Eds. ; ACS Symposium Series No 353,1987;p 49.

(b) Andzelm,J.;Wimmer,E.;Salahub,D.R. in Spin Density Functional Approach to the Chemistry of Transition Metal Clusters , Salahub,D.R.; Zerner,M.C. ,Eds. ACS Symposium Series No 394,1989; p 229.

(c) DGauss available from CRAY

(d) DMol available from Biosym

(e) DMon is developed by Salahub et al. at University of Montreal,Canada

(f) NUMOL is developed by Becke at Queens University,Kingston,Canada

31. Ziegler,T.;Rauk,A. Theor.Chim.Acta. 1977,46,1

32. (a) Becke,A.D. J.Chem.Phys. 1982,76,6037

(b) Becke,A.D. J.Chem.Phys. 1983,78,4787

(c) Becke,A.D;Dickson,R.M. J.Chem.Phys. 1988,89,2993

(d) Becke,A.D. Int.J.Quantum Chem. 1989,S23,599

(e) Becke,A.D.;Dickson,R.M. J.Chem.Phys. 1990,92,3610

(f) The program outlined in Ref. 50d-e is called NUMOL.

33. (a)L.Laaksonen ;D.Sundholm.;P.Pyykkö. Computer Phys. Rept. 4,313 (1986)

(b)D.Heinemann;B.Fricke;D.Kolb Phys.Rev. A 38,4998 (1988)

(c) D.Heinemann;A.Rosén;B.FrickeChem.Phys.Lett. 166,627 (1990)

34. (a) J.G.Snijders;E.J.Baerends Mol.Phys. 33,1651 (1977)

(b) J.Andzelm;E.Radzio; D.R.SalahubJ.Chem.Phys. 83,4573 (1985)

(c) W.Ravenek, Ph.D. Thesis, Nijmegen 1983.

(d) W.Ravenek ;E.J.Baerends J.Chem.Phys. 81,865 (1984)

384

35. (a) J.G.Snijders ; E.J.Baerends , Mol.Phys. 36 ,1789 .(1978)
    (b) .J.G.Snijders ;E.J.Baerends ; P.Ros ,Mol.Phys. 38 ,1909 (1979)
    (c) H.Gollisch; L.Fritsche ,Phys.Status.Solidi B86 ,145 (1978)
    (d) D.D.Koelling ; P.N.Harmon J.Phys. C10,3107 (1977)
    (e) D.E.Ellis;A.Rosen Z.Phys. A 283, 3 (1977)
    (f) C.Y.Yang ;S.Rabii Phys.Rev. A 12, 362 (1977)
36. (a) C.Satoko Chem.Phys.Lett. 83,111 (1981)
    (b) F.W.Averill ;G.S.Painter Phys.Rev. B 32,2141 (1985)
    (c) C.Satoko Phys.Rev. 30,1754 (1984)
    (d) J.Harris ;R.O.Jones ;J.E. Mueller J.Chem.Phys. 75,3904 (1981)
    (e) L.Martins ;J.Buttet;R.Car Phys.Rev.Lett. 53,655 (1984)
    (f) B.Delley J.Chem.Phys. ,in press
37. J.W.Bruno;H.A.Stecher;L.R.Mors;D.C.Sonnenberg;T.J.MarksJ.Am.Chem.Soc.
    7275,108 (1986)
38. J.A.Connor;M.T.ZafaraniMoatter;J.Bickerton;N.I.Saied;S.Suradi;R.Carson;G.A.
    Altackhin;H.A.Skinner Organometallics 1,1166 (1982)
39. F.Ungvary Organomet.Chem. 36,363 (1872)
40. T.Ziegler;V.Tschinke;A.Becke ,Polyhedron 6,685 (1987).
41. (a) T.Ziegler;J.G.Snijders;E.J.Baerends, J.Chem.Phys 74,5737 (1981)
    (b) T.Ziegler,;V.Tschinke;E.J.Baerends;J.G.Snijders;W.RavenekJ .Phys.Chem.
    93,3050 (1989)
    (c) T.Ziegler NATO ASI B87,421 (1981)
    (d) T.Ziegler NATO ASI C176,189 (1986)
    (e) T.Ziegler;J.G.Snijders,E.J.Baerends ACS Symposium Series 395,322 (1989)
    (f) T.Ziegler;V.Tschinke ACS Symposium Series 428 (1990)
42. W.J. Hehre, L. Radom, P.V.R. Schleyer and J.A. Pople, Ab initio Molecular
    Theory, John Wiley   & Sons, New York, 1986.
43. (a)K. Faegri Jr. and J. Almlöf, Chem. Phys. Lett., 107, 121(1984).
    (b) H.P. Lüthi, P.E.M. Siegbahn and J. Almlöf, J. Phys. Chem., 89,   2156(1985).
    (c)H.P. Lüthi, J.H. Ammeter, J. Almlöf and K. Faegri,Jr., J. Chem. Phys., 77,
    2002(1982).
    (d)J. Almlöf, K. Faegri Jr., B.E.R. Schilling and H.P. Lüthi, Chem Phys. Lett., 106,
    266(1984).
44. K.Faegri;J.Almlöf Chem.Phys.Lett. 107,121 (1984)
45. L.Fan;T.Ziegler J.Chem.Phys. in press
46. J.Labanowski ; J.Andzelm, Eds. "Density Functional Methods in Chemistry",
    Springer-Verlag,Heidelberg 1991.
47. T.Ziegler J.Am.Chem.Soc. 105,7543 (1983)
48. (a) T.ZieglerJ.Am.Chem.Soc 106,5901(1984)
    (b) T.Ziegler, J.Am.Chem.Soc. 107,4453 (1985)
49. (a) Th.A.van der Knaap;F.Bickelhaupt;J.G.Kraaykamp;G.van Koten
    J.C.P.Bernards; H.T.Edzes ; W.S.Weeman ; E.de Boer;E.J.Baerends

Organometallics 3,1804 (1984)

(b)F.M.Bickelhaupt;E.J.Baerends;W.Ravenek Inorg.Chem. 29, 350 (1990)

(c) T.Ziegler ;A.Rauk Inorg.Chem.18,1558 (1979)

(d) T.Ziegler Inorg.Chem. ,24,1547 (1985)

(e)A.P.Hitchcock,A.P.;N.G.Hao;N.H.Werstiuk;M.G.McGlinchey,M.G.;
T.Ziegler Inorg. Chem. 21,793 (1982)

(f) T.Ziegler Inorg.Chem. 25,2723 (1986)

(g) R.L.DeKock ;E.J.Baerends ;R.Hengelmolen, Organometallics 3,289 (1984)

(h) T.Ziegler Organometallics 4,675 (1985)

(i) C.Famiglietti;E.J.Baerends Chem.Phys. 62,407 (1981)

50.   T.Ziegler ;A.Rauk Inorg.Chem. 18,1755 (1979)

51.   (a)D.Antolovic;E.R.DavidsonJ.Chem.Phys. 88,4967 (1988)

(b)L.Versluis.; T.ZieglerJ.Am.Chem.Soc. 111,2018 (1989)

52.   We base our rationalization of the trend in DEc on the assumptions that the correlation energy in atoms and molecules is roughly proportional to the number of opposite-spin electron pairs.

53.   V.Tschinke;T.Ziegler Theor.Chim.Acta , in press

54.   M.Buijse;E.J.Baerends J.Chem.Phys

55.   T.Ziegler;A.Rauk;E.J.Baerends Theor.Chim.Acta , 43,261 (1977)

56.   G.L.Gutsev ;T.Ziegler J.Comp.Chem. ,submitted

57.   J.Baker,; R.H.Nobes ;L.Radom,L. J.Comp.Chem. 3,349 (1986)

58.   T.T.Chen ; W.D.Smith ;J.Simons Chem.Phys.Lett. 26,296 (1974)

59.   P.Pulay Mol.Phys. 17,197 (1967)

60.   (a) L.Fan;L.Versluis;T.Ziegler;E.J.Baerends;W.Ravenek Int. J. Quantum Chem. S22,173 (1988)

(b) A.St-Amant ;R.Fournier ;D.R.Salahub Int. J.Quantum Chem. S23 (1989)

61.   R. Fournier J.Chem.Phys. 92,5422 (1990)

62.   L.Versluis ;T.Ziegler Inorg.Chem. 29,4530 (1990)

63.   T.Ziegler;L.Versluis;V.Tschinke J.Am.Chem.Soc. 108,612 (1986)

64.   D.C.Roe Organometallics 6, 942 (1987)

65.   L.Versluis;T.Ziegler , Organometallics, 9, 2985 (1990)

66.   T.Ziegler ;V.Tschinke ;L.Fan ;A.D.Becke,A. J.Am.Chem.Soc. 111,9177 (1990)

67.   J.F.Harrod ;V.Tschinke ;T.Ziegler . Organometallics 9,897.(1990)

68.   A.P.Masters ;T.S.Soerensen;T.Ziegler , Organometallics, 8,1088 (1989)

69.   L.Versluis ;T.Ziegler ,J.Am.Chem.Soc. 112,6163 (1990)

70.   E.Folga.;T.Ziegler Can.J.Chem., submitted.

71.   V.Tschinke .;E.Folga.;T.Ziegler, ,Organometallics,submitted

THEORETICAL MODELS FOR ORGANOMETALLIC REACTIONS

MARGARETA R.A. BLOMBERG, PER E.M. SIEGBAHN,
MATS SVENSSON and JAN WENNERBERG
Institute of Theoretical Physics
University of Stockholm
Vanadisvägen 9
S-11346 Stockholm
Sweden

ABSTRACT. Quantum chemical methods are applied to calculate potential energy surfaces for the reaction between transition metal systems and simple hydrocarbons like methane, ethane and ethene. Naked metal atoms are used to model metal complexes, making it possible to make clear cut comparisons between different metals and also between different classes of reactions. The use of simple models also allows for the application of sophisticated quantum chemical methods, including an accurate treatment of electron correlation effects, thus ensuring that reliable relative energies are obtained. It is shown that the variation in both binding energies and activation energies can be understood from the electronic spectra of the metal atoms. The effects of using more realistic model systems are also investigated. Finally, the results from a methodological investigation of calculated binding energies in an experimentally rather well known system, $Ni(CO)_4$, are discussed. For this system, which belongs to a class of molecules that are unusually difficult to treat accurately by quantum chemical methods, the discrepancy between calculated and experimental binding energies is about 10 percent.

1. Introduction

The large present interest in organometallic thermochemistry is motivated by its importance for the understanding of the factors that govern chemical transformations, as well as for the description of chemical bonding mechanisms. Although the amount of experimental information relating to the strength of transition metal-ligand bonds is rapidly growing, quantum chemical studies can still play an important role in this area. Using quantum chemical methods accurate relative energies can be calculated to compare the stability of different molecular conformations.

J. A. Martinho Simões (ed.), Energetics of Organometallic Species, 387–421.

One strength of the quantum chemical approach is that the differences in stabilities can be interpreted in terms of electronic structures. Furthermore, theoretical methods offer the opportunity to study model systems, which can not be prepared in actual experiments, and therefore different factors influencing the stability of certain compounds can be isolated by investigating them one at a time. The use of model systems also leads to the possibility to determine trends in stability or reactivity, by using the same model system for different transition metals. Another advantage of theoretical methods is that all the steps in a chemical reaction can be followed, thus determining not only the overall thermochemistry of the reaction but also activation energies, offering the possibility to compare the reactivity of different species as well as the energetics of different reaction paths. On the other hand one has to make certain that the approximations inherent in the theoretical methods are not introducing artifacts which make the conclusions unreliable. Recent method development has made it possible to apply accurate *ab initio* quantum chemical methods to studies of transition metal chemistry on a larger scale and in this article some results mainly dealing with the reactivity of metal complexes towards hydrocarbons are summarized. In these studies simple models for the metal complexes are used. An example of an accurate study of metal-ligand binding energies in an actual metal complex will also be given.

In section 2 below the choice of models are discussed and in section 3 the quantum chemical methods used are briefly described. In section 4 a summary is given for some of the results obtained in studies on alkane activation by transition metal complexes. In section 5 the energetics of π-coordination of different ligands to transition metals is reported, including a comparison of alkene coordination to different metals as well as a summary of an accurate study of the strength of the different metal-carbonyl bonds in $Ni(CO)_4$. In section 6 finally some concluding remarks are given.

2. Models

The interaction of two molecules can be described by a potential energy surface. To study a chemical reaction by quantum chemical tools implies that the relative energies of the most important stationary points on this surface are determined. To use simple models of the reacting molecules is a computational advantage, since it reduces the number of degrees of freedom of the potential surfaces. Furthermore, as mentioned in the introduction above, the use of models is also a means for obtaining a deeper knowledge than if only very realistic systems were studied.

The smallest model of a metal complex is a naked metal atom and this is the main model used in the presently described studies on hydrocarbon activation by transition metal complexes. Information about the reactivity of different naked metal atoms allows a detailed understanding of the basic reaction mechanisms for groups of reactions. This type of information also makes it possible to compare different classes of reactions, for example the difference between C-H and C-C activation can be completely understood from naked atom model studies as will be further discussed in section 4 below. Furthermore, the purely atomic contribution from the metal centers is the most important part of the reactivity of the metal complexes. When the basic mechanisms for the reactivity of the naked metal atoms towards a certain class of molecules are understood the effects of different types of ligands on the metal can be studied and a full understanding of the reactivity of the metal complexes can finally be obtained.

The reactivity of naked metal atoms have also been studied experimentally. Recently, measurements of reactivities of neutral metal atoms have been performed [1-3], whereas earlier studies were mainly concerned with the reactivities of metal cations. The reactivity of transition metal cations with hydrocarbons has been studied intensively during the last decades [4-6]. A comparison between these types of experiments and the present model calculations offers both a calibration of the calculations and a contribution to the interpretation of the experimental results.

3. Methods

To obtain reliable relative energies for the chemical reactions studied it is necessary to perform accurate calculations. The use of simple models is therefore a computational advantage also in the sense that accurate quantum chemical methods can be used. Based on previous experience the choice of methods, basis sets and geometries are not expected to lead to errors of more than a few kcal/mol for relative energies in the studies on hydrocarbon coordination and activation discussed below.

The first requirement for obtaining reliable results is that reasonably large atomic basis sets are used. The generalized contraction scheme [7] yielding a minimal basis description of the core orbitals and a double to triple zeta description of the valence shells is used. For the metal atoms the primitive basis sets of Wachters [8] and Huzinaga [9] are normally used and the addition of one diffuse d function leads to a triple zeta description of the valence d shell. Further, two valence p functions are added together with one contracted f function. For carbon and hydrogen the primitive basis of Huzinaga [10] is normally used and on carbon one d function is added. The active hydrogen has a triple zeta s shell and one p function added.

In the description of both organometallic reactions and π-coordination to metals the effects of electron correlation are extremely important. For example, the results for the thermochemistry of the insertion of a nickel atom into a C-H or C-C bond are changed from the Hartree-Fock value by 50-60 kcal/mol by the inclusion of near-degeneracy and dynamical correlation effects in the calculations [11]. Simple calculations at the Hartree-Fock level will thus not at all be enough for a study of this type of reactions. The situation for the description of the coordination of ethylene or carbonyl to transition metals is very similar [12,13]. It has further been shown that it is important to include the correlation effects of all valence electrons in the interacting systems, not only those directly involved in the bonding [11,14], which leads to a large number of correlated electrons in many cases. The methods therefore have to be size-extensive and also applicable to cases with relatively large near-degeneracies. Two different types of methods have been developed that fulfill these requirements, the coupled pair functional (CPF) method [15], based on a single and double excitation configuration expansion, and the single and double excitation coupled-cluster (CCSD) method, including a perturbational estimate of connected triple excitations, denoted CCSD(T)[16]. Both these methods are restricted to a single reference state and are therefore efficient and straight forward to use. It has been shown that these methods give accurate values for relative energies of the types of systems discussed here. Most of the correlated calculations are therefore performed using a variant of the CPF method, the Modified Coupled Pair Functional (MCPF) method [17]. In a few cases, where several reference states are required, for example low-spin coupled open shell states which can not be described by a single reference determinant, another variant of the CPF method is used, the multi-reference Average Coupled Pair Functional (MR-ACPF) method [18]. All valence electrons, on the metals and on the ligands, are correlated in all calculations described.

Relativistic effects are accounted for using first order perturbation theory including the mass-velocity and Darwin terms [19]. Particularly for the second row metals the inclusion of relativistic effects is necessary for obtaining reliable results. For the third row metals a more sophisticated treatment of relativistic effects is needed, and these metals have therefore not yet been included in our studies.

In the determination of stationary points on the potential energy surfaces only the most important parameters are optimized and for example the internal methyl structure is kept frozen in all calculations. The geometries for the insertion products and the transition state structures in the alkane activation reactions are optimized for several of the atoms and for the rest of the atoms extrapolated from the optimized structures. Most of the optimizations are performed at the SCF level and test calculations showed that MCPF optimizations lead to very small energy changes. For the metal-alkene interaction geometry optimizations are performed on the MCPF level for all systems.

4. Alkane Activation

Selective C-H and C-C activation of saturated alkanes by transition metal catalysts is an important step in the transformation of the abundant, but inert alkanes into more useful products. Therefore a large number of studies, both experimental and theoretical have been performed to elucidate the mechanisms of alkane activation by transition metal complexes. Two fundamentally different mechanisms for transition metal activation of hydrocarbons in solution have been established by experiment, *1.* oxidative addition to a coordinatively unsaturated metal center and *2.* σ-bond metathesis to a metal alkyl or hydride. The main part (section 4.1) of this summary will be concerned with the oxidative addition mechanism, but a short report will also be given in section 4.2 on some preliminary results from a study on the σ-bond metathesis mechanism.

4.1 OXIDATIVE ADDITION

It was not until 1982 that direct intermolecular insertion of a metal center into unactivated carbon-hydrogen bonds was first observed [20-22]. These first observations involved iridium and rhodium complexes but subsequently insertion into alkane carbon-hydrogen bonds was observed also for rhenium, iron and osmium complexes [23] One important question in this context which will be addressed below is, which are the differences between the transition metals in their ability to activate saturated hydrocarbons? The analogous insertion of a metal center into unactivated carbon-carbon bonds has not yet been observed, despite the ready cleavage of carbon-carbon bonds by heterogeneous catalysts. This observation leads to a second question and this is why the C-H bond is more easily broken than the C-C bond, although the C-H bond is stronger, which will be addressed below.

For the comparison between the C-C and C-H activation reactions we have chosen to study the oxidative addition of a metal atom into the carbon-carbon bond of ethane (reaction 1) and into the carbon-hydrogen bond of methane (reaction 2).

$$M + \begin{matrix} CH_3 \\ | \\ CH_3 \end{matrix} \rightarrow M \begin{matrix} CH_3 \\ CH_3 \end{matrix} \qquad (1)$$

$$M + \begin{matrix} H \\ | \\ CH_3 \end{matrix} \rightarrow M \begin{matrix} H \\ CH_3 \end{matrix} \qquad (2)$$

The thermochemistry and activation energy for reactions (1) and (2) are calculated and the results for two different metals, nickel and palladium are summarized and discussed in section 4.1.1 below. For the comparison between different metals the energetics of reaction (2) is calculated for the neutral atoms of the whole second row of transition metals and the results are reported in section 4.1.2 below. A comparison between neutral metal atoms and cations is also performed, and some preliminary results from the calculation of the energetics of reaction (2) for the second row transition metal cations are given in section 4.1.3. Finally, some ligand effects on the reactivity of the metal atoms are discussed in section 4.1.4.

4.1.1. Comparison Between C-H and C-C Activation. The intermolecular oxidative addition of transition metals to alkane C-H bonds are observed in homogenous media, while the corresponding reaction for unactivated C-C bonds has not been observed. As can be seen from Table 1 and Figure 1, the barrier for the C-C insertion is calculated to be 14-16 kcal/mol higher than that for the C-H insertion, in agreement with these observations. It is also clear from Table 1 that the difference between the two reactions can not be explained by differences in the reaction energies. In fact, the calculated reaction energies for the metal insertion are not very different for the two reactions and, in particular, they are not in favour of the C-H insertion. The C-C insertion is slightly more exothermic than the C-H insertion.

Table 1. *Calculated energies for the insertion of nickel and palladium atoms into the C-C bond of C_2H_6 and into the C-H bond of CH_4. The relative energies (in kcal/mol) are given relative to the low-spin states of the atoms. Positive values for the reaction energies, ΔE, means that the addition reaction is endothermic. $\Delta E^{\ddagger}(add)$ is the barrier for the addition reaction and $\Delta E^{\ddagger}(elim)$ is the barrier for the elimination reaction.*

	Ni		Pd	
	C_2H_6	CH_4	C_2H_6	CH_4
ΔE	-11	-3	+7	+9
$\Delta E^{\ddagger}(add)$	+32	+18	+31	+15
$\Delta E^{\ddagger}(elim)$	+43	+21	+24	+6

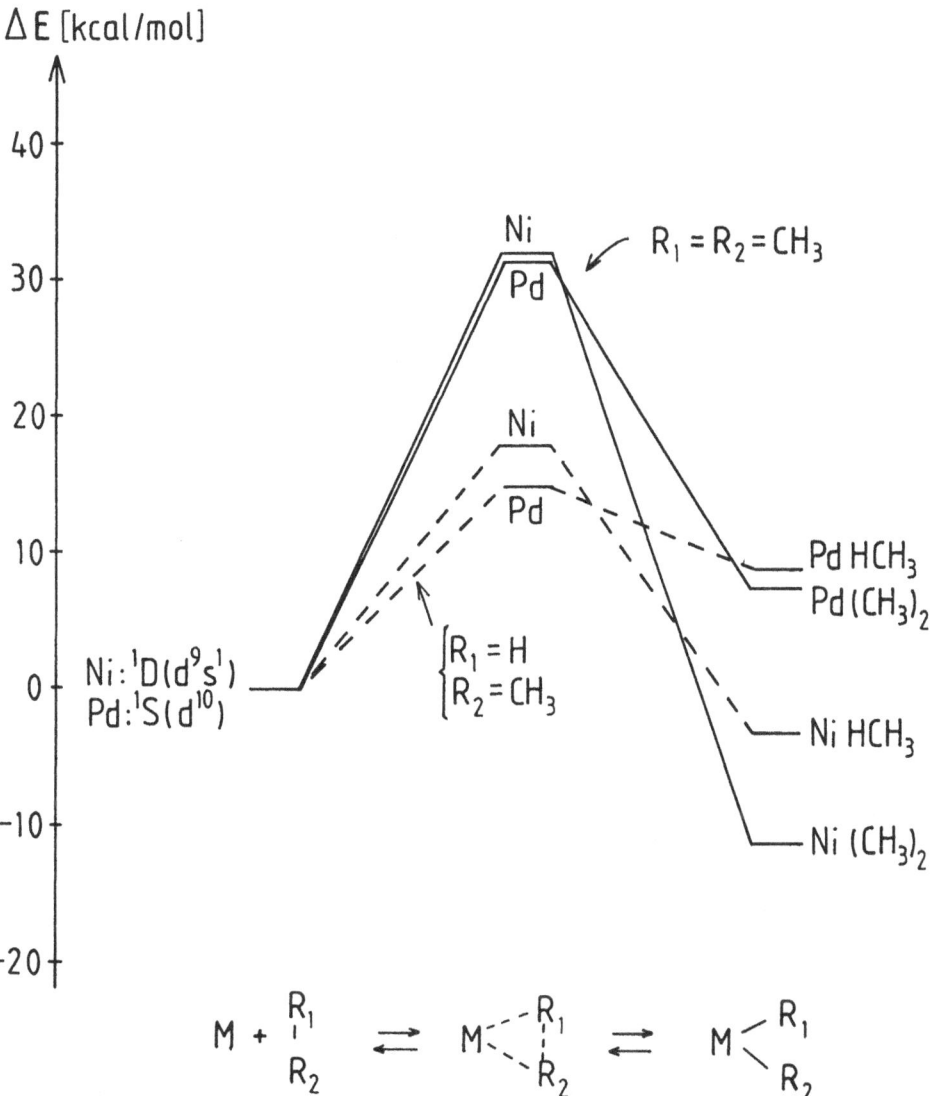

Figure 1. *Insertion of* Ni *and* Pd *into the* C-H *bond of methane (dashed curves) and into the* C-C *bond of ethane (solid curves). Calculated energies relative to the lowest low-spin state of the metal atom and free hydrocarbon.*

Instead, the difference between the two reactions can be explained by the reaction mechanism to be discussed below.

For the nickel atom the activation energy for the C-C insertion is 32 kcal/mol and for the C-H insertion 18 kcal/mol (see Table 1). The results obtained for palladium are quite similar, 32 kcal/mol for C-C insertion and 15 kcal/mol for C-H insertion. These values are obtained using the lowest low-spin state of the metal atom as reference point. In Ref. 11 several other first and second row metals were studied and it was found that for all metals the difference between the C-H and C-C activation barriers were in the range 14-20 kcal/mol.

The fact that the difference in barrier height for the C-H insertion and the C-C insertion is so constant from metal to metal suggests that the explanation for this difference should be found in the properties of hydrogen and methyl rather than in the properties of the metal. A simple explanation of this kind involves the difference in directionality between the H- and the CH_3-bond [11,24,25]. Since the spherically symmetrical hydrogen atoms can bind to the metal atom and to each other at the same time in the transition state structure, the metal insertion into the H_2 molecule forming the bent MH_2 complex has a very low or no barrier [26]. The sp^3 hybridized carbon in a methyl group, on the other hand, has one optimal binding direction and when the metal-carbon bond is starting to form in the MR_1R_2 complex the methyl groups have to rotate into a position that is no longer optimal for the R_1-R_2 bond. This gives rise to a higher barrier for metal insertion into C-C and C-H bonds compared to the H-H bond, and also to a higher barrier for the insertion into a C-C bond compared to the insertion into a C-H bond, since in the first case two methyl groups are involved and in the latter case only one.

Considering the reverse reactions, the elimination of ethane and methane from the MR_1R_2 complexes, two factors are important. First, as can be seen in Table 1, the C-H coupling has a lower barrier than the C-C coupling. This result is, of course, explained by the same reaction mechanisms as discussed above for the addition reactions. Secondly, the reaction energies obviously also affect the elimination barrier. With a large exothermicity for the addition reaction there will clearly be a large activation energy for the elimination reaction. Previous calculations show that processes involving H-H bonds have essentially no barriers, which is due to the spherical nature of the H atoms as discussed above. From this fact it should not be concluded, however, that the H-H coupling occurs more easily than the C-H and C-C couplings, since the metal dihydrides are often much stronger bound than the metal methyl complexes. For palladium, for example, the dihydride is bound relative to the ^1S ground state of palladium by about 8 kcal/mol [11] and the activation barrier for C-H coupling is 6 kcal/mol. Thus, about the same activation energy is needed for H-H coupling and C-H coupling.

4.1.2. Comparison Between Different Metals for C-H *Activation.* Only a few transition metals are represented among those metal complexes which have been observed to insert into C-H bonds in saturated hydrocarbons via an oxidative addition mechanism. The first observations of alkane C-H insertion in solution were made in 1982 for iridium complexes, and the active intermediates were supposed to be coordinatively unsaturated fragments of the general formula Cp*IrL (L=CO,PR$_3$)[20,21]. Shortly afterwards the analogous rhodium fragment (Cp*RhL) was also found to be active [22] and later on also the ClRhL$_2$ (L=PPh$_3$) fragment [27]. Other metals proven to be effective in alkane oxidative addition by metal complexes are iron, rhenium and osmium [23]. Furthermore, the Cp*ML (M=Ir,Rh) and CpML (M=Ir,Rh) fragments have been shown to insert into methane and other alkane C-H bonds in matrices at low temperature (10-20 K) [28-30]. Apparently rhodium is the only second row metal active in homogenous oxidative addition to alkanes and an important question to answer, is why rhodium is more efficient in this process than the rest of the second row metals.

To address this question the oxidative addition of all second row transition metal atoms to methane were studied. The thermochemistry of reaction (2) was calculated and the barrier heights were determined. From several experimental investigations on both iridium and rhodium complexes it has been inferred that a molecularly bound metal-alkane intermediate is formed in the C-H activation reaction [31-33]. Therefore the energy of a molecularly bound metal-methane complex was also calculated for all metal atoms. Thus four regions on the potential surfaces were investigated: the metal atom plus free methane **1**, a molecular precursor complex with the η^2-structure **2**, the transition state **3** and the insertion product MHCH$_3$ complex **4**. The energies are given relative to the ground state of the metal atom, which in many cases does not have the same spin state as the product complex **4**. Therefore the lowest atomic state with the same spin as the product was also calculated. For the η^2-complex **2** two different spin states were investigated for most cases, one corresponding to the atomic ground state and one corresponding to the product complex **4**. The resulting energies are summarized in Table 2 and the electronic structures are presented in Table 3.

$$M + CH_4 \qquad M \overset{\text{..-H}}{\underset{\text{-.H}}{\diagdown}} CH_2 \qquad M \overset{\text{H}}{\underset{\text{---}}{\text{...}}} CH_3 \qquad M \overset{\text{H}}{\diagdown} CH_3$$

$$\quad 1 \qquad\qquad\qquad 2 \qquad\qquad\qquad 3 \qquad\qquad\qquad 4$$

The results show that for yttrium to technetium the reaction barriers for ground

Table 2. *Calculated energies for the insertion of neutral metal atoms into the C-H bond of* CH_4*, relative to the ground state of the metal atom and free methane (kcal/mol). Positive values for* ΔE *means that this structure lies above the metal plus methane asymptote. The calculated energy of the lowest low-spin atomic state is also given.*

	M + CH_4		η^2-complex[a]		Transition state		Product complex	
	State	ΔE	State	ΔE	State	ΔE	State	ΔE
Y	$^2D(d^1s^2)$	0	$^2A_1$	19	$^2A'$	42	$^2A'$	-7
Zr	$^3F(d^2s^2)$	0	$^3A_2$	22	$^3A"$	33	$^3A"$	-12
Nb	$^6D(d^4s^1)$	0	$^6A_1$	3	$^4A"$	32	$^4A"$	-10
	$^4F(d^3s^2)$	4	$^4A_1$	12				
Mo	$^7S(d^5s^1)$	0	$^7A_1$	16	$^5A'$	48	$^5A'$	13
	$^5S(d^5s^1)$	32	$^5A_1$	34				
Tc	$^6S(d^5s^2)$	0	$^6A_1$	28	$^4A'$	49	$^4A'$	27
	$^4D(d^6s^1)$	42	$^4B_1$	42				
Ru	$^5F(d^7s^1)$	0	$^3A_1$	8	$^3A'$	24	$^3A'$	9
	$^3F(d^7s^1)$	20	$^5B_2$	9				
Rh	$^4F(d^8s^1)$	0	$^2A_2$	6	$^2A'$	14	$^2A'$	-5
	$^2D(d^9)$	15	$^4A_1$	9				
Pd	$^1S(d^{10})$	0	$^1A_1$	-4	$^1A'$	16	$^1A'$	9

a. *The energy of the* η^2*-complex is for all metals calculated in the geometry optimized for the low-spin rhodium* η^2*-complex.*

state insertion into the C-H bond of methane are larger than about 30 kcal/mol. The lowest barriers are found for rhodium (14 kcal/mol) and palladium (16 kcal/mol).

Table 3. *Mulliken populations for the neutral* M + CH$_4$ *systems. Charge* (q$_M$) *and 4d population on the metal atom.*

	η^2-complex[a]			Transition state			Insertion product		
	State	q$_M$	4d	State	q$_M$	4d	State	q$_M$	4d
Y	$^2A_1$	-0.3	1.3	$^2A'$	+0.1	1.3	$^2A'$	+0.5	1.0
Zr	$^3A_2$	-0.3	2.4	$^3A"$	+0.1	2.6	$^3A"$	+0.5	2.3
Nb	$^6A_1$	-0.1	4.2	$^4A"$	+0.1	3.9	$^4A"$	+0.4	3.5
	$^4A_1$	-0.2	3.6						
Mo	$^7A_1$	-0.1	4.9	$^5A'$	+0.1	5.1	$^5A'$	+0.3	4.7
	$^5A_1$	-0.1	4.9						
Tc	$^6A_1$	-0.1	5.9	$^4A'$	+0.1	6.0	$^4A'$	+0.3	5.9
	$^4B_1$	-0.1	6.1						
Ru	$^3A_1$	0.0	7.8	$^3A'$	+0.1	7.6	$^3A'$	+0.2	7.1
	$^5B_2$	-0.1	6.9						
Rh	$^2A_2$	0.0	8.6	$^2A'$	+0.1	8.4	$^2A'$	+0.1	8.2
	$^4A_1$	-0.1	7.9						
Pd	$^1A_1$	0.0	9.9	$^1A'$	+0.1	9.4	$^1A'$	+0.2	9.2

a. The geometry for the η^2-complex is taken from the low-spin rhodium η^2-complex.

For rhodium the product complex is bound by 5 kcal/mol, while palladium has an unstable product complex, unbound by 9 kcal/mol. Also for ruthenium, with an intermediate barrier of 24 kcal/mol, the product complex is unstable by 9 kcal/mol. It should further be noticed that for palladium the lowest point on the atomic ground state potential curve is the molecular η^2-complex, bound by about 4 kcal/mol, making palladium unique among the second row atoms in that it is the only one that has an η^2-complex which is bound relative to the ground state of the atom.

To simplify the comparison between the metals, the reaction energies (binding energies and barrier heights) for all second row metals are plotted in Fig. 2. From this figure it can be seen that rhodium has the lowest barrier for C-H addition and palladium has the lowest barrier for C-H elimination.

The trends for the calculated binding energies can be understood from the binding mechanisms and the electronic spectra of the metal atoms. In the product complex **4** two covalent bonds are formed and the bonding electron configuration on the metal therefore has to have two open shells. The main bonding atomic configuration for the systems studied here is found to be $4d^{n+1}5s^1$, i.e. the two covalent bonds are formed by one $5s$ and one $4d$ electron from the metal. The corresponding state on the atom with the lowest energy will have a high spin coupling of these two electrons, which will often be referred to as the high-spin state below, while the product complex will be referred to as the low-spin state. The position of these states in the atomic spectra is an important factor determining the binding energy of the insertion product. To the left in the periodic table the $4d^n5s^2$ state will have the same spin as the product complex and will therefore contribute to the bonding. To the right in the periodic table the $4d^{n+2}$ state will have the same spin as the product complex and will therefore contribute to the bonding. Another factor influencing the binding energy in the insertion product is the amount of loss of exchange energy upon bond formation [34], which is largest for the atoms with many open shells, i.e. the atoms in the middle of the row.

The barrier for the reaction between methane and the transition metal atoms is a result of a crossing between two surfaces. Therefore the energy in the transition state region **3** is determined by two important factors. Before the barrier the C-H bond starts to break and methane prepares for the bonding towards the metal. In this region the interaction between methane and the metal atom is essentially repulsive and the metal adopts the state which is least repulsive. Since the $5s$ electrons are the most diffuse electrons in the metal, the lower the number of $5s$ electrons in the configuration the smaller is the repulsion and thus the configurations with zero $5s$ electrons give the lowest repulsion. For the low-spin coupled state with one $5s$ electron, the repulsion can also be reduced by the formation of two sd hybrids, one pointing towards the C-H bond and the other perpendicular to the line connecting the metal with the C-H bond. By placing two electrons in the latter of these, rather than one in each, the repulsion is significantly decreased. After the barrier the two bonds are formed and the strength of the bonding in the product complex determines this part of the potential surface. The barrier height for the addition reaction is thus reduced both by a low repulsion in the entrance region and by strong bond-formation in the product region. The last of these two factors is responsible for the fact that the general shape of the curve for the barriers in

399

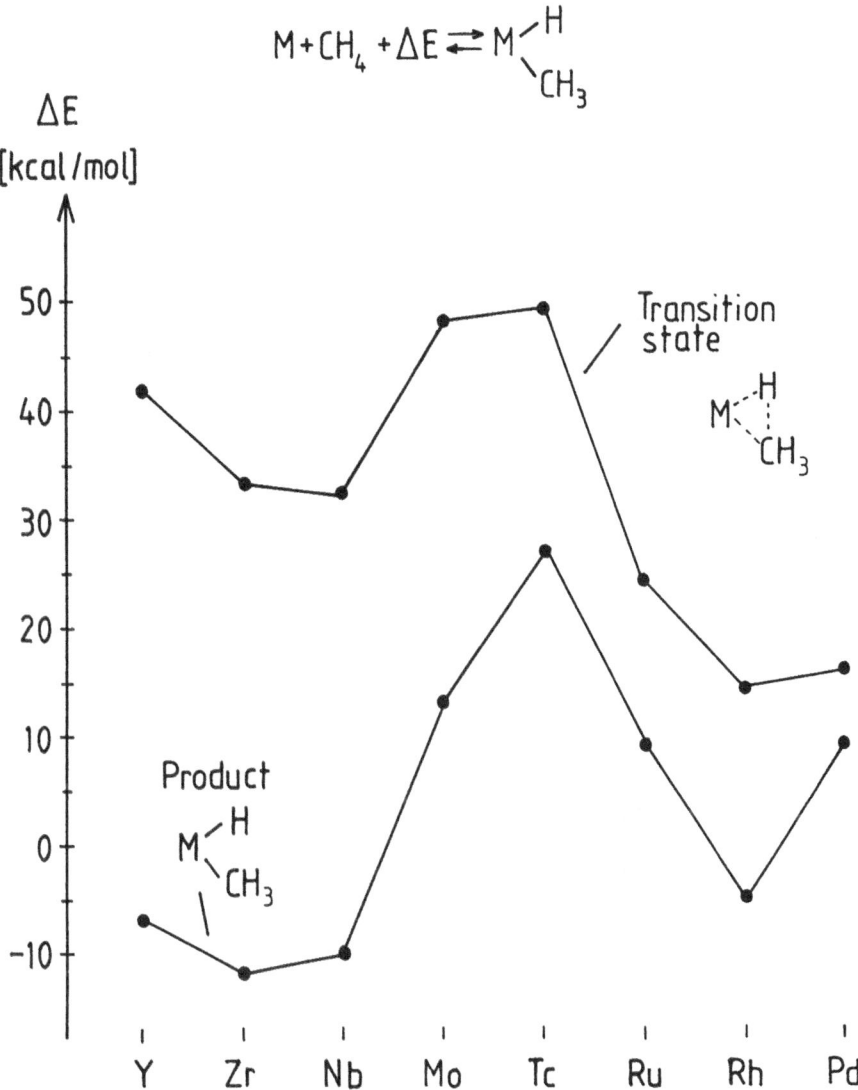

Figure 2. *Insertion of the second row metal atoms into the C-H bond of methane. Calculated energies for the transition state structures (upper curve) and the product complexes (lower curve) relative to the ground state of the metal atom and free methane. The upper curve corresponds to the addition barrier and a large positive value for ΔE corresponds to a high activation barrier. The lower curve corresponds to the binding energy of the product complex and a large positive value for ΔE corresponds to a strongly endothermic addition reaction, while a negative value for ΔE corresponds to an exothermic addition reaction.*

Figure 2 is rather similar to the curve for the binding energies. The first factor makes the barriers to the right in the periodic table lower than to the left since the atoms to the right have low lying s^1 and s^0 states of the same spin as the product complex.

In the region of the η^2-complex **2** a low-lying $4d^{n+2}$ state is crucial to reduce the repulsion. The bonding force is achieved by an admixture of the low-spin coupled $4d^{n+1}5s^1$ state, which by means of the sd hybridization discussed above, causes a distortion of the electron distribution around the metal atom. Through this distortion the metal nucleus is partly unshielded and the electrons in the C-H bond will experience an attractive force. It is therefore important that both the $4d^{n+2}$ and the $4d^{n+1}5s^1$ states are low-lying for obtaining bound η^2-complexes, and reasonably bound η^2-complexes only exist on the low-spin potential surfaces of the metals to the right in the periodic table. It should be noted that the calculations in this region of the potential surfaces are performed for one single geometry for all metals, the one optimized for the low-spin state of the rhodium system. For those metals which were found to have unbound η^2-complexes at this particular geometry there might be a minimum on the potential surface for longer metal-methane distances, but these minima have to be very weakly bound, probably not by more than 1 kcal/mol.

The main motivation for the present study is to contribute to the understanding of the mechanisms of C-H activation in homogenous catalysis where the active catalysts are ligated complexes. Below we will discuss how to best make comparisons between the naked atom model presently used and the conclusions drawn from the experimental information about the ligated metal complexes. Comparisons will also be made between the present calculations and experimental information about the reactivity of naked metal atoms.

The conclusion that the rhodium atom has the lowest barrier for insertion into the C-H bond in methane is in accord with the fact that rhodium complexes, to our knowledge, are the only second row metal complexes experimentally observed to activate alkane C-H bonds in solution [22,27] and in matrices [28-30]. It is highly unlikely that this is just a coincidence and in the section on ligand effects below the differences and similarities between ligated metal complexes and naked metal atoms are examined more in detail. Comparing the potential surface for the ClRh(PH$_3$)$_2$ complex with the results for both the neutral rhodium atom and the cation it is found that the best agreement with the ligated complex is obtained for the neutral atom. Thus the fact that the neutral rhodium atom for a large part of the potential surface gives the best correspondence to the ligated Rh(I)-complex shows that the oxidation state of the metal is of minor importance in modelling a metal complex.

As will be seen in the section on ligand effects below, the best overall agreement with the potential surface for the ClRh(PH$_3$)$_2$ complex is further obtained if the low-spin coupled excited state of the rhodium atom is used as reference point rather than the high-spin coupled ground state. However, such a similarity can not be expected for all metal atoms. In particular, for the metals with very large splittings between the spin states it is not expected that the atomic excited low-spin state should give potential surfaces as close to the ligated metal complexes as is found in the rhodium case. The reactivity of a certain metal in complexes is probably best estimated by considering both the ground state and the low-spin reference points. It is interesting to note that among the second row metals, rhodium will have the lowest barrier for insertion into the C-H bond of methane whichever of these two approaches is followed. Using the high-spin ground state as reference point leads to an insertion barrier of 14 kcal/mol for rhodium, followed by palladium and ruthenium with barrier heights of 16 and 24 kcal/mol, respectively. Following the low-spin surface, rhodium is the only metal without a barrier. The second lowest barrier is obtained for ruthenium with a value of 4 kcal/mol.

Turning now to the experiments performed for naked metal atoms the comparison between the calculated results and experiment is obviously more direct since no modelling is involved. For neutral metal atoms, however, there are rather few experiments performed to elucidate their reactivity towards saturated hydrocarbons. In most of these experimental studies [1a,2,35,36] no reaction have been found between neutral metal atoms and alkanes, which is in agreement with the relatively high barriers presently calculated for the methane insertion for the ground state atoms. In a recent study, however, Klabunde et al [1b] investigated the reactivity of all transition metal atoms towards methane and in contrast to previous results found several second row atoms to be strongly interacting with methane. On the basis of the presently reported results we conclude that there must be some other explanation for these observations than the insertion of ground state metal atoms into the C-H bond of methane, in particular since the experiments were performed at low temperature (10-20 K). Perhaps excited states of the metal atoms are involved. It should also be noted that in this experiment, among the second row metals, the insertion product MHCH$_3$ was only identified for the case of rhodium. Rhodium, having the lowest barrier and a stable product complex, is actually the second row transition metal for which insertion into the C-H bond of methane is most likely to be observed for the naked metal atom. However, an activation energy of about 14 kcal/mol is needed and a spin transition, to reach from the quartet ground state of the metal atom to the doublet state of the insertion product.

4.1.3. Comparison Between Neutral Metal Atoms and Cations for C-H Activation.
To demonstrate the tendencies in the metal reactivity when the oxidation state is changed, calculations were also performed for reaction (2) for all second row

transition metal cations. For the η^2-complexes **2** the geometries were optimized and the calculated binding energies are listed in Table 4. In contrast to the neutrals most cations have fairly strongly bound η^2-complexes on the atomic ground state potential surfaces. To get a quick estimate of the activation energies and binding energies for the C-H insertion reaction, calculations were also performed in the transition state region **3** and in the insertion product region **4**, using the geometries from the corresponding neutral systems, slightly modified in the following way. For yttrium to thechnetium the H-M-C bend angle of the insertion product, and the M-C bond distances for both structures, were taken from the $M(CH_3)_2^+$ complexes in ref. 37. The results are listed in Table 5. The product complexes are in most cases less bound for the cations than for the neutrals, while the energies in the transition state regions, calculated in this way, for most metals are a few kcal/mol lower for the cations than for the neutrals. These calculated activation energies are preliminary and are likely to decrease somewhat when the true transition state geometries are determined, which is currently done. Still, the present results indicate that for most of the cations there are substantial activation barriers for the insertion into the C-H bond of methane.

The calculated binding energies for the M^+-CH_4 complexes **2** are in the range 0-17 kcal/mol, and are listed in Table 4. An η^2-structure, with two hydrogens pointing towards the metal, was assumed, and the M-CH_4 distances were optimized. The bonding in this region is dominated by the charge-induced dipole interaction. As can be seen from Table 4, the binding energies correlate strongly with the distance to the metal, a short bond-distance giving a larger binding energy than a long bond-distance, which is in accord with the behaviour of the dominating bonding force. The bond-distance, in turn, is determined by several factors, and one of these is the electronic configuration on the metal. The radius of the $4d$ orbital is smaller than that of the $5s$ orbital, and metals with a $4d^{n+1}$ ground state of the cation therefore have shorter bond distances than those with a $4d^n 5s^1$ ground state, as can be seen from Table 4. It can also be seen from Table 4 that for metals with a $4d^n 5s^1$ ground state of the cation, the more compact $4d^{n+1}$ state is mixed in, if the two states are of the same spin. This is the case for zirconium and also for the triplet state of the yttrium system. For the yttrium cation the ground state is actually the $^1S(5s^2)$ state, but due to the large repulsion of the two s-electrons there is no minimum in the η^2-region for this state and the triplet state correlating with the $^3D(4d^1 5s^1)$ state is the ground state of the η^2-complex. For the cations to the right in the periodic table the $4d^n 5s^1$ and $4d^{n+1}$ states have different spin and can therefore not mix. As can be seen from Table 4 for the case of technetium, which is the only one of these metals having a $4d^n 5s^1$ grounds state, there is no mixing in of the excited $4d^{n+1}$ state.

Table 4. *Calculated energies (in kcal/mol) for the η^2-complexes of the M^+-CH_4 systems, relative to the ground state M^+ + CH_4 asymptote. The optimized M-C bond distances are also given, as well as the charge and 4d population on the metal.*

	M^+		η^2-complex				
	Ground state	State	ΔE	$R(M\text{-}C)^a$	q_M	4d	
Y^+	$^1S(s^2)^b$	$^3A_2$	-0.4^b	5.7	+0.8	1.2	
Zr^+	$^4F(d^2s^1)$	$^4B_1$	-10.4	5.1	+0.7	2.6	
Nb^+	$^5D(d^4)$	$^5B_2$	-11.8	5.0	+0.7	3.9	
Mo^+	$^6S(d^5)$	$^6A_1$	-8.8	5.3	+0.8	5.0	
Tc^+	$^7S(d^5s^1)$	$^7A_1$	-4.5	5.9	+0.9	5.0	
Ru^+	$^4F(d^7)$	$^4B_1$	-13.3	4.9	+0.9	6.9	
Rh^+	$^3F(d^8)$	$^3A_2$	-13.0	5.0	+0.8	7.9	
Pd^+	$^2D(d^9)$	$^2A_1$	-17.0	4.8	+0.8	9.0	

a. In a_0. b. The $^3D(d^1s^1)$ state is 3.4 kcal/mol above the ground state.

The binding energies of the insertion products and the activation energies are listed in Table 5. For three of the metals, ruthenium, rhodium and palladium, the calculated energy in the transition state region is lower than the energy in the insertion product region, indicating that there are no minima for the insertion products for these metals. The only metals for which the insertion product is below the ground state M^+ + CH_4 asymptote are yttrium and zirconium, with binding energies of 7 and 9 kcal/mol. The yttrium cation has a $5s^2$ ground state with a low lying $4d^15s^1$ state and the zirconium cation has a $4d^25s^1$ ground state. All the other metals have fairly unbound insertion products and all, except technetium, have $4d^{n+1}$ ground states. These results indicate that, in line with the results for the neutrals, the $4d^n5s^1$ state is the best bonding state also for the cations. In

Table 5. *Calculated energies for the insertion of metal cations into the* C-H *bond of* CH_4, *relative to the ground state of the cation and free methane (kcal/mol). Positive values for* ΔE *means that this structure lies above the metal plus methane asymptote. The geometries used in these calculations are taken from the corresponding neutral systems and slightly modified as described in the text.*

	M^+	Transition state		Product complex	
	Ground state	State	ΔE	State	ΔE
Y^+	$^1S(s^2)$	$^1A'$	34	$^1A'$	-7
Zr^+	$^4F(d^2s^1)$	$^2A'$	23	$^2A'$	-9
Nb^+	$^5D(d^4)$	$^3A"$	30	$^3A"$	10
Mo^+	$^6S(d^5)$	$^4A'$	47	$^4A'$	39
Tc^+	$^7S(d^5s^1)$	$^5A'$	23	$^5A'$	14
Ru^+	$^4F(d^7)$	$^4A"$	16	$^4A"$	20
Rh^+	$^3F(d^8)$	$^3A'$	11	$^1A'$	24
Pd^+	$^2D(d^9)$	$^2A'$	15	$^2A'$	20

particular molybdenum and rhodium have very unstable insertion products, unbound by 39 and 24 kcal/mol. These insertion products are almost 30 kcal/mol less bound than the corresponding insertion products for the neutral atoms. For these two cations the $4d^n5s^1$ states are as much as 37 and 49 kcal/mol above the $4d^{n+1}$ ground states, while the neutrals have $4d^{n+1}5s^1$ ground states. The technetium cation has a $4d^55s^1$ ground state, but due to the large loss of exchange energy upon bond formation the product complex is fairly unbound, by 14 kcal/mol. Technetium and yttrium are the only metals for which the cations are not less bound than the neutrals in the insertion product region. The calculated activation energies for the insertion reaction are in the range of 20 to 50 kcal/mol for the cations, compared to 15 to 50 kcal/mol for the neutrals.

For the gas phase reactions of naked transition metal cations with alkanes the experimental information is abundant. Most of the metal cations investigated are very reactive towards alkanes. The metal insertion in an alkane C-H or C-C bond is postulated to be the first step in these reactions, even though the insertion product complexes have never been observed. Therefore a comparison between these experiments and the present calculations should be relevant. The C-H insertion step is believed to have no barrier in the alkane activation reactions actually observed [5,6,38]. However, most experimental information about the reactivity of transition metal cations with methane does not contradict the presently calculated large activation energies for insertion of second row cations into the C-H bond of methane. While cations experimentally are found to be very reactive with larger alkanes, most metal cations do not react with methane at low collision energies. For example, the second row metal cations yttrium[39], niobium [40], molybdenum[41], ruthenium, rhodium and palladium [42,43] have been found to be unreactive with methane, while all these metals do react with larger alkanes. The only second row cation found to react with methane is zirconium [44].

Even if the main experimental information on methane activation by cations could be interpreted to support the present results, there are discrepancies between the present preliminary calculated high activation energies for the cation insertion into methane and some of the experimental information on alkane activation. First, as mentioned above, in the case of zirconium methane activation has actually been observed. Therefore a determination of the transition state geometry for this particular case was performed, and the barrier height thus calculated is as high as 18 kcal/mol. One possible explantion for this discrepancy between theory and experiment could be that there is a mistake in the interpretation of the experimental result, in the sense that the reacting cations actually did have enough energy (electronic or kinetic) to surmount such a barrier. Another explanation could be that there is a different, low energy reaction pathway available which has not yet been found in the calculations. This latter possibility is presently explored. A second reason for questioning the high calculated activation energies for the cation insertion into methane is the expected similarity between methane and ethane. Since ethane reacts with several second row cations, and therefore should have no barrier for the C-H insertion reaction, the barrier height for insertion into methane is expected to be very low. There are different ways this discrepancy between the theoretical and experimental predictions could be resolved. One possibility is that there is no similarity between methane and ethane for this step of the reaction. The expected similarity in barrier heights is based on the similarity in the binding energy of the η^2-complex of methane and ethane. But in the transition state region there might be energy lowering mechanisms available for ethane which are not available for methane, e.g. agostic bonding, leading to a substantially lower activation energy for ethane activation than for methane activation. Another possibility is that the barriers actually are rather similar for ethane and methane, and that both are fairly high as predicted by the present preliminary calculations. If this is the case there has to be enough energy available for the reactants in the experiments to surmount the high barriers, and the difference in reactivity between methane and ethane could be caused by the fact that the final products are more endothermic in the case of methane than in the case of ethane. Finally, there is still the possibility of a different, low energy reaction pathway available for the insertion of cations into the C-H bond of both methane and ethane, which has not yet been found in the calculations. As mentioned above this possibility is presently explored.

In summary the cations are more strongly bound to methane in the η^2-region than the neutral atoms are. This stronger bond in the η^2-region, which is expected to shift the potential surfaces of the cations to lower energies compared to the neutrals [5,6,38], is only one of the factors determining the differences between neutral atoms and cations in their reactivity towards methane. Another factor is the differences in the atomic spectra of the neutrals and the cations. The cations, for example, in general have lower lying $5s^0$ states, which should contribute to lowering the barriers. On the other hand, the best states for forming the two covalent bonds in the final insertion products are the $5s^1$ and $5s^1 5p^1$ states, which are generally less available for the cations than for the neutrals, leading to less bound insertion products for the cations. This weaker bonding, in turn, should contribute to increasing the barrier heights of the cations. In conclusion, there are several minor differences between the neutrals and the cations in their interaction with methane, but the present results do not indicate large differences between the second row transition metal neutral atoms and cations in their reactivity with methane.

4.1.4. Ligand Effects on C-H Activation. To examine the differences and similarities between ligated metal complexes and naked metal atoms calculations were performed also for the reaction between methane and the rhodium complex $ClRh(PH_3)_2$. This complex is chosen to model the $ClRh(PPh_3)_2$ complex, which is believed to be active in an experimentally observed C-H insertion process [27]. The same four points on the potential surface as for the naked metal atoms were calculated for the $ClRh(PH_3)_2$ complex. The results are summarized in Table 6, where also the reaction energies for the naked rhodium atom have been included, using two different reference points, the high-spin ground state and the low-spin excited state corresponding to the state of the insertion product. Another aspect on the comparison between metal atoms and metal complexes is the oxidation state

Table 6. *Calculated relative energies, in kcal/mol, for the reaction of different* Rh *fragments with methane.*

L_nM	$L_nM + CH_4$		η^2-complex		Transition state		Product complex	
	State	ΔE	State	ΔE	State	ΔE	State	ΔE
$ClRh(PH_3)_2$	$^1A_1$	0	$^1A_1$	-12	$^1A_1$	-2	$^1A_1$	-10
Rh	^4F	0	$^2A_2$	6	$^2A'$	14	$^2A'$	-5
Rh	^2D	0	$^2A_2$	-9	$^2A'$	-1	$^2A'$	-19
Rh^+	^3F	0	$^3A_2$	-13	$^3A'$	11	$^1A'$	24

of the metal. The neutral naked metal atom has the oxidation state zero, while in the reacting metal complexes higher oxidation states are common. One simple way to obtain correct oxidation states for the naked atom model is to use cations instead of neutral atoms, with a charge corresponding to the oxidation state of the metal in a particular type of complexes. Rh^+ could thus be used to model Rh(I) in the $ClRh(PH_3)_2$ complex. In Table 6 we have therefore also included the reaction energies for the naked rhodium cation using the geometries of the neutral rhodium atom.

Comparing the $ClRh(PH_3)_2$ complex with the different naked atom models in Table 6 it can be seen that the best overall agreement with the ligated complex is obtained for the neutral atom following the low-spin surface. In the region of the η^2-complex both the cation and the low-spin rhodium atom give results close to the full complex. For the cationic system, the η^2-complex is bound by 13 kcal/mol and the attraction between rhodium and the methane molecule is dominated by the charge-induced dipole interaction. For the neutral rhodium atom the low-spin coupling of the valence electrons is crucial for the formation of an η^2-complex, the low-spin η^2-complex is bound by 9 kcal/mol relative to the low-spin asymptote, while on the high spin surface there is no minimum for the assumed η^2-structure. For the ligated complex the binding energy of the η^2-structure is 12 kcal/mol and the binding can be viewed as a combination of the type of binding in the neutral atomic low-spin complex and the pure electrostatic binding of the cationic complex.

In the region of the insertion product the ligated complex is bound by 10 kcal/mol. For the neutral rhodium atom the product complex is bound by as much as 19 kcal/mol relative to the low-spin asymptote and by 5 kcal/mol relative to the high-spin ground state, while for the cation there is no minimum for the product complex and the energy in this region is 24 kcal/mol above the asymptote. Thus, for the product complex only the neutral atom give results reasonably close to the ligated complex. Also in the region of the transition state the low-spin coupled neutral atom gives the best correspondence to the full complex with a relative energy of -1 kcal/mol compared to -2 kcal/mol for the ligated complex.

Thus the fact that the naked rhodium atom gives a potential surface for the interaction with methane that is quite similar to the $ClRh(PH_3)_2$ complex can be interpreted as if the ligands in metal complexes only have very small effects on the reactivity. This interpretation is supported by the results in Ref. 45, where potential surfaces were calculated for the reaction between methane and different metal complexes. Two different rhodium complexes were investigated, CpRhCO and $CpRhPH_3$, and for both these complexes the potential surfaces obtained are quite similar to the present one for the $ClRh(PH_3)_2$ complex and thus also quite similar to the one for the naked rhodium atom. Further, in Ref. 45 a ruthenium complex, $Ru(CO)_4$, was also investigated and the potential surface for the reaction between methane and this complex has large similarities to the one presently obtained for the naked ruthenium atom. The calculated insertion barrier for the $Ru(CO)_4$ complex is about 20 kcal/mol and the insertion product is found to be unbound by about 5 kcal/mol [45]. Relative to the ground state of the ruthenium atom the presently calculated barrier for methane insertion is 24 kcal/mol and the product complex is unbound by 9 kcal/mol. It would, however, be premature to draw the general conclusion that the ligands on a metal complex have no influence. There are several studies, both experimental [46,47] and theoretical [48,49], showing that the choice of ligands or the position of a ligand can have a large effect on the stability and reactivity of metal complexes. The most interesting conclusion to be drawn in this context is that the difference between the rhodium and ruthenium complexes is similar to the difference between the rhodium and ruthenium atoms. For both the ligated complexes and the naked atoms ruthenium has a higher methane insertion barrier than rhodium, and rhodium has the most stable product complex.

One important ligand effect to be mentioned here is the drastic change in the splitting between different spin states compared to the naked metal atom. The low-spin (singlet) state of $ClRh(PH_3)_2$ is calculated to be only 3 kcal/mol higher in energy than the ground triplet state. Such a small calculated splitting between a triplet and a singlet state indicates that the actual ground state of the metal fragment might very well be the singlet state, since usually a more accurate description of correlation effects will decrease the energy of the singlet state relative to the triplet state. This splitting between the high-spin ground state and the low-spin excited state of the ligated rhodium complex of 3 kcal/mol or less should be compared to the splitting of 15 kcal/mol between the high-spin ($^4F(4d^85s^1)$) and low-spin ($^2F(4d^85s^1)$) states of the rhodium atom. This is an example of the general trend that the ligands in metal complexes stabilize the low-spin states, and therefore the low-spin states of metal fragments are expected to be either the ground state or, at least, to have an excitation energy that is much lower than for the free metal atom.

4.2. σ-BOND METATHESIS

The σ-bond metathesis reaction for H-H or C-H activation is described by the general formula:

$$L_nMR_1 + RH \rightarrow L_nMR + R_1H \tag{4}$$

A number of high valent, electrophilic metal systems have been observed to intermolecularly activate alkanes by this mechanism, both in solution [50,51] and in gas phase [52,53]. The systems found to be active in solution are metallocene derivatives of the type Cp_2^*MR, with M=Sc,Y or Lu, and R=H or alkyl. The system studied in gas phase is the similar cationic complex $Cp_2ZrCH_3^+$. All these complexes are considered as d^0 systems and the oxidative addition path for alkane activation should not be available, since the metals are already in their highest possible oxidation states. The experimental information indicate that reaction (4) occurs via a four-center transition state structure **5**.

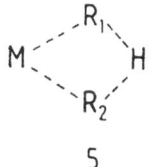

5

We have chosen to study three different cases of reaction (4):

$$L_nMH + H_2 \rightarrow L_nMH + H_2 \tag{5}$$

$$L_nMCH_3 + H_2 \rightarrow L_nMH + CH_4 \tag{6}$$

$$L_n MCH_3 + CH_4 \rightarrow L_n MCH_3 + CH_4 \tag{7}$$

Four different metals have been investigated, Sc, Y, Ti and Zr and the Cp and Cp* ligands are modelled by hydrogens. This modelling is justified by test calculations on reaction (5) for yttrium, yielding very similar relative energies for Cp and H ligands, using fixed geometries. The complexes used to model the solution experiments are thus $H_2 MR$, with M=Sc, Y and R=H, CH_3. The gas phase experiments are modelled by the $H_2 MR^+$ complexes, with M=Ti, Zr and R=H, CH_3. A four center transition state **5** is determined for each of the reactions and the activation energies are summarized in Table 7.

The most important result in Table 7 is that for each metal the activation energy increases with the number of methyl groups in the four center transition state structure **5**. Reaction (5) with no methyl group involved has the lowest barrier, varying between 1 and 10 kcal/mol for the different metals, and reaction (7) with two methyl groups in the transition state structure has the highest barrier for all metals, varying between 12 and 28 kcal/mol. Reaction (6) with only one methyl group has an intermediate barrier for all metals, varying between 5 and 13 kcal/mol. This trend has the same origin as the difference between C-H and C-C activation by the oxidative addition mechanism discussed above, namely the difference in directionality between the hydrogen atom and the methyl group. The spherically symmetric hydrogens can bind in several directions at the same time, while the sp^3 hybridized carbon in the methyl group has one optimal binding direction. There-

Table 7. *Calculated activation energies in kcal/mol for the different σ-bond metathesis reactions.*

	$L_n MH + H_2$ (5)	$L_n MCH_3 + H_2$ (6)	$L_n MCH_3 + CH_4$ (7)
$H_2 ScR$	10	13	28
$H_2 YR$	9	12	25
$H_2 TiR^+$	1	7	12
$H_2 ZrR^+$	1	5	12

fore, the energy of the transition state structure **5** is higher the more methyl groups that are involved, since for this structure each R-group has to bind in two directions.

For the comparison with the gas phase experimental results it should be noted that only reaction (6) has been investigated, and only for the zirconium case [52,53]. Hydrogen activation was observed to occur in this experiment, while the reverse reaction did not occur [52]. The presently calculated activation energy for the hydrogen activation by the zirconium-methyl complex is only 5 kcal/mol, making it very plausible that the observed reaction actually occurs via this transition state structure. The fact that the reverse of reaction (6) did not occur was interpreted as a stronger Cp_2Zr^+-H bond than the corresponding Cp_2Zr^+-CH_3 bond. In contrast to these observations, the present calculations indicate that reaction (6) is approximately thermoneutral for the zirconium case. To resolve this discrepancy between theory and experiment the effects of the Cp ligands on the Zr-H and Zr-CH_3 binding energies are currently investigated.

In solution-experiments reaction (5) is observed to occur very fast for the scandium case [51] and reaction (7) is observed for both scandium and yttrium [50,51]. The yttrium complex reacts about 250 times faster than the scandium complex [51]. The calculated barrier for reaction (5) for the scandium case is reasonably low, 10 kcal/mol, and the four center structure **5** is therefore a possible transition state for the observed reaction. For reaction (7), however, the calculated barriers are very high, 25 kcal/mol for yttrium and 28 kcal/mol for scandium. The difference obtained between yttrium and scandium, a 3 kcal/mol lower barrier for yttrium, is in accord with the observed difference in reaction rate between the two metals. However, the high barriers indicate that the methane activation reaction (7) is not very likely to occur via the four center structure **5**. In fact, from experimental measurement of the reaction rate as a function of temperature, the activation enthalpy is determined to be 11.6 kcal/mol for the corresponding lutetium reaction [54]. Since the reaction rate for lutetium is in between yttrium and scandium the presently calculated barrier heights for the four center transition state does not agree with the measured activation energy for reaction (7).

From the populations given in Table 8 it can be seen that the metals in these complexes are quite far from d^0. For scandium and yttrium the d occupations in both the L_nMR complexes and the transition state structures are rather close to one, which corresponds to the neutral atomic ground state occupation, d^1s^2, for both metals. For the neutral titanium and zirconium atoms the ground state occupations are d^2s^2 and as can be seen from Table 8 the d populations in the L_nMR complexes and the transition state structures are rather close to two. However, the metal d electrons in the L_nMR complexes are involved in covalent bonding to the ligands, and an oxidative addition alkane activation mechanism for the metals in

Table 8. *Mulliken population analysis for the different σ-bond metathesis structures.*

| | L_nMR | | | | Transition state | | | |
| | L_nMH | L_nMCH_3 | $L_nMH + H_2$ | $L_nMCH_3 + H_2$ | | $L_nMCH_3 + CH_4$ | |
	q_M d	q_M d	q_M d	q_M	d	q_M	d
H_2ScR	0.85 0.81	0.99 0.82	0.83 0.93	0.84	0.84	0.80	0.92
H_2YR	0.98 0.88	1.24 0.85	0.96 1.02	1.01	1.00	1.02	0.98
H_2TiR^+	1.04 1.72	1.06 1.76	0.82 1.89	0.82	1.91	0.84	1.93
H_2ZrR^+	1.32 1.68	1.54 1.68	1.15 1.89	1.17	1.88	1.15	1.86

these complexes is thus still unlikely to be favorable. It should further be noted that for both yttrium and zirconium the neutral naked metal atoms have high barriers (42 kcal/mol and 33 kcal/mol, respectively, see Table 2) for the oxidative addition mechanism.

5. π-Coordination to Transition Metals

Two important π-acceptor ligands in organotransition metal complexes are carbon monoxide and ethylene. Both these ligands have empty π^* orbitals which interact with filled d orbitals on the metal, thus delocalizing electron density from the metal to the ligand. There is also a σ interaction, between a filled orbital on the ligand (the σ-lone pair on CO and the π orbital on the olefin) and empty orbitals of σ symmetry on the metal. This type of metal ligand interaction was first described by Dewar, Chatt and Duncanson [55]. It has turned out to be a challenge to quantum chemical methods to calculate accurate values for the binding energies of π-coordinated ligands. Below we summarize the results from two different studies, one concerning the variation in ethene binding energies with different metals and the other dealing with the different individual carbonyl binding energies in the $Ni(CO)_4$ complex.

5.1 ETHENE BINDING ENERGIES

The coordination of alkenes to transition metal centers constitute the starting point in several important catalytic reactions, e.g. the hydrogenation of unsaturated hydrocarbons and the polymerization of ethene and other olefins. The polymerization of ethene, catalyzed by transition metal compounds, is a very important industrial process, but still the reaction mechanisms are not well understood. For example, the mechanism for the carbon-carbon bond-forming polymer chain propagation step is still controversial. It is expected that the strength of the metal-ethene bond influences the C-C bond-forming step, which is described as insertion of the ethene moiety into the metal-alkyl bond. A starting point in an investigation of the mechanisms for ethene polymerization is therefore to determine the variation in the metal-ethene binding energies over different metals. Below we summarize the results from such a study.

The binding energies of ethene to all second row transition metal atoms were determined and the results are summarized in Table 9 and Figure 3. For molybdenum and technetium the ground states were found to be repulsive and no other state was found to be bound relative to the ground state of the metal atom. For these metals the binding energies are therefore set to zero. The general shape of the curve for the binding energy of ethylene to the second row transition metals (Fig. 3) is strikingly similar to the corresponding curve for the methane insertion product (Fig. 2), with the lowest binding energy in the middle of the row. The explanation for the low binding energies in the middle of the row for the methane insertion product is the large loss of exchange energy upon covalent bond formation. In the ethene case, where the bonding is described as donation and back-donation of electrons in the σ- and π-systems, the variation of the amount of exchange energy along the transition row also plays an important role for the shape of the curve in Fig. 3.

The dominating bonding force between ethene and the metal is the π-interaction, which in the present calculations, performed in C_{2v} symmetry, occurs in the b_2 representation. All states of interest therefore have to have the $4d_{b_2}$ orbital occupied on the metal. Ruthenium, rhodium and palladium have this orbital doubly occupied, leading to a better donation to the π^* orbital of ethene, and these metals therefore have the largest binding energies, 22-28 kcal/mol. All the other metals, with less electrons, have this orbital singly occupied and it would cost too much in atomic excitation energy to make this orbital doubly occupied. Therefore the binding energy due to π-interaction is smaller for these metals, yielding binding energies of 10-19 kcal/mol for yttrium, zirconium and niobium. The interaction in the σ-system is easiest analyzed in terms of repulsion between the metal electrons and the doubly occupied π orbital on ethene. All metals except molybdenum and technetium can minimize this repulsion by placing all the σ electrons in an orbital

Table 9. *Calculated relative energies (in kcal/mol) for the coordination of* C_2H_4 *to the neutral metal atoms. Mulliken populations for the metals are also given, together with the dominating configuration for the* MC_2H_4 *complexes.*

	State	Conf	q_M	5s	5p	4d	ΔE
Y	$^2B_2$	$\sigma^2 d_{b_2}^1$	+0.01	1.58	0.43	0.93	-18
Zr	$^3B_1$	$\sigma^2 d_{b_2}^1 d_{a_2}^1$	-0.08	1.47	0.36	2.21	-10
Nb	$^6A_1$	$\sigma^1 d_{b_2}^1 d_{a_2}^1 d_{b_1}^1 d_{a_1}^1$	-0.01	0.79	0.15	4.03	-19
Mo	$^7A_1$	$\sigma^{1+1} d_{b_2}^1 d_{a_2}^1 d_{b_1}^1 d_{a_1}^1$					0
Tc	$^6A_1$	$\sigma^{2+1} d_{b_2}^1 d_{a_2}^1 d_{b_1}^1 d_{a_1}^1$					0
Ru	$^3A_2$	$\sigma^1 d_{b_2}^2 d_{a_1}^2 d_{b_1}^2 d_{a_2}^1$	+0.14	0.38	0.10	7.32	-22
Rh	$^2A_2$	$\sigma^2 d_{b_2}^2 d_{a_1}^2 d_{b_1}^2 d_{a_2}^1$	+0.08	0.51	0.12	8.24	-28
Pd	$^1A_1$	$\sigma^2 d_{b_2}^2 d_{a_1}^2 d_{b_1}^2 d_{a_2}^2$	+0.05	0.33	0.11	9.45	-27

that is more or less sd_σ-hybridized to point away from the ethene electrons. For some of the metals this is possible since there is only one electron in the σ orbitals and for other metals there are two electrons which are low-spin coupled, allowing a double occupation of one orbital. In the case of ruthenium and rhodium this means that the ground state of the MC_2H_4 system correlate with an excited low-spin coupled state of the metal atom, but since the low-spin coupling is not very costly for these metals, these states are still strongly bound relative to the ground states of the atoms. Ruthenium, rhodium and palladium also have low lying $4d^{n+2}$ states, which can be mixed in to lower the repulsion. On the other hand, for both molybdenum and technetium one electron has to be placed in the sd_σ-hybrid pointing towards ethene. This is necessary for molybdenum because the two electrons in the σ-system are high-spin coupled and for technetium because there are three electrons in the σ-system. Therefore, for these two metals the repulsion in the σ-system is

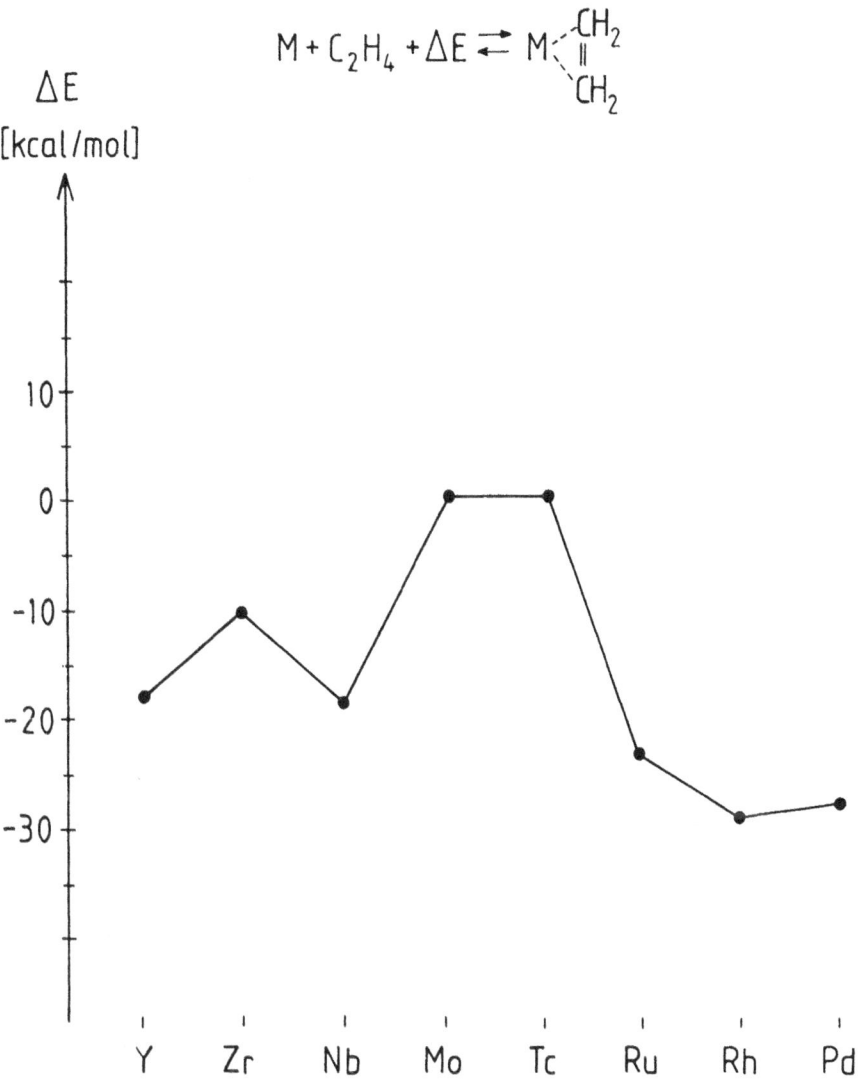

Figure 3. *Coordination of ethene to the second row metal atoms. Calculated energies relative to the ground state of the metal atoms and free ethene.*

too large to make them bound. There are, however, bound low-spin coupled states for these metals, but since the low-spin coupling is very costly, due to the large loss of exchange energy, these states are not bound relative to the ground state of the metal atoms.

5.2 CARBONYL BINDING ENERGIES

The transition metal-carbonyl bond is of fundamental importance in both organometallic chemistry and surface chemistry. It has therefore been extensively studied, both experimentally and theoretically, and the bonding mechanism is essentially understood. A puzzling observation, which still needs further investigation, however, is the irregular trend in the experimentally determined carbonyl binding energies for $Ni(CO)_x \rightarrow Ni(CO)_{x-1} + CO$ for x=1-4. The binding energies are obtained from a combination of photoelectron spectroscopy measurements [56] and appearance potential measurements [57] on the negative ions $Ni(CO)_x^-$, x=1-3. These experiments give a total Ni-CO binding energy of 120 kcal/mol for $Ni(CO)_4$, thus yielding an average Ni-CO binding energy of 30 kcal/mol. The first and the fourth carbonyl groups are bound by 29 and 25 kcal/mol, respectively. This is reasonably close to the average. Surprisingly enough, however, the second carbonyl obtains a much larger binding energy of 54 kcal/mol, and the third carbonyl a very small binding energy of 13 kcal/mol. Previous theoretical investigations [14] did not agree with this trend in binding energies, and below we present the results from a recent reinvestigation of the $Ni(CO)_4$ binding energies including a comparison between different quantum chemical methods.

Three different methods were applied to calculate the different CO binding energies in $Ni(CO)_4$, the Modified Coupled Pair Functional method (MCPF), the Coupled Cluster method, with and without the effect of triple excitations (denoted CCSD(T) and CCSD, respectively) and the Multi-Reference Average Coupled Pair Functional (MR-ACPF) method. The MR-ACPF method was only applied to the first and second carbonyl binding energies, since for the larger complexes this method would lead to too long configuration expansions, which could not presently be handled. The MCPF and the coupled cluster methods are based on one reference state, leading to shorter configuration expansions, and could therefore be applied also to the largest complexes. The results are summarized in Table 10 and only a few comments will be made in this context.

The most reliable of the applied methods should be the MR-ACPF method, since this is the only one that explicitly treats the near degeneracy effects. The results for the first and second carbonyl energy show that of the single reference methods in particular the CCSD(T) method gives results very close to the multi-reference calculations. For the first carbonyl a binding energy of 34 kcal/mol is obtained

Table 10. *Calculated Ni-CO binding energies for the last* CO *in* Ni(CO)$_x$, $x = 1-4$, *in kcal/mol.*

	NiCO	Ni(CO)$_2$	Ni(CO)$_3$	Ni(CO)$_4$	Total	Corrected for BSSE
MCPF	34	31	35	24	124	109
CCSD	19	39	30	23	111	97
CCSD(T)	34	43	35	30	142	125
MR-ACPF	34	39				
exp1[a]	29±15	54±15	13±10	25±2	121	
exp2[b]					140	
exp3[c]	36±4	51±4	29±2			

a Ref. 56. b Ref. 58. c Ref. 59.

for both methods and for the second carbonyl 39 kcal/mol is obtained using the MR-ACPF method and 43 kcal/mol using the CCSD(T) method. For the total carbonyl binding energy in Ni(CO)$_4$ the most reliable experimental result is 140 kcal/mol [58]. The calculated result for the total carbonyl energy at the CCSD(T) level, 142 kcal/mol, agree very well with this experimental value. However, this good agreement is somewhat fortuitous since the calculations suffer from a basis set superposition error (BSSE) of 17 kcal/mol, leading to an underestimate of about 10 percent of the total carbonyl binding energy in Ni(CO)$_4$ at the CCSD(T) level, when the basis set superposition error is corrected for. The CCSD calculations give a result for the total carbonyl binding energy that is 30 kcal/mol lower than the CCSD(T) value, showing that the triple excitations are quite important for the binding energies when the coupled cluster method is applied. The MCPF method gives a somewhat larger error than the CCSD(T) method for the total carbonyl binding energy, leading to an underestimate of about 20 percent, when the basis set superposition error is corrected for, which is still reasonably good taking into account that the correlation energy of as many as 50 electrons has to be calculated.

Concerning the trend in binding energy for the different carbonyl groups the results from the different theoretical methods are rather coherent. The results for the first and the last carbonyls agree fairly well with the experimental values, while for the second carbonyl the calculated values, varying between 31 and 43 kcal/mol for the different methods, are consistently lower than the experimental value of 54 kcal/mol and for the third carbonyl the calculated values, varying between 30 and 35 kcal/mol, are substantially larger than the experimental value of 13 kcal/mol. For the binding energy of the third carbonyl in $Ni(CO)_4$ a recent experiment [59] gives a value of 29 kcal/mol, in much better agreement with the calculated values. For the second carbonyl, however, this new experiment gives a value of 51 kcal/mol, maintaining a large discrepancy between theory and experiment for this particular case.

In conclusion, for a system as demanding as $Ni(CO)_4$, involving fairly large near degeneracies and dynamic correlation effects from 50 electrons, the deviation between experiment and the best calculated results for the total binding energy is about 10 percent. For many important systems in organometallic chemistry this value should represent an upper limit for the expected uncertainty in calculated relative energies.

6. Concluding Remarks

Recent development of quantum chemical methods has drastically increased the possibility to treat problems in organometallic chemistry. There are now methods available which are applicable to systems with fairly large near degeneracies, which are often present in transition metal systems and which treat electron correlation effects in a size-extensive manner, making it possible to correlate many electrons, which is often necessary for transition metal systems. The most important feature of the new methods is that they are simple to use, since they are based on a single reference configuration and can therefore be applied to whole classes of systems, thus determining differences and similarities between different metals and between different groups of reactions. The study of trends in interaction energies has turned out to be crucial for the elucidation of both binding and reaction mechanisms. In this paper we have briefly reviewed some results obtained from applying this approach to the study of reactions between transition metals and simple hydrocarbons. We foresee an increased importance of quantum chemical studies in the area of transition metal chemistry.

References

1. *a.* K.J. Klabunde and Y. Tanaka, J. Am. Chem. Soc. **105**, 3544 (1983). *b.* K.J. Klabunde, Gi Ho Jeong and A.W. Olsen, in *Selective Hydrocarbon Activation: Principles and Progress*, J.A. Davies, P.L. Watson, A.Greenberg and J.F.

Liebman, Ed., VCH Publishers, New York, 1990, pp 433-466.

2. D. Ritter and J.C. Weisshaar, J. Am. Chem. Soc. **112**, 6425 (1990).

3. S.A. Mitchell and P.A. Hackett, J. Chem. Phys. **93**, 7822 (1990).

4. J.C. Weisshaar, Advances in Chem. Phys. **81**, in press.

5. P.B. Armentrout and J.L. Beauchamp, Acc. Chem. Res. **22**, 315 (1989).

6. P.B. Armentrout, in *Selective Hydrocarbon Activation: Principles and Progress*, J.A. Davies, P.L. Watson, A.Greenberg and J.F. Liebman, Ed., VCH Publishers, New York, 1990, pp 467-533.

7. *a.* J. Almlöf and P.R. Taylor, J. Chem. Phys. **86**, (1987) 4070, *b.* R.C. Raffenetti, J. Chem. Phys. **58**, (1973) 4452.

8. A.J.H. Wachters, J. Chem. Phys. **52**, (1970) 1033.

9. S. Huzinaga, J. Chem. Phys. **66**. (1977) 4245.

10. S. Huzinaga, J. Chem. Phys. **42** (1965) 1293.

11. M.R.A. Blomberg, P.E.M. Siegbahn, U. Nagashima and J. Wennerberg. J. Am. Chem. Soc. **113**, 476 (1991).

12. M. Blomberg, U. Brandemark, L. Pettersson, P. Siegbahn and M. Larsson. in Molecular Properties: Proceedings of the CCP1 Study Weekend. Cambridge 25-27 March 1983. ed. by R.D. Amos and M.F. Guest (Daresbury)

13. P.-O. Widmark. B.O. Roos and P.E.M. Siegbahn. J. Phys. Chem. **89** (1985) 2180.

14. *a.* M. Blomberg. J. Johansson, P. Siegbahn and J. Wennerberg. J. Chem. Phys. **88** (1988) 4324. *b.* M.R.A. Blomberg, U.B. Brandemark, P.E.M. Siegbahn. J. Wennerberg and C.W. Bauschlicher, Jr.. J. Am. Chem. Soc. **110** (1988) 6650.

15. R. Ahlrichs, P. Scharf and C. Erhardt. J. Chem. Phys. **82** (1985) 890.

16. Raghavachari, G. W. Trucks, J. A. Pople, and M. Head-Gordon. Chem. Phys. Lett. **157** (1989) 479. The coupled cluster calculations are performed using the TITAN set of electronic structure programs, written by T. J. Lee. A. P. Rendell and J. E. Rice.

17. D.P. Chong and S.R. Langhoff, J. Chem. Phys. **84** (1986) 5606.

18. R.J. Gdanitz and R. Ahlrichs. Chem. Phys. Lett. **143**, (1988) 413. In some applications the internally contracted variant of the ACPF method is used. This program is written by Per Siegbahn and is based on the reference: H.-J. Werner and P.J. Knowles, J. Chem. Phys. **89**, (1988) 5803.

19. R.L. Martin, J. Phys. Chem. **87**, 750 (1983); see also R.D. Cowan and D.C. Griffin, J. Opt. Soc. Am. **66**, 1010 (1976).

20. *a.* A.H. Janowicz and R.G. Bergman, J. Am. Chem. Soc. **104**, 352 (1982), *b.* A.H. Janowicz and R.G. Bergman, J. Am. Chem. Soc. **105**, 3929 (1983).

21. *a.* J.K. Hoyano and W.A.G. Graham, J. Am. Chem. Soc. **104**, 3723 (1982), *b.* J.K. Hoyano, A.D. McMaster and W.A.G. Graham, J. Am. Chem. Soc. **105**,

7190 (1983).

22. W.D. Jones and F.J. Feher, J. Am. Chem. Soc. **104**, 4240 (1982),

23. *Perspectives in the Selective Activation of C-H and C-C Bonds in Saturated Hydrocarbons*, ed B. Meunier and B. Chaudret, Scientific Affairs Division - NATO, Brussels, 1988.

24. M. Blomberg, U. Brandemark, L Pettersson and P. Siegbahn, Int. J. Quantum Chem. **23**, 855 (1983)

25. *a.* J.J. Low and W.A. Goddard III, J. Am. Chem. Soc. **106**, 8321 (1984), *b.* J.J. Low and W.A. Goddard III, Organometallics **5**, 609 (1986), *c.* J.J. Low and W.A. Goddard III, J. Am. Chem. Soc. **106**, 6928 (1984), *d.* J.J. Low and W.A. Goddard III, J. Am. Chem. Soc. **108**, 6115 (1986).

26. *a.* M.R.A. Blomberg and P.E.M. Siegbahn, J. Chem. Phys. **78**, (1983) 986, *b.* M.R.A. Blomberg and P.E.M. Siegbahn, J. Chem. Phys. **78**, (1983) 5682.

27. T. Sakakura, T. Sodeyama, K. Sasaki, K. Wada and M. Tanaka, J. Am. Chem. Soc. **112**, 7221 (1990).

28. A.J. Rest, I. Whitwell, W.A.G. Graham, J.K. Hoyano and A.D. McMaster, J. Chem. Soc. Chem. Commun., 624 (1984).

29. D.M. Haddleton, A. McCamley and R.N. Perutz, J. Am. Chem. Soc. **110**, 1810 (1988).

30. S.T. Belt, F.-W. Grevels, W.E. Klotzbücher, A. McCamley and R.N. Perutz, J. Am. Chem. Soc. **111**, 8373 (1989).

31. J.M. Buchanan, J.M. Stryker and R.G. Bergman, J. Am. Chem. Soc. **108**, 1537 (1986).

32. R.A. Periana and R.G. Bergman, J. Am. Chem. Soc. **108**, 7332 (1986).

33. B.H. Weiller, E.P. Wasserman, R.G. Bergman, C.B. Moore and G.C. Pimentel, J. Am. Chem. Soc. **111**, 8288 (1989).

34. E.A. Carter and W.A. Goddard III, J. Phys. Chem. **92**, 5679 (1988).

35. M.R. Zakin, D.M. Cox and A. Kaldor, J. Chem. Phys. **89**, (1988) 1201.

36. P. Fayet, A. Kaldor and D.M. Cox, J. Chem. Phys. **92**, (1990) 254.

37. M. Rosi, C.W. Bauschlicher, Jr., S.R. Langhoff and H. Partridge, J. Phys. Chem. **94**, 8656 (1990).

38. P.A.M. van Koppen, M.T. Bowers, J.L. Beachaump and D.V. Dearden, in *Bonding Energetics in Organometallic Compounds*, T.J. Marks, Ed., ACS Symposium Series, Washington, DC 1990, pp 34-54.

39. Y. Huang, M.B. Wise, D.B. Jacobson and B.S. Freiser, Organometallics **6**, 346 (1987).

40. S.W. Buckner, T.J. MacMahon, G.D. Byrd and B.S. Freiser, Inorg. Chem. **28**, 3511 (1989).

41. J.B. Schilling and J.L. Beachamp, Organometallics **7**, 194 (1988).

42. M.A. Tolbert, M.L. Mandich, L.F. Halle and J.L. Beachamp, J. Am. Chem. Soc. **108**, 5675 (1986).

43. G.D. Byrd and B.S. Freiser, J. Am. Chem. Soc. **104**, 5944 (1982).

44. T.J. MacMahon, Y.A. Ranasinghe and B.S. Freiser J. Phys. Chem. **95**, 7721 (1991).

45. T. Ziegler, V. Tschinke, L. Fan and A.D. Becke, J. Am. Chem. Soc. **111**, 9177 (1989).

46. T. Yamamoto, A. Yamamoto and S. Ikeda, J. Am. Chem. Soc. **93**, (1971) 3350.

47. T. Sakakura, T. Sodeyama, K. Sasaki, K. Wada and M. Tanaka, J. Am. Chem. Soc. **112**, 7221 (1990).

48. M.R.A. Blomberg, J. Schüle and P.E.M. Siegbahn, J. Am. Chem. Soc. **111**, 6156 (1989),

49. M.R.A. Blomberg, P.E.M. Siegbahn and M. Svensson, New J. of Chem. in press

50. P.L. Watson, J. Am. Chem. Soc. **105** (1983) 6491.

51. M.E. Thompson, S.M. Baxter, A.R. Bulls, B.J. Burger, M.C. Nolan, B.D. Santarsiero, W.P. Schaefer and J.E. Bercaw, J. Am. Chem. Soc. **109** (1987) 203.

52. C.S. Christ, J.R. Eyler and D.E. Richardson, J. Am. Chem. Soc. **110** (1988) 4038.

53. C.S. Christ, J.R. Eyler and D.E. Richardson, J. Am. Chem. Soc. **112** (1990) 96.

54. P.L. Watson, in *Selective Hydrocarbon Activation: Principles and Progress*, J.A. Davies, P.L. Watson, A.Greenberg and J.F. Liebman, Ed., VCH Publishers, New York, 1990, pp 79-112.

55. *a.* M.J.S. Dewar, Bull. Soc. Chim. Fr. 18C , 71 (1951), *b.* J. Chatt and L.A. Duncanson, J. Chem. Soc. 2939 (1953).

56. A.E. Stevens, C.S. Feigerle and W.C. Lineberger, J. Am. Chem. Soc. **104**, (1982) 5026.

57. R.N. Compton and J.A.D. Stockdale, Int. J. Mass Spectrom. Ion Phys. **22**, (1976) 47.

58. A.K. Fischer, F.A. Cotton and G. Wilkinson, J. Am. Chem. Soc., **79**, (1957) 2044.

59. Lee S. Sunderlin and Robert R. Squires, private communication.

List of Posters and Contributed Oral Presentations

Posters

G. J. M. Gruter, B. L. M. van Baar, M. Hogenbirk, O. S. Akkerman, and
F. Bickelhaupt
*An Interface for the Introduction of Air and Moisture Sensitive
Compounds into a Magnetic Sector Mass Spectrometer*

J. P. Leal, N. Marques, A. Pires de Matos, A. M. Galvão, M. J.
Calhorda, and J. A. Martinho Simões
Uranium-Ligand Bond Dissociation Enthalpies

N. Papadopoulos, C. Hasiotis, G. Kokkinidis, and G. Papanastasiou
*Determination of Kinetic Parameters of an Homogeneous Chemical Reaction
Following an Electrochemical Reaction, Using 3D Electrochemistry*

J. S. Chickos, D. G. Hesse, and J. F. Liebman
Estimation of the Heat Capacities of Solid and Liquid Organic Compounds

L. S. Sunderlin, D. Wang, and R. R. Squires
Metal Carbonyl Anion Bond Strengths

M. Tilset and A. Pedersen
Chemical and Electrochemical Oxidation of $CpCo(PPh_3)Me_2$

B. Salih and N. Balcioglu
Energetics of Mass Spectral Isomerization of 1,5-cyclodecadiene

J. Catalan and J. C. del Valle
Experimental Acidity of Carbazole

S. T. Kapellos, A. Mavridis, and J. F. Harrison
Electrostatic, High Spin Bonding in MNH_2^+ (M=Sc, Ti, V, Cr, and Mn)

J. Marçalo and A. Pires de Matos
*$D(X_nM-L)/D(H-L)$ and $\Delta H_f^o(X_nML)/\Delta H_f^o(HL)$ Correlations in Organolanthanide,
Actinide, and Early Transition Metal Compounds*

D. Clemmer and P. Armentrout
*Understanding M^+-NH_2, M^+-OH, and M^+-NO Bond Energies: Covalent and
Dative Interactions*

M. Azzaro, S. Breton, M. Decouzon, S. Géribaldi
*FT-ICR Study of Gas-Phase Reactions of Sc^+, Y^+, and Lu^+ with Silane and
Alkoxysilanes*

H. J. Arnold, J. P. Fitzgerald, and J. P. Collman
Heterometallic Metal-Metal Bonded Dimers

A. R. Dias, H. P. Diogo, M. E. Minas da Piedade, J. A. Simoni, and
J. A. Martinho Simões
Energetics of Zr-H, Zr-C, Zr-I, and Zr-OR Bonds in the Complexes
Zr(Cp)₂(Cl)L

P. Dias, A. R. Dias, J. A. Simoni, C. Teixeira, and J. A. Martinho
Simões
Photocalorimetry. A Microcalorimetric System for Probing the
Energetics of Light-Induced Reactions in Solution

H. Alt, P. Dias, A. R. Dias, R. Santos, and J. A. Martinho Simões
Energetics of Titanium-Phosphorus Bonds in the Complexes
Ti(Cp)₂(CO)(PR₃) (R=Me, Et) and Ti(Cp)₂(PMe₃)₂

T. Burkey
New Results on an Old Photochemical Reaction

Oral Presentations

J. S. Chickos, D. G. Hesse, and J. F. Liebman
Estimations of Sublimation Enthalpies

G. J. M. Gruter, B. L. M. van Baar, M. Hogenbirk, O. S. Akkerman, and
F. Bickelhaupt
An Interface for the Introduction of Air and Moisture Sensitive
Compounds into a Magnetic Sector Mass Spectrometer

S. T. Kapellos, A. Mavridis, and J. F. Harrison
Low Spin Bonding in MNH₂⁺ (M=Sc, Ti, V, Cr, and Mn)

C. G. Screttas and M. Micha-Screttas
Correlating Spectroscopic Electronegativities with Thermochemical Data

M. Azzaro, S. Breton, M. Decouzon, S. Géribaldi
FT-ICR Study of Gas-Phase Reactions of Sc⁺, Y⁺, and Lu⁺ with Silane and
Alkoxysilanes

H. J. Arnold, J. P. Fitzgerald, and J. P. Collman
Heterometallic Metal-Metal Bonded Dimers

B. Wayland
Thermochemistry of Rhodium Dimers

A. S. Miller
Acidities of Metal Hydrides

List of Participants

Jorge M. M. Abalroado
Departamento de Engenharia Química, Instituto Superior Técnico,
1096 Lisboa Codex, Portugal

Peter B. Armentrout
Department of Chemistry, University of Utah,
Salt Lake City, Utah 8411, U.S.A.

Hilary J. Arnold
Department of Chemistry, Stanford University,
Stanford, California 94305, U.S.A.

Murat Azik
Department of Chemistry, Hacettepe University, Beytepe Campus,
06532 Ankara, Turkey

Jack L. Beauchamp
Noyes Laboratory, California Institute of Technology,
Pasadena, California 91125, U.S.A.

Margareta R. A. Blomberg
Institute of Theoretical Physics, University of Stockholm,
Vanadisvägen, S-11346 Stockholm, Sweden

Sylvie Breton
Laboratoire de Chimie Physique Organique, Université de Nice-Sophia
Antipolis, Parc Valrose, 06034 Nice Cedex, France

Theodore J. Burkey
Department of Chemistry, Memphis State University,
Memphis, Tennessee 38152, U.S.A.

Maria J. Calhorda
Centro de Tecnologia Química e Biológica, Rua da Quinta Grande, 6,
Apartado 127, Oeiras, Portugal

Teresa Cañada
Instituto de Química Física Rocasolano, CSIC, Calle Serrano 119,
E-28006 Madrid, Spain

Alexander S. Carson
Department of Physical Chemistry, The University of Leeds,
Leeds LS2 9JT, U.K.

James S. Chickos
Department of Chemistry, College of Arts and Sciences, University of
Missouri-St. Louis, 8001 Natural Bridge Rd., St. Louis,
Missouri 63121-4499, U.S.A.

David E. Clemmer
Department of Chemistry, University of Utah,
Salt Lake City, Utah 84112, U.S.A.

Joseph A. Connor
Chemical Laboratory, The University, Canterbury, Kent CT2 7NH, U.K.

Alberto Romão Dias
Departamento de Engenharia Química, Instituto Superior Técnico,
1096 Lisboa Codex, Portugal

Palmira B. Dias
Departamento de Engenharia Química, Instituto Superior Técnico,
1096 Lisboa Codex, Portugal

Hermínio P. Diogo
Departamento de Engenharia Química, Instituto Superior Técnico,
1096 Lisboa Codex, Portugal

Maria Teresa Leal da Silva Duarte
Departamento de Engenharia Química, Instituto Superior Técnico,
1096 Lisboa Codex, Portugal

Ender Erdik
Department of Chemistry, Ankara University, Ankara, Turkey

Eduardo Peris Fajarnes
Departament de Química Inorgànica, Universitat de València,
Doctor Moliner 50, 46100 Burjassot, València, Spain

Adelino M. Galvão
Departamento de Engenharia Química, Instituto Superior Técnico,
1096 Lisboa Codex, Portugal

Ayten Göçmen
Department of Chemistry, Istanbul Technical University, I.T.Ü. Fen-Ed.
Fak. Kimya Böl., Maslak, 80626 Instanbul, Turkey

Jorge M. Gonçalves
Departamento de Química, Faculdade de Ciências da Universidade do
Porto, Praça Gomes Teixeira, 4000 Porto, Portugal

Gert-Jan Gruter
Scheikundig Laboratorium, Vrije Universiteit, De Boelelaan 1083,
1081 HV Amsterdam, The Netherlands

William D. Jones
Department of Chemistry, University of Rochester,
Rochester, New York 14627, U.S.A.

George Kalatzis
Inorganic Chemistry Laboratory, Department of Chemistry, University of
Athens, Panepistimiopolis, Kouponia, 157 01 Athens, Greece

S. T. Kapellos
Physical Chemistry Laboratory, Department of Chemistry, University of
Athens, Panepistimiopolis, Kouponia, 157 71 Athens, Greece

Ünel Köklü
Department of Chemistry, Istanbul Technical University, Fen-Ed. Fak.
Kimya Böl., Maslak, 80626 Instanbul, Turkey

Juan C. del Valle Lázaro
Departamento de Química Física Aplicada, C II. 203, Universidad
Autonoma de Madrid, E-28049 Madrid, Spain

João Paulo Leal
Departamento de Química, Laboratório Nacional de Engenharia e
Tecnologia Industrial, 2686 Sacavém Codex, Portugal

Joaquim Marçalo
Departamento de Química, Laboratório Nacional de Engenharia e
Tecnologia Industrial, 2686 Sacavém Codex, Portugal

Tobin J. Marks
Department of Chemistry, Northwestern University, 2145 Sheridan Road,
Evanston, Illinois 60201, U.S.A.

António Pires de Matos
Departamento de Química, Laboratório Nacional de Engenharia e
Tecnologia Industrial, 2686 Sacavém Codex, Portugal

M. Agostinha R. Matos
Departamento de Química, Faculdade de Ciências da Universidade do
Porto, Praça Gomes Teixeira, 4000 Porto, Portugal

Amy E. Stevens Miller
PL/LID, 1102F Greiner St., Hanscom AFB, MA 01731-5000, U.S.A.

Manuel João Monte
Departamento de Química, Faculdade de Ciências da Universidade do
Porto, Praça Gomes Teixeira, 4000 Porto, Portugal

Peter Mulder
Center for Chemistry and the Environment, Gorlaeus Laboratories,
Leiden University, 2300 RA Leiden, The Netherlands

Helena I. Seguro Nogueira
Departamento de Química, Universidade de Aveiro, 3800 Aveiro, Portugal

Rafael Notario
Instituto de Quimica Física Rocasolano, CSIC, Calle Serrano 119,
E-28006 Madrid, Spain

Nikos Papadopoulos
Laboratory of Physical Chemistry, Department of Chemistry, University
of Thessaloniki, Thessaloniki 54006, Greece

Astrid Pedersen
Department of Chemistry, University of Oslo, P. O. Box 1033, Blindern,
N-0315 Oslo 3, Norway

Maria Angeles Ubeda Picot
Departament de Química Inorgànica, Universitat de València,
Doctor Moliner 50, 46100 Burjassot, València, Spain

Manuel E. Minas da Piedade
Departamento de Engenharia Química, Instituto Superior Técnico,
1096 Lisboa Codex, Portugal

Geoffrey Pilcher
Department of Chemistry, University of Manchester,
Manchester M13 9PL, U.K.

David E. Richardson
Department of Chemistry, University of Florida,
Gainesville, Florida 32611, U.S.A.

Bekir Salih
Department of Chemistry, Faculty of Engineering, Hacettepe University,
Beytepe Campus, 06532 Ankara, Turkey

Luís M. N. B. F. Santos
Departamento de Química, Faculdade de Ciências da Universidade do
Porto, Praça Gomes Teixeira, 4000 Porto, Portugal

Rui M. Borges dos Santos
Departamento de Engenharia Química, Instituto Superior Técnico,
1096 Lisboa Codex, Portugal

C. G. Screttas
National Hellenic Research Foundation, 48 Vassileos Constantinov Av.,
Athens 116 35, Greece

Manuel A. V. Ribeiro da Silva
Departamento de Química, Faculdade de Ciências da Universidade do
Porto, Praça Gomes Teixeira, 4000 Porto, Portugal

Maria das Dores Ribeiro da Silva
Departamento de Química, Faculdade de Ciências da Universidade do
Porto, Praça Gomes Teixeira, 4000 Porto, Portugal

José A. Martinho Simões
Departamento de Engenharia Química, Instituto Superior Técnico,
1096 Lisboa Codex, Portugal

José de Alencar Simoni
Instituto de Química, Universidade Estadual de Campinas,
13081 Campinas, S. Paulo, Brazil

Henry A. Skinner
Department of Chemistry, University of Manchester,
Manchester M13 9PL, U.K.

Robert R. Squires
Department of Chemistry, Purdue University,
West Lafayette, Indiana 47906, U.S.A.

Lee S. Sunderlin
Department of Chemistry, Purdue University, West Lafayette,
Indiana 47907, U.S.A.

Clementina Teixeira
Departamento de Engenharia Química, Instituto Superior Técnico,
1096 Lisboa Codex, Portugal

Mats Tilset
Department of Chemistry, University of Oslo, P. O. Box 1033, Blindern,
N-0315 Oslo 3, Norway

Tito da Silva Trindade
Departamento de Química, Universidade de Aveiro, 3800 Aveiro, Portugal

Chris Tsiamis
Department of General and Inorganic Chemistry, Aristotelian University,
Thessaloniki 54008, Greece

Canan Ünaleroglu
Department of Chemistry, Hacettepe University, Beytepe Campus,
06532 Ankara, Turkey

Luís Filipe Coelho Veiros
Departamento de Engenharia Química, Instituto Superior Técnico,
1096 Lisboa Codex, Portugal

Robin Walsh
Department of Chemistry, University of Reading, Whiteknights,
P.O. Box 224, Reading RG6 2AD, U.K.

Bradford B. Wayland
Department of Chemistry, University of Pennsylvania,
Philadelphia, Pennsylvania 19104-6323, U.S.A.

Danial D. M. Wayner
Division of Chemistry, National Research Council, 100 Sussex Drive,
Ottawa, Ontario K1A OR6, Canada

Tom Ziegler
Department of Chemistry, University of Calgary,
Calgary, Alberta T2N 1N4, Canada

Subject Index

Ab initio 299-301, 324, 329, 331, 357, 358, 362, 363, 365-377, 387-418
Acidity 253-266
   solvation effects 263-266
   substituent effects 259, 260-266
Actinometry 87
Agostic bond *(see also Metal-alkane σ complexes* and *Metal-silane σ complexes)* 84, 90, 331, 378, 406
Alkane dehydrogenation 303, 304
Aluminium compounds 14, 15
Antimony compounds 4, 21, 22
η^2-Arene coordination 55-63
Arsenic compounds 4, 21, 22, 29

Backbonding 83, 86
Bismuth compounds 4, 21, 22, 29
Bond dissociation enthalpy
   absolute values in solution 36
   anchor 36, 199-201
   C-H in alkanes 174-176
   Co-C in co-enzyme B12 318, 319
   Co-CH 193
   Co-CH(Me)Ph 217
   Co-CH2Ph 211-213
   Co^+-CH4 347, 349
   Co^+-C2H4 311, 350
   Co^+-C2H6 341, 349
   Co^+-C3H6 309-311, 350
   Co^+-C3H8 349
   Co^+-CO 311
   Co^--CO 279, 281
   Co-H 123, 260, 281, 329, 367
   Co^+-H2O 346
   Co^+-ligand 311
   Co-Me 367
   Co-Mn 191
   Co-Re 191
   Cr-alkane 80-82, 85, 86, 91, 92
   Cr-amine 82, 83
   Cr-arene 48, 91, 92, 200
   Cr-CO 80, 81, 85, 86, 190, 279, 366
   Cr^+-CO 342, 345
   Cr^--CO 279, 281, 283
   Cr-cy-C6H12 91
   Cr-dichloroethane 91, 92
   Cr-H 123
   Cr^+-H2O 346
   Cr-hexene 91
   Cr-ligand 243

The manufacturer's authorised representative in the EU is Springer
Nature Customer Service Centre GmbH, Europaplatz 3, 69115 Heidelberg,
Germany. If you have any concerns regarding our products, please
contact ProductSafety@springernature.com

Printed and bound by CPI Group (UK) Ltd, Croydon, CR0 4YY
23/04/2026
02095624-0007